BIOMEMBRANES

A Multi-Volume Treatise

Volume 5 • 1996

ATPases

BIOMEMBRANES

A Multi-Volume Treatise

ATPases

Editor: A. G. LEE
Department of Biochemistry
University of Southampton
Southampton, England

VOLUME 5 • 1996

JAI PRESS INC.

Greenwich, Connecticut *London, England*

55 Old Post Road No. 2
Greenwich, Connecticut 06836

JAI PRESS LTD.
38 Tavistock Street
Covent Garden
London WC2E 7PB
England

ISBN: 1-55938-662-2

Manufactured in the United States of America

Transferred to digital printing 2008

CONTENTS

LIST OF CONTRIBUTORS

Karlheinz Altendorf — Universität Osnabrück
Osnabruck, Germany

Krister Bamberg — Center for Ulcer Research and Education
V.A. Wadsworth Medical Center
Los Angeles

Dennis Bayle — Center for Ulcer Research and Education
V.A. Wadsworth Medical Center
Los Angeles

Adrian R. Beaudoin — Département de Biologie
Université de Sherbrooke
Quebec

Philippe Champeil — Department of Cellular and Molecular Biology
Centre d'Etudes de Saclay
Gif-sur-Yvette, France

Flemming Cornelius — Institute of Biophysics
University of Aarhus
Denmark

Saibel Dey — Department of Biochemistry and Molecular Biology
Wayne State University School of Medicine

Dexian Dou — Department of Biochemistry and Molecular Biology
Wayne State University School of Medicine

Wolfgang Epstein — Department of Molecular Genetics and Cell Biology, The University of Chicago

Pilar Eraso — Departamento de Bioquímica, Instituto de Investigaciones Biomédicas del C.S.I.C., Madrid

Masamitsu Futai — Institute of Scientific and Industrial Research, Osaka University, Osaka, Japan

Danilo Guerini — Department of Biochemistry, Swiss Federal Institute of Technology

Joachim Krebs — Department of Biochemistry, Swiss Federal Institute of Technology

A.G. Lee — Department of Biochemistry, University of Southampton

Michael E. Maguire — Department of Pharmacology, School of Medicine, Case Western Reserve University

Ronald N. McElhaney — Department of Biochemistry, University of Alberta

Robert K. Nakamoto — Department of Molecular Physiology and Biological Physics, University of Virginia

Nathan Nelson — Roche Institute of Molecular Biology, Roche Research Center, Nutley, New Jersey

Maryse Picher — Département de Biologie, Université de Sherbrooke, Quebec

Francisco Portillo
Departamento de Bioquímica
Instituto de Investigaciones Biomédicas del C.S.I.C.
Madrid

Barry P. Rosen
Department of Biochemistry and Molecular Biology
Wayne State University School of Medicine

George Sachs
Center for Ulcer Research and Education
V.A. Wadsworth Medical Center
Los Angeles

Ramón Serrano
Departamento de Biotechnología
Universidad Politécnica
Valencia, Spain

Jean Sévigny
Département de Biologie
Université de Sherbrooke
Quebec

Jai Moo Shin
Center for Ulcer Research and Education
V.A. Wadsworth Medical Center
Los Angeles

Tao Tao
Department of Pharmacology
School of Medicine
Case Western Reserve University

PREFACE

The quantity of information available about membrane proteins is now too large for any one person to be familiar with anything but a very small part of the primary literature. A series of volumes concentrating on molecular aspects of biological membranes therefore seems timely. The hope is that, when complete, these volumes will provide a convenient introduction to the study of a wide range of membrane functions.

Volume 5 of *Biomembranes* covers an important group of membrane proteins, the ATPases. The P-type ATPases couple the hydrolysis of ATP to the movement of ions across a membrane and are characterized by the formation of a phosphorylated intermediate. Included are the plasma membrane and muscle sarcoplasmic reticulum Ca^{2+}-ATPases, the (Na^{+}-K^{+})-ATPase, the gastric (H^{+}-K^{+})-ATPase, the plasma membrane H^{+}-ATPase of fungi and plants, the Mg^{2+}-transport ATPases of *Salmonella typhimurium*, and the K^{+}-ATPase of *Escherichia coli*, KdpB. The other important classes of ATPase in eukaryotic systems are the vacuolar H^{+}-ATPases and the F_0F_1 ATP synthase, and, in bacteria, the anion-translocating ATPases, responsible for resistance to arsenicals and antimonials, and the (Na^{+}-Mg^{2+})-ATPase of *Acholeplasma*. Finally, eukaryotic systems contain a variety of ecto-nucleotidases important, for example, in hydrolysis of extracellular ATP released as a cotransmitter from cholinergic and adrenergic nerve terminals. Volume 5 of *Biomembranes* explores structure–function relationships for these membrane-bound ATPases.

As editor, I wish to thank all the contributors for their efforts and the staff of JAI Press for their professionalism in seeing everything through to final publication.

A.G. Lee
Editor

STRUCTURE OF THE SR/ER Ca^{2+}-ATPASE

A. G. Lee

I. THE Ca^{2+}-ATPASE

The Ca^{2+}-ATPase of endoplasmic reticulum (ER) or sarcoplasmic reticulum (SR) is one of the P-type ion pumps characterized by cation-activated phosphorylation of an Asp residue. The Ca^{2+}-ATPase consists of a single polypeptide chain, unlike other ATPases which exist as αβ-heterodimers (e.g., the (Na^+–K^+)-ATPase); reports that interaction between the Ca^{2+}-ATPase and a 53 kD glycoprotein in the SR membrane are important for coupling of ATP hydrolysis to Ca^{2+} transport appear to be unfounded (Grimes et al., 1991). Sequences of many ER/SR Ca^{2+}-ATPases are now available, and show some conservation with respect to other P-type ion

Biomembranes
Volume 5, pages 1–42.

ISBN: 1-55938-662-2.

pumps (Green and Stokes, 1992). As described below, analysis of hydropathy plots suggests the presence of 10 transmembrane α-helices. Most of the extramembranous residues are in two cytoplasmic regions separated by transmembrane helices M3 and M4 (Figure 1). The first (the β-strand domain in Figure 1) has a predicted anti-parallel β-structure, and is linked to the membrane by α-helices S2 and S3 (Green and Stokes, 1992). It contains a trypsin-cleavage site and, since cleavage at this site affects ATPase activity, the domain has also been referred to as the transduction domain. The second region, between transmembrane helices M4 and M5, is larger and has been subdivided into three domains. The phosphorylation domain and nucleotide-binding domain are both predicted to consist of alternating β-strands and α-helices. The pattern is similar to that found in kinases where ATP, bound to the second of two parallel β-sheet domains, phosphorylates a substrate bound to the other domain. The phosphorylation domain contains the residue (Asp-351) phosphorylated by ATP. The nucleotide-binding domain contains a

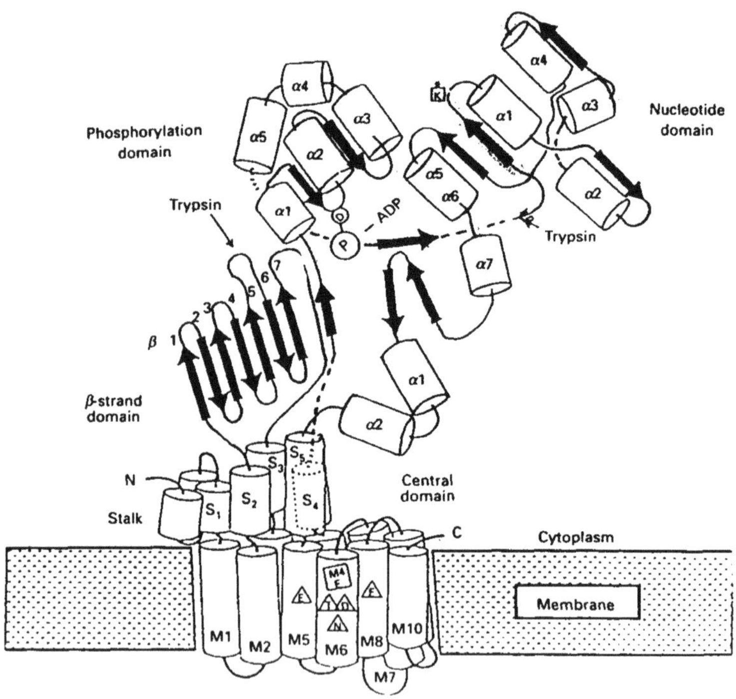

Figure 1. Diagrammatic representation of the Ca^{2+}-ATPase. Surface-exposed regions of the ATPase, as defined by binding of mAbs and antipeptide antibodies, are shaded. Also shown are Lys-515 labeled by FITC (K*) and two trypsin-cleavage sites (Mata et al., 1992).

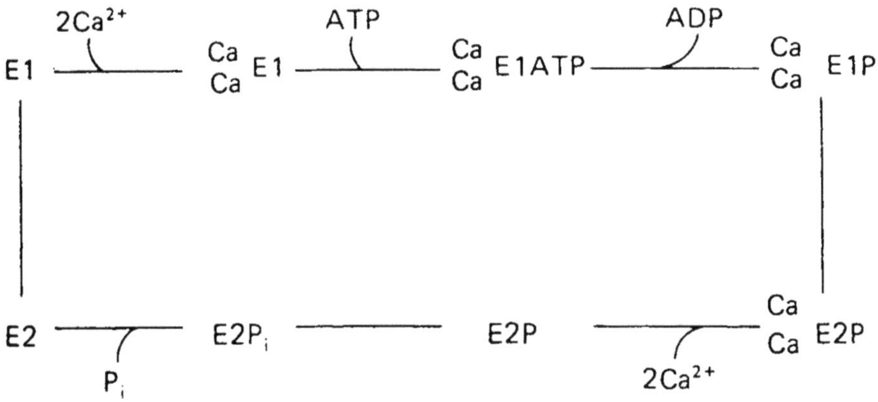

Figure 2. A simplified reaction scheme for the Ca^{2+}-ATPase.

conserved sequence around Lys-515 which can be labeled in an ATP-protectable manner by fluorescein isothiocyanate (FITC) and is thus believed to contain the binding site for ATP. Finally, at the C-terminal side of this second region is a central or hinge domain which, since it is labeled by γ-phosphate affinity labels (see below), must be close to the phosphorylation site in the three-dimensional structure of the ATPase.

The mechanism of the Ca^{2+}-ATPase is usually described in terms of the E2–E1 model developed from the Post–Albers scheme for the (Na^+–K^+)-ATPase (de Meis, 1981; Figure 2). In the E1 conformation, the ATPase has two outward-facing binding sites of high affinity. Following the binding of MgATP, the ATPase is phosphorylated to give Ca_2E1P, which can undergo a change in conformation to a state (Ca_2E2P) in which the two Ca^{2+} binding sites are of low affinity and inward facing. Following loss of Ca^{2+} to the inside of the SR, the ATPase can dephosphorylate and recycle to E1. Thus, binding of Ca^{2+} to high-affinity exterior-facing sites on the ATPase causes a change in chemical reactivity for the ATPase, from being reactive to P_i and H_2O in the E2 state, to being reactive with ATP and ADP in the E1 state. Only in the Ca^{2+}-bound form can the ATPase be phosphorylated by ATP and undergo the series of conformational changes leading to the occlusion of Ca^{2+} and then its translocation and release to the luminal spaces of SR. The key to understanding the mechanism of the ATPase, therefore, appears to be understanding how the switch in reactivity of the ATPase (ATP reactive to P_i reactive) is linked to the change in orientation or accessibility (cytoplasmic or luminal) of the Ca^{2+} sites.

II. ISOENZYMES OF THE Ca^{2+}-ATPASE

Three genes encoding the Ca^{2+}-ATPase of endoplasmic or sarcoplasmic reticulum have been identified, SERCA1, SERCA2, and SERCA3. The SERCA1 gene encodes the Ca^{2+}-ATPase of sarcoplasmic reticulum, expressed mainly in fast-twitch skeletal muscle (Brandl et al., 1986) and the SERCA2 gene encodes the isoforms expressed in slow-twitch skeletal, cardiac or smooth muscle, brain, stomach mucosa, liver, kidney, and other tissues (MacLennan et al., 1985). SERCA3 is expressed at high levels in large intestine and spleen, at intermediate levels in brain, stomach, uterus, skeletal muscle, and heart, and at low levels in some other tissues; SERCA3 protein has potential sites for phosphorylation by cAMP-dependent protein kinase (Burk et al., 1989).

The SERCA1 primary transcript is processed by one of two alternative routes. In adult muscle, the penultimate exon is retained, but in the neonate it is spliced out. The adult isoform codes for a protein with a C-terminal G instead of the last eight amino acids (DPEDERRK) of the neonatal isoform (Brandl et al., 1986). The function of these differences is unknown. Splicing of the SERCA2 primary transcript differs between tissues. In cardiac muscle, the donor splicing site of the penultimate exon is recognized, where it is fused to the last exon of the gene. In smooth muscle and in non-muscle cells, this donor site is not recognized and transcription stops at a polyadenylation site located before the last exon of the gene. This alternative splicing generates two mRNAs that differ at their 3′ ends. The last 4 amino acids of SERCA2a are replaced by 49 amino acids in SERCA2b, so that the former encodes a protein of 110 kD and the later a protein of 115 kD (Lytton and MacLennan, 1988). Heart and slow-twitch skeletal muscle express mainly the SERCA2a isoform, whereas smooth muscle and brain express the SERCA2b isoform. Similar alternative splicing is observed in the crustacean *Artemia* (Escalante and Sastro, 1993). The C-terminal extension of the Ca^{2+}-ATPase found in smooth muscle and brain is hydrophobic, and it has been suggested that it could constitute an extra transmembrane domain. Comparison of the *Artemia* and vertebrate sequences show that the sequence of the extra domain is poorly conserved, suggesting that it cannot be involved in the enzymology of the ATPase; a possible role in regulating interactions with other cell components has been suggested (Escalante and Sastro, 1993). Immunological studies of SERCA2a and SERCA2b expressed in COS cells have shown that the carboxyl-termini of the two isoforms are located on opposite sides of the membrane, consistent with a 10-helix model for the SERCA2a isoform (see following) and an 11-helix model for the SERCA2b isoform (Campbell et al., 1992; see also Bayle et al., 1995).

Regulation of the Ca^{2+}-ATPases has been observed at the level of gene transcription. Thus, chronic low frequency stimulation of fast-twitch skeletal muscle leads to switching to slow-twitch muscle. This is accomplished by a gradual reduction in the level of the SERCA1 gene and an increase in the level of the SERCA2a gene; this follows a change in the expression of myosin isoforms and accompanies an increase in the level of expression of phospholamban and a decrease in the expression of calsequestrin (Leberer et al., 1989).

III. ORGANIZATION OF THE Ca^{2+}-ATPASE IN THE MEMBRANE

The most detailed information about the organization of the Ca^{2+}-ATPase in the membrane has come from electron crystallographic studies of two- and three-dimensional crystals. Incubation of SR with vanadate (an analog of phosphate) in the absence of Ca^{2+} leads to the formation of 2D crystals with a dimeric unit cell (P2). If, however, SR is incubated in the presence of lanthanide ions or high concentrations of Ca^{2+}, then a monomeric unit cell is observed (P1; Taylor et al., 1988). The P1-type membrane crystal is presumably related to the E1 state of the ATPase and the P2-type membrane crystal is related to the E2 state.

In-plane projections of the crystalline arrays observed in the presence of vanadate and praseodymium are shown in Figure 3 and show a pear-shaped profile for the Ca^{2+}-ATPase, very similar in the P1 and P2 crystals, the difference being the presence of dimers in the vanadate-induced crystals. A three-dimensional reconstruction of the vanadate-induced crystals with uranyl acetate staining is shown in Figure 4. The reconstruction shows that interactions between ATPase molecules are of two types, occurring at different heights above the membrane surface. Molecules are linked to form dimers by a bridge which is located 42 Å above the bilayer surface. Projecting lobes located about 28 Å above the membrane surface then link dimers into ribbons (Taylor et al., 1986). These conclusions are in agreement with studies of three-dimensional microcrystals of the Ca^{2+}-ATPase grown from solution in the detergent $C_{12}E_8$ in the presence of Ca^{2+} (Stokes and Green, 1990b; Figure 5). The extramembranous domain of the ATPase is observed almost exclusively on one side of the membrane (corresponding to the cytoplasmic side in SR vesicles), and consists of the pear-shaped head (65 × 40 × 50 Å) centered about 35 Å above the cytoplasmic surface of the membrane and connected to it by a stalk of 28 Å diameter (Stokes and Green, 1990b). The pear shape is produced by a smaller lobe protruding from the stalk centered some 30 Å above the membrane surface, leaving a gap of ca. 16 Å between it and the membrane surface.

The greatest detail about the structure of the Ca^{2+}-ATPase has come from studies of vanadate-induced crystals of the ATPase embedded in ice using cryoelectron microscopy (Toyoshima et al., 1993). As found previously (see Figure 4), ribbons of dimer are observed, produced by two strands of Ca^{2+}-ATPase molecules running in opposite directions along a helical track (Figure 6). The thickness of the membrane, as defined by the distance between the two nearly continuous bands shown in Figures 6 and 7c is ca. 32 Å (Toyoshima et al., 1993). This probably corresponds to the distance between phospholipid phosphate groups across the bilayer, but is rather smaller than the separation of ca. 38 Å measured for pure lipid bilayers at room temperature. The thickness of the bilayer is, however, known to decrease with decreasing temperature and this may account for the observed result.

The total height of the ATPase molecule is 120 Å and it extends some 75 Å above the membrane surface. The cytoplasmic region has been likened to the head and

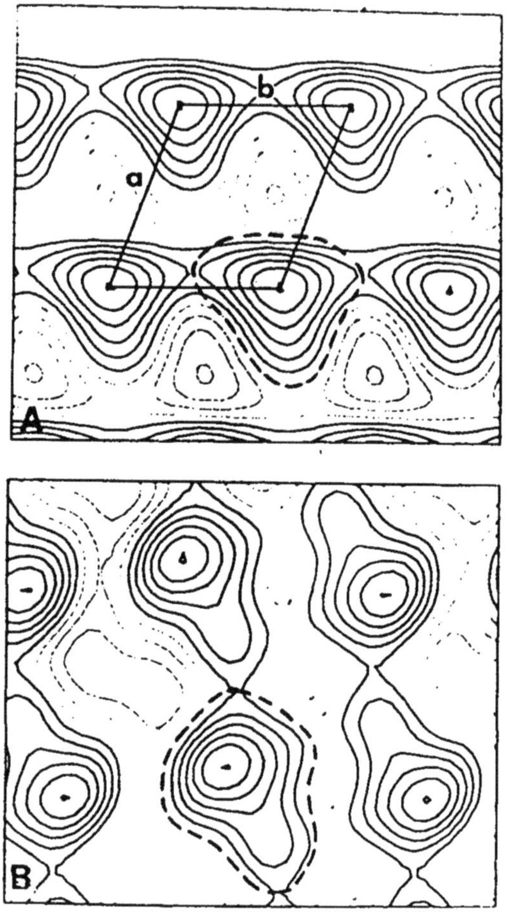

Figure 3. Electron density maps of (**A**) gadolinium-induced P1 crystals and (**B**) vanadate-induced P2 crystals. One molecule of the Ca^{2+}-ATPase is outlined with a dashed line and shows a pear-shaped profile in projection. In (**A**) the unit cell is marked, with dimensions a = 6.2 nm and b = 5.4 nm. (Reproduced with permission from Dux et al., 1985.)

neck of a bird (Toyoshima et al., 1993). The head, responsible for the formation of dimer ribbons, is connected to the stalk which is 25 Å long and composed of two segments. At the top of the stalk, the two segments separate to form a cavity (Figures 7 and 8). A cavity or groove is also visible in the head region, near the base of the "beak" (Figure 8). The transmembrane part of the ATPase consists of three segments (A, B, and C in Figures 7 and 8) clearly resolved in the hydrophobic core of the membrane. The largest segment, segment A, is composed of two parts, one oriented vertically (A2), the other (A1) being inclined so that the two parts are separated at

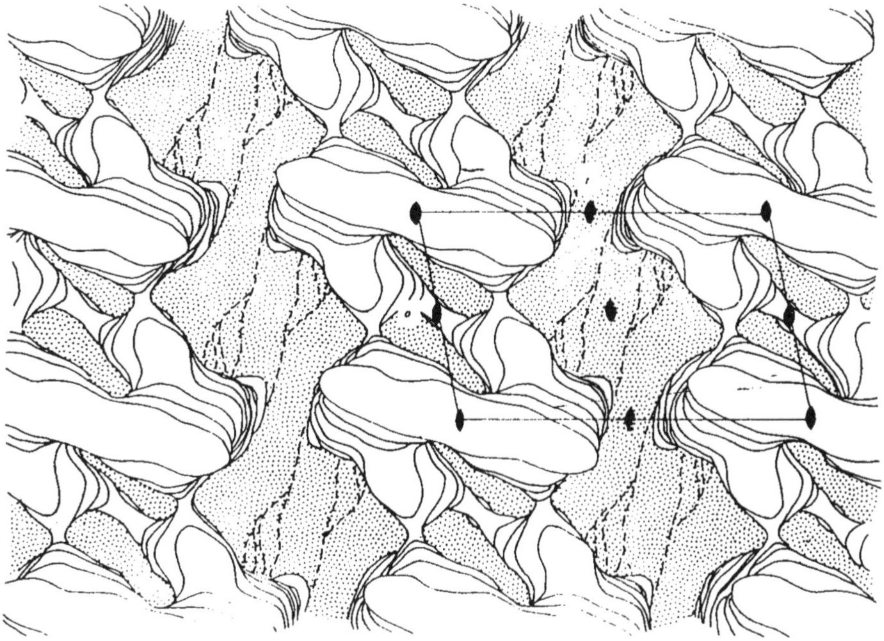

Figure 4. A three-dimensional reconstruction of vanadate-induced crystals. The reconstruction is viewed from the cytoplasmic side. Ca^{2+}-ATPase monomers are connected by a bridge to form dimers. Dimers are connected to form ribbons through a projecting lobe. The stippling shows the approximate location of the lipid bilayer surface. (Reproduced with permission from Taylor et al., 1986.)

the luminal surface (Figure 7c). On the cytoplasmic side, A1 continues into the stalk, whereas A2 is displaced from the stalk and is connected to segment B at both surfaces of the membrane (Figure 8). Whereas segment B merges with A2 on the cytoplasmic side, the 40° inclination of segment B means that it is ca. 40 Å away from A2 on the luminal side; a small luminal domain (marked by asterisks in Figures 7a, 7c, and 7d) links B and A2. C, the third transmembrane segment, is curved and extends ca. 20 Å into the cytoplasm. Toyoshima et al. (1993) and Stokes et al. (1994) have suggested that segment B contains the seventh transmembrane helix (M7) and one other helix, probably helix 9, and that the small luminal domain contains the M7–M8 loop, so that M8 runs through segment A2. Stokes et al. (1994) also suggest that segment C contains M1 and M10, and that M4 and M5 are likely to be located in the A1 segment with M6 and M8 located in segment A2. The gap between A1 and A2 could then correspond to the opening of the ATPase from which Ca^{2+} is released into the lumen of the SR (Toyoshima et al., 1993).

Information about the structure of the ATPase has also been obtained using fluorescence energy transfer to determine distances between defined residues on

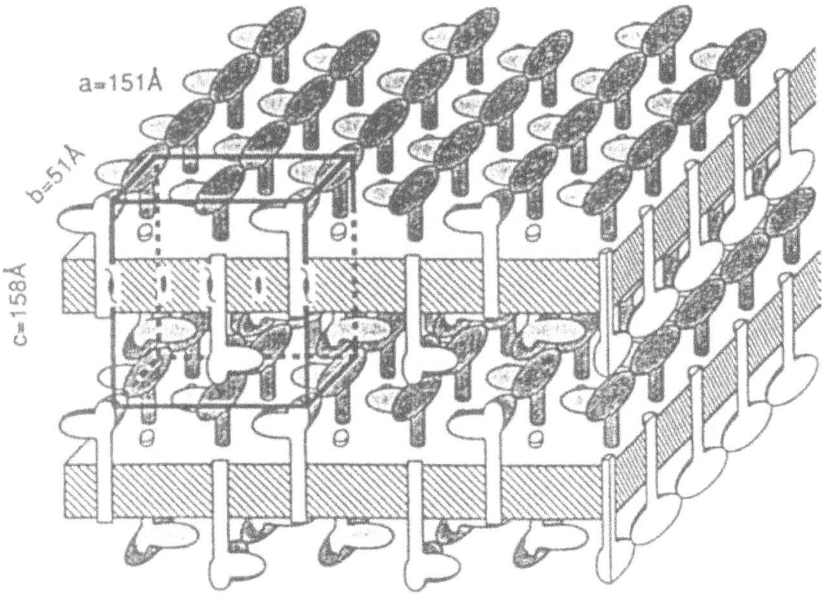

Figure 5. Packing of Ca^{2+}-ATPase molecules in microcrystals grown from $C_{12}E_8$ in the presence of Ca^{2+}. The Ca^{2+}-ATPase molecules are shown as two ellipsoids on a narrow cylinder. A unit cell is shown at the left. Contacts between ATPase molecules occur along the axis of stacking in the crystal (c), between the tops of the Ca^{2+}-ATPase heads and along the axis of ribbons (b) between the sides of the ATPase heads. (Reproduced with permission from Stokes and Green, 1990b, by copyright permission of the Biophysical Society.)

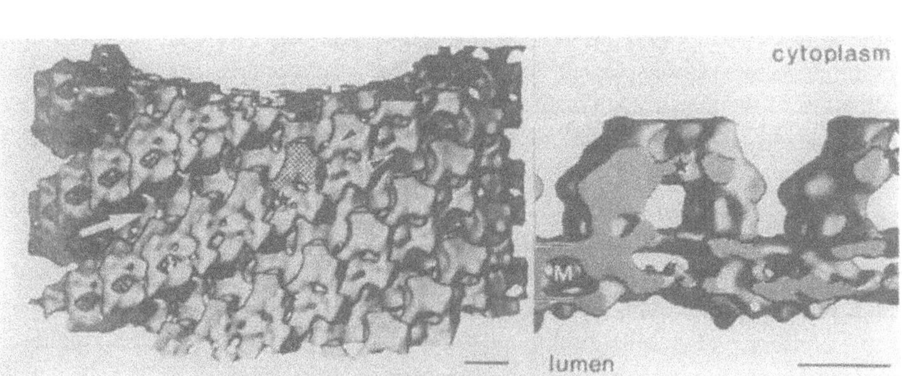

Figure 6. Surface model of the Ca^{2+}-ATPase viewed along the membrane surface. The two leaflets of the lipid bilayer (M) are shown. The Ca^{2+}-ATPase molecules are linked on the cytoplasmic surface into dimer ribbons. The small domains on the luminal surface (marked by white asterisks) also appear to make a link between dimer ribbons. The bar represents 50 Å. (Reproduced with permission from Toyoshima et al., 1993, copyright Macmillan Magazines Limited.)

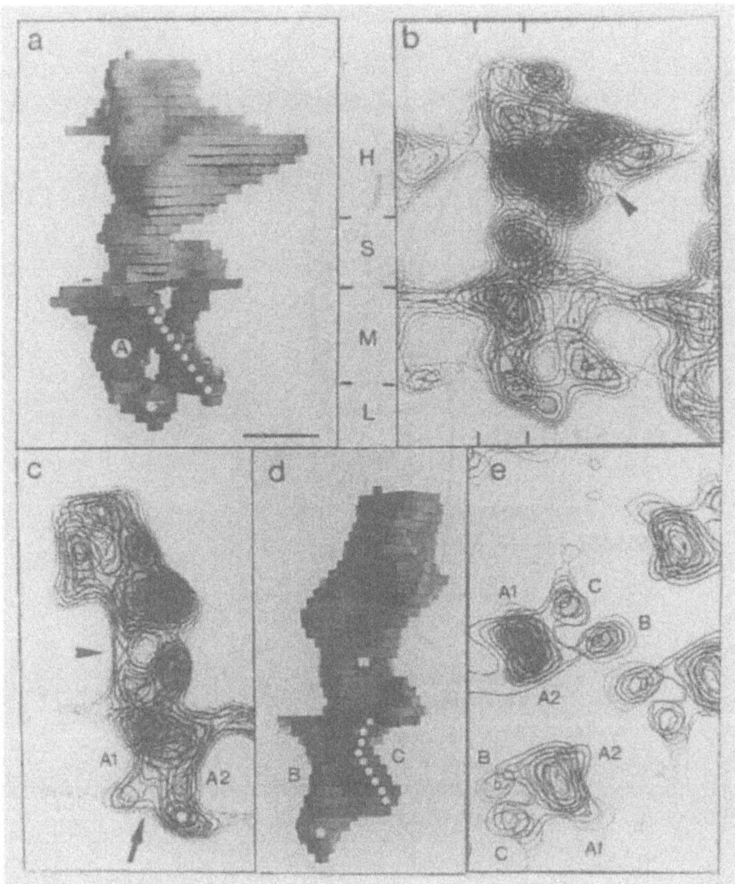

Figure 7. Three-dimensional structure of the Ca^{2+}-ATPase. Views perpendicular to the dimer ribbon are shown in (**a**) and (**b**), along the ribbon in (**c**) and (**d**), and normal to the membrane in (**e**). Equivalent views of a wooden model (**a**) and a stack of transparent sections (**b**) show the head portion (H), the stalk (S), the transmembrane region (M), and the luminal domain (L). The two nearly continuous densities flanking M represent the phospholipid headgroup regions of the bilayer. The line of white dots in (**a**) represents the segment B shown in Figure 8. (**c**) is a view at right angles to that shown in (**b**), showing the structure of the A segment (Figure 8) and the stalk. The arrowhead indicates the cavity at the top of the stalk region, and the arrow shows the separation of the two parts of segment A at the luminal surface. (**d**) a view of the wooden model shown in (**a**) looking from right to left. The line of white dots represents the segment C shown in Figure 8. (**e**) stacks of sections cut parallel to the membrane surface showing the three-transmembrane segments. The scale bar in (**a**) corresponds to 25 Å. (Reproduced with permission from Toyoshima et al., 1993, copyright Macmillan Magazines Limited.)

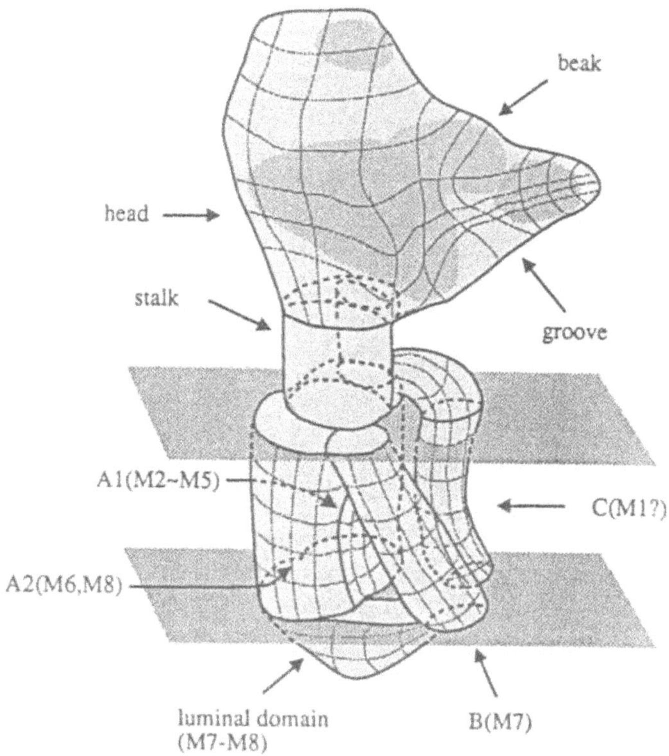

Figure 8. Proposed structure of the Ca^{2+}-ATPase. (Reproduced with permission from Toyoshima et al., 1993, copyright Macmillan Magazines Limited.)

the ATPase (Scott, 1985; Teruel and Gomez-Fernandez, 1986; Herrmann et al., 1986; Gutierrez-Merino et al., 1987; Squier et al., 1987; Birmachu et al., 1989; Munkonge et al., 1989; Bigelow and Inesi, 1991; Baker et al., 1993; Corbalan-Garcia et al., 1993; Mata et al., 1993; Stefanova et al., 1993b). The heights of particular residues on the ATPase above the bilayer surface have been determined by labeling the ATPase with a suitable donor fluorophore and reconstituting it into bilayers of phospholipid containing phosphatidylethanolamine (PE) labeled in the headgroup region with a suitable acceptor fluorophore. A complication is that the definition of the surface of the membrane in fluorescence studies is likely to be different to that in electron microscopy. It has been shown that the fluorescence properties of phosphatidylethanolamine labeled with the dansyl group in the headgroup region are consistent with a conformation in which the dansyl group is folded back and penetrates into the bilayer (Ghiggino et al., 1981). It, therefore, seems likely that the phospholipid–water interface as defined by fluorescence probes will correspond to the level of the glycerol backbone of the phospholipid. The surface of the membrane as defined by electron microscopy is likely to correspond to the phospholipid headgroup region. Since a phosphatidylcholine headgroup extends approximately 15 Å from the glycerol backbone region in crystals of the phospholipid (Pearson and Pascher, 1979), definitions of the membrane surface by fluorescence and electron microscopy may differ by as much as 15 Å.

Lys-515 in the nucleotide binding domain of the ATPase can be labeled with FITC. Since binding of ATP inhibits labeling with FITC, it is assumed that Lys-515 is part of the ATP binding site, the hydrophobic fluorescein group presumably occupying the adenine-binding region. The distance between FITC and the phospholipid–water interface has been measured as ca. 80 Å, putting Lys-515 on the top surface of the ATPase (Gutierrez-Merino et al., 1987; Munkonge et al., 1989; Figure 9). The height above the phospholipid–water interface of Cys-670/Cys-674 in the hinge region of the ATPase labeled with IAEDANS has been determined as 54 Å (Baker et al., 1993). Corbalan-Garcia et al. (1993) obtained a significantly smaller height for Cys-670/Cys-674 above the membrane surface (39 Å), but under the labeling conditions used, Cys residues other than Cys-670/Cys-674 may have been labeled. As described below, the hinge region is believed to make up part of the ATP-binding site of the ATPase. It has been suggested by Toyoshima et al. (1993) that the groove on the cytoplasmic domain of the ATPase located ca. 40 Å above the surface of the membrane (Figure 8) could represent the ATP-binding site. The hinge region of the ATPase could, therefore, constitute the opening of the groove. If the binding site for ATP were arranged with the γ-phosphate of the ATP binding near the mouth of the groove and the adenine binding in more hydrophobic buried regions, the binding site would then extend to ca. 69 Å above the surface of the membrane (an ATP molecule is ca. 15 Å in length). If Lys-515 were part of the ATP-binding site, then these estimates would locate Lys-515 closer to the membrane surface than the estimates made using FITC-labeled ATPase. However, it remains possible that Lys-515 is not a residue that actually lines the ATP-binding site, but is located some way from the true binding site, competition between FITC and ATP on the ATPase arising from the bulky nature of the fluorescein ring system.

Sites at or near the Ca^{2+}-binding sites on the ATPase can be labeled with the fluorescent carbodiimide N-cyclohexyl-N′-(4-dimethylamino-1-naphthyl)carbodiimide (NCD-4); the labeled sites were located ca. 20 Å from the phospholipid–water interface (Munkonge et al., 1989). Originally it was suggested that the sites were in the cytoplasmic region of the ATPase, but the energy-transfer measurements would be equally consistent with a location within the transmembranous region of the ATPase; such a location would be consistent with the observed quenching of fluorescence by spin-labeled fatty acids which will partition into the phospholipid bilayer (Munkonge et al., 1989). As described below, experiments using site-directed mutagenesis have suggested that Ca^{2+} binding involves residues in postulated transmembranous α-helices (Clarke et al., 1989). Lanthanide ions have also

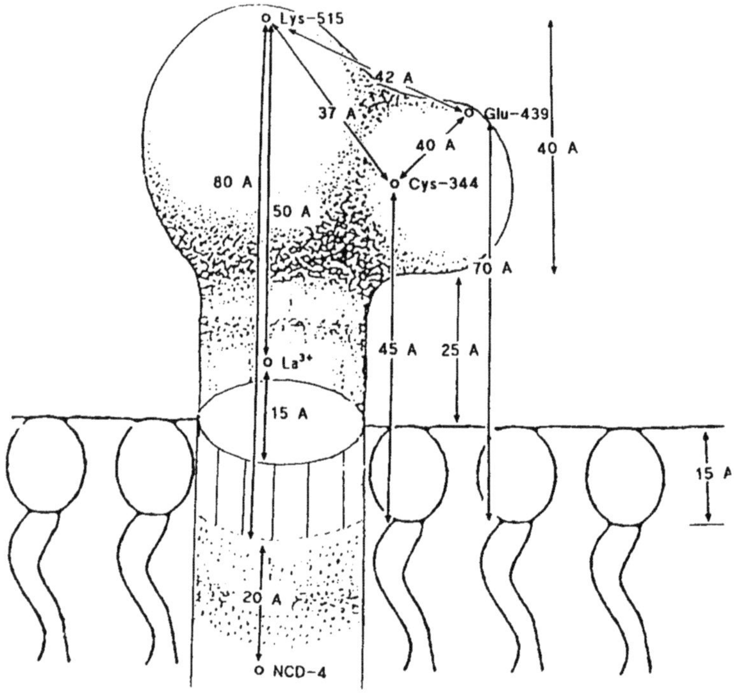

Figure 9. Location of residues on the ATPase as defined by fluorescence energy transfer. (**A**) Positions of Lys-515, Cys-344 and Glu-439 are given on the structure as deduced by Stokes and Green (1990a) from studies of negatively stained crystals of the ATPase. Also shown are the locations of sites labeled by N-cyclohexyl-N″-(d-dimethylamino-1-naphthyl)carbodiimide (NCD-4), believed to be associated with the binding of Ca^{2+}, and a possible location for site(s) binding lanthanides (Mata et al., 1993). (**B**) A scaled perspective drawing showing distances measured on the ATPase (Baker et al., 1993). (Continued)

been used to locate Ca^{2+} ion binding sites, but it is unclear whether the lanthanides bind at the 'true' Ca^{2+}-binding sites or at some other metal ion binding site(s) on the ATPase (Fujimori and Jencks, 1990; Imamura and Kawakita, 1991; Ogurusu et al., 1991; Henao et al., 1992). Scott, (1985) using Tb^{3+} as the probe for Ca^{2+}, has suggested a separation of 47 Å between the Tb^{3+}-binding site and Lys-515, which, with a location for Lys-515 80 Å above the surface, would put the Tb^{3+}-binding site ca. 30 Å above the surface. X-ray diffraction studies have located binding sites for lanthanides ca. 12 Å above the phospholipid polar headgroup region of the bilayer (Asturias and Blasie, 1991); with a thickness of the headgroup region of 15 Å, the site locations estimated by X-ray diffraction and fluorescence would be in close agreement. These experiments suggest that lanthanides bind in the stalk region of

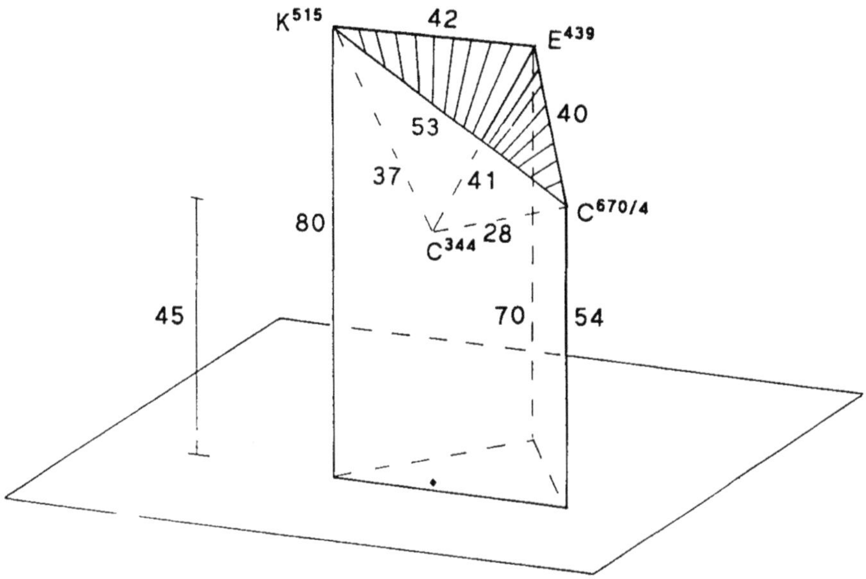

Figure 9. (Continued)

the ATPase—it is unclear whether these binding sites have any role in the normal function of the ATPase.

Cys residues 670 and 674 can be labeled with IAEDANS. Structure predictions suggest that this region of the ATPase will be α-helical giving a separation between the two Cys residues of 6.1 Å. These two residues will, therefore, be located close enough in the three-dimensional structure of the ATPase to be treated as a single site; a distance of 53 Å has then been estimated to FITC at Lys-515 (Bigelow and Inesi, 1991; Baker et al., 1993). Bigelow and Inesi (1991) have determined the separation between IAEDANS and two Cys residues designated MAL A and MAL B labeled with maleimide derivatives as 37- and 28-Å. The measured separation between MAL A and FITC at Lys-515 is the same as that measured between Cys-344 and Lys-515 (Mata et al., 1993) suggesting that MAL A is Cys-344 and thus giving the Cys-344 to IAEDANS distance as 28 Å. These measurements suggest a location for Cys-344 (and thus of the residue phosphorylated by ATP, Asp-351) toward the center of the ATPase molecule. The height of Cys-670/Cys-674 above the phospholipid–water interface has been estimated by measuring resonance energy transfer between IAEDANS-labeled ATPase and FITC-PE and, as described above, has been found to be 54 Å. The separation between bound Pr^{3+} and IAEDANS at Cys-670 and Cys-674 has been estimated as 18 Å (Squier et al., 1987). It has been suggested that the binding site for lanthanides is located ca. 15 Å above the phospholipid polar headgroup region or ca. 30 Å above the glycerol backbone region (Mata et al., 1993), giving a Pr^{3+}-IAEDANS separation of ca. 24 Å, in reasonable agreement with the direct estimate.

Cys-344, close to the residue (Asp-351) phosphorylated by ATP, and Glu-439 have also been labeled and distances estimated, as shown in Figure 9 (Baker et al., 1993; Mata et al., 1993; Stefanova et al., 1993b). As shown, if Lys-515 is on the top surface of the ATPase, this would define the larger lobe as the nucleotide-binding domain, with Cys-344 on the smaller lobe, representing the phosphorylation domain (Mata et al., 1993). Location of Lys-515 on the top surface of the ATPase would be consistent with the considerable surface exposure of the protein in this region as defined by the binding of antibodies to the native ATPase (Colyer et al., 1989; Mata et al., 1992). Studies of the binding of antipeptide antibodies raised to the phosphorylation domain of the ATPase also suggest that much of this domain is also surface exposed (Mata et al., 1992).

Stahl and Jencks (1987) have suggested a conformational change on the ATPase following binding of ATP to the ATPase in the presence of Ca^{2+}, serving to relocate the nucleotide-binding and phosphorylation domains on the ATPase, bringing the γ-phosphate of ATP close to Asp-351. Distance measurement using fluorescence energy transfer gave no evidence for any major relocation of Cys-344 in the phosphorylation domain on binding Ca^{2+} or vanadate, either with respect to Lys-515 in the nucleotide-binding domain or with respect to the phospholipid–water interface. Similarly, binding of Ca^{2+} or vanadate had no effect on the distances between Lys-515 or Glu-439 and the phospholipid–water interface (Gutierrez-Merino et al., 1987; Stefanova et al., 1992b) or on the separation between the sites labeled by NCD-4 and the phospholipid–water interface (Munkonge et al., 1989). It appears that the conformational differences between the E1 and E2 states of the ATPase and between Ca^{2+}-bound and -free forms must be localized in small regions of the structure. This is also consistent with the observation that most monoclonal antibodies binding to the native ATPase have no significant effect on activity (Colyer et al., 1989).

IV. THE TRANSMEMBRANE REGION OF THE Ca^{2+}-ATPASE

A hydropathy plot for the Ca^{2+}-ATPase of SR using the scale of Hudecek and Anzenbacher (1988) is shown in Figure 10. Four clear transmembrane helices are found in the N-terminal region. In the C-terminal region the plots are less clear, but six helices of lengths of ca. 20 amino acids can be identified (Figure 11). Other ATPases of this class also contain four clear transmembrane helices at the N-terminus, but at the C-terminus the number could be four to six transmembrane helices; the exact number is still uncertain. Boundaries of transmembrane helices are often difficult to determine from hydropathy plots, but further information can be obtained by comparison of Ca^{2+}-ATPase sequences (Figure 12). Transmembrane helix M2 has been started at Phe-88 since the *Plasmodium falciparum* sequence contains an insert at this position (Kimura et al., 1993). Similarly, the *Plasmodium*

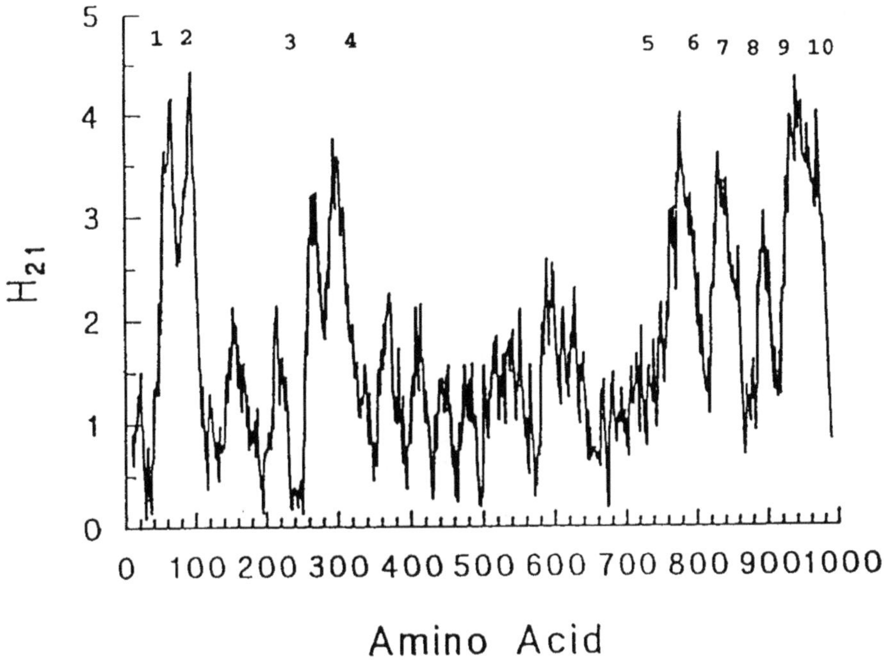

Figure 10. A hydropathy plot for the Ca^{2+}-ATPase of SR using the scale of Hudecek and Anzenbacher (1988) with a sliding window of 21 residues. Transmembrane helices are taken to correspond to regions for which $H_{21} \geq 2.5$.

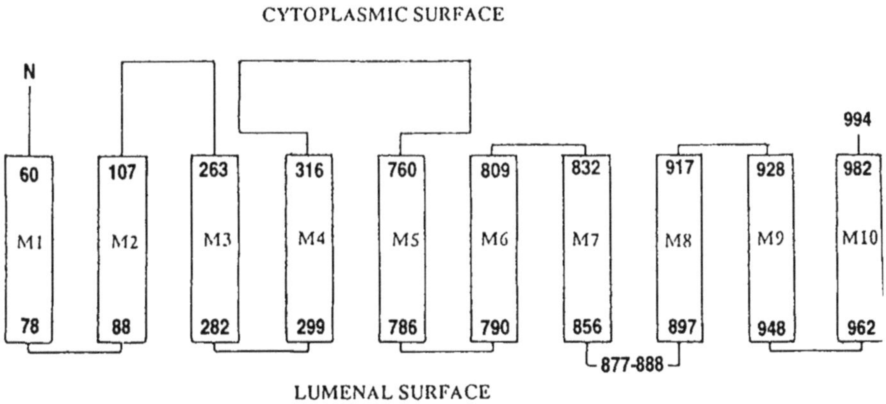

Figure 11. Numbering of the putative ten-transmembrane helices for the Ca^{2+}-ATPase of SR.

M1

60

1	L	L	V	R	I	L	L	L	A	A	C	I	S	F	V	L	A	W	F
2	L	L	V	R	I	L	L	L	A	A	C	I	S	F	V	L	A	W	F
3	L	L	V	R	I	L	L	L	A	A	L	V	S	F	V	L	A	W	F
4	L	L	V	K	I	L	L	L	A	A	I	I	S	F	V	L	A	L	F
5	L	L	V	K	I	L	L	L	A	A	F	I	S	F	V	L	T	L	L
6	L	L	V	K	I	L	L	L	A	A	F	V	S	F	A	L	T	L	L
7	L	L	V	K	I	L	L	L	A	A	I	I	S	L	V	L	A	L	F
8	T	L	V	K	I	L	L	G	A	A	F	I	S	F	V	L	A	Y	V
9	A	M	C	V	V	L	I	L	A	A	A	L	S	F	G	T	T	D	W

M2

88

1	F	V	E	P	F	V	I	L	L	I	L	I	A	N	A	I	V	G	V	W
2	F	V	E	P	F	V	I	L	L	I	L	V	A	N	A	I	V	G	V	W
3	F	V	E	P	L	V	I	M	L	I	L	V	A	N	A	I	V	G	V	W
4	F	V	E	P	L	V	I	L	L	I	L	I	A	N	A	V	V	G	V	W
5	F	I	E	P	L	V	I	V	L	I	L	I	L	N	A	A	V	G	V	W
6	F	I	E	P	V	V	I	L	M	I	L	I	L	N	A	A	V	G	V	W
7	Y	V	E	P	F	V	I	L	L	I	L	I	A	N	A	V	V	G	V	W
8	Y	V	E	P	L	V	I	L	W	I	L	V	L	N	A	I	V	G	V	W
9	-	-	-	-	-	V	I	S	A	I	I	V	L	N	I	T	V	G	F	I

M3

263

1	V	I	S	L	I	C	V	A	V	W	L	I	N	I	G	H	F	N	D	P
2	V	I	S	L	I	C	I	A	V	W	I	I	N	I	G	H	F	N	D	P
3	A	I	S	V	I	C	V	A	V	W	V	I	N	I	G	H	F	A	D	P
4	V	I	S	V	I	C	V	A	V	W	A	I	N	I	G	H	F	N	D	P
5	I	I	F	V	I	C	V	T	V	W	I	I	N	F	K	H	F	S	D	P
6	I	I	F	I	I	C	V	H	V	W	I	I	N	F	K	H	F	S	D	P
7	V	I	S	V	I	C	V	A	V	W	A	I	N	I	G	H	F	N	D	P
8	A	I	G	V	V	C	L	V	V	W	A	I	N	Y	K	Y	F	L	S	W
9	K	L	T	V	L	A	Y	I	L	F	C	I	A	I	I	L	A	I	I	V

Figure 12. Comparison of sequences of putative transmembrane helices for the Ca^{2+}-ATPases. (1) SERCA1 fast-twitch rabbit skeletal muscle (Brandl et al., 1986); (2) SERCA2, slow-twitch rabbit skeletal muscle (MacLennan et al., 1985); (3) SERCA3 rat kidney (Burk et al., 1989); (4) *Drosophila melanogaster* (Magyar and Varadi, 1990); (5) *Plasmodium falciparum* (Kimura et al., 1993); (6) *Plasmodium yoelii* (Murakami et al., 1990); (7) *Artemia* (Palmero and Sastre, 1989); (8) tomato (Wimmers et al., 1992); (9) cta3 protein of yeast (Ghislain et al., 1990). For M4, M5, M6, and M8 the sequence (10) of human plasma membrane Ca^{2+}-ATPase (Verma et al., 1988) is also given. (Continued)

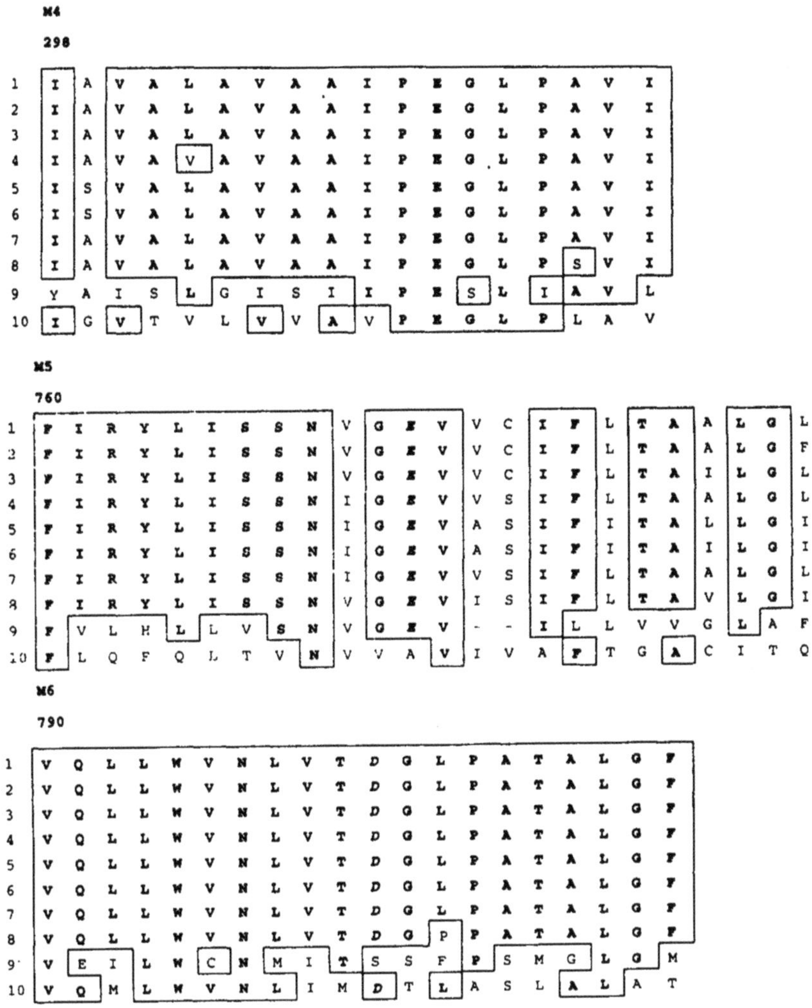

Figure 12. (Continued)

falciparum, *Plasmodium yoelii*, and tomato sequences contain inserts close to Phe-856 (Murakami et al., 1990; Wimmers et al., 1992; Kimura et al., 1993) making this a likely end of M7. Inserts in the *Plasmodium falciparum* and tomato sequences (Wimmers et al., 1992; Kimura et al., 1993) at Pro-897 make this a likely start for M8. A triple mutant at Ile-Thr-Thr-317 led to a functionally inactive mutant, but one which was stably incorporated into the membrane, suggesting that this sequence represents the end of M4 and the start of stalk S4 (Vilsen et al., 1991).

M7

832

1	W	L	F	F	R	Y	M	A	I	G	G	Y	V	G	A	A	T	V	G	A	A	A	W	W	F
2	W	L	F	F	R	Y	L	A	I	G	C	Y	V	G	A	A	T	V	G	A	A	A	W	W	F
3	W	L	F	F	R	Y	L	A	I	G	V	Y	V	G	L	A	T	V	A	A	A	T	W	W	F
4	W	L	F	F	R	Y	M	A	I	G	F	Y	V	G	A	A	T	V	G	A	A	A	W	W	F
5	L	T	L	L	R	Y	I	I	I	G	T	Y	V	G	I	A	T	V	S	I	F	V	Y	W	F
6	L	T	L	L	R	Y	I	V	I	G	T	Y	V	G	I	A	T	V	S	I	F	I	Y	W	Y
7	W	L	F	F	R	Y	M	A	I	G	T	Y	V	G	A	A	T	V	G	A	A	A	H	W	F
8	W	V	F	F	R	Y	M	A	I	G	G	Y	V	G	I	A	T	V	G	I	F	I	V	W	Y
9	S	L	I	N	D	M	M	V	Y	G	F	F	L	G	V	V	S	L	M	T	W	V	V	I	M

M8

896

1	P	M	T	M	A	L	S	V	L	V	T	I	E	M	C	N	A	L	N	S	L
2	P	M	T	M	A	L	S	V	L	V	T	I	E	M	C	N	A	L	N	S	L
3	P	T	T	M	A	L	S	V	L	V	T	I	E	M	C	N	A	L	N	S	V
4	A	M	T	M	A	L	S	V	L	V	T	I	E	M	L	N	A	M	N	S	L
5	A	S	T	L	S	L	S	V	L	V	L	I	E	M	F	N	A	L	N	A	L
6	A	S	T	L	S	L	S	V	L	V	L	I	E	M	F	N	A	L	N	A	L
7	P	M	T	M	A	L	S	V	L	V	T	I	E	M	L	N	I	L	N	S	L
8	A	M	T	L	S	L	S	V	L	V	A	I	E	M	F	N	S	L	N	A	L
9	A	V	V	T	F	C	I	L	I	M	A	V	E	V	K	N	F	D	N	S	L
10	H	Y	T	I	V	F	N	T	F	V	L	M	Q	L	F	N	E	I	N	A	R

M9

928

1	W	V	N	I	W	L	L	G	S	I	C	L	S	M	S	L	H	F	L	I	L
2	W	E	N	I	W	L	V	G	S	I	C	L	S	M	S	L	H	F	L	I	L
3	W	L	N	P	W	L	L	G	A	V	V	M	S	M	A	L	H	F	L	I	L
4	W	C	N	L	W	L	I	G	S	M	A	L	S	F	T	L	H	F	V	I	L
5	W	R	N	M	Y	L	V	L	A	T	I	G	S	L	L	L	H	V	L	I	L
6	W	R	N	M	Y	L	V	L	A	T	I	G	S	L	F	L	H	C	L	I	I
7	W	S	N	I	W	L	I	S	A	I	C	L	S	M	T	L	H	F	V	I	L
8	W	R	N	P	W	L	L	V	A	M	S	L	S	F	A	L	H	S	V	I	L
9	V	E	N	K	F	L	A	W	A	I	A	L	A	A	V	S	V	F	P	T	I

M10

962

1	L	D	L	T	Q	W	L	M	V	L	K	I	S	L	P	V	I	G	L	D	E
2	L	N	V	T	Q	W	L	M	V	L	K	I	S	L	P	V	I	L	M	D	E
3	L	S	G	R	Q	W	G	V	V	L	Q	M	S	L	P	V	I	L	L	D	E
4	L	S	A	E	E	W	I	T	V	M	K	F	S	I	P	V	V	L	L	D	E
5	L	S	A	Y	D	W	F	L	V	F	L	W	S	F	P	V	I	I	L	D	E
6	L	T	L	H	D	W	F	L	V	F	L	W	S	F	P	V	I	I	I	D	E
7	L	T	L	T	E	W	I	V	V	L	K	I	S	F	P	V	E	L	L	D	E
8	L	S	L	Y	E	W	L	L	V	I	L	L	S	A	P	V	I	L	I	D	E
9	Y	I	G	W	E	W	G	V	V	A	V	A	V	M	F	Y	F	F	Y	V	E

Figure 12. (Continued)

Immunological studies have been used to clarify the number and arrangement of the transmembrane helices. An even number of transmembrane helices is indicated since antipeptide antibodies raised to the N- and C-terminal sequences of the Ca^{2+}-ATPase have both been shown to bind to the ATPase in sealed SR vesicles, demonstrating a cytoplasmic location for both (Matthews et al., 1989). The cytoplasmic location of the N-terminus has also been shown by chemical labeling experiments (Reithmeier and MacLennan, 1981). It has also been demonstrated using antipeptide antibodies that both the N- and C-terminii of the (Na^+–K^+)-ATPase (Antolovic et al., 1991) are located on the cytoplasmic side of the membrane, again giving an even number of transmembrane helices. A monoclonal antibody whose epitope is located between M1 and M2 has been located at the extracellular surface of the Ca^{2+}-ATPase in erythrocytes (Feschenko et al., 1992), corresponding to a luminal location for the M1–M2 loop for the Ca^{2+}-ATPase of SR. A monoclonal whose epitope is located at the M3–M4 boundary has been found to bind to the (Na^+–K^+)-ATPase at extracellular sites (Kano et al., 1990), corresponding to a luminal location for the M3–M4 boundary for the Ca^{2+}-ATPase of SR.

The differences between the 8- and 10-helical models occur largely in the C-terminal half of the structure. In the 10-helical model, the C-terminal region up to residue 760 is very largely buried in the membrane, with the largest exposed region being a loop between residues 856 and 897, located in the lumen of the SR (Figure 11). In the 8-helical models, either the region between residues 856 and 897 is predicted to constitute a loop on the cytoplasmic surface of the SR with a large luminal loop between residues 794 and 836, or a large lumenal loop is predicted between residues 856 and 928 (Matthews et al., 1990). Antipeptide- and monoclonal-antibody binding studies have shown a luminal location for the 877–888 loop (Clarke et al., 1990a; Matthews et al., 1990). Treatment of the ATPase with proteinase K produces a 30 kD fragment, resistant to proteolysis, containing both the C-terminus and residues 877–888; the molecular weight would be consistent with a fragment from ca. residue 720 to the C-terminus. This suggests that much of this segment is transmembranous, with the loops connecting the transmembranous regions being relatively short and, therefore, protected from proteinase K cleavage (Matthews et al., 1990). These observations are consistent with the 10-helical model shown in Figure 11.

The similarity in hydropathy plots for the members of the P-type ATPase family suggests that they all have the same number of transmembrane α-helices. The topology of the Mg^{2+}-ATPase of *Salmonella typhimurium* was studied by constructing fusions with BlaM and LacZ, utilizing the observations that lactamase confers penicillin resistance when located extracytoplasmically and that LacZ is only functional when expressed cytoplasmically. All fusions were consistent with a 10-helix model, except for that with the BlaM protein fused at Pro-766 (Smith et al., 1993). However, a study of the (Na^+–K^+)-ATPase involving addition of antibody epitopes to defined regions of the ATPase was consistent with an 8-helix model; both the N- and C-termini were found to be cytoplasmic, but the evidence was against a transmembrane helix at the C-terminus, equivalent to M10 in the Ca^{2+}-ATPase (Canfield and Levenson, 1993; Figure 11).

The 10-helical model for the Ca^{2+}-ATPase is consistent with fluorescence quenching data. It has been shown that a maximum of 85% of the fluorescence of the tryptophan residues can be quenched by hydrophobic quenchers, consistent with the location of 10- of the 13-tryptophan residues (77%) in hydrophobic regions of the ATPase (Froud et al., 1986). Furthermore, it has been shown that binding of fatty acids to the membrane quenches up to 35% of the tryptophan fluorescence (Froud et al., 1986). Since the quenching mechanism involves close contact between the carboxyl group and the tryptophan residue quenched, and since the fatty acids are located in the membrane with their carboxyl groups at the lipid–water interface, this locates 5 (38%) of the 13 tryptophans at or close to the ends of the transmembrane helices. This is also consistent with experiments with water-soluble quenchers. KI is able to quench fluorescence by 33% and the tryptophan residues quenched by KI are also accessible for quenching from the lipid phase since the hydrophobic quencher $C_2Cl_4Br_2$ at 0.05 mM can quench 67% of the fluorescence, but addition of KI to the ATPase quenched with $C_2Cl_4Br_2$ results in little extra quenching (Froud et al., 1986).

Because the ions transported by the ATPases are positively charged, residues with negatively charged carboxyls (glutamic and aspartic acids) are likely to play a critical role in binding and transporting these cations. In terms of the 10-helix model, four negatively charged residues are found to be conserved in all the ER/SR Ca^{2+}-ATPase, except for the cta3 protein of yeast (Figure 12): Glu-309, Glu-771, Asp-800, and Glu-908 in M4, M5, M6, and M8, respectively. In cta3, Asp in M5 is replaced by Ser. Asp-800 in M6 is conserved in all the P-type ATPases and M4 contains a sequence PEXL found in all the ATPases except the H^+-ATPase of *N. crassa*; presumably the charged Glu residue is essential for the transport of all large cations, but not for H^+. The importance of the four negatively charged residues has been shown by site-specific mutagenesis, since replacement of the residues generally leads to loss of Ca^{2+}-specific phosphorylation of the ATPase by ATP and loss of Ca^{2+} inhibition of phosphorylation by P_i (Clarke et al., 1990b; Vilsen and Andersen, 1992b). However, mutagenesis experiments have also led to the suggestion that Glu-908 is not directly involved in Ca^{2+}-binding (Andersen and Vilsen, 1994, 1995). A chimeric ATPase containing helices M1–M4 of the (Na^+–K^+)-ATPase and helices M5–M10 of the Ca^{2+}-ATPase showed Ca^{2+} binding (Luckie et al., 1992); although the chimera contains only three of the Ca^{2+}-ATPase transmembrane helices believed essential for Ca^{2+} binding, as described above, M4 from the (Na^+–K^+)-ATPase contains the conserved region believed to be involved in Ca^{2+} binding and can thus, presumably, substitute for M4 from the Ca^{2+}-ATPase.

For an Asp-800 → Asn mutant of the Ca^{2+}-ATPase, Ca^{2+}-dependent phosphorylation by ATP was not observed, but Ca^{2+} inhibition of phosphorylation by P_i did occur, although 0.5 mM Ca^{2+} was required, suggesting a marked reduction in affinity for one of the two Ca^{2+} sites. For a Glu-309 → Gln mutant, again no Ca^{2+}-dependent phosphorylation by ATP was observed, but Ca^{2+} inhibition of phosphorylation by P_i occurred normally at 10 μM Ca^{2+}. In a Glu-771 → Gln mutant, Ca^{2+}-dependent phosphorylation by ATP was observed at high concentrations of Ca^{2+}. It was suggested that the Ca^{2+}-binding sites were organized with Glu-309 contributing to the second, outer most Ca^{2+}-binding site, and Asp-800 contributing to the first, inner most Ca^{2+}-binding site (Andersen and Vilsen, 1992); this, however, would not be consistent with sequence comparisons with the plasma membrane Ca^{2+}-ATPase described below. Subsequent studies have suggested that only a single Ca^{2+} ion can bind to the Glu-309 → Gln mutant, and that access to the Ca^{2+}-binding site is possible only from the luminal side, this binding being responsible for the observed Ca^{2+}-inhibition of phosphorylation by P_i (Skerjanc et al., 1993). Since phosphorylation of the Ca^{2+}-ATPase by P_i in the native ATPase is only inhibited by Ca^{2+} on the cytoplasmic side of the membrane (where it can bind to high affinity sites on the E1 conformation) and not by Ca^{2+} on the luminal side of the membrane, this implies a change in the accessibility of the Ca^{2+} sites as well as a change in affinity. It is clear that interpretation of these results without a full structure for the ATPase will be difficult.

Mutation of Gly residues in M4, M5, and M6 led to a complex pattern of results, some mutants exhibiting reduced Ca^{2+} affinity and others slowing a decreased rate of dephosphorylation of the phosphorylated ATPase (Andersen et al., 1992). Mutation of residues around Glu-309 in M4 had no effect on Ca^{2+} binding to the ATPase, but did prevent ATP hydrolysis and locked the phosphorylated ATPase in an ADP-insensitive state; these experiments make it clear that the M4 domain is involved in the Ca_2E1P–E2P conformation change as well as in Ca^{2+} binding (Rice et al., 1993).

It is likely that the four negatively charged residues in the ER/SR Ca^{2+}-ATPases are distributed two at each of the two Ca^{2+} binding sites. The plasma membrane Ca^{2+}-ATPase binds only 1 Ca^{2+} and contains only two of these acidic residues, corresponding to Glu-309 and Asp-800 in helices 4 and 6, respectively (Figures 12). It is, therefore, logical to suggest that in the Ca^{2+}-ATPase of SR, Glu-309 and Asp-800 (M4 and M6) make up one Ca^{2+}-binding site and Glu-771 and Glu-908 (M5 and M8) make up the other. Such an arrangement of these negatively charged residues is only possible in the 10-helix model for the ATPase. In a 7-helix model, for example, only one of the key residues (Glu-309) would be located within the transmembrane region (Moller et al., 1991).

The four helices M4, M5, M6, and M8 could be organized on the corners of a square with M4 and M6 making up one side and M5 and M8 the other (as in Figures 13 and 14) or with M4 and M8 making up one side and M5 and M6 the other (Lee et al., 1993). The required oxygen ligands around the Ca^{2+} can be provided by the negatively charged residues and the conserved Asn residues on helices 5, 6, and 8 (together with backbone oxygens), this introducing a marked bend into the helices (Figure 13). Eisenman and Dani (1987) have suggested that one role for proline

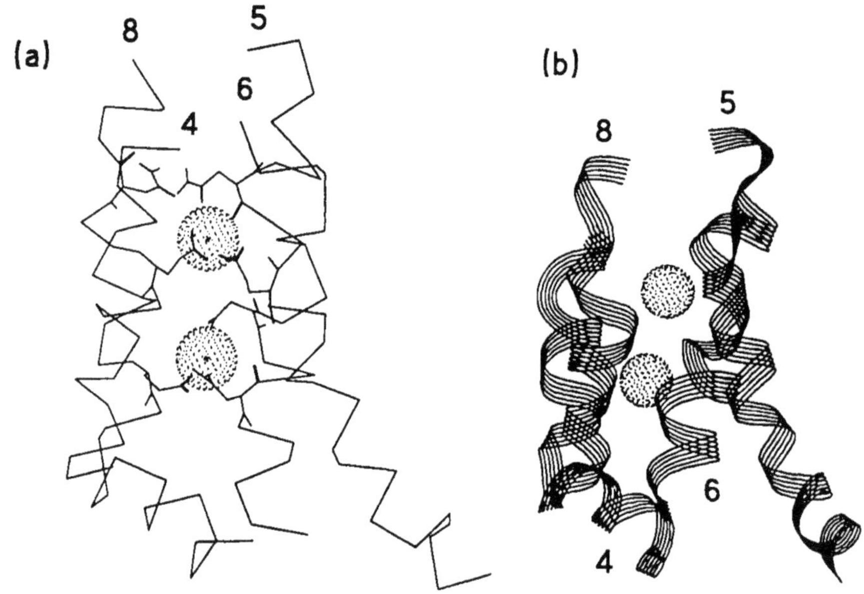

Figure 13. Possible arrangement of transmembrane α-helices making up the two Ca^{2+}-binding sites on the Ca^{2+}-ATPase showing (**a**) the α-carbon backbone and residues within 5Å of the two bound Ca^{2+} ions viewed parallel to the bilayer normal, with the cytoplasmic surface at the top and the luminal surface at the bottom. (**b**) A ribbon diagram of (**a**) showing the bent α-helical structure.

residues in ion channels is to provide a nonhydrogen-bonded carbonyl-oxygen for liganding to a cation. In the model shown in Figure 13, the carbonyl oxygen of Pro-308 in M4 could provide an extra ligand for the Ca^{2+}. As described later, mutation of Pro-308 leads to a reduction in Ca^{2+} affinity (Vilsen et al., 1989).

Evidence about the possible modes of binding of the two Ca^{2+} ions to the ATPase comes from kinetic studies. The two Ca^{2+} ions bind sequentially, since occupation of the second, outer site, prevents dissociation of Ca^{2+} from the first, inner site (Orlowski and Champeil, 1991a). Two possible models are shown in Figure 15. The first model proposes binding of the two Ca^{2+} ions in a channel-like structure. The second model envisages a conformation change on the ATPase following binding of the first Ca^{2+} ion. In the second model, in the absence of Ca^{2+}, only a single, inner site is available for binding Ca^{2+}. Following binding of Ca^{2+} to this initial site to give CaE1, the ATPase undergoes a conformational change to give CaE1′ with the appearance of the second Ca^{2+}-binding site. Binding of Ca^{2+} to this second, outer, site then gives Ca_2E1′. Evidence in favor of the second model comes from fluorescence studies. It has been shown that the tryptophan fluorescence

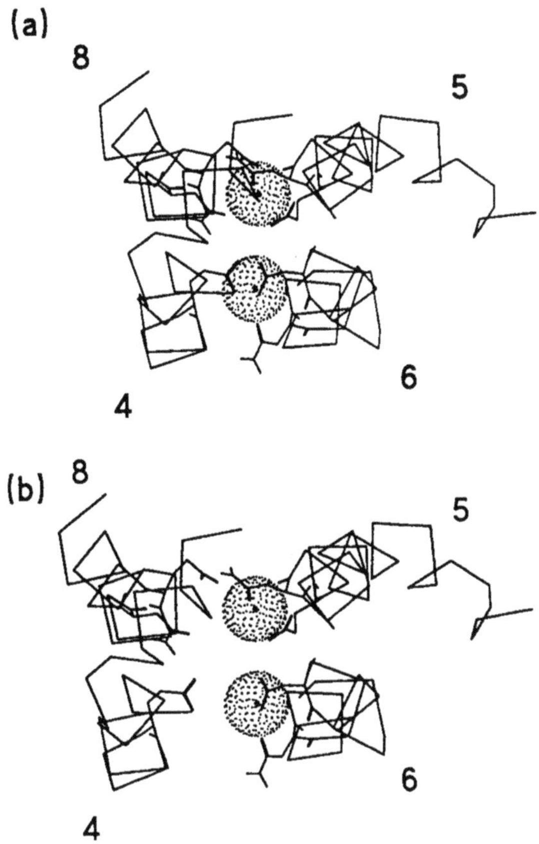

Figure 14. (**a**) The suggested arrangement of transmembrane α-helices shown in Figure 13, viewed from the cytoplasmic surface of the membrane. (**b**) A rearrangement of helices M4 and M8 relative to helices M6 and M5 that would lead to low affinity binding of Ca^{2+}.

intensity for the ATPase changes on binding Ca^{2+} and that, under a wide variety of conditions, changes in tryptophan fluorescence intensity accurately follow changes in the occupancy of the Ca^{2+}-binding sites on the ATPase (Henderson et al., 1994a,1994b). The change in tryptophan fluorescence intensity on binding Ca^{2+} is unlikely to follow directly from occupation of the two Ca^{2+}-binding sites on the ATPase; this would require equal changes in fluorescence for binding at the two Ca^{2+}-binding sites and there is no reason to expect an equal distribution of the 13 tryptophan residues in the ATPase about the two sites. However, either model shown in Figure 15 could account for a tryptophan fluorescence change that reflects Ca^{2+} occupancy if the equilibrium constant for the CaE1–CaE1′ step is equal to 1.

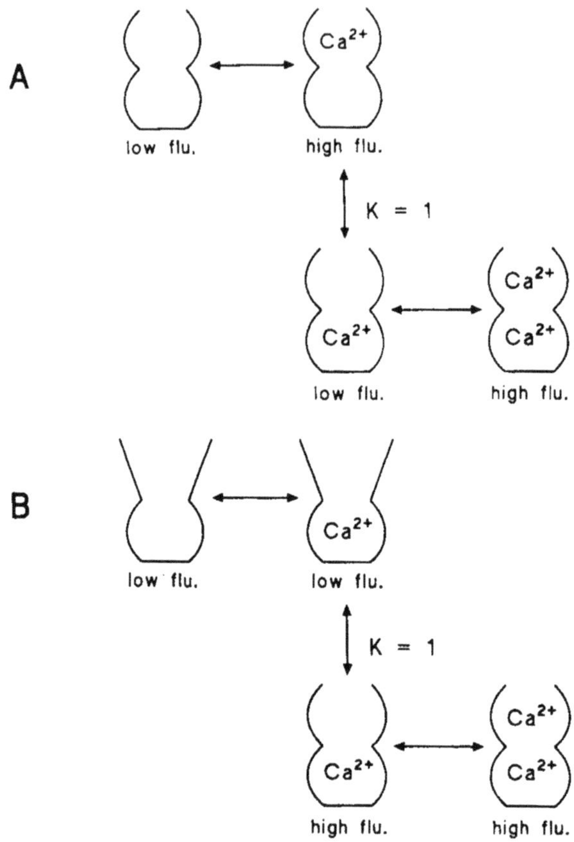

Figure 15. Two possible models for Ca^{2+} binding to the ATPase. (**A**) binding of Ca^{2+} in a channel like structure and (**B**) binding involving a conformational change following binding of the first Ca^{2+} ion to the ATPase. Binding of Ca^{2+} is sequential: $E1 + Ca^{2+} \rightleftharpoons CaE1 \rightleftharpoons CaE1' + Ca^{2+} \rightleftharpoons Ca_2E1'$.

For the first model it is assumed that the change in tryptophan fluorescence monitors occupancy of the outer of the two Ca^{2+}-binding sites. With an equilibrium constant of 1 for the CaE1–CaE1′ step, the relative fluorescence changes on binding one- and two-Ca^{2+} ions will be in the proportion 0.5:1, as required. For the second model, it is proposed that the states CaE1′ and Ca_2E1' are states of high fluorescence, and again with the equilibrium constant for the CaE1–CaE1′ step equal to one, the relative fluorescence changes on binding one- and two-Ca^{2+} ions will be in the proportion 0.5:1. The later model provides a more natural explanation for the observed cooperativity of Ca^{2+} binding, but either would be consistent with the equilibrium binding data. The later model is in better agreement with kinetic data as described in Henderson et al. (1994b).

In any sequential Ca^{2+}-binding model in which one Ca^{2+} ion binds "above" another, there is an obvious problem in ensuring access to the inner binding site. Free access to the first site requires that the second site not be fully formed in the initial conformation as in the second model presented in Figure 15. Conformational changes following the binding of Ca^{2+} to the first site could then result in distortion of M4 and M6 bringing M5 and M8 together (Figure 13) and so creating the second site. The model shown in Figures 13 and 14 also provides a possible explanation for the change in nature of the sites from high affinity and exposed to the outside (the E1 conformation) to low affinity and accessible to the inside in the phosphorylated form (E2P). Moving M4 and M8 away from M5 and M6 would remove one of the two COO^- groups from each binding site, resulting in a large decrease in affinity, at the same time allowing unrestricted release of Ca^{2+} into the lumen of the SR (Figure 14). Experiments by Orlowski and Champeil (1991b) have shown that dissociation of Ca^{2+} from Ca_2E2P is nonsequential, unlike the dissociation of Ca^{2+} from Ca_2E1 (Henderson et al., 1994b).

Information about the possible packing of the 10-transmembrane α-helices can be obtained from sequence comparisons. Figure 16 shows the sequences of M1–M10 in the form of helical wheels, with the amino acids arranged as an ideal α-helix (100° rotation per residue) viewed down the long axis from the N-terminal end. Many of the helices show a nonrandom distribution of conserved residues. Thus, in M1 conserved residues are located in an arc on one side of the helix. This is also shown in the view in Figure 17 where the α-helix is seen as a cylinder cut along the long axis of the helix and flattened. In this view, the most conserved residues are located in a central cluster. It is considered that regions of highest conservation correspond to regions of protein–protein contact, since it is likely that protein–lipid interactions will be structurally less demanding. A similar distribution is observed for M2. For M3, there is only one residue, Ile-274, conserved among all the sequences, although six others, located in an arc on one face of the helix, are conserved in all but cta3 of yeast (Figure 16).

For M4, conservation is particularly marked around Glu-309 (Figures 16 and 17). M4 is also of interest in containing two proline residues, one of which, Pro-308, is present in all sequences including that of the cta3 protein of yeast and the plasma membrane Ca^{2+}-ATPase, while the other, Pro-312, is present in the plasma membrane Ca^{2+}-ATPase, but not the cta3 protein of yeast. There has been much speculation about the importance of proline residues in transmembrane α-helices (von Heijne, 1991). It has been suggested that the kinked helices formed by Pro residues are oriented with their convex or open sides (defined by the proline itself and its +1, –3 and –4 neighbors) toward the protein interior and their concave sides towards the surrounding lipids. It has also been suggested that charged residues tend to be found on the convex face of proline-kinked transmembrane helices, resulting in the simultaneous burial of the kink region and the charged residues in

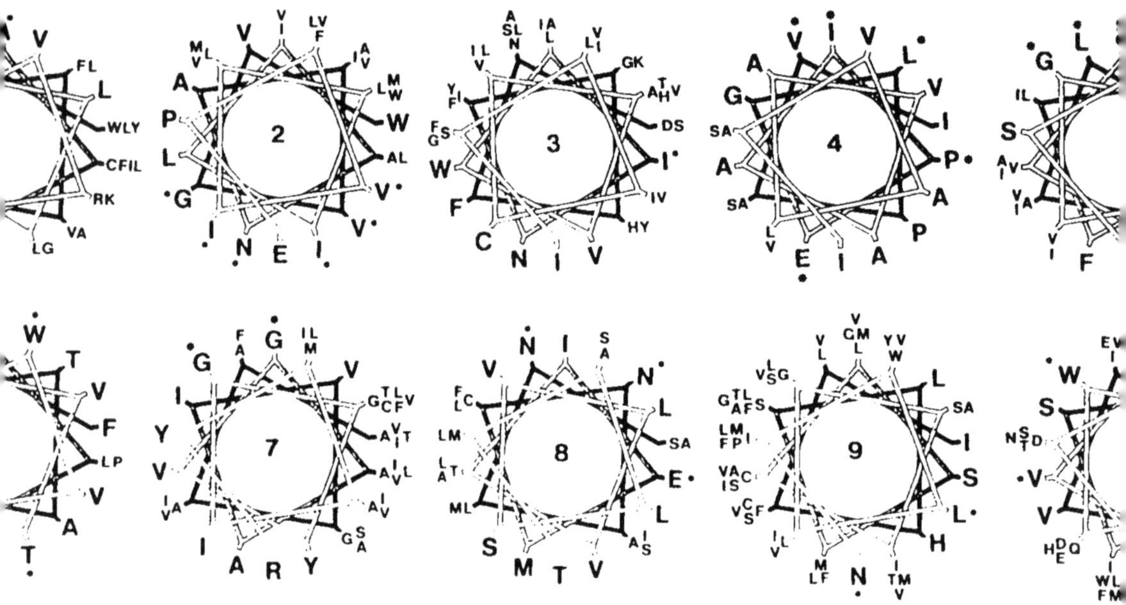

/heel diagrams of the putative 10-transmembrane helices for the Ca^{2+}-ATPase. Residues conserved amc
1–8 given in Figure 12 are shown in large capitals (data for the frog muscle (Vilsen and Andersen, 1992) is
arisons). Those also conserved in cta3 of yeast (sequence 9 of Figure 12) are marked *. Residues not conserv
als.

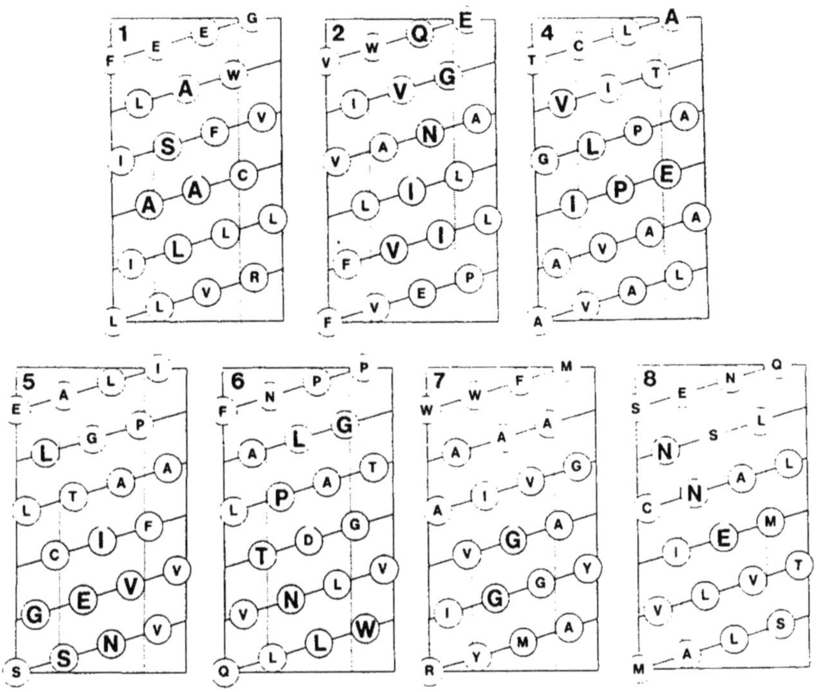

Figure 17. Putative transmembrane helices for the Ca^{2+}-ATPase. Helices M1, M2, M4, M5, M6, M7, and M8 are viewed as a cylinder cut along the axis of the helix. Residues conserved in all the sequences given in Figure 12 including cta3 protein of yeast are shown in large capitals.

the protein interior (von Heijne, 1991). The two proline residues around Glu-309 are located in such a way as to create a very marked kink in this helix, with Glu-309 on the convex side. M5 shows a high degree of conservation around Glu-771 (Figures 16 and 17). Mutation of conserved proline residues 308 in M4 and 803 in M6 led to a reduced affinity for Ca^{2+} (Vilsen et al., 1989). In contrast, mutation of Pro-312 in M4 had little effect on Ca^{2+} affinity, but did lead to loss of accumulation of Ca^{2+}, the phosphorylated ATPase being locked in an ADP-sensitive form, presumably because of a decrease in the rate of the $Ca_2E1P \rightarrow Ca_2E2P$ step (Vilsen et al., 1989).

Mutation of hydrophobic residues around the conserved Glu-309 led to abolition of Ca^{2+} transport (Clarke et al., 1993). Interestingly, mutation (Clarke et al., 1993) of those residues shown in Figure 15 as being conserved in all the Ca^{2+}-ATPases, including the cta3 protein of yeast, also led to a reduction in Ca^{2+} transport despite being located on the opposite side of the helix to Glu-309.

The conserved Asn-768 is of interest since Asn residues are often found at Ca^{2+}-binding sites (McPhalen et al., 1991). Conservation in an arc around Asp-800 in M6 is clear, including conserved Asn and Pro residues at positions 796 and 803, respectively. M7 is unusually long (25 residues in Figure 12), consistent with the data of Toyoshima et al. (1993) discussed above. Only two residues in M7 are conserved among all the SR/ER Ca^{2+}-ATPases, Gly-841 and Gly-845, located above each other in the helical structure, suggesting a tight packing of this part of the helix surface with the other helices. M8 shows a lower degree of conservation than M4, M5 and M6, but two Asn residues (911 and 914) are conserved. M9 and M10 show relatively low degrees of conservation.

The high degree of conservation in M4 and M6 would suggest a structure in which these helices are largely in contact with other helices, rather than with phospholipid. On the other hand, helices such as M1, M2, M3, M7, and M9 where nonconserved hydrophobic residues are found on one face of the helix, are likely to be organized with the nonconserved face exposed to lipid and the conserved face in contact with other helices. The low degree of conservation of M10 would be consistent with a relatively lipid-exposed location for this helix. The observation that it is the hydrophobic residues in M10 that vary does, however, suggest a trans-membrane orientation for this α-helix, rather than a location, for example, on the surface of the membrane.

The first transmembrane α-helix contains a motif, (R or K)ILLL, that is also found in the T cell antigen receptor and believed to determine retention of the receptor within the endoplasmic reticulum (see Magyar and Varadi, 1990). This sequence is absent from the plasma membrane Ca^{2+}-ATPases, and it has therefore been suggested that this sequence could serve as an internal signal sequence.

It has been found that the Ca^{2+}-ATPase can be inhibited by sesquiterpene lactones, such as thapsigargin and trilobolide, and by dihydroquinones, such as BHQ. Both inhibit by shifting the E1/E2 equilibrium towards E2 with a decrease in the rate of the E2 $\rightarrow$ E1 step (Khan et al., 1995; Wictome et al., 1992, 1995). The hydrophobicity of these molecules makes a binding site in the transmembrane region of the ATPase likely. A chimera of Met-1 to Thr-355 and Lys-712 to Ala-994 of the Ca^{2+}-ATPase flanking Leu-379 to Lys-724 of the (Na^{+}–K^{+})-ATPase showed Ca^{2+} binding and inhibition by thapsigargin or BHQ (Sumbilla et al., 1993). A chimera of the Ca^{2+}-ATPase in which Met-1 to Ile-163 were replaced by the corresponding region of the (Na^{+}–K^{+})-ATPase retained sensitivity to thapsigargin (Ishii and Takeyasu, 1993). The binding site for thapsigargin would then seem to lie between M3 and M10.

Since effects of BHQ and trilobolide on the ATPase are identical, it is likely that bind to similar sites on the ATPase and that the -OH groups of the two molecules are important in the binding. However, the separation between -OH groups found in the crystal structure of trilobolide (3.64 Å) is very different to that predicted by modeling for BHQ (5.52 Å) making it unlikely that they interact with the same residues on the Ca^{2+}-ATPase. Since polar derivatives of BHQ have no effect on the ATPase, the binding site is likely to be located within the transmembrane region of the ATPase. Helix M5 contains a cluster of Tyr and Ser residues (Y^{763}-XXSS) which

(a) (b)

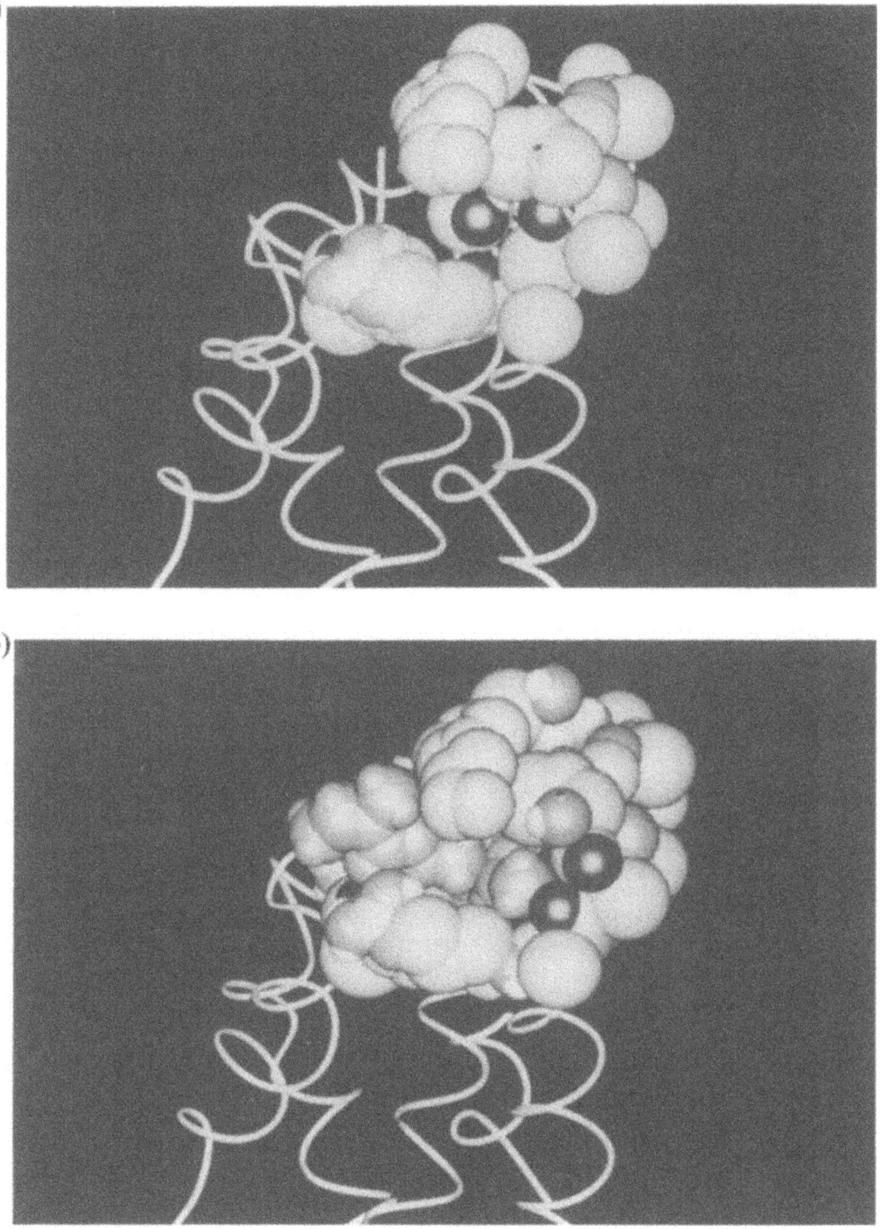

Figure 18. Possible arrangement of transmembrane helices M4–M8 showing a cluster of Tyr and Ser residues in helices M5 and M8 which could constitute a binding site for trilobolide, a sesquiterpene lactone. (**a**) shows residues around the potential binding site for trilobolide and (**b**) shows trilobolide in the site.

modeling shows could provide suitable ligands for hydrogen bonding with trilobolide or BHQ, with possible further hydrogen bonding to Tyr-837 of M7 (Figure 18). These residues are not found in plasma membrane Ca^{2+}-ATPase and the plasma membrane Ca^{2+}-ATPase is insensitive to thapsigargin or BHQ.

V. CYTOPLASMIC DOMAINS OF THE Ca^{2+}-ATPase

Serrano and Portillo (1990) have identified six motifs conserved in cytoplasmic domains of all the ATPases they surveyed (Figure 19). Motif I, DXSX(I or L)TGES, is found in all the available Ca^{2+}-ATPase sequences, except that in some E replaces D and, in the cta3 protein of yeast, the third residue is A instead of the usual S. It has been suggested that this motif, in the β-strand or transduction domain, is involved in the hydrolysis of the phosphorylated intermediate. A conserved sequence, (I or L)CSDKTGTLTXN, is found around the residue (Asp-351) phosphorylated by ATP and is found in all the available Ca^{2+}-ATPase sequences, except for that of the cta3 protein of yeast where I replaces L and the final N is replaced by G. The third conserved sequence KGA is found in all sequences and contains the

```
                   I                II                 III
               * * *****      ********** *             ***
SR        176  DQSILTGES  348 ICSDKTGTLTTNQ  514  VKGAP
LCA            EQSSLTGES      ICSDKTGTLTTNQ       VKGAP
P.falc         EQSMLTGES      ICSDKTGTLTTNQ       CKGAP
Spo            DEALLTGES      ICSDKTGTITQGK       AKGAV

                IV            V             VI
               ** *         * ***         *   *  *
SR        601  DPPRK   623  MITGDN   676  ARVEPSHK
LCA            DPPRE        VITGDN        SRAEPRHK
P.falc         DPPRK        MITGDN        CRTEPKHK
Spo            DPPRT        MLTGDH        ARCAPQTK

                         VII
                 **** ** * ****  * *
SR        699  AMTGDGVNDAPALKKAEIGIA
LCA            AMTGDGVNDAPALKLADIGIA
P.falc         AMTGDGVNDAPALKSADIGIA
Spo            AMTGDGVNDSPSLKQANVGIA
```

Figure 19. Conserved motifs in the cytoplasmic regions of the Ca^{2+}-ATPase. * denotes the residues identified by Serrano and Portillo (1990) as conserved in the P-type ATPase family. SR, fast-twitch rabbit skeletal muscle (Brandl et al., 1986); LCA, tomato (Wimmers et al., 1992); P.falc, *Plasmodium falciparum* (Kimura et al., 1993); Spo, cta3 protein from *Schizosaccharomyces pombe* (Ghislain et al., 1990).

residue (Lys-515) labeled by FITC; since, as described above, labeling is competitive with binding of ATP, this is presumed to be part of the adenine-binding site region. The fourth conserved sequence, DPXR, believed to be part of the adenine-binding region of the ATP-binding site, is also found in all sequences. The fifth conserved sequenced, MXTGD, is also believed to be part of the adenine-binding region, and is again found in all sequences except for the plant sequence where M is replaced by V. Finally, two conserved motifs are observed in the hinge region at the C-terminal end of the nucleotide-binding domain. Of the sequence AXXXPXXK found in a variety of ATPases, only P and K are found conserved in all the Ca^{2+}-ATPases (Figure 17). A conserved sequence, TGDGXNDXPXLKKAXXGXA, is also found in all sequences except for the KK sequence which is replaced by KL, KS, and KQ, respectively in the plant, *Plasmodium falciparum* and yeast sequences. The hinge region of the ATPase is presumed to be involved in intramolecular changes necessary to bring the phosphorylation and nucleotide domains together allowing phosphorylation of Asp-351. The close similarity between all the ion pumps in these catalytic regions argues for a similar mechanism of energy transduction in all the ATPases.

Chemical labeling has been used to identify residues which may be part of the ATP-binding site of the ATPase. Labeling of Lys-515 with fluorescein isothiocyanate (FITC) is competitive with binding of ATP and labeling leads to loss of ATPase activity although hydrolysis of acetyl phosphate is still possible, suggesting that Lys-515 is part of the nucleotide-binding region (Figure 20; Mitchinson et al., 1982; Pick and Karlish, 1982). Adenosine triphosphopyridoxal labels Lys-684 in

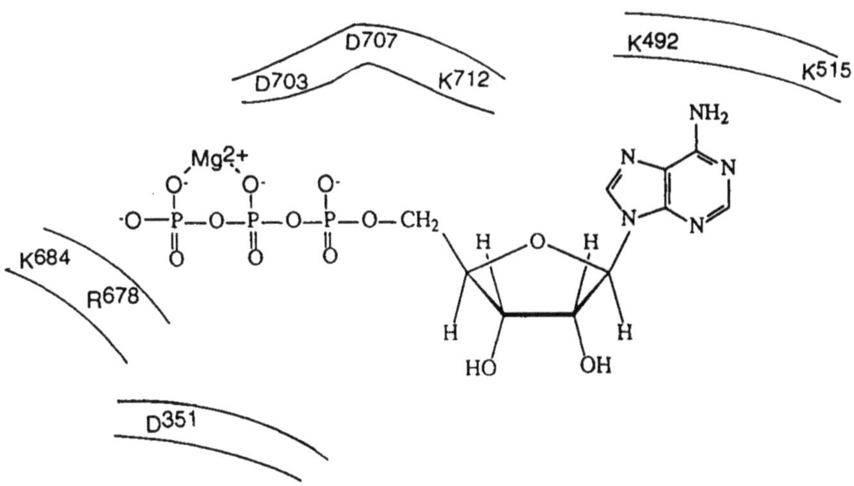

Figure 20. Residues implicated by chemical labeling experiments in binding MgATP to the Ca^{2+}-ATPase of SR.

the presence of Ca^{2+} and both Lys-684 and Lys-492 in the absence of Ca^{2+} (Yamamoto et al., 1988, 1989), consistent both with a change in the relative positions of Lys-684 and Lys-492 on binding Ca^{2+} and with a location for these two residues in the ATP binding site (Yamamoto et al., 1989). Lys-492 has also been shown to be labeled by 2′,3′-O-(2,4,6-trinitrophenyl)-8-azido-ATP (McIntosh et al., 1992) and Lys-492 and Arg-678 have been reported to be cross-linked by glutaraldehyde (McIntosh and Ross, 1992), each in an ATP-protectable manner, again consistent with Lys-492 being part of the ATP binding site. Lys-492 has been labeled with pyridoxal phosphate, and protection studies suggest that it may be located close to the α-phosphoryl group of ATP (Yamagata et al., 1993). Lys-492 is a conserved residue and the equivalent residues in lamb kidney (Na^{+}–K^{+})-ATPase (Lys-480) and pig gastric (H^{+}–K^{+})-ATPase (Lys 497) can also be labeled with pyridoxal phosphate (Tamura et al., 1989; Hinz and Kirley, 1990). In dog kidney (Na^{+}–K^{+})-ATPase, it has been shown that FITC can react with Lys-501, Lys-480, or Lys-766, but with only one per ATPase molecule; it has therefore been suggested that these residues (equivalent to Lys-515, Lys-492, and Lys-758 in the Ca^{2+}-ATPase) are clustered around the binding site for fluorescein (Xu, 1989). The pH dependence of inactivation of Lys-492 by (2,4,6-trinitrophenyl)-8-azido-ATP has been found to be consistent with a pK of 7.5 for this lysine, about 3 orders of magnitude lower than that for lysine in solution, suggesting an unusual environment in the ATPase (Seebregts and McIntosh, 1989). This residue has also been shown to be the most reactive with the succinimidyl ester of 7-amino-4-methylcoumarin-3-acetic acid (Stefanova et al., 1993a). Further, since labeling with the succinimidyl ester is unaffected by labeling of Lys-515 with FITC, Lys-515 and Lys-492 are presumably not in close proximity on the ATPase (Stefanova et al., 1993a).

It has been shown by McIntosh and Woolley (1994) that the γ-phosphate of an ATP analogue (2′3′-O-(2,4,6-trinitrophenyl)-8-azido-ATP) covalently linked to Lys-492 on the ATPase is able to phosphorylate Asp-351, albeit at a slow rate. This implies that Lys-492 and Asp-351 should be separated by about 14 Å in the Ca-bound complex. Further, it implies that Lys-492 must be close to the adenyl moiety of the nucleotide in the complex (McIntosh and Woolley, 1994).

Ohta et al. (1986) have reported that 5′-p-fluorosulfonyl benzoyladenosine labels a lysine in the (Na^{+}–K^{+})-ATPase equivalent to Lys-712 in the Ca^{2+}-ATPase, suggesting this Lys is also in the ATP-binding site; mutation of this residue in the Ca^{2+}-ATPase has no effect on activity, so that its role cannot be essential (Maruyama et al., 1989). Two Asp residues in the (Na^{+}–K^{+})-ATPase, conserved residues equivalent to Asp-703 and Asp-707 in the Ca^{2+}-ATPase are labeled by adenosine 5′-[N-[4-[N-(2-chloroethyl)-N-methylamino] benzyl]-γ-amidotriphosphate], so that these residues are also presumably part of the ATP-binding site (Dzhandzhugazyan et al., 1988).

Further information about the structure of the cytosolic domain has come from antibody binding experiments (Mata et al., 1992). As shown in Figure 1, a relatively large number of antibodies bind to the phosphorylation and nucleotide binding domains, suggesting that these domains are relatively exposed in the three-dimensional structure of the ATPase. It was found that residues 510–515 were surface exposed, suggesting that these residues are not directly involved in binding ATP. Antibodies against fluorescein failed to bind to the ATPase labeled with FITC at Lys-515, suggesting a buried location for the fluorescein, despite the relatively exposed location of Lys-515 itself (Mata et al., 1989). The observation that the first tryptic-cleavage site (T_1) on the ATPase is located at Arg-505 is also consistent with surface exposure in this region. Interestingly, whereas the region beyond Lys-515 (residues 516–519) is highly conserved in all the P-type ATPases, residues 510–515 are not.

Competitive-binding experiments have shown that antibodies binding to residues 510–515 bind competitively with antibodies binding to residues 662–666, but that binding of antibodies which bind to residues 580–586 are not competitive with either of the former antibodies (Tunwell et al., 1991). The polypeptide chain of the ATPase between residues 510 and 666 would thus seem to be in the form of a large loop, bringing the two ends close together and away from residues 580–588. The loop could start close to the first site of trypsin cleavage (Arg-505) which presumably is surface exposed. The observation that antibodies binding to residues 510–515 or 662–666 (but not 580–586) inhibit the ATPase suggest that either the region where the loop comes together plays an essential role in the function of the ATPase or that binding of antibodies to this region inhibit important conformational changes on the ATPase.

All the residues implicated in binding ATP are indicated in Figure 20. In phosphotransferases, Mg is found bridging the β- and γ-phosphates, as shown. Ion pairing between phosphate oxygens and guanidinium groups of arginine and ammonium groups of lysine are common. Rather little is known about the coordination around Mg bound to proteins, but for MgGTP bound to the *Ha ras* oncogene product, p21, apart from two phosphate oxygens, oxygens of Ser, Thr, and possibly Asp residues make up the inner coordination sphere. Girardet et al. (1993) have suggested that residues in the region 696–707 in the hinge region may be involved in binding Mg^{2+}.

Close proximity of residues from the phosphorylation and nucleotide-binding domains have been shown by cross-linking experiments (Gutowski-Eckel et al., 1993). The ATPase has been reacted with an analog of ATP activated at the γ-phosphate to form a mixed anhydride with Asp-351. This then, is the target for nucleophilic attack by a neighboring amino acid side-chain, possibly a Lys residue. This second step was only observed in the presence of Ca^{2+}, reflecting a conformational change in the region around Asp-351. Following digestion, a cross-linked double-peptide was obtained, one chain running from Ala-327 to Met-361 and the second from Ile-624 to Met-700 (Gutowski-Eckel et al., 1993).

Site-directed mutagenesis has not yet defined the roles of individual residues around the ATP-binding site, but a number of mutations in the conserved regions have been shown to have a block in the $Ca_2E1P \rightarrow Ca_2E2P$ transition (MacLennan and Toyofuku, 1992).

The region C-terminal of Asp-351 is of interest in the regulation of the Ca^{2+}-ATPase. Slow-twitch muscles like heart express an isoform (SERCA2) of the Ca^{2+}-ATPase whose activity is regulated by phospholamban. Phospholamban is a membrane protein with a single hydrophobic domain at the C-terminus (residues 31–52) and a hydrophilic N-terminal domain (residues 1–30) containing residues whose phosphorylation reduces interaction with the Ca^{2+}-ATPase. Fast-twitch skeletal muscle contains no phospholamban but co-expression of phospholamban and the fast-twitch isoform of the Ca^{2+}-ATPase (SERCA1) in COS cells showed that phospholamban is capable of regulating SERCA1 in identical fashion to SERCA2 (Toyofuku et al., 1993). Cross-linking experiments have suggested that phospholamban interacts with the Ca^{2+}-ATPases at a region just C-terminal of Asp-351, and, using ^{125}I-labeled phospholamban, radioactivity was recovered in Lys-397 and Lys-400 (James et al., 1989; Vorherr et al., 1992). These two residues occur in a region conserved in the SERCA1 and SERCA2 isoforms of the Ca^{2+}-ATPase, but not in the SERCA3 isoform (Burk et al., 1989) or in more distantly related Ca^{2+}-ATPases such as those in *Plasmodium* (Kimura et al., 1993), *Artemia* (Palmero and Sastre, 1989), or plants (Wimmers et al., 1992). Phospholamban has been shown not to affect the Ca^{2+} affinity of the SERCA3 isoform of the Ca^{2+}-ATPase and the construction of chimeric Ca^{2+}-ATPases between SERCA2 and SERCA3 has shown that the region between residues 336 and 412 is essential for interaction with phospholamban (Toyofuku et al., 1993). Presumably, binding of the positively charged phospholamban would be to the negatively charged residues found in this region of the ATPase (e.g., Glu-392, Glu-304, and Asp-399). It was found that a second region, between residues 467–762 in the nucleotide-binding/hinge region, was also essential for interaction (Toyofuku et al., 1993).

A number of charged residues are conserved in the stalk region of all the Ca^{2+}-ATPases, including the plasma membrane Ca^{2+}-ATPase. These include the motif ^{108}EXXE in S2, ^{243}EX(E or D)XXXXXXK in S3, and ^{728}KXXX(E or D)XXXXDD and ^{751}R in S5. Mutation of residues in S3 led to relatively mild changes in the kinetics of the ATPase, with some decrease in the rate of dephosphorylation of the phosphorylated ATPase (Andersen and Vilsen, 1993); mutations at the M5-S5 boundary led to a phenotypic variant of the Ca^{2+}-ATPase in which hydrolysis of ATP no longer led to net accumulation of Ca^{2+} (Andersen, 1995).

Evidence in favor of Ca^{2+}-binding sites on the luminal side of the SR membrane has come from studies of the effects of luminal Ca^{2+} on the level of phosphorylation of the ATPase by P_i; these studies imply that Ca^{2+} must be able to bind to luminal sites on both the unphosphorylated (E2) and the phosphorylated (E2P) ATPase (Suko et al., 1981; Froud and Lee, 1986; Jencks et al., 1993; Myung and Jencks, 1995). It has been suggested that the sites on the unphosphorylated ATPase are distinct from those on the phosphorylated ATPase (Jencks et al., 1993). Only a small number of charged residues are predicted on the luminal side of the membrane which might interact with Ca^{2+}. For loops between helices 3–4, 7–8, and 9–10 no charged residues appear to be conserved between all the Ca^{2+}-ATPases. For loops between helices 1–2 and 5–6 negatively charged residues (E or D) are consistently found at positions corresponding to ^{79}E and ^{785}E in all the Ca^{2+}-ATPases.

As shown in Figure 1, antibody binding studies show that the region around the second tryptic-cleavage site (T_2) is exposed on the ATPase in the presence of Ca^{2+} (East et al., 1992). Apart from the N-terminus itself, no antibodies binding to the native ATPase between the N-terminus and the T_2 site were obtained, suggesting that this region of the ATPase is largely buried in the native structure (Mata et al., 1992). Mutation of conserved residues in the β-strand domain have been shown to lead to a reduction in the rate of the $Ca_2E1P \rightarrow Ca_2E2P$ transition (MacLennan and Toyofuku, 1992). Residues affecting the rate of this transition have also been found in the hinge domain and in the stalk region (MacLennan and Toyofuku, 1992).

VI. STRUCTURE AND MECHANISM

A number of the features of the structure of the Ca^{2+}-ATPase described above are important in thinking about the mechanism of the ATPase. The first is the wide separation between the Ca^{2+}-binding sites, located within the transmembrane region of the ATPase, and the ATP-binding site in the cytoplasmic region of the ATPase. This separation necessitates coupling over long distances between the phosphorylation domain and the Ca^{2+}-binding sites. Such a large separation may be required to combine the enzyme-like properties of the phosphorylation domain and the machine-like properties of the Ca^{2+}-binding domain. In an enzyme, the transition state of the reacting substrate binds more strongly to the enzyme than the ground state of the substrate, as the structure of the active site of an enzyme is designed to match the transition-state configuration of the substrate rather than the ground state. The substrate deforms to fit the active site. In contrast, at the transport site, the transported ion undergoes no change, but the site does; it is the site that deforms and has to be conformationally flexible (Krupka, 1993).

It may be that phosphorylation is particularly suited to provide the link between catalysis and transport, since strong binding of the oxygens of a covalently bound phosphate group to residues in a neighboring domain could result in significant conformational changes, linked to changes at the ion binding sites. Nevertheless, it appears that conformational changes on the Ca^{2+}-ATPase are rather limited in their magnitude. Fluorescence energy transfer measurements fail to detect any significant changes in the structure of the ATPase between Ca^{2+}-bound and vanadate-bound states (Gutierrez-Merino et al., 1987; Stefanova et al., 1993b). Only a very small number of antibodies binding to the ATPase have any effect on function, indicating that little of the surface of the ATPase undergoes significant structural change during the reaction cycle (Colyer et al., 1989). Circular dichroism (Nakamoto and Inesi, 1986) also detects little difference between Ca^{2+}- and vanadate-bound forms of the Ca^{2+}-ATPase although changes in the proportion of α-helix have been detected by infrared spectroscopy (Arrondo et al., 1987). It has been shown that vanadate inhibits trypsin cleavage at the T_2 site, but since the effect is not observed in detergent solution, this could reflect a change in organization of the ATPase within the plane of the membrane, rather than a major conformational change in the region of the T_2-cleavage site (Andersen and Jorgensen, 1985). It appears, therefore, that Ca^{2+} transport involves only minor changes in the overall structure of the ATPase, changes being localized to small regions of the ATPase. This is consistent with the measured value of the E1/E2 equilibrium constant which is close to 1 (Henderson et al., 1994a), implying that the free energy difference between E1 and E2 is close to zero ($\Delta G° = -RT\ln K_{equil}$), and thus that the E1 and E2 conformations are likely to have rather similar structures. As described above, relatively small movements of helices M4, M5, M6, and M8 could lead both to a change in accessibility of the Ca^{2+}-binding sites and a change in their affinity for Ca^{2+}. How such changes could be linked to changes in the phosphorylation domain, and the nature of these later changes, remains to be determined.

ACKNOWLEDGMENTS

I thank Dr. J. M. East and Dr. I. Matthews with whom many of the ideas discussed in this review were developed, and the BBSRC and Wellcome Trust for financial support.

REFERENCES

Andersen, J. P. (1995). Functional consequences of alterations to amino acids at the M5S5 boundary of the Ca^{2+}-ATPase of sarcoplasmic reticulum -Mutation Tyr763→Gly uncouples ATP hydrolysis from Ca^{2+} transport. J. Biol. Chem. 270, 908–914.

Andersen, J. P., & Jorgensen, J. L. (1985). Conformational states of sarcoplasmic reticulum Ca^{2+}-ATPase as studied by proteolytic cleavage. J. Memb. Biol. 88, 187–198.

Andersen, J. P., & Vilsen, B. (1992). Functional consequences of alterations to Glu-309, Glu-771 and Asp-800 in the Ca^{2+}-ATPase of sarcoplasmic reticulum. J. Biol. Chem. 267, 19383–19387.

Andersen, J. P., & Vilsen, B. (1994). Amino acids Asn(796) and Thr(799) of the Ca^{2+}-ATPase of sarcoplasmic reticulum bind Ca^{2+} at different sites. J. Biol. Chem. 269, 15931–15936.

Andersen, J. P., & Vilsen, B. (1995). Structure-function relationships of cation translocation by Ca^{2+}-and Na^{+},K^{+}-ATPases studied by site-directed mutagenesis. FEBS Lett. 359, 101–106.

Andersen, J. P., Vilsen, B., & MacLennan, D. H. (1992). Functional consequences of alterations to Gly310, Gly770, and Gly801 located in the transmembrane domain of the Ca^{2+}-ATPase of sarcoplasmic reticulum. J. Biol. Chem. 267, 2767–2774.

Andersen, J. P., & Vilsen, B. (1993). Functional consequences of substitution of the seven-residue segment LysIleArgAspGlnMetAla240 located in the stalk helix S3 of the Ca^{2+}-ATPase of sarcoplasmic reticulum. Biochemistry 32, 10015–10020.

Antolovic, R., Bruller, H. J., Bunk, S., Linder, D., & Schoner, W. (1991). Epitope mapping by amino-acid-sequence-specific antibodies reveals that both ends of the alpha subunit of Na/K-ATPase are located on the cytoplasmic side of the membrane. Eur. J. Biochem. 199, 195–202.

Arrondo, J. L. R., Mantsch, H. H., Mullner, N., Pikula, S., & Martonosi, A. (1987). Infrared spectroscopic characterization of the structural changes connected with the $E_1 \rightarrow E_2$ transition in the Ca^{2+}-ATPase of sarcoplasmic reticulum. J. Biol. Chem. 262, 9037–9043.

Asturias, F. J., & Blasie, J. K. (1991). Location of high-affinity metal binding sites in the profile structure of the Ca^{2+}-ATPase in the sarcoplasmic reticulum by resonance X-ray diffraction. Biophys. J. 59, 488–502.

Baker, K. J., East, J. M., & Lee, A. G. (1993). Localization of the hinge region of the Ca^{2+}-ATPase of sarcoplasmic reticulum using resonance energy transfer. Biochim. Biophys. Acta, in press.

Bayle, D., Weeks, D., & Sachs, G. (1995). The membrane topology of the rat sarcoplasmic and endoplasmic reticulum calcium ATPases by *in vitro* translation scanning. J. Biol. Chem. 270, 25678–25684.

Bigelow, D. J., & Inesi, G. (1991). Frequency-domain fluorescence spectroscopy resolves the location of maleimide-directed spectroscopic probes within the tertiary structure of the Ca^{2+}-ATPase of sarcoplasmic reticulum. Biochemistry 30, 2113–2125.

Birmachu, W., Nisswandt, F. L., & Thomas, D. D. (1989). Conformational transitions in the calcium adenosinetriphosphatase studied by time-resolved fluorescence resonance energy transfer. Biochemistry 28, 3940–3947.

Brandl, C. J., Green, N. M., Korczak, B., & MacLennan, D. H. (1986). Two Ca^{2+}-ATPase genes: Homologies and mechanistic implications of deduced amino acid sequences. Cell 44, 597–607.

Burk, S. E., Lytton, J., MacLennan, D. H., & Shull, G. E. (1989). cDNA cloning, functional expression, and mRNA tissue distribution of a third organellar Ca^{2+} pump. J. Biol. Chem. 264, 18561–18568.

Campbell, A. M., Kessler, P. D., & Fambrough, D. M. (1992). The alternative carboxyl termini of avian cardiac and brain sarcoplasmic reticulum/endoplasmic reticulum Ca^{2+}-ATPases are on opposite sides of the membrane. J. Biol. Chem. 267, 9321–9325.

Canfield, V. A., & Levenson, R. (1993). Transmembrane organization of the Na,K-ATPase determined by epitope addition. Biochemistry 32, 13782–13786.

Clarke, D. M., Loo, T. W., Inesi, G., & MacLennan, D. H. (1989). Location of high affinity Ca^{2+}-binding sites within the predicted transmembrane domain of the sarcoplasmic reticulum Ca^{2+}-ATPase. Nature 339, 476–478.

Clarke, D. M., Loo, T. W., & MacLennan, D. H. (1990a). The epitope for monoclonal antibody A20 (amino acids 870–890) is located on the luminal surface of the Ca^{2+}-ATPase of sarcoplasmic reticulum. J. Biol. Chem. 265, 17405–17408.

Clarke, D. M., Loo, T. W., & MacLennan, D. H. (1990b). Functional consequences of alterations to polar amino acids located in the transmembrane domain of the Ca^{2+}-ATPase of sarcoplasmic reticulum. J. Biol. Chem. 265, 6262–6267.

Clarke, D. M., Loo, T. W., Rice, W. J., Andersen, J. P., Vilsen, B., & MacLennan, D. H. (1993). Functional consequences of alterations to hydrophobic amino acids located in the M4 transmembrane sector of the Ca^{2+}-ATPase of sarcoplasmic reticulum. J. Biol. Chem. 268, 18359–18364.

Colyer, J., Mata, A. M., Lee, A. G., & East, J. M. (1989). Effects on ATPase activity of monoclonal antibodies raised against (Ca^{2+} + Mg^{2+})-ATPase from rabbit skeletal muscle sarcoplasmic reticulum and their correlation with epitope location. Biochem. J. 262, 439–447.

Corbalan-Garcia, S., Teruel, J.A., & Gomez-Fernandez, J. C. (1993). Intramolecular distances within the Ca^{2+}-ATPase from sarcoplasmic reticulum as estimated through fluorescence energy transfer between probes. Eur. J. Biochem. 217, 737–744.

de Meis, L. (1981). The Sarcoplasmic Reticulum, Wiley, New York.

Dux, L., Taylor, K. A., Ting-Beall, H. P., & Martonosi, A. (1985). Crystallization of the Ca^{2+}-ATPase of sarcoplasmic reticulum by calcium and lanthanide ions. J. Biol. Chem. 260, 11730–11743.

Dzhandzhugazyan, K. N., Lutsenko, S. V., & Modyanov, N. N. (1988). In: The Na,K-Pump, Part A: Molecular Aspects (Skou, J. C., Norby, J. G., Maunsbach, A. B., & Esmann, M., eds.) pp. 181–188. Alan R. Liss, New York.

East, J. M., Matthews, I., Tunwell, R. E. A., Mata, A. M., & Lee, A. G. (1992). Studies of the structure and function of sarcoplasmic reticulum (Ca^{2+}-Mg^{2+})-ATPase using immunological approaches. Biochem. Soc. Trans. 20, 550–554.

Eisenman, G., & Dani, J. A. (1987). An introduction to molecular architecture and permeability of ion channels. Ann. Rev. Biophys. Chem. 16, 205–226.

Escalante, R., & Sastro, L. (1993). Similar alternative splicing events generate two sarcoplasmic reticulum or endoplasmic reticulum Ca^{2+}-ATPase isoforms in the crustacean *Artemia franciscana* and in vertebrates. J. Biol. Chem. 268, 14090–14095.

Feschenko, M. S., Zvaritch, E. I., Hofmann, F., Shakhparonov, M. I., Modyanov, N. N., Vorherr, T., & Carafoli, E. (1992). A monoclonal antibody recognizes an epitope in the first extracellular loop of the plasma membrane Ca^{2+} pump. J. Biol. Chem. 267, 4097–4101.

Froud, R. J., & Lee, A. G. (1986). A model for the phosphorylation of the Ca^{2+} + Mg^{2+}-activated ATPase by phosphate. Biochem. J. 237, 207–215.

Froud, R. J., East, J. M., Rooney, E. K., & Lee, A. G. (1986). Binding of long-chain alkyl derivatives to lipid bilayers and to (Ca^{2+}-Mg^{2+})-ATPase. Biochemistry 25, 7535–7544.

Fujimori, T., & Jencks, W. P. (1990). Lanthanum inhibits steady-state turnover of the sarcoplasmic reticulum calcium ATPase by replacing magnesium as the catalytic ion. J. Biol. Chem. 265, 16262–16270.

Ghiggino, K. P., Lee, A. G., Meech, S. R., O'Connor, D. V., & Phillips, D. (1981). Time-resolved emission spectroscopy of the dansyl fluorescence probe. Biochemistry 20, 5381–5389.

Ghislain, M., Goffeau, A., Halachmi, D., & Eilam, Y. (1990). Calcium homeostasis and transport are affected by disruption of cta3, a novel gene encoding Ca^{2+}-ATPase in *Schizosaccharomyces pombe*. J. Biol. Chem. 265, 18400–18407.

Girardet, J. L., Bally, I., Arlaud, G., & Dupont, Y. (1993). Localization of a putative magnesium-binding site within the cytoplasmic domain of the sarcoplasmic reticulum Ca^{2+}-ATPase. Eur. J. Biochem. 217, 225–231.

Green, N. M., & Stokes, D. L. (1992). Structural modelling of P-type ion pumps. Acta Physiol. Scand. 146, 59–68.

Grimes, E. A., Burgess, A. J., East, J. M., & Lee, A. G. (1991). Reconstitution experiments provide no evidence for a role for the 53-kDa glycoprotein in coupling Ca^{2+} transport to ATP hydrolysis by the (Ca^{2+}-Mg^{2+})-ATPase in sarcoplasmic reticulum. Biochim. Biophys. Acta 1064, 335–342.

Gutierrez-Merino, C., Munkonge, F., Mata, A. M., East, J. M., Levinson, B. L., Napier, R. M., & Lee, A. G. (1987). The position of the ATP binding site on the (Ca^{2+}-Mg^{2+})-ATPase. Biochim. Biophys. Acta 897, 207–216.

Gutowski-Eckel, Z., Mann, K., & Baumert, H. G. (1993). Identification of a cross-linked double-peptide from the catalytic site of the Ca^{2+}-ATPase of sarcoplasmic reticulum formed by the Ca^{2+}- and pH-dependent reaction with ATP γ-P-imidazolidate. FEBS Lett. 324, 314–318.

Henao, F., Orlowski, S., Merah, Z., & Champeil, P. (1992). The metal sites on sarcoplasmic reticulum membranes that bind lanthanide ions with the highest affinity are not the ATPase Ca^{2+} transport sites. J. Biol. Chem. 267, 10302–10312.

Henderson, I. M. H., Khan, M. Y., East, J. M., & Lee, A. G. (1994a). Binding of calcium to the (Ca^{2+}-Mg^{2+})-ATPase of sarcoplasmic reticulum: Equilibrium studies. Biochem. J. 297, 615–624.

Henderson, I. M. H., Starling, A. P., Wictome, M., East, J. M., & Lee, A. G. (1994b). Binding of calcium to the (Ca^{2+}-Mg^{2+})-ATPase of sarcoplasmic reticulum: Kinetic studies. Biochem. J. 297, 625–636.

Herrmann, T. R., Gangola, P., & Shamoo, A. E. (1986). Estimation of inter-binding-site distances in sarcoplasmic reticulum (Ca^{2+} + Mg^{2+})-ATPase using Eu(III) luminescence energy transfer. Eur. J. Biochem. 158, 555–560.

Hinz, H. R., & Kirley, T. L. (1990). Lysine 480 is an essential residue in the putative ATP site of lamb kidney (Na,K)-ATPase. J. Biol. Chem. 265, 10260–10265.

Hudecek, J., & Anzenbacher, P. (1988). Secondary structure prediction of liver microsomal cytochrome P-450; proposed model of spatial arrangement in a membrane. Biochim. Biophys. Acta 955, 361–370.

Imamura, Y., & Kawakita, M. (1991). Analysis of the binding sites of Gd^{3+}- on Ca^{2+}-transporting ATPase of the sarcoplasmic reticulum through its effects on fluorescence of tryptophan residues and a covalently attached fluorescent probe N-(1-anilinonaphthyl-4)maleimide. J. Biochem. Tokyo 110, 214–219.

Ishii, T., & Takeyasu, K. (1993). The amino-terminal 200 amino acids of the plasma membrane Na^+–K^+-ATPase α subunit confer ouabain sensitivity on the sarcoplasmic reticulum Ca^{2+}-ATPase. Proc. Natl. Acad. Sci. USA 90, 8881–8885.

James, P., Inui, M., Tada, M., Chiesi, M., & Carafoli, E. (1989). Nature and site of phospholamban regulation of the Ca^{2+} pump of sarcoplasmic reticulum. Nature 342, 90–92.

Jencks, W. P., Yang, T., Peisach, D., & Myung, J. (1993). Calcium ATPase of sarcoplasmic reticulum has four binding sites for calcium. Biochemistry 32, 7030–7034.

Kano, I., Satoh, K., Nagai, F., Ushiyama, K., Nakao, T., Hara, Y., & Kano, K. (1990). Antigenic determinant of a monoclonal antibody: Extracellular domain at the M3–M4 junction of the alpha-subunit of Na,K-ATPase. Can. J. Biochem. 68, 1262–1267.

Khan, Y. M., Wictome, M., East, J. M., & Lee, A. G. (1995). Interactions of dihydroxybenzenes with the Ca^{2+}-ATPase: Separate binding sites for dihydroxybenzenes and sesquiterpene lactones. Biochemistry 34, 14385–14393.

Kimura, M., Yamaguchi, Y., Takada, S., & Tanabe, K. (1993). Cloning of a Ca^{2+}-ATPase gene of *Plasmodium Falciparum* and comparison with vertebrate Ca^{2+}-ATPases. J. Cell Sci. 104, 1129–1136.

Krupka, R. M. (1993). Coupling mechanisms in ATP-driven pumps. Biochim. Biophys. Acta 1183, 114–122.

Leberer, E., Hartner, K. T., Brandl, C. J., Fujii, J., Tada, M., MacLennan, D. H., & Pette, D. (1989). Slow/cardiac sarcoplasmic reticulum Ca^{2+}-ATPase and phospholamban mRNAs are expressed in chronically stimulated rabbit fast-twich muscle. Eur. J. Biochem. 185, 51–54.

Lee, A. G., East, J. M., Giles, I. G., Henderson, I. M. H., Starling, A. P., & Warmerdam, J. Q. M. (1993). In: The Temperature Adaptation of Biological Membranes (Cossins, A., ed.), pp. 13–29. Portland Press, London.

Loomis, C. R., Martin, D. W., McCaslin, D. R., & Tanford, C. (1982). Phosphorylation of calcium adenosinetriphosphatase by inorganic phosphate: Reversible inhibition at high magnesium ion concentrations. Biochemistry 21, 151–156.

Luckie, D. B., Lemas, V., Boyd, K. L., Fambrough, D. M., & Takeyasu, K. (1992). Molecular dissection of functional domains of the E1E2-ATPase using sodium and calcium pump chimeric molecules. Biophys. J. 62, 220–227.

Lytton, J., & MacLennan, D. H. (1988). Molecular cloning of cDNAs from human kidney coding for two alternatively spliced products of the cardiac Ca^{2+}-ATPase gene. J. Biol. Chem. 263, 15024–15031.

MacLennan, D. H., Brandl, C. J., Korczak, B., & Green, N. M. (1985). Amino-acid sequence of a Ca^{2+} + Mg^{2+}-dependent ATPase from rabbit muscle sarcoplasmic reticulum, deduced from its complementary DNA sequence. Nature 316, 696–700.

MacLennan, D. H., & Toyofuku, T. (1992). Structure–function relationships in the Ca^{2+} pump of the sarcoplasmic reticulum. Biochem. Soc. Trans. 20, 559–562.

Magyar, A., & Varadi, A. (1990). Molecular cloning and chromosomal localization of a sarco/endoplasmic reticulum-type Ca^{2+}-ATPase of *Drosophila melanogaster*. Biochem. Biophys. Res. Commun. 173, 872–877.

Maruyama, K., Clarke, D. M., Fujii, J., Inesi, G., Loo, T. W., & MacLennan, D. H. (1989). Functional consequences of alterations to amino acids located in the catalytic center (isoleucine 348 to threonine 357) and nucleotide-binding domain of the Ca^{2+}-ATPase of sarcoplasmic reticulum. J. Biol. Chem. 264, 13038–13042.

Mata, A. M., Lee, A. G., & East, J. M. (1989). Probing the nucleotide-binding site of sarcoplasmic reticulum (Ca^{2+}-Mg^{2+})-ATPase with anti-fluorescein antibodies. FEBS Lett. 253, 273–275.

Mata, A. M., Matthews, I., Tunwell, R. E. A., Sharma, R. P., Lee, A. G., & East, J. M. (1992). Definition of surface exposed and trans-membranous regions of the (Ca^{2+}-Mg^{2+})-ATPase of sarcoplasmic reticulum using anti-peptide antibodies. Biochem. J. 286, 567–580.

Mata, A. M., Stefanova, H. I., Gore, M. G., East, J. M., & Lee, A. G. (1993). Localization of Cys-344 on the (Ca^{2+}-Mg^{2+})-ATPase of sarcoplasmic reticulum using fluorescence energy transfer. Biochim. Biophys. Acta 1147, 6–12.

Matthews, I., Colyer, J., Mata, A. M., Green, N. M., Sharma, R. P., Lee, A. G., & East, J. M. (1989). Evidence for the cytoplasmic location of the N- and C-terminal segments of sarcoplasmic reticulum (Ca^{2+}-Mg^{2+})-ATPase. Biochem. Biophys. Res. Commun. 161, 683–688.

Matthews, I., Sharma, R. P., Lee, A. G., & East, J. M. (1990). Transmembranous organisation of (Ca^{2+}-Mg^{2+})-ATPase from sarcoplasmic reticulum: Evidence for the luminal location of residues 877–888. J. Biol. Chem. 265, 18737–18740.

McIntosh, D., & Ross, D. (1992). Glutaraldehyde cross-links Lys-492 and Arg-678 at the active site of sarcoplasmic reticulum Ca^{2+}-ATPase. FASEB J. 6, A120.

McIntosh, D. B., Woolley, D. G., & Berman, M. C. (1992). 2′,3′-O-(2,4,6-trinitrophenyl)-8-azido-AMP and -ATP photolabel Lys-492 at the active site of the sarcoplasmic reticulum Ca^{2+}-ATPase. J. Biol. Chem. 267, 5301–5309.

McIntosh, D. B., & Woolley, D. G. (1994). Catalysis of an ATP analogue untethered and tethered to Lys-492 of sarcoplasmic reticulum Ca^{2+}-ATPase. J. Biol. Chem., in press.

McPhalen, C. A., Strynadka, N. C. J., & James, M. N. G. (1991). Calcium-binding sites in proteins: A structural perspective. Adv. Protein Chemistry 42, 77–144.

Mitchinson, C., Wilderspin, A. F., Trinnaman, B. J., & Green, N. M. (1982). Identification of a labelled peptide after stoichiometric reaction of fluorescein isothiocyanate with the Ca^{2+}-dependent adenosine triphosphatase of sarcoplasmic reticulum. FEBS Lett. 146, 87–92.

Moller, J. V., le Maire, M., & Champeil, P. (1991). Location of amino acid residues critical for Ca^{2+} transport by sarcoplasmic reticulum Ca^{2+}-ATPase in light of membrane protein folding proposed for Na^{+},K^{+}-ATPase. In: The Sodium Pump: Recent Developments (Kaplan, J. H., & De Weer, P., eds.) pp. 259–264. Rockefeller University Press, New York.

Munkonge, F., East, J. M., & Lee, A. G. (1989). The positions of the sites labeled by N-cyclo-N′-(4-dimethylamino-l-naphthyl)-carbodiimide on the (Ca^{2+}-Mg^{2+})-ATPase. Biochim. Biophys. Acta 979, 113–120.

Murakami, K., Tanabe, K., & Takada, S. (1990). Structure of a *Plasmodium yoelii* gene-encoded protein homologous to the Ca^{2+}-ATPase of rabbit skeletal muscle sarcoplasmic reticulum. J. Cell Sci. 97, 487–495.

Myung, J., & Jencks, W. P. (1995). There is only one phosphoenzyme intermediate with bound calcium on the reaction pathway of the sarcoplasmic reticulum calcium ATPase. Biochemistry 34, 3077–3083.

Nakamoto, R. K., & Inesi, G. (1986). Retention of ellipticity between enzymatic states of the Ca^{2+}-ATPase of sarcoplasmic reticulum. FEBS Lett. 194, 258–262.

Ogurusu, T., Wakabayashi, S., & Shigekawa, M. (1991). Functional characterization of lanthanide binding sites in the sarcoplasmic reticulum Ca^{2+}-ATPase—do lanthanide ions bind to the calcium transport site? Biochemistry 30, 9966–9973.

Ohta, T., Nagano, K., & Yoshida, M. (1986). The active site structure of Na^{+}/K^{+}-transporting ATPase: Location of the 5′-(p-fluorosulfonyl)benzoyladenosine binding site and soluble peptides released by trypsin. Proc. Natl. Acad. Sci. USA 83, 2071–2075.

Orlowski, S., & Champeil, P. (1991a). Kinetics of calcium dissociation from its high-affinity transport sites on sarcoplasmic reticulum ATPase. Biochemistry 30, 352–361.

Orlowski, S., & Champeil, P. (1991b). The two calcium ions initially bound to nonphosphorylated sarcoplasmic reticulum Ca^{2+}-ATPase can no longer be kinetically distinguished when they dissociate from phosphorylated ATPase towards the lumen. Biochemistry 30, 11331–11342.

Palmero, I., & Sastre, L. (1989). Complementary DNA cloning of a protein highly homologous to mammalian sarcoplasmic reticulum Ca^{2+}-ATPase from the crustacean *Artemia*. J. Mol. Biol. 210, 737–748.

Pearson, R. H., & Pascher, I. (1979). The molecular structure of lecithin dihydrate. Nature 281, 499–501.

Pick, U., & Karlish, S. J. (1982). Regulation of the conformation transition in the Ca^{2+}-ATPase from sarcoplasmic reticulum by pH, temperature, and calcium ions. J. Biol. Chem. 257, 6120–6126.

Reithmeier, R. A., & MacLennan, D. H. (1981). The NH_2 terminus of the (Ca^{2+} + Mg^{2+})-adenosine triphosphatase is located on the cytoplasmic surface of the sarcoplasmic reticulum membrane. J. Biol. Chem. 256, 5957–5960.

Rice, W. J., Clarke, D. M., Loo, T. W., & MacLennan, D. H. (1993). Functional roles of amino acids in transmembrane sequence M4 of SERCA1. Biophys. J. 64, A333.

Scott, T. L. (1985). Distances between the functional sites of the (Ca^{2+} + Mg^{2+})-ATPase of sarcoplasmic reticulum. J. Biol. Chem. 260, 14421–14423.

Seebregts, C. J., & McIntosh, D. B. (1989). 2′,3′-O-(2,4,6-trinitrophenyl)-8-azido-adenosine mono-, di-, and triphosphates as photoaffinity probes of the Ca^{2+}-ATPase of sarcoplasmic reticulum. Regulatory/superfluorescent nucleotides label the catalytic site with high efficiency. J. Biol. Chem. 264, 2043–2052.

Serrano, R., & Portillo, F. (1990). Catalytic and regulatory sites of yeast plasma membrane H^+-ATPase studied by directed mutagenesis. Biochim. Biophys. Acta 1018, 195–199.

Skerjanc, I. S., Toyofuku, T., Richardson, C., & MacLennan, D. H. (1993). Mutation of glutamate-309 to glutamine alters one Ca^{2+}-binding site in the Ca^{2+}-ATPase of sarcoplasmic reticulum expressed in Sf9 cells. J. Biol. Chem. 268, 15944–15950.

Smith, D. L., Tao, T., & Maguire, M. E. (1993). Membrane topology of a P-type ATPase. The MgtB magnesium transport protein of *Salmonella typhimurium*. J. Biol. Chem. 268, 22469–22479.

Squier, T. C., Bigelow, D. J., de Ancos, J. G., & Inesi, G. (1987). Localization of site-specific probes on the Ca^{2+}-ATPase of sarcoplasmic reticulum using fluorescence energy transfer. J. Biol. Chem. 262, 4748–4754.

Stahl, N., & Jencks, W. P. (1987). Reactions of the sarcoplasmic reticulum calcium adenosinetriphosphatase with adenosine 5′-triphosphate and Ca^{2+} that are not satisfactorily described by an E1–E2 model. Biochemistry 26, 7654–7667.

Stefanova, H. I., Mata, A. M., East, J. M., Gore, M. G., & Lee, A. G. (1993a). Reactivity of lysyl residues on the (Ca^{2+}-Mg^{2+})-ATPase to 7-amino-4-methylcoumarin-3-acetic acid succinimidyl ester. Biochemistry 32, 356–362.

Stefanova, H. I., Mata, A. M., Gore, M. G., East, J. M., & Lee, A. G. (1993b). Labeling the (Ca^{2+}–Mg^{2+})-ATPase of sarcoplasmic reticulum at Glu-439 with 5-(bromomethyl)-fluorescein. Biochemistry 32, 6095–6103.

Stokes, D. L., & Green, N. M. (1990a). Comparison of frozen-hydrated and negatively stained crystals of Ca^{2+}-ATPase suggest a shape for the intramembranous domain. Biochem. Soc. Trans. 18, 841–843.

Stokes, D. L., & Green, N. M. (1990b). Three-dimensional crystals of CaATPase from sarcoplasmic reticulum. Symmetry and molecular packing. Biophys. J. 57, 1–14.

Stokes, D. L., Taylor, W. R., & Green, N. M. (1994). Structure, transmembrane topology and helix packing of P-type ion pumps. FEBS Lett. 346, 32–38.

Suko, J., Plank, B., Preis, P., Kolassa, N., Hellmann, G., & Conca, W. (1981). Formation of magnesium-phosphoenzyme and magnesium-calcium-phosphoenzyme in the phosphorylation of adenosine triphosphatase by orthophosphate in sarcoplasmic reticulum: Models of a reaction sequence. Eur. J. Biochem. 119, 225–236.

Sumbilla, C., Lu, L., Lewis, D. E., Inesi, G., Ishii, T., Takeyasu, K., Feng, Y., & Fambrough, D. M. (1993). Ca^{2+}-dependent and thapsigargin-inhibited phosphorylation of Na^+,K^+-ATPase catalytic domain following chimeric recombination with Ca^{2+}-ATPase. J. Biol. Chem. 268, 21185–21192.

Tamura, S., Tagaya, M., Maeda, M., & Futai, M. (1989). Pig gastric (H^+–K^+)-ATPase. J. Biol. Chem. 264, 8580–8584.

Taylor, K. A., Dux, L., & Martonosi, A. (1986). Three-dimensional reconstruction of negatively stained crystals of the Ca^{2+}-ATPase from muscle sarcoplasmic reticulum. J. Mol. Biol. 187, 417–427.

Taylor, K. A., Dux, L., Varga, S., Ting-Beall, H. P., & Martonosi, A. (1988). Analysis of two-dimensional crystals of Ca^{2+}-ATPase in sarcoplasmic reticulum. Methods Enzymol. 157, 271–289.

Teruel, J. A., & Gomez-Fernandez, J. C. (1986). Distances between the functional sites of sarcoplasmic reticulum (Ca^{2+} + Mg^{2+})-ATPase and the lipid/water interface. Biochim. Biophys. Acta 863, 178–184.

Toyofuku, T., Kurzydlowski, K., Tada, M., & MacLennan, D. H. (1993). Identification of regions in the Ca^{2+}-ATPase of sarcoplasmic reticulum that affect functional association with phospholamban. J. Biol. Chem. 268, 2809–2815.

Toyoshima, C., Sasabe, H., & Stokes, D. L. (1993). Three-dimensional cryo-electron microscopy of the calcium ion pump in the sarcoplasmic reticulum membrane. Nature 362, 469–471.

Tunwell, R. E., Conlan, J. W., Matthews, I., East, J. M., & Lee, A. G. (1991). Definition of surface-exposed epitopes on the (Ca^{2+}–Mg^{2+})-ATPase of sarcoplasmic reticulum. Biochem. J. 279, 203–212.

Verma, A. K., Filoteo, A. G., Stanford, D. R., Wieben, E. D., Penniston, J. T., Strehler, E. E., Fischer, R., Heim, R., Vogel, G., Mathews, S., Strehler-Page, M. A., James, P., Vorherr, T., Krebs, J., & Carafoli, E. (1988). Complete primary sequence of a human plasma membrane Ca^{2+} pump. J. Biol. Chem. 263, 14152–14159.

Vilsen, B., Andersen, J. P., Clarke, D. M., & MacLennan, D. H. (1989). Functional consequences of proline mutations in the cytoplasmic and transmembrane sectors of the Ca^{2+}-ATPase of sarcoplasmic reticulum. J. Biol. Chem. 264, 21024–21030.

Vilsen, B., Andersen, J. P., & MacLennan, D. H. (1991). Functional consequences of alterations to hydrophobic amino acids located at the M4S4 boundary of the Ca^{2+}-ATPase of sarcoplasmic reticulum. J. Biol. Chem. 266, 18839–18845.

Vilsen, B., & Andersen, J. P. (1992a). Deduced amino acid sequence and E1–E2 equilibrium of the sarcoplasmic reticulum Ca^{2+}-ATPase of frog skeletal muscle. FEBS Lett. 306, 213–218.

Vilsen, B., & Andersen, J. P. (1992b). CrATP-induced Ca^{2+} occlusion in mutants of the Ca^{2+}-ATPase of sarcoplasmic reticulum. J. Biol. Chem. 267, 25739–25743.

von Heijne, G. (1991). Proline kinks in transmembrane α-helices. J. Mol. Biol. 218, 499–503.

Vorherr, T., Chiesi, M., Schwaller, R., & Carafoli, E. (1992). Regulation of the calcium ion pump of sarcoplasmic reticulum: Reversible inhibition by phospholamban and by the calmodulin binding domain of the plasma membrane calcium ion pump. Biochemistry 31, 371–376.

Wictome, M., Michelangeli, F., Lee, A. G., & East, J. M. (1992). The inhibitors thapsigargin and 2,5-di(tert-butyl)-1,4-benzohydroquinone favour the E2 form of the (Ca^{2+}-Mg^{2+})-ATPase. FEBS Lett. 304, 109–113.

Wictome, M., Khan, Y. M., East, J. M., & Lee, A. G. (1995). Binding of sesquiterpene lactone inhibitors to the Ca^{2+}-ATPase. Biochem. J. 310, 859–868.

Wimmers, L. E., Ewing, N. N., & Bennett, A. B. (1992). Higher plant Ca^{2+}-ATPase: Primary structure and regulation of mRNA abundance by salt. Proc. Natl. Acad. Sci. USA 89, 9205–9209.

Xu, K. Y. (1989). Any of several lysines can react with 5′-isothiocyanatofluorescein to inactivate sodium and potassium ion activated adenosinetriphosphatase. Biochemistry 28, 5764–5772.

Yamagata, K., Daiho, T., & Kanazawa, T. (1993). Labeling of lysine 492 with pyridoxal 5′-phosphate in the sarcoplasmic reticulum Ca^{2+}-ATPase. J. Biol. Chem. 268, 20930–20936.

Yamamoto, H., Tagaya, M., Fukui, T., & Kawakita, M. (1988). Affinity labeling of the ATP-binding site of Ca^{2+}-transporting ATPase of sarcoplasmic reticulum by adenosine triphosphopyridoxal: Identification of the reactive lysyl residue. J. Biochem. (Tokyo) 103, 452–457.

Yamamoto, H., Imamura, Y., Tagaya, M., Fukui, T., & Kawakita, M. (1989). Ca^{2+}-dependent conformational change of the ATP-binding site of Ca^{2+}-transporting ATPase of sarcoplasmic reticulum as revealed by an alteration of the target-site specificity of adenosine triphosphopyridoxal. J. Biochem. (Tokyo) 106, 1121–1125.

KINETIC CHARACTERIZATION OF SARCOPLASMIC RETICULUM Ca^{2+}-ATPASE

Philippe Champeil

Biomembranes
Volume 5, pages 43–76.

ISBN: 1-55938-662-2.

I. INTRODUCTION

Sarcoplasmic reticulum Ca^{2+}-ATPase is a membranous enzyme which couples the energetically down-hill ATP hydrolysis to the up-hill transport of Ca^{2+} from the muscle cytoplasm, in which the free Ca^{2+} concentration is in the submicromolar range, to the luminal compartment of sarcoplasmic reticulum in which the free Ca^{2+} concentration is in the submillimolar or millimolar range. This enzyme belongs to the class of P-type ATPases, that is, ATPases whose catalytic cycle involves formation of a covalent phosphoenzyme (Pedersen and Carafoli, 1987). It has been studied in great detail from the functional point of view (see Inesi et al., 1988 and many other chapters in Volume 157 of *Methods in Enzymology*, as well as Jencks, 1989; Andersen and Vilsen, 1990; and Inesi et al., 1992 for more recent descriptions). Its study has often benefited from those of the other P-type ATPases, and has sometimes stimulated them in turn. It now appears that this enzyme is the prototype of a whole family of intracellular Ca^{2+} pumps, the so-called SERCA-ATPases (sarcoplasmic or endoplasmic reticulum calcium ATPases), which play a major role in the regulation of cytosolic free Ca^{2+} levels in most cell types (e.g., Lytton and Nigam, 1992). The purpose of the present review is to introduce the reader to a vast literature, by describing experimental results which have contributed to establishing some aspects of the mechanism of ATP hydrolysis and Ca^{2+} transport by sarcoplasmic reticulum Ca^{2+}-ATPase. Points of controversy will be mentioned only briefly. We hope this description of the current knowledge and the current uncertainties about the elementary steps involved in ATP hydrolysis and Ca^{2+} transport will be helpful for those in various fields: in the molecular study of structure/function relationships of this particular enzyme, to help detailed interpretation of the effects of chemical labeling or directed mutagenesis; in a more physiological or pharmacological perspective, to help elucidation of the basis for the regulatory role of any new agent modulating intracellular Ca^{2+} transport; in the study of the less extensively documented SERCA-ATPases, to provide a starting point for their functional characterization. A few examples of such analyses will be described.

II. OVERALL REACTION FOR ATP HYDROLYSIS AND Ca^{2+} TRANSPORT

Hasselbach and Makinose found that in the presence of oxalate, $^{45}Ca^{2+}$ was actively pumped into muscle microsomes due to transient activation of microsomal ATPase activity and precipitation of Ca^{2+}–oxalate crystals in the microsomal inner compartment (Hasselbach and Makinose, 1961). The microsomal fraction responsible for Ca^{2+} uptake was soon shown to be derived from sarcoplasmic reticulum. In the initial experiments of these authors, termination of Ca^{2+} uptake was induced by sample cooling followed by centrifugation. Subsequently, calcium-loaded vesicles were removed from the suspension by precipitating them with $HgCl_2$ (Hasselbach and Makinose, 1963), and these confirmatory measurements established that the coupling ratio between Ca^{2+} transport and Ca^{2+}-dependent ATP hydrolysis was close to two (Figure 1). Filtration through nitrocellulose filters was later introduced as a convenient way to separate membranes from the uptake medium and measure trapped Ca^{2+} (Martonosi and Feretos, 1964). From the beginning, Ca^{2+} uptake and ATPase activity were found to be associated with ATP–ADP exchange activity (i.e., formation of radioactive ATP from [^{14}C]- or [^{32}P]-ADP in the medium), suggesting that formation of an intermediate phosphorylated protein was part of the ATP hydrolysis mechanism. This phosphoenzyme was soon observed (Yamamoto and Tonomura, 1968; Makinose, 1969; Martonosi, 1969; Inesi et al., 1970). The γ-phosphate of ATP appeared to bind to an aspartyl residue, through an acid-stable acyl-phosphate bond (e.g., Bastide et al., 1973). The Ca^{2+} dependence of phosphoenzyme formation matched that of ATP hydrolysis, Ca^{2+} transport and ATP–ADP exchange: the enzyme's apparent affinity for Ca^{2+} was in the submicromolar range, consistent with physiological Ca^{2+} levels (Figure 2).

The complete cycle of Ca^{2+} transport and ATP hydrolysis by sarcoplasmic reticulum ATPase was found to be reversible. Steady-state Ca^{2+} depletion by sarcoplasmic reticulum vesicles resulting from ATP hydrolysis induced continuous formation of [^{32}P]ATP from ADP and [^{32}P]-labeled inorganic phosphate (P_i) (Makinose, 1971), ADP- and P_i-dependent release of previously accumulated $^{45}Ca^{2+}$ was found to be coupled to ATP synthesis with a ratio again close to two to one

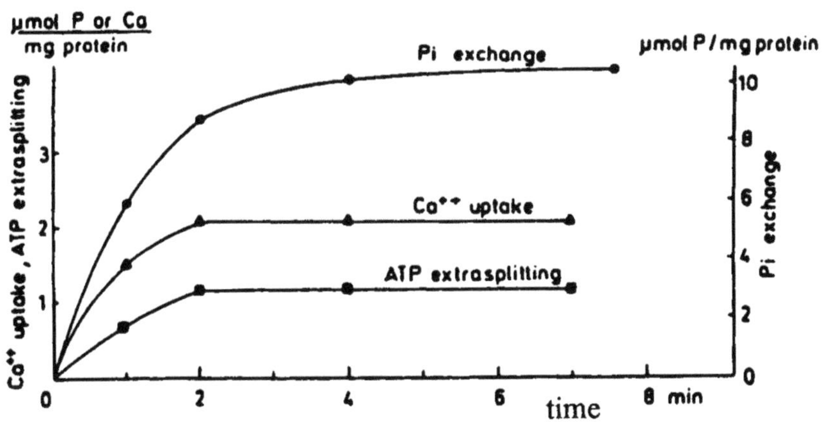

Figure 1. Simultaneous measurement of Ca^{2+} uptake, Ca^{2+}-dependent ATP hydrolysis, and Ca^{2+}-dependent phosphate exchange between ATP and ADP, in the presence of oxalate. [ATP] = [Mg] = [oxalate] = 5×10^{-3} M, [ADP] = 2×10^{-3} M, [Ca] = 0.12×10^{-4} M, 20°C (redrawn from Hasselbach and Makinose, 1963, with permission).

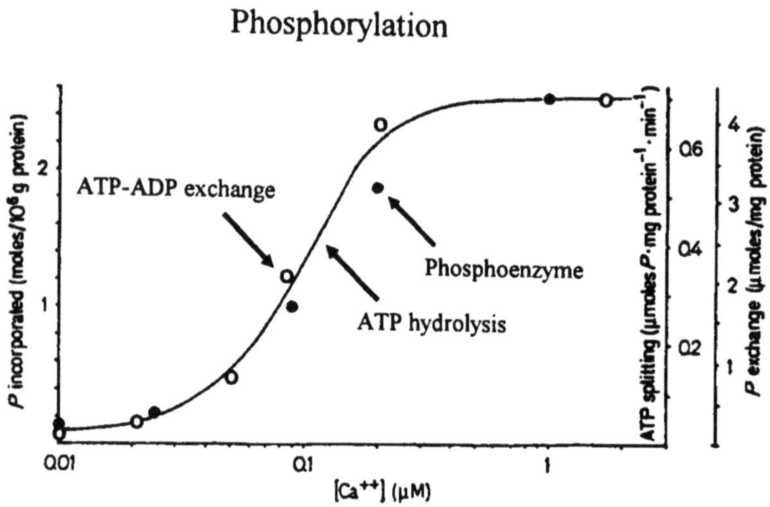

Figure 2. Identical dependency of phosphoprotein formation (closed circles), ATP extra-splitting (continuous line) and phosphate exchange (open circles) on the concentration of free calcium ions (from Makinose, 1969, with permission).

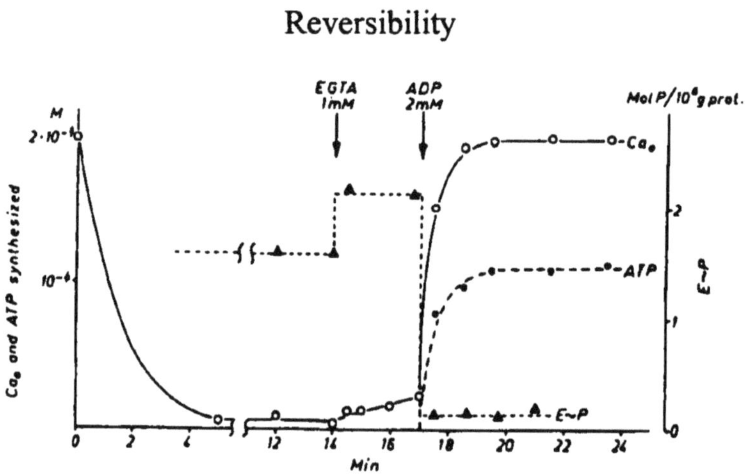

Figure 3. Reversal of the sarcoplasmic reticulum calcium pump, and phosphoenzyme formation from [^{32}P]-orthophosphate. Before the actual measurement, loading of the vesicles was first performed in the presence of acetylphosphate, glucose, hexokinase, and P_i (from Makinose, 1972, with permission).

(Makinose and Hasselbach, 1971), and in the presence of a Ca^{2+} gradient, formation of phosphoenzyme from $[^{32}P]P_i$ was evidenced (Makinose, 1972; see Figure 3). In the absence of a Ca^{2+} gradient, phosphoenzyme formation from P_i was also demonstrated at low enough Ca^{2+} concentration, both through direct $[^{32}P]$ measurements (Masuda and de Meis, 1973) and through P_i–water exchange measurements (Kanazawa and Boyer, 1973). This "gradient-independent" phosphoenzyme

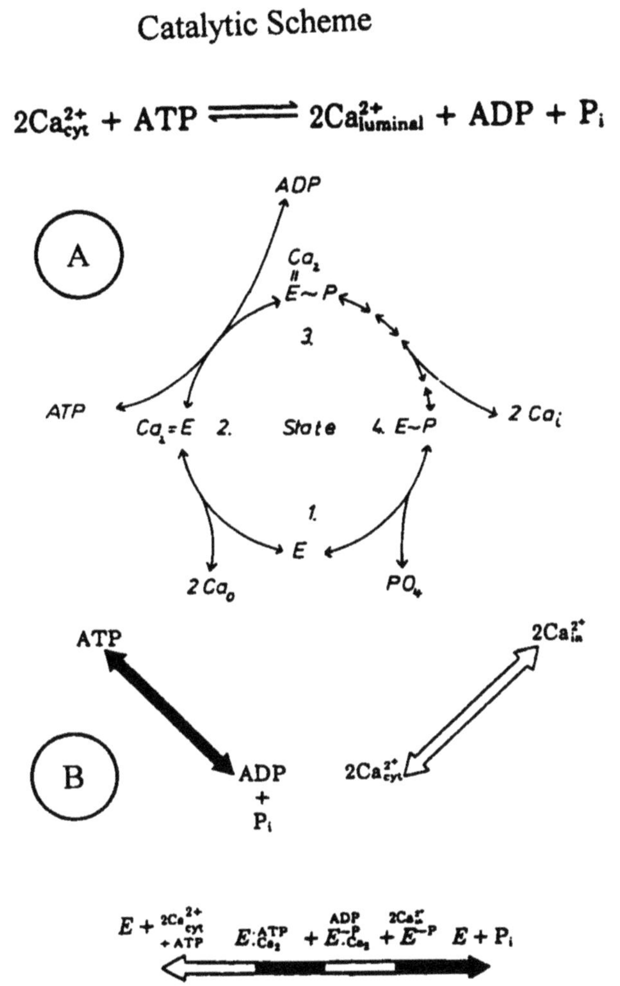

Figure 4. Reaction scheme for the reaction catalyzed by sarcoplasmic reticulum ATPase (**A**, from Makinose, 1973, with permission). Under physiological conditions, hydrolysis of ATP is energetically down-hill, while calcium uptake is up-hill. According to this scheme, each process is divided into two, and the ATPase alternates the steps of the chemical and vectorial reaction (**B**, from Jencks, 1989, with permission).

was found to subsequently react with ADP if a high concentration of Ca^{2+} was present on the luminal side (de Meis and Carvalho, 1974; Knowles and Racker, 1975). As early as 1973, a simple scheme for the ATPase catalytic cycle was proposed (Makinose, 1973; see Figure 4). This scheme describes the presence or absence of Ca^{2+} at the transport sites as the switch allowing the ATPase catalytic site to become reactive to ATP or to P_i (states 2 and 1, respectively). Conversely, phosphorylated ATPase switches from an ADP-sensitive state to a water-reactive state when Ca^{2+} dissociates towards the lumen of sarcoplasmic reticulum (from state 3 to state 4). This scheme, which is still a useful summary of the cycle, has been elegantly described by Jencks as the basis for the tight coupling observed between the vectorial events and the chemical events catalyzed by the Ca^{2+} pump, as follows: the chemical reaction at the catalytic site is divided into phosphorylation and dephosphorylation, the vectorial reaction is divided into the binding of Ca^{2+} on one side of the membrane and its dissociation on the other, and neither reaction can occur unless the other one also occurs because the cycle alternates the steps of the chemical and vectorial reactions (Jencks, 1989).

III. ELEMENTARY STEPS

A. Phosphorylation by ATP

Figure 5 illustrates typical time-resolved measurements of the initial events of the catalytic cycle (Verjovski-Almeida et al., 1978). Starting from an enzyme pre-equilibrated with Ca^{2+}, addition of ATP with a rapid mixing device induces fast formation of phosphoenzyme, as measured after acid quenching (Panel A and schematic drawing in Figure 5). At high (millimolar) ATP concentrations, phosphorylation occurs within a few milliseconds, and cannot be resolved with the usual rapid mixing techniques. P_i is not produced during this phase, implying that dephosphorylation steps are slower (Panel B). A remarkable feature of these initial events is illustrated in Panel C. Here, the phosphorylation reaction was triggered by ATP addition in the presence of $^{45}Ca^{2+}$, it was then quenched by addition of EGTA, a Ca^{2+} chelator, instead of acid, and the vesicles were subsequently filtered on a nitrocellulose filter to measure the amount of $^{45}Ca^{2+}$ taken up. A "burst" of $^{45}Ca^{2+}$ internalization was observed, in a two-to-one ratio with the amount of phosphoenzyme. This burst results from the fact that EGTA addition stopped further phosphorylation from ATP, but did not prevent the already phosphorylated ATPase molecules from completing their catalytic and transport cycle before filtration was performed. This burst implies that as soon as phosphoenzyme was formed, the addition of EGTA no longer induced dissociation of Ca^{2+} towards the cytosolic side: bound Ca^{2+} became inaccessible from the outer medium because of phosphorylation. Phosphorylation at the catalytic site is, therefore, not just a simple chemical reaction; it is also associated with restriction of the accessibility of the Ca^{2+} ions bound to the transport site.

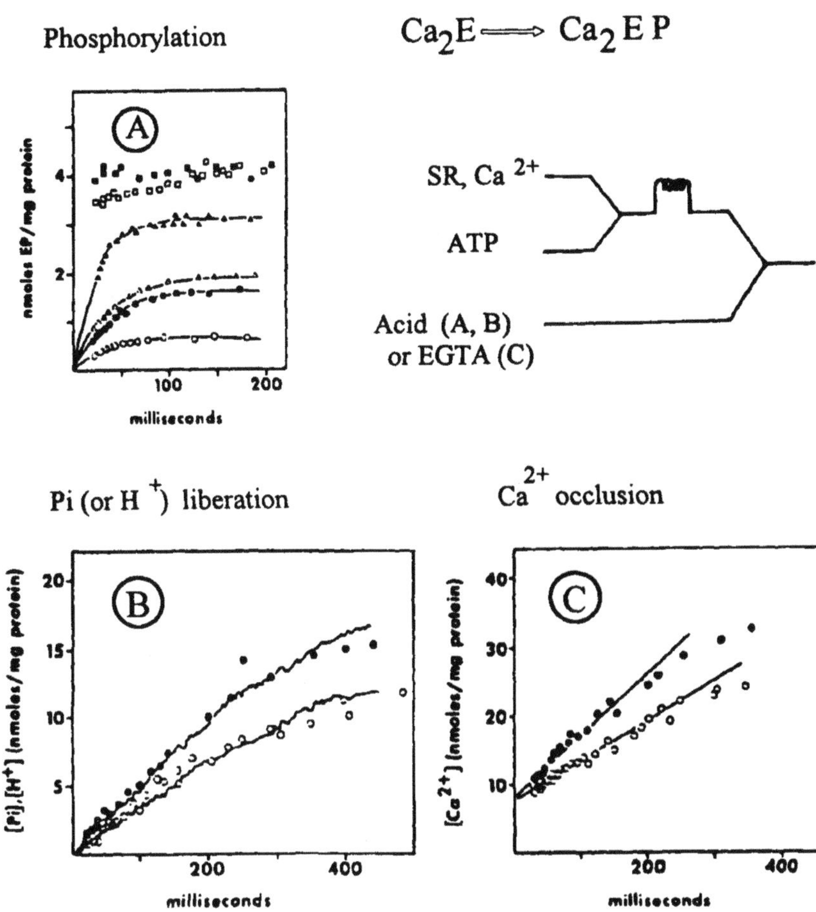

Figure 5. Phosphorylation from ATP and associated events. Panel **A**: phosphoenzyme formation, in the presence (from bottom to top) of 1-, 2-, 2.9-, 5-, 100-, or 1000-μM ATP. Panel **B**: production of phosphate (circles) and medium acidification (continuous trace), in the presence of 100- (open symbols) or 1000- (closed symbols) μM ATP. Panel **C**: trapped calcium under the conditions of Panel **B**, after EGTA quenching of the reaction and subsequent filtration. The intercept corresponds to twice the amount of phosphoenzyme formed (from Verjovski-Almeida et al., 1978, with permission).

Because phosphorylation from ATP is faster than ATPase dephosphorylation, it was initially accepted that there was a practically 1:1 relation between the amount of phosphoenzyme measured and the amount of ATPase in the sarcoplasmic reticulum membranes (Meissner et al., 1973). A few years later, the situation appears to be less clear-cut for various reasons: impurities in radiolabeled ATP might bias the measurements, the rate constants for the various elementary steps might not be very different from each other under usual assay conditions where ATP concentrations are kept moderate (micromolar or submillimolar; e.g., Carvalho et al., 1976; Henao et al., 1991), and partially denatured protein is probably part of the total ATPase polypeptides, even in purified SR fractions (Barrabin et al., 1984; Coll and Murphy, 1984; Gafni and Boyer, 1984). In addition, the overall equilibrium constant for the phosphorylation step itself (from E. ATP to E–P. ADP) is not always high (Shigekawa and Kanazawa, 1982; Petithory and Jencks, 1986). Maximal levels of phosphoenzyme and hence, the level of active ATPases, can be obtained by phosphorylation with Ca–ATP in the presence of high KCl (McIntosh et al., 1992).

The proportion of ATPase with bound substrate (ATP in Figure 5a) obviously controls the apparent rate of phosphorylation. As ATP binds with a second-order rate constant which is not very large, typically 10^6–10^7 $M^{-1}s^{-1}$, it might also contribute to the limitation of this apparent rate at micromolar concentrations of ATP (e.g., Lacapère and Guillain, 1993). Phosphorylation per se is believed to be fast, including that for low affinity substrates (e.g., ITP or acetylphosphate; see Scofano et al., 1979; Bodley and Jencks, 1987). It has, in fact, been suggested that a conformational change following substrate binding, and not phosphoryl transfer itself, controls the apparent maximal rate of ATPase phosphorylation (Petithory and Jencks, 1986). At a high enough concentration of ATP, the apparent rate of phosphorylation is faster that the rate of Ca^{2+} dissociation from unphosphorylated ATPase (Rauch et al., 1978; Sumida et al., 1978), and this makes it possible to trigger single-cycle turnover by simultaneously adding ATP and a high concentration of EGTA to initially Ca^{2+}-saturated ATPase. Exploitation of this favorable circumstance will be described below. It is also worth noting that the substrate for ATPase phosphorylation from ATP is believed to be Mg–ATP, ADP being released from the phosphoenzyme to which Mg^{2+} remains bound (Vianna, 1975; Makinose and Boll, 1979; Ronzani et al., 1979; Shigekawa et al., 1983). Yet, replacement of Mg–ATP by other metal–ATP complexes generally also leads to phosphoenzyme formation (e.g., Yamada and Ikemoto, 1980), but the phosphoenzyme thus formed has a stability different from that of Mg–ATP-derived phosphoenzyme (references in Section IIIB). In particular, the phosphoenzyme-like complex formed from Cr–ATP is very stable, and keeps occluding Ca^{2+} at the transport sites for hours, a potentially useful characteristic (Serpersu et al., 1982; Vilsen and Andersen, 1986).

As was mentioned in the introduction, ADP-induced reversal of ATPase phosphorylation gives rise to ATP–ADP exchange, which may be much faster than ATP hydrolysis itself (see Figure 1). This exchange may also be associated with exchange of Ca^{2+} between luminal and cytosolic compartments (e.g., Waas and Hasselbach, 1981; Takenaka et al., 1982; Inao and Kanazawa, 1986; Soler et al., 1990).

B. Ca^{2+} Dissociation Toward the Luminal Side

Initial attempts to study Ca^{2+} dissociation toward the luminal side of the membrane took advantage of the possibility to use preparations derived from sarcoplasmic reticulum vesicles, but with high passive permeability for Ca^{2+}. In such cases, Ca^{2+} released on the luminal side because of ATP utilization should freely diffuse out of the leaky vesicles, and the amount of membrane-bound Ca^{2+} should thus be reduced. Such Ca^{2+} release was indeed found (Ikemoto, 1975). However, the breakthrough was to realize that Ca^{2+} release occurred only if the subsequent steps were slowed down enough to permit steady-state accumulation of Ca^{2+}-free states of ATPase, state 4 or state 1 (Watanabe et al., 1981). One way to do this is to add DMSO (de Meis et al., 1980); another is to use high Mg^{2+} concentrations at alkaline pH (Nakamura and Tonomura, 1982; Takisawa and Makinose, 1983; Andersen et al., 1985). Steady-state Ca^{2+} release was thus observed, either by $^{45}Ca^{2+}$ filtration techniques or by optical techniques with a Ca^{2+} dye (Figure 6a). Time-resolved measurements based on the same rationale were then designed, again using leaky preparations derived from sarcoplasmic reticulum, and these measurements allowed estimation of the rate of Ca^{2+} dissociation from phosphorylated ATPase towards the luminal side of the membrane (Champeil and Guillain, 1986). More refined versions of the protocol made it possible to measure the rate of this dissociation under conditions where steady-state Ca^{2+} release was low: for this purpose, the rapid filtration experiments included a chase by $^{40}Ca^{2+}$ of the released $^{45}Ca^{2+}$ (Wakabayashi et al., 1986; Orlowski and Champeil, 1991b). However, because of the technical limitations of rapid filtration measurements and the possible concerns with the leakiness of membrane preparations, such experiments are not appropriate under experimental conditions where Ca^{2+} dissociation is expected to be very fast, for example, at a high temperature.

A conceivable alternative to the measurement of the rate of Ca^{2+} dissociation itself is the measurement of the rate with which phosphoenzyme switches from an ADP-sensitive form (the one with bound Ca^{2+}, state 3) to an ADP-insensitive form (the Ca^{2+}-free form, state 4). These measurements are technically demanding, as they generally involve phosphorylation from [γ-^{32}P]ATP first, followed by addition of EGTA or nonradioactive ATP to stop further [^{32}P] incorporation, then addition of ADP, after various incubation periods, to dephosphorylate ADP-sensitive phosphoenzyme, and finally acid quenching, to determine the amount of ADP-insensitive phosphoenzyme. Nevertheless, such measurements were successfully performed in some cases, generally at a low temperature, and they contributed to establishing the validity of the scheme depicted in Figure 4a (e.g., Shigekawa and Dougherty, 1978; Shigekawa and Akowitz, 1979; Nakamura and Tonomura, 1982; Takisawa and Makinose, 1983; Andersen et al., 1985). However, when such experiments are performed at room temperature, various processes with similar rate constants may contribute to the observed kinetics, including ATP dissociation after its synthesis from ADP-sensitive phosphoenzyme and ADP-induced acceleration

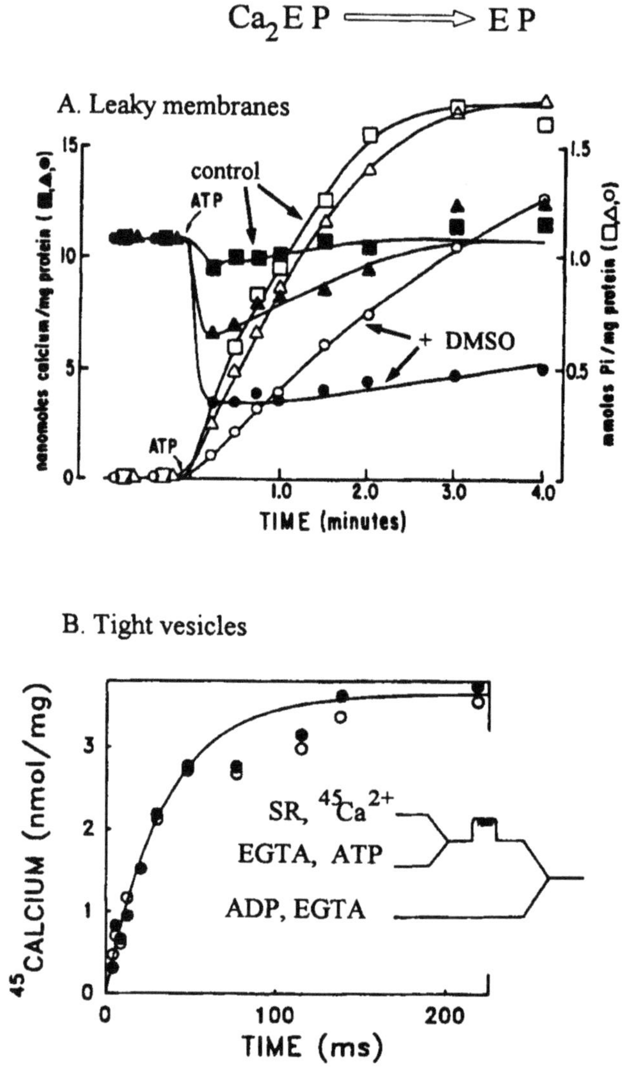

Figure 6. Calcium release toward the luminal side of the membrane. Panel **A**: effect of DMSO on steady-state release of calcium from purified (thus leaky) ATPase membranes during ATP hydrolysis. At zero time, 1 mM ATP was added, in the presence of 0% (squares), 20% (triangles) or 30% (circles) DMSO (from Watanabe et al., 1981). Panel **B**: calcium internalization into native (tight) sarcoplasmic reticulum vesicles. A single turnover was initiated by simultaneously adding ATP and EGTA to vesicles preincubated with $^{45}Ca^{2+}$; after the indicated period, further internalization was stopped by adding ADP, and the sample was filtered (from Hanel and Jencks, 1991, with permission). Open and closed symbols correspond to experiments performed with empty vesicles or with vesicles passively loaded with nonradioactive calcium, respectively.

of the transition to the ADP-insensitive form of phosphoenzyme. As a result, the pattern of phosphoenzyme disappearance upon addition of ADP is difficult to interpret, especially if P_i production is not measured simultaneously (Inesi et al., 1982; Pickart and Jencks, 1982; Froehlich and Heller, 1985; Hobbs et al., 1985; Wang, 1986).

Under such conditions, the rate of Ca^{2+} dissociation toward the luminal side of the membrane was therefore studied using a different protocol, with tight vesicles, by combining rapid mixing with $^{45}Ca^{2+}$ filtration. In this protocol, $^{45}Ca^{2+}$ internalization in the vesicle lumen during single turnover was monitored; the essential step, after triggering single turnover by adding EGTA and ATP to Ca^{2+}-saturated ATPase, was to add ADP and EGTA at various times, before vesicle filtration, to release to the outer medium those $^{45}Ca^{2+}$ ions bound to phosphoenzyme (state 3) and not yet released in the vesicle lumen (Figure 6b). These experiments are difficult, and two slightly different versions of this protocol gave conflicting results as to whether dissociation of the two transported Ca^{2+} ions was sequential or not (Inesi, 1987; Hanel and Jencks, 1991). This issue remains controversial (Orlowski and Champeil, 1991b; McIntosh et al., 1991; Ross et al., 1991) and might benefit from recent findings with Cr–ATP-dependent Ca^{2+} occlusion (Chen et al., 1991; Vilsen and Andersen, 1992a).

Ca^{2+} dissociation from phosphorylated ATPase toward the luminal side of the membrane, as well as the corresponding switch in chemical reactivity of the phosphoenzyme, is accelerated by ATP (Champeil and Guillain, 1986; Wakabayashi et al., 1986); it depends on pH only moderately, but is slowed down by K^+, which contributes to make the ADP-sensitive phosphoenzyme predominant at steady-state in the presence of K^+ (e.g., Shigekawa and Akowitz, 1979). Compared to Mg^{2+}, other metallic cofactors of ATP (which remain bound to phosphoenzyme) also slow it down (e.g., Wakabayashi and Shigekawa, 1987; Fujimori and Jencks, 1990). It should finally be mentioned that Ca^{2+} dissociation from phosphorylated ATPase in leaky preparations is associated with binding of protons to the ATPase (Yamaguchi and Kanazawa, 1984, 1985), consistent with electrogenic H^+/Ca^{2+} exchange being mediated by the Ca^{2+} pump (Chiesi and Inesi, 1980; Levy et al., 1990; Nishie et al., 1990; Bamberg et al., 1993). Measurement of the rate of turnover-dependent H^+ binding using pH-sensitive dyes is, therefore, an alternative to the measurement of the rate of Ca^{2+} dissociation (Yamaguchi and Kanazawa, 1985), but again only with leaky preparations under conditions where a significant amount of Ca^{2+}-free phosphoenzyme accumulates at steady-state.

C. ATPase Dephosphorylation

One of the most useful tools for studying phosphoenzyme hydrolysis is [^{18}O]-exchange between P_i and water, "medium P_i–water exchange" (Boyer et al., 1977; Ariki and Boyer, 1980): every time the chemical bond in phosphoenzyme is hydrolyzed, the P_i pool, initially enriched in [^{18}O], gains one [^{16}O]-atom from water

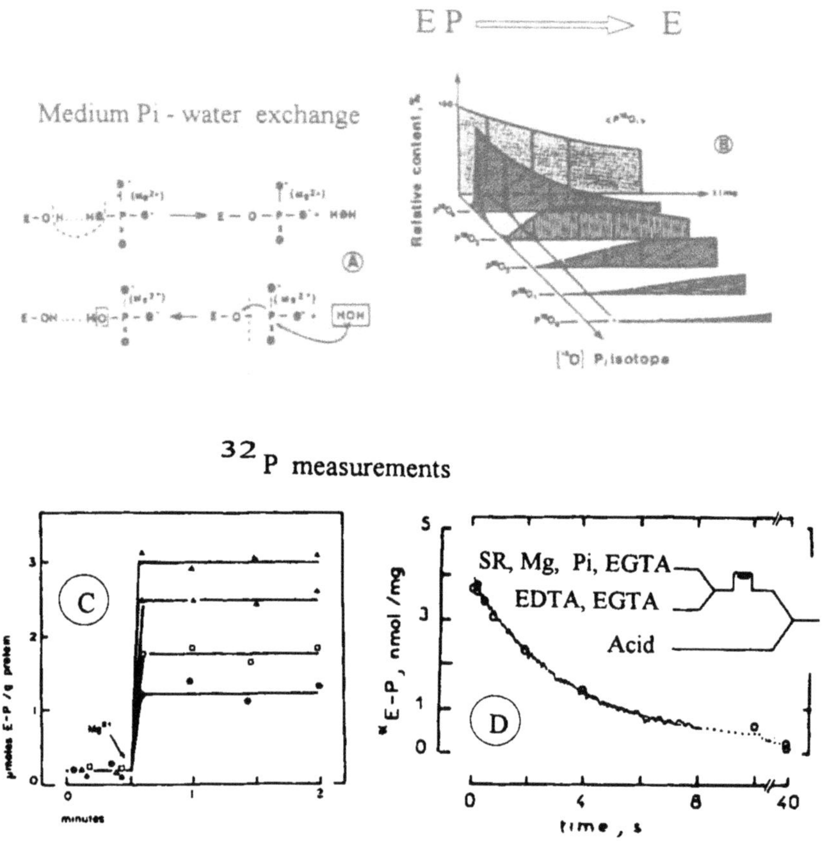

Figure 7. Formation and hydrolysis of phosphoenzyme from P_i. Panels **A** and **B**: mechanism of oxygen exchange between [^{16}O]-water and [^{18}O]-P_i (from Champeil et al., 1985, drawn after the data in Ariki and Boyer, 1980). Panel **C**: dependence on Mg^{2+} of phosphorylation from P_i, in the absence of Ca^{2+} and K^+ at pH 6. At the time indicated by the arrow, 10 mM Mg^{2+} was added. The various symbols correspond to different concentrations of P_i (Masuda and de Meis, 1973). Panel **D**: ATPase dephosphorylation after Mg^{2+} chelation in the presence of 15% DMSO, as measured either by chemical quenching procedures with [^{32}P] and a multimixing apparatus (open circles, see schematic representation of the protocol) or by stopped-flow fluorescence (continuous lines) (from Champeil et al., 1985, with permission).

(Figures 7a and 7b). The medium P_i–water exchange technique is unique in that it gives information about the rate of phosphoenzyme hydrolysis even under conditions where the amount of phosphoenzyme derived from P_i is low at steady-state, for example, under physiological conditions at neutral pH in the presence of K^+ and ATP (McIntosh and Boyer, 1983). Unfortunately, adequate equipment for medium P_i–water exchange measurements is rarely available. The rate of phosphoenzyme hydrolysis can also be deduced from rapid quenching chemical measurements. In the latter measurements, "gradient-independent" phosphoenzyme (see above), initially formed from [^{32}P]P_i, is generally allowed to hydrolyze by mixing with a Mg^{2+} chelator, which removes the phosphorylation co-substrate (Figures 7c and 7d), or by mixing with excess P_i, which dilutes out the specific radioactivity of newly-formed phosphoenzyme. [^{32}P]P_i tracer can also be added to nonradioactive phosphoenzyme at equilibrium, and the rate of [^{32}P] incorporation is then measured, as under such conditions this incorporation is limited by phosphoenzyme hydrolysis (Boyer et al., 1977). Under some conditions, dephosphorylation can also be monitored by stopped-flow measurements of the changes in ATPase intrinsic fluorescence (e.g., Lacapère et al., 1981; Champeil et al., 1985; see Figure 7d). Under conditions where the equilibrium amount of phosphoenzyme derived from P_i is low, estimation with [^{32}P] measurements of the true rate of dephosphorylation is possible only with two-step protocols: phosphoenzyme is first formed from [^{32}P]P_i under ionic conditions allowing formation of maximal amounts of this species (e.g., at pH 6 in the absence of K^+), and its hydrolysis is measured only after a "jump" to the desired pH, ionic, or ligand conditions, with the tentative assumption that this jump instantaneously modifies the dephosphorylation rate (e.g., de Meis et al., 1980; Inesi et al., 1982). Using such protocols, moderately fast dephosphorylation rates can also be studied with rapid filtration techniques, the dead time of which is a few tens of milliseconds (Champeil and Guillain, 1986).

Compared to physiological conditions, the rate of phosphoenzyme hydrolysis is slowed down at acidic pH, in the absence of K^+, in the absence of ATP, and of course at low temperatures. At alkaline pH, it is also slowed down by Mg^{2+}, possibly because of competition between Mg^{2+} and Ca^{2+} for luminal sites on phosphorylated ATPase (de Meis et al., 1980). This rate is also dramatically reduced in the presence of DMSO or other organic co-solvents (de Meis et al., 1980). Mg^{2+} is not the sole metallic cofactor permitting phosphorylation from P_i (Mintz et al., 1990). The equilibrium constant for covalent bond formation from the noncovalent complex (E–P_i) has been the subject of some dispute. As the maximal amount of phosphoenzyme formed from P_i in the absence of Ca^{2+} gradient was much smaller than the amount of active ATPase present, this equilibrium constant was initially considered to be close to one at neutral pH (Kolassa et al., 1979). Yet, later on, it was suggested that at least in the absence of K^+, the equilibrium constant for phosphorylation was well above one both at pH 6 and pH 7, and that the less than complete maximal phosphorylation from P_i observed at pH 7 was the result of the fact that besides being a co-substrate, Mg^{2+} also acted as an inhibitor for phosphorylation from P_i (Loomis et al., 1982; Champeil et al., 1985).

The dephosphorylation step is not just an event restricted to the catalytic site: reorientation of the transport sites partly occurs at this step, from a luminal orientation in phosphoenzyme (state 4) to a mainly non-luminal orientation in the unphosphorylated Ca^{2+}-free form of ATPase (state 1). Evidence for this will be discussed below.

D. Binding of Ca^{2+} to Unphosphorylated ATPase

The high affinity binding of two $^{45}Ca^{2+}$ ions to the transport sites of sarcoplasmic reticulum Ca^{2+}-ATPase was initially deduced from filtration or column chromatography experiments (Fiehn and Migala, 1971; Meissner, 1973; Ikemoto, 1975; Dupont, 1980; Inesi et al., 1980). This binding soon appeared to be associated with long-range conformational changes, as detected by spectroscopic probes located in various parts of the polypeptide chain (e.g., Champeil et al., 1976; Dupont, 1976; Murphy, 1976). Such long-range conformational changes are consistent with the role of Ca^{2+} binding as a switch for the chemical reactivity of the catalytic site. As regards the rate of the Ca^{2+}-induced transition, from state 1 to state 2, analysis of a wide-range of experiments allowed L. de Meis to infer that this rate was relatively slow in the absence of substrate, but was accelerated in the presence of ATP and could, therefore, contribute to overall rate-limitation, depending on the exact conditions (Carvalho et al., 1976; Souza and de Meis, 1976). This hypothesis accounted for the observation that in empty or leaky vesicles, the level of turnover-dependent ATPase phosphorylation from ITP (or ATP at low concentration) was much less than stoichiometric, and phosphorylation from P_i was possible: in the presence of ITP (or at low concentrations of ATP) the P_i-reactive form of ATPase (state 1) piled up because of too slow transition to its NTP-reactive form (state 2). The first estimation of the rate of this transition came from stopped-flow measurements of the rate of the associated changes in ATPase tryptophan fluorescence (Dupont, 1978). Direct measurement of the rate of Ca^{2+} binding was made possible only later, when it was realized that *perfusion* of adsorbed vesicles with $^{45}Ca^{2+}$ for controlled periods could be used for this purpose (Dupont, 1982, 1984; see Figure 8b). Meanwhile, it had been shown that at least under selected experimental conditions, the rate of the changes in ATPase tryptophan fluorescence after addition of Ca^{2+} to Ca^{2+}-deprived ATPase indeed reflected the rate with which this ATPase became reactive to ATP (Guillain et al., 1981; see Figure 8a). It was recently suggested that the rise in tryptophan fluorescence could indeed reflect the formation of ATPase forms with not just one, but two metal ions at the transport site (Orlowski and Champeil, 1993; Champeil, 1993); it could also reflect a critical conformational change, following Ca^{2+} binding (Henderson et al., 1994a, 1994b). Chemical quenching measurements of the rate of Ca^{2+} binding were also designed and carried out, exploiting single-site turnover of ATPase in the presence of ATP and EGTA and the notion that binding of the two Ca^{2+} ions to their transport sites was required to permit phosphoenzyme formation and Ca^{2+} internalization (Petithory and Jencks, 1988b).

The description of $^{45}Ca^{2+}$ dissociation from its binding sites on unphosphorylated Ca^{2+}-ATPase, towards the cytoplasmic side of the membrane, has led to interesting results. In this case, dissociation is clearly sequential, suggesting that the binding pocket for Ca^{2+} in the Ca_2E species (state 2) is narrow enough to prevent dissociation of the more deeply bound ion if the superficial position is occupied, either by

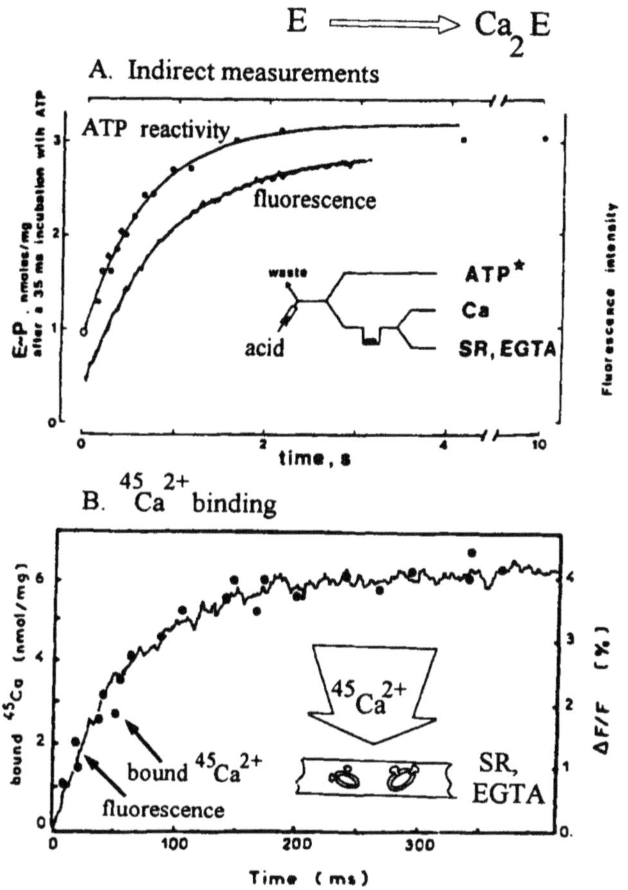

Figure 8. Calcium binding-induced transition of nonphosphorylated ATPase. Panel **A** shows the rise in tryptophan fluorescence following addition of calcium to Ca^{2+}-deprived sarcoplasmic reticulum vesicles under certain conditions (continuous line); the reactivity of ATPase to ATP under the same conditions (closed circles) has the same time dependence. In the latter chemical measurement, the reaction mixture was quenched with acid a few milliseconds after addition of ATP (from Guillain et al., 1981, with permission). Panel **B**: under different experimental conditions, resulting in a faster transition, ^{45}Ca-binding (circles) to calcium-deprived ATPase, adsorbed on a nitrocellulose filter, was directly measured. This binding was also correlated with the changes in tryptophan fluorescence (continuous line) (from Dupont, 1984, with permission).

Ca^{2+} or by its analog, Sr^{2+}; in addition, Ca^{2+} movements inside the half-occupied pocket have been suggested to be fast (Dupont, 1982, 1984; Inesi, 1987; Petithory and Jencks, 1988a; Orlowski and Champeil, 1991a; Fujimori and Jencks, 1992b). In contrast, the detailed mechanism for the *binding* of the two Ca^{2+} ions to Ca^{2+}-ATPase is not yet completely elucidated. Ca^{2+} binding is generally found to occur with positive cooperativity, but even the pH dependence of this apparent cooperativity has been disputed until recently (e.g., Forge et al., 1993a). Under several conditions, the time course of Ca^{2+} binding is thought to be biphasic, but does this imply sequential binding? Controversy has focused on the location in the reaction sequence of the conformational rearrangement which accompanies Ca^{2+} binding: does this conformational change follow (e.g., Inesi et al., 1980; Petithory and Jencks, 1988b) or precede (e.g., de Meis and Vianna, 1979) Ca^{2+} binding itself (e.g., Dupont, 1978; Champeil et al., 1983; Dupont, 1984; Inesi, 1987; Stahl and Jencks, 1987; Nakamura, 1989; Wakabayashi and Shigekawa, 1990; Forge et al., 1993a, 1993b; Henderson et al., 1994a, 1994b, and the preceding chapter in this book)? One of the underlying important questions is: at which step in the cycle do the Ca^{2+} transport sites change from a luminal back to a cytoplasmic orientation? This will be examined in the next section.

E. Implications for the Reaction Mechanism

In the original formulation of the $^{*}E$–E (or E_2–E_1) model for Ca^{2+}-ATPase (de Meis and Vianna, 1979), Ca^{2+}-free nonphosphorylated ATPase was thought to be in equilibrium between two different slowly-interconverting forms, the $^{*}E$ (or E_2) form which was supposed to expose its transport sites on the luminal side, and the E (or E_1) form which was supposed to expose its transport sites on the cytosolic side. If this were the case, binding of Ca^{2+} from the cytosolic side of the membrane should be affected by the presence of Ca^{2+} on the luminal side, because of competition. In contrast to this prediction, binding of $^{45}Ca^{2+}$ to Ca^{2+}-ATPase, as well as related spectroscopic signals, were found to be similar with loaded and non-loaded sarcoplasmic reticulum vesicles, within experimental error (Myung and Jencks, 1991; Henderson et al., 1994a; Orlowski and Champeil, unpublished results). The conclusion, already stated at the end of Section IIIC, to be derived from these results, is that phosphorylation at the catalytic site, and not a conformational change, controls the vectorial specificity for Ca^{2+} binding to Ca^{2+}-ATPase, and Ca^{2+}-free nonphosphorylated ATPase probably no longer exposes its transport sites to the luminal medium. Nevertheless, it is not possible to account for all kinetic results by assuming that high affinity Ca^{2+}-binding sites are available on the cytoplasmic side of every Ca^{2+}-free nonphosphorylated ATPase molecule: some kind of moderately slow transition probably limits the availability at the cytoplasmic surface of the transport sites of a fraction of the ATPase molecules (e.g., Wakabayashi and Shigekawa, 1990). To reconcile the above facts, one might propose that a form of ATPase exists (the one generally referred to as $^{*}E$ or E_2), in which the Ca^{2+}-binding pocket faces neither the lumen nor the cytoplasm, but is occluded. It could be the nonphosphorylated counterpart of the form of phosphorylated ATPase with occluded Ca^{2+}, mentioned above in Section IIIA, except that the occluded binding pocket would now contain protons instead of Ca^{2+} ions (Pick and Karlish, 1982; McIntosh et al., 1991; Forge et al., 1993a). In this respect, the Ca^{2+}-ATPase would, in fact, be completely analogous to the model "E_2–E_1" enzyme, Na^+/K^+ ATPase, in which, in the absence of Na^+, the transport sites of unphosphorylated ATPase are also occluded from the medium and normally occupied by K^+.

The results of Myung and Jencks (1991) discussed previously, imply that a majority of unphosphorylated, Ca^{2+}-free ATPase molecules do not expose their transport sites toward the luminal side of the membrane. Yet, they probably do so in a transient way, and with a low probability. This is the easiest way to account for the fact that in the absence of Ca^{2+} on the cytosolic side, nonphosphorylated Ca^{2+}-ATPase mediates a slow Ca^{2+} efflux from previously loaded sarcoplasmic reticulum vesicles, and that this efflux is inhibited by Ca^{2+} binding to the ATPase transport sites (Boland et al., 1975; Sorensen, 1983; Gould et al., 1987; de Meis et al., 1990). The various forms of Ca^{2+}-ATPase must, therefore, probably be viewed as dynamic conformations (Orlowski and Champeil, 1991a), in which the transport sites do not necessarily have a rigidly fixed orientation, but in which the binding pocket is separated from the luminal or the cytosolic medium by two gates whose probability of opening differs in the various forms. In the occluded forms, the opening probability is small for both gates. In the other ATPase forms, these gates generally open in a mutually exclusive, phosphorylation-dependent way, ensuring communication of the binding pocket with either one side of the membrane or the other, and tight coupling of the pump. However, this rule is not absolute under all conditions, and pump slippage or uncoupled fluxes may occur, as for instance in the case of Ca^{2+} efflux through nonphosphorylated ATPase.

On the basis of experimental results suggesting that Ca^{2+} binding to low affinity sites on the luminal side of nonphosphorylated ATPase influences phosphoenzyme formation from P_i even though it does not influence binding of Ca^{2+} on the cytosolic side (Chaloub et al., 1979; Suko et al., 1981; Jencks et al., 1994), an alternative view was developed for the mechanism of Ca^{2+} transport. In this view, transport does not result from reorientation of the Ca^{2+}-binding pocket, but results from the movement of the two Ca^{2+} ions from a pair of cytoplasmic-facing sites to a pair of luminal-facing sites (Jencks et al., 1994). Future work will have to confirm or disprove this appealing possibility. Structural predictions suggested that a number of negatively charged groups were located on the luminal side of the Ca^{2+}-ATPase, and these groups could be candidates for low affinity Ca^{2+}-binding sites (MacLennan et al., 1985).

IV. MODULATORS OF ATPASE ACTIVITY

A. Ca^{2+} Analogs

The Ca^{2+} analogs which are transported by the Ca^{2+} pump are few. Sr^{2+} is the one whose interaction with Ca^{2+}-ATPase has been characterized most completely. Although Sr^{2+} has an affinity for unphosphorylated Ca^{2+}-ATPase much lower than that of Ca^{2+}, the kinetics of most of the elementary steps in the cycle are identical when Ca^{2+} or Sr^{2+} are transported (Fujimori and Jencks, 1992a). Just like Ca^{2+}, Sr^{2+} is transported with a stoichiometry of two to one with respect to hydrolyzed ATP (e.g., Berman and King, 1990; Hasselbach and Migala, 1985). Nevertheless, the cooperativity of Sr^{2+} binding to Ca^{2+}-ATPase is poor (Fujimori and Jencks, 1992b; Orlowski and Champeil, 1993). The established fact that a preformed Sr^{2+} gradient is unable to support ATP synthesis (Guimaraes-Motta et al., 1984) might be partly due to the low affinity of luminal Sr^{2+} for its sites on phosphorylated ATPase (Fujimori and Jencks, 1992a). However, as there is evidence that Sr^{2+} and Ca^{2+} do compete for these luminal sites at the concentrations attained inside actively loaded vesicles, it might also be due to the inability of Sr^{2+} to promote the back conversion of phosphoenzyme from a P_i-reactive to an ADP-reactive form, which would then imply the existence of an intermediate form between state 4 and state 3 in Figure 4a (Guimaraes-Motta et al., 1984). Mn^{2+} is the second metal ion which has been shown to be transported by the pump (Chiesi and Inesi, 1980; Costa and Madeira, 1986). However, Mn^{2+} may also replace Mg^{2+} at the catalytic site (e.g., Ogurusu et al., 1991), which makes the effects of Mn^{2+} as a Ca^{2+} analog more difficult to interpret. As discussed in the preceding chapter, lanthanide ions are poor analogs of Ca^{2+} with respect to their binding to Ca^{2+}-ATPase.

B. ATP Analogs and Other Substrates

The Ca^{2+}-ATPase site for nucleotides does not have a high selectivity for ATP. For instance, ITP, GTP, acetylphosphate, and paranitrophenylphosphate are hydrolyzed and support Ca^{2+} uptake. ATP analogs like AMPPCP and AMPPNP compete with ATP binding with comparable affinities, and trinitrophenyl derivatives (TNPATP, TNPADP, TNPAMP, as well as photoactivatable azido-derivatives of these nucleotides) bind with even higher affinity (e.g., Seebregts and McIntosh, 1989; McIntosh et al., 1992). It was mentioned in Section IIIA above, that at high substrate concentrations, the maximal rate of ATPase phosphorylation from low affinity substrates like acetylphosphate (Bodley and Jencks, 1987) or ITP (Scofano et al., 1979) was not very different from that obtained using ATP. However, when such poor substrates are used, other steps in the ATPase cycle are slower than when ATP is used. As noted above, not only the Ca2+-induced transition from state 1 to state 2, but also the two other steps in the cycle, Ca2+ dissociation from phosphoenzyme and phosphoenzyme hydrolysis, are accelerated in the presence of ATP. The result is a non-Michaelian profile of ATPase activity as a function of the

ATP concentration. What is the reason for the activating effects of ATP on phosphoenzyme processing? Part of the reason is that after release of ADP from the phosphorylated catalytic site, ATP may bind to this site, and the unusual number of four phosphate groups in the site destabilizes the phosphoenzyme and stimulates its processing (McIntosh and Boyer, 1983; Cable et al., 1985; Bishop et al., 1987; Champeil et al., 1988; Seebregts and McIntosh, 1989). In fact, metal-free ATP is a very efficient activator for dephosphorylation (Champeil et al., 1988); Mg-ATP has also been suggested to stimulate phosphoenzyme turnover (Stefanova et al., 1987). It is difficult, however, to completely exclude the possible existence of a second, regulatory site for nucleotides (e.g., Suzuki et al., 1990; Coll and Murphy, 1991). In particular, it is a puzzling observation that addition of a high concentration of unlabeled ATP stimulates covalent phosphorylation of Ca^{2+}-ATPase from a previously formed noncovalent [γ-^{32}P]ATP-enzyme precursor (Shigekawa and Kanazawa, 1982). Analysis of the stimulation by nucleotides is made even more complicated by possible interactions between ATPase monomers, either nonrandom interactions in stable oligomers or random collisions in the native sarcoplasmic reticulum membrane (see following).

C. Ca^{2+}-precipitating Agents

We would like to briefly comment on the use of Ca^{2+}-precipitating agents in the study of Ca^{2+} transport and ATP hydrolysis. In the initial studies reported in Section II above, the use of oxalate made it possible to make a major step forward. Ca^{2+}-precipitating agents, oxalate or P_i, are still used, for example to discriminate between Ca^{2+} transport mediated by membranous fragments derived from endoplasmic reticulum or from plasma membranes: only the former membranes are thought to contain the anion transporter required to permit oxalate or P_i to passively follow Ca^{2+} uptake. Once in the organelle lumen, oxalate is thought to permit precipitation of Ca^{2+}-oxalate because of the high concentration of luminal Ca^{2+}, and this removes the inhibitory effect of accumulated Ca^{2+} on further transport. Yet, it has been convincingly shown that seeding of the very first Ca^{2+}-oxalate crystals and/or delayed precipitation of the low affinity complex formed limits the efficiency of this precipitating ion. As a result, the luminal free Ca^{2+} concentration during steady-state Ca^{2+} uptake probably remains much higher than could be expected on the basis of the solubility product of Ca^{2+}-oxalate, and the ATPase activity of tight vesicles in the presence of oxalate is lower than that of ionophore-treated membranes, as the latter are completely free of inhibition by accumulated Ca^{2+} (Feher and Briggs, 1980; Madeira, 1984; Feher and Lipford, 1985).

D. Lipids, Detergents, and Protein–protein Interactions

The lipid environment of Ca^{2+}-ATPase is, of course, a major determinant of its activity, as are also the possible protein–protein interactions. Detergents have been used to replace lipids and obtain information not only about structural features of

Ca^{2+}-ATPase, but also about its functional capabilities. As this has been reviewed elsewhere (le Maire and Møller, 1986; Møller et al., 1986; Hidalgo, 1987; Andersen, 1989) it will not be covered here extensively; only a few points will be recalled. The activity level of membranous Ca^{2+}-ATPase does not appear to be the mere consequence of the average fluidity of the lipid phase; although the membrane must be of adequate thickness and fluid enough to permit conformational rearrangement of the enzyme during its catalytic cycle, interaction between the protein and individual lipids probably plays a major role in determining the pumping activity, despite the relatively short residence time of these lipids close to the protein (e.g., Caffrey and Ferguson, 1981; East et al., 1984, 1985; Navarro et al., 1984; Lentz et al., 1985; Gould et al., 1987; Squier et al., 1988; Mahaney et al., 1992; Starling et al., 1993). Detergent-embedded, completely delipidated monomeric ATPases also display various degrees of steady-state activity in various detergents, even with detergents which, on the long term, do not inactivate the ATPase irreversibly (e.g., Lund et al., 1989). The role of residual lipids in detergent-solubilized protein has been shown to be critical for both structure and stability of Ca^{2+}-ATPase (McIntosh and Ross, 1985; Vilsen and Andersen, 1987). Monomeric ATPase, either detergent-solubilized or reconstituted with excess lipid, possesses all the functional capabilities of native membranous ATPase—possibly with different rates for the individual steps in the reaction cycle (e.g., Dean and Tanford, 1978; le Maire et al., 1978; Kosk-Kosicka et al., 1983; Martin, 1983; Vilsen and Andersen, 1986; McIntosh and Ross, 1988; Heegard et al., 1990). Nevertheless, individual polypeptide chains in sarcoplasmic reticulum membranes are probably so close that they probably do interact with each other in the native membrane. These random or nonrandom interactions might result in modulation of some of the intermediate steps during catalysis and transport.

V. TOOLS FOR STUDYING NANOGRAM AMOUNTS OF MUTATED ATPASE

Section III describes the various tools which have been developed for studying, in detail, the relative contribution of individual steps in the cycle to overall catalysis by Ca^{2+}-ATPase. However, it should be realized that many of these measurements require a substantial amount of ATPase. For instance, when $^{45}Ca^{2+}$ binding is measured by filtration, 50–300 μg of ATPase protein is routinely used for each filter. It would be desirable to apply similar methods to the study of ATPases modified by site-directed mutagenesis, but the relatively low amount of ATPase which can be produced in transient expression systems (Maruyama and MacLennan, 1988) is a serious obstacle. Expression systems producing larger amounts of ATPase will be welcome (e.g., Hussain et al., 1992; Skerjanc et al., 1993); nevertheless, specific tools for studying low amounts of ATPase have been successfully developed.

The first tool which was found useful for detection of very low amounts of active ATPase was the measurement of ATP-dependent accumulation of Ca^{2+}-oxalate crystals in ATPase-containing microsomes (Maruyama and MacLennan, 1988). This accumulation proved to be detectable under conditions where the concentrations of expressed ATPase had to be determined quantitatively through immunoreactivity tests. Its study made it possible to discriminate between active and inactive expressed ATPases, and also to some extent to study overall regulation properties, for example, the sensitivity to Ca^{2+} or inhibitors (Campbell et al., 1991; Lytton et al., 1992). A major step forward, then, was to miniaturize protocols for the measurement of ATPase phosphorylation from either ATP or P_i (Clarke et al., 1989). This was obtained by detecting ^{32}P-labeled phosphoenzyme by autoradiography after electrophoretic separation of the phosphorylated ATPase, under electrophoresis conditions where the acyl-phosphate bond was sufficiently stable. To measure phosphorylation from ATP in the presence of Ca^{2+}, low concentrations of [γ-^{32}P]ATP were used, to keep the signal-to-noise ratio high. Conversely, to measure phosphoenzyme formed from [^{32}P]P_i in the absence of Ca^{2+}, DMSO was included in the medium to increase the ATPase apparent affinity for P_i (de Meis et al., 1980) and to permit the use of moderate concentrations of P_i.

This was the starting point for more detailed studies of the catalytic cycle of expressed ATPases. For instance, experiments were designed and performed to recognize the ability of the phosphorylated expressed ATPase to experience the transition from an ADP-sensitive to an ADP-insensitive form: in these experiments, phosphorylation from ATP was followed by addition of EGTA and ADP before acid quenching (Andersen et al., 1989). Other two-step protocols were ingeniously adapted from the literature, e.g., phosphorylation from [γ-^{32}P]ATP followed by a chase with nonradioactive ATP, to measure phosphoenzyme turnover under steady-state conditions (Sumbilla et al., 1993). Even $^{45}Ca^{2+}$ occlusion in the presence of CrATP was measured (Luckie et al., 1992; Vilsen and Andersen, 1992b), as well as ATPase hydrolysis itself (Vilsen et al., 1991). Special mention should also be made of remarkable early experiments in which the ability of mutated and expressed ATPases to bind Ca^{2+} and experience the conformational change from state 1 to state 2 was examined (Figure 9). In mutants lacking this ability, calcium no longer permitted the ATPase to be phosphoryled from ATP; concomitantly, although the phosphorylation site was intact, as judged from the fact that phosphorylation from P_i was possible, the presence of Ca^{2+} no longer inhibited phosphorylation from P_i (Clarke et al., 1989). It is of interest that similar results, derived from standard phosphorylation experiments, were obtained in parallel with Ca^{2+}-ATPase chemically derivatized with a specific carbodiimide reagent. In this case, the large amount of modified ATPase available made it possible to directly prove that Ca^{2+} binding to ATPase was inhibited as a consequence of derivatization (Inesi et al., 1990; Sumbilla et al., 1991). With the recent advent of high-yield expression systems, similar measurements with mutated ATPases will be possible (Skerjanc et al., 1993). Tools are therefore, now becoming available to describe functional consequences of various mutations (e.g., Andersen and Vilsen, 1992, 1993; MacLennan et al., 1992; Sumbilla et al., 1993).

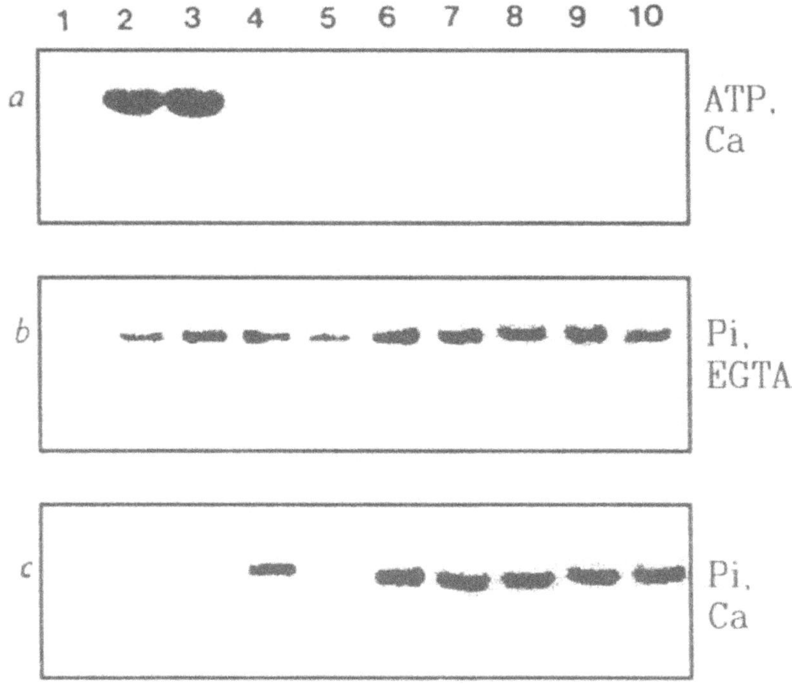

Figure 9. Phosphorylation of expressed wild-type and mutant ATPases, (**a**) from [γ-^{32}P]ATP in the presence of Ca^{2+}, (**b**) from [^{32}P]P_i in the absence of Ca^{2+}, and (**c**) from [^{32}P]P_i in the presence of Ca^{2+}. After acid quenching of the reaction, proteins were applied to a polyacrylamide gel for electrophoresis followed by autoradiography. Wild-type ATPase was loaded onto lane 2, and mutant ATPases onto lanes 3–10. The mutant ATPases in lanes 4 and 7–10 were not phosphorylated by ATP in the presence of Ca^{2+}, and their phosphorylation from P_i was not prevented by Ca^{2+}, which is a strong inhibitor of the reaction with P_i in the native or wild-type ATPase (from Clarke et al., 1989, with permission; see also Figure 9 in Inesi et al., 1992).

VI. SPECIFIC EXAMPLES OF RATE-LIMITATION

This final section will mention a few selected fields in which such a detailed analysis of the individual steps in the ATPase catalytic cycle has been attempted, as well as the difficulty and/or the limitations of such an approach.

Understanding how a particular perturbing agent exerts its stimulatory or inhibitory effect on Ca^{2+}-ATPase is one of these fields. For instance, many studies have focused on the interaction of Ca^{2+}-ATPase with amphiphilic substances, and from detailed kinetic studies, the particular target of the perturbing drug was revealed (e.g., McIntosh and Davidson, 1984; Hara and Kanazawa, 1986; Kawashima et al., 1990; Wakabayashi et al., 1988). It should be stressed that depending on the experimental conditions, either one or the other of all four individual steps depicted in Figure 4 may become rate-limiting. As a result, the perturbing effect of one particular molecule on one of these steps may or may not manifest itself, depending on whether this step contributes to rate limitation or not, under the conditions of the experiment (e.g., Champeil et al., 1986; de Foresta et al., 1992).

A different and interesting case has been illustrated by a detailed study which attempted to find which individual step in the cycle was modified in the presence of a small concentration of La^{3+} (Fujimori and Jencks, 1990). In this study, all four of the individual steps depicted in Figure 4 were found to have similar rate constants when they were measured in the presence of this concentration of La^{3+}, added immediately before the actual test. Nevertheless, steady-state activity in the presence of La^{3+} was much lower than in its absence. The way out of this paradox was to realize that the small amount of La–ATP formed was responsible for the observed steady-state inhibition, but that the inhibitory effect of La–ATP as an analog of Mg–ATP was developing slowly, because only a small fraction of the ATPase molecules bound La–ATP and became inhibited at each new catalytic cycle — a feature reminiscent of what had been recognized as the reason for the inhibitory role of Ca–ATP (Nakamura, 1984; Yamada et al., 1986; Lund and Møller, 1988; Orlowski et al., 1988).

In some cases, more than just one step was found to be modulated by the drug considered. Among the few relatively specific inhibitors of Ca^{2+}-ATPase which have been described (e.g., Kass et al., 1989; Seidler et al., 1989; Wictome et al., 1992), a remarkable example is given by thapsigargin, the recently discovered potent inhibitor of all SERCA ATPases. This inhibitor, which acts at subnanomolar concentrations, affects both ATP binding (although in a noncompetitive way) and Ca^{2+} binding-related steps (see Lytton et al., 1991; Kijima et al., 1991; Sagara and Inesi, 1991; Inesi and Sagara, 1992; Wictome et al., 1992; DeJesus et al., 1993; Sumbilla et al., 1993). Another example is intramolecular cross-linking of the Ca^{2+}-ATPase active site with glutaraldehyde, which dramatically lowers ATP binding affinity, blocks phosphorylation from P_i in the absence of Ca^{2+}, but still permits phosphorylation from ATP or small substrates like acetylphosphate in the presence of Ca^{2+}, although blocking Ca^{2+} release to the vesicle lumen after this phosphorylation (McIntosh et al., 1991; Ross et al., 1991). Such simultaneous inhibition of different steps in the cycle argues in favor of long-range interactions as being the basis for the catalytic properties of Ca^{2+}-ATPase (Inesi et al., 1992).

Ultimately, one would like to correlate the type of perturbation of Ca^{2+}-ATPase activity observed under specific conditions with the ATPase structural modifications giving rise to this functional perturbation. In particular, this is the purpose of a number of directed mutagenesis experiments, the effects of which are currently interpreted in terms of modification of the Ca^{2+} binding sites, or of the phosphorylation site, or of the sites modulating the rate of one of the transitions depicted in Figure 4. The results of such experiments have been described in the preceding chapter of this book, and will not be further discussed in detail. We only wish to emphasize here that such studies of modified ATPases must take into account the fact that the overall behavior resulting from one particular modification of ATPase is the combined result of its effects on the various steps in the cycle, so that interpretation is not always easy. A remarkable example was provided by a study of chimeric proteins (Toyofuku et al., 1992). In this study, as Ca^{2+} transport mediated by SERCA 3 is activated by lower Ca^{2+} concentrations than Ca^{2+} transport mediated by SERCA 1 or SERCA 2, chimeric proteins between SERCA 2 and SERCA 3 were constructed, and tested for their sensitivity to Ca^{2+}. The result was that the nucleotide binding/hinge domain of the ATPase played a crucial role in determining the isoform-specific Ca^{2+}-sensitivity of Ca^{2+}-ATPase. As this domain is widely believed to be part of the cytoplasmic globule of the ATPase cytosolic, whereas the Ca^{2+}-binding sites are likely to be membranous, this result emphasizes that the Ca^{2+}-dependence of ATPase activity reflects not only the affinity of Ca^{2+} for the transport sites of nonphosphorylated ATPase, but also the relative contribution of the different catalytic steps.

The same conclusion was illustrated in another study of the effect of a phospholamban antibody (PlAb) on the Ca^{2+}-dependence of skeletal (SERCA 1) and cardiac (SERCA2) Ca^{2+}-ATPase (Cantilina et al., 1993). Phospholamban is thought to interact with the cytosolic loop of cardiac Ca^{2+}-ATPase (James et al., 1989; Toyofuku et al., 1993), resulting in a reduction of its apparent affinity for Ca^{2+}. The phospholamban antibody shifted the Ca^{2+} concentration dependence of Ca^{2+} transport by cardiac (but not skeletal) sarcoplasmic reticulum towards lower Ca^{2+} concentrations, but it did not affect equilibrium Ca^{2+} binding to either skeletal or cardiac ATPase. The results were completely accounted for by assuming that the antibody exerted its effect by stimulating both the forward and reverse rates of the Ca^{2+}-triggered enzyme isomerization (from state 1 to state 2, in terms of the scheme in Figure 4), without changing the overall equilibrium constant for Ca^{2+} binding. Again, an effect on the apparent Ca^{2+} dependence of the transport velocity was obtained without modification of the equilibrium properties of the Ca^{2+}-binding sites, due to the interplay between the various kinetic constants in the ATPase catalytic cycle.

ACKNOWLEDGMENTS

We are specially grateful to the late Drs. C.M Gary-Bobo and F. Bastide for introducing us to the study of sarcoplasmic reticulum Ca^{2+}-ATPase, to F. Guillain for his collaboration over the past years, and to S. Orlowski, B. de Foresta, M. le Maire, D. McIntosh, and A.G. Lee for critically reading this manuscript.

REFERENCES

Andersen, J. P. (1989). Monomer–oligomer equilibrium of sarcoplasmic reticulum Ca^{2+}-ATPase and the role of subunit interaction in the Ca^{2+} pump mechanism. Biochim. Biophys. Acta 988, 47–72.

Andersen, J. P., & Vilsen, B. (1990). Primary ion pumps. Opinion in Cell Biol. 2, 722–730.

Andersen, J. P., & Vilsen, B. (1992). Structural basis for the E_1/E_1 P–E_2/E_2 P conformation changes in the sarcoplasmic reticulum Ca^{2+}-ATPase studied by site-specific mutagenesis. Acta Physiol. Scand. 146, 151–159.

Andersen, J. P., & Vilsen, B. (1993). Functional consequences of substitution of the seven-residue segment LysIleArgAspGlnMetAla240 located in the stalk helix S3 of the Ca^{2+}-ATPase of sarcoplasmic reticulum. Biochemistry 32, 10015–10020.

Andersen, J. P., Lassen, K., & Møller, J. V. (1985). Changes in Ca^{2+} affinity related to conformational transitions in the phosphorylated state of soluble monomeric Ca^{2+}-ATPase from sarcoplasmic reticulum. J. Biol. Chem. 260, 371–380.

Andersen, J. P., Vilsen, B., Leberer, E., & MacLennan, D. H. (1989). Functional consequences of mutations in the β-strand sector of the Ca^{2+}-ATPase of sarcoplasmic reticulum. J. Biol. Chem. 264, 21018–21023.

Ariki, M., & Boyer, P. D. (1980). Characterization of medium inorganic phosphate–water exchange catalyzed by sarcoplasmic reticulum vesicles. Biochemistry 19, 2001–2004.

Bamberg, E., Butt, H. J., Eisenrauch, A., & Fendler, K. (1993). Charge transport of ion pumps on lipid bilayer membranes. Quat. Rev. Biophys. 26, 1–25.

Barrabin, H., Scofano, H., & Inesi, G. (1984). Adenosinetriphosphatase site stoichiometry in sarcoplasmic reticulum vesicles and purified enzyme. Biochemistry 23, 1542–1548.

Bastide, F., Meissner, G., Fleischer, S., & Post, R. L. (1973). Similarity of the active site of phosphorylation of the adenosine triphosphatase for transport of sodium and potassium ions in kidney to that for transport of calcium ions in the sarcoplasmic reticulum of muscle. J. Biol. Chem. 248, 8385–8391.

Berman, M. C., & King, S. B. (1990). Stoichiometries of calcium and strontium transport coupled to ATP and acetylphosphate hydrolysis by skeletal sarcoplasmic reticulum. Biochim. Biophys. Acta 1029, 235–240.

Bishop, J. E., Al-Shawi, M. K., & Inesi, G. (1987). Relationship of the regulatory nucleotide site to the catalytic site of the sarcoplasmic reticulum Ca^{2+}-ATPase. J. Biol. Chem. 262, 4658–4663.

Bodley, A., & Jencks, W. P. (1987). Acetylphosphate as a substrate for the Ca^{2+}-ATPase of sarcoplasmic reticulum. J. Biol. Chem. 262, 13997–14004.

Boland, A. R., Jilka, R. L., & Martonosi, A. N. (1975). Passive Ca^{2+} permeability of phospholipid vesicles and sarcoplasmic reticulum membranes. J. Biol. Chem. 250, 7501–7510.

Boyer, P. D., de Meis, L., Carvalho, M. G. C., & Hackney, D. D. (1977). Dynamic reversal of enzyme carboxyl group phosphorylation as the basis of the oxygen exchange catalyzed by the sarcoplasmic reticulum adenosine triphosphatase. Biochemistry 16, 136–140.

Cable, M. B., Feher, J. J., & Briggs, F. N. (1985). Mechanism of allosteric regulation of the Ca,Mg-ATPase of sarcoplasmic reticulum: Studies with 5′-adenylyl methylenediphosphate. Biochemistry 24, 5612–5619.

Caffrey, M., & Feigenson, G. W. (1981). Fluorescence quenching in model membranes. 3. Relationship between calcium adenosinetriphosphatase enzyme activity and the affinity of the protein for phosphatidylcholines with different acyl chain characteristics. Biochemistry 20, 1949–1961.

Campbell, A. M., Kessler, P. D., Sagara, Y., Inesi, G., & Fambrough, D. M. (1991). Nucleotide sequences of avian cardiac and brain SR/ER Ca^{2+}-ATPases and functional comparisons with fast twitch Ca^{2+}-ATPase: Calcium affinities and inhibitor effects. J. Biol. Chem. 266, 16050–16055.

Cantilina, T., Sagara, Y., & Inesi, G. (1993). Comparative studies of cardiac and skeletal sarcoplasmic reticulum ATPases: Effect of a phospholamban antibody on enzyme activation by Ca^{2+}. J. Biol. Chem. 268, 17018–17025.

Carvalho, M. G. C., Souza, D. G., & de Meis, L. (1976). On a possible mechanism of energy conservation in sarcoplasmic reticulum membranes. J. Biol. Chem. 251, 3629–3636.

Chaloub, R. M., Guimaraes-Motta, H., Verjovski-Almeida, S., de Meis, L., & Inesi, G. (1979). Sequential reactions in P_i utilization for ATP synthesis by sarcoplasmic reticulum. J. Biol. Chem. 254, 9464–9468.

Champeil, P. (1993). Discrimination between sarcoplasmic reticulum ATPase species with one or two Ca^{2+} or Sr^{2+} ions bound to the transport sites, as deduced from the fluorescence of fluorescein-isothiocyanate bound to Lys-515 and of tryptophan residues. Biophys. J. 64, A353.

Champeil, P., & Guillain, F. (1986). Rapid filtration study of the phosphorylation-dependent dissociation of calcium from transport sites of purified sarcoplasmic reticulum ATPase and ATP modulation of the catalytic cycle. Biochemistry 25, 7623–7633.

Champeil, P., Bastide, F., Taupin, C., & Gary-Bobo, C. M. (1976). Spin-labelled sarcoplasmic reticulum vesicles: Ca^{2+}-induced spectral changes. FEBS Lett. 63, 270–272.

Champeil, P., Gingold, M. P., Guillain, F., & Inesi, G. (1983). Effect of magnesium on the calcium-dependent transient kinetics of sarcoplasmic reticulum ATPase, studied by stopped flow fluorescence and phosphorylation. J. Biol. Chem. 258, 4453–4458.

Champeil, P., Guillain, F., Vénien, C., & Gingold, M. P. (1985). Interaction of magnesium and inorganic phosphate with calcium-deprived sarcoplasmic reticulum adenosinetriphosphatase as reflected by organic solvent-induced perturbation. Biochemistry 24, 69–81.

Champeil, P., le Maire, M., Andersen, J. P., Guillain, F., Gingold, M. P., Lund, S., & Møller, J. V. (1986). Kinetic characterization of the normal and detergent-perturbed reaction cycles of the sarcoplasmic reticulum calcium pump. J. Biol. Chem. 261, 16372–16384.

Champeil, P., Riollet, S., Orlowski, S., Guillain, F., Seebregts, C. J., & McIntosh, D. B. (1988). ATP regulation of sarcoplasmic reticulum Ca^{2+}-ATPase: Metal-free ATP and 8-bromo-ATP bind with high affinity to the catalytic site of phosphorylated ATPase and accelerate dephosphorylation. J. Biol. Chem. 263, 12288–12294.

Chen, Z., Coan, C., Fielding, L., & Cassafer, G. (1991). Interaction of CrATP with the phosphorylation site of the sarcoplasmic reticulum ATPase. J. Biol. Chem. 266, 12386–12394.

Chiesi, M., & Inesi, G. (1980). Adenosine 5'-triphosphate-dependent fluxes of manganese and hydrogen ions in sarcoplasmic reticulum vesicles. Biochemistry 19, 2912–2918.

Clarke, D. M., Loo, T. W., Inesi, G., & MacLennan, D. H. (1989). Location of high affinity Ca^{2+}-binding sites within the predicted transmembrane domain of the sarcoplasmic reticulum. Nature 339, 476–478.

Coll, R. J., & Murphy, A. J. (1984). Purification of the Ca^{2+}-ATPase of sarcoplasmic reticulum by affinity chromatography. J. Biol. Chem. 259, 14248–14254.

Coll, R. J., & Murphy, A. J. (1991). Kinetic evidence for two nucleotide binding sites on the CaATPase of sarcoplasmic reticulum. Biochemistry 30, 1456–1461.

Costa, A. G., & Madeira, V. M. C. (1986). Magnesium and manganese ions modulate Ca^{2+} uptake and its energetic coupling in sarcoplasmic reticulum. Arch. Biochem. Biophys. 249, 199–206.

Dean, W. L., & Tanford, C. (1978). Properties of a delipidated, detergent-activated Ca^{2+}-ATPase. Biochemistry 17, 1683–1690.

de Foresta, B., Henao, F., & Champeil, P. (1992). Kinetic characterization of the perturbation by dodecylmaltoside of sarcoplasmic reticulum Ca^{2+}-ATPase. Eur. J. Biochem. 209, 1023–1034.

DeJesus, F., Girardet, J. L., & Dupont, Y. (1993). Characterization of ATP binding inhibition to the sarcoplasmic reticulum Ca^{2+}-ATPase by thapsigargin. FEBS Lett. 332, 229–232.

de Meis, L., & Carvalho, M. G. C. (1974). Role of the Ca^{2+} concentration gradient in the adenosine 5'-triphosphate-inorganic phosphate exchange catalyzed by sarcoplasmic reticulum. Biochemistry 13, 5032–5038.

de Meis, L., & Vianna, A. L. (1979). Energy interconversion by the Ca^{2+}-dependent ATPase of the sarcoplasmic reticulum. Annu. Rev. Biochem. 48, 275–292.

de Meis, L., Martins, O. B., & Alves, E. A. (1980). Role of water, hydrogen ion, and temperature on the synthesis of adenosinetriphosphate by the sarcoplasmic reticulum adenosine triphosphatase in the absence of a calcium ion gradient. Biochemistry 19, 4252–4261.

de Meis, L., Suzano, V. A., & Inesi, G. (1990). Functional interactions of catalytic site and transmembrane channel in the sarcoplasmic reticulum ATPase. J. Biol. Chem. 265, 18848–18851.

Dupont, Y. (1976). Fluorescence studies of the sarcoplasmic reticulum calcium pump. Biochem. Biophys. Res. Commun. 71, 544–550.

Dupont, Y. (1978). Transient kinetics of sarcoplasmic reticulum (Ca^{2+}+Mg^{2+})-ATPase studied by fluorescence. Nature 273, 396–398.

Dupont, Y. (1980). Occlusion of divalent cations in the phosphorylated calcium pump of sarcoplasmic reticulum. Eur. J. Biochem. 109, 231–238.

Dupont, Y. (1982). Low-temperature studies of the sarcoplasmic reticulum calcium pump: Mechanism of calcium binding. Biochim. Biophys. Acta 688, 75–87.

Dupont, Y. (1984). A rapid-filtration technique for membrane fragments or immobilized enzymes: Measurements of substrate binding or ion fluxes with a few-millisecond time resolution. Anal. Biochem. 142, 505–510.

East, J. M., Jones, O. T., Simmonds, A. C., & Lee, A. G. (1984). Membrane fluidity is not an important physiological regulator of the (Ca^{2+}–Mg^{2+})-dependent ATPase of sarcoplasmic reticulum. J. Biol. Chem. 259, 8070–8071.

East, J. M., Melville, D., & Lee, A. G. (1985). Exchange rates and numbers of annular lipids for the calcium and magnesium ion dependent adenosinetriphosphatase. Biochemistry 24, 2615–2623.

Feher, J. J., & Briggs, F. N. (1980). The effect of calcium oxalate crystallization kinetics on the kinetics of calcium uptake and calcium ATPase activity of sarcoplasmic reticulum vesicles. Cell Calcium 1, 105–118.

Feher, J. J., & Lipford, G. B. (1985). Calcium oxalate and calcium phosphate capacities of cardiac sarcoplasmic reticulum. Biochim. Biophys. Acta 818, 373–385.

Fiehn, W., & Migala, A. (1971). Calcium binding to sarcoplasmic membranes. Eur. J. Biochem. 20, 245–248.

Forge, V., Mintz, E., & Guillain, F. (1993a). Ca^{2+} binding to sarcoplasmic reticulum ATPase revisited: (I) mechanism of affinity and cooperativity modulation by H^+ and Mg^{2+}. J. Biol. Chem. 268, 10953–10960.

Forge, V., Mintz, E., & Guillain, F. (1993b). Ca^{2+} binding to sarcoplasmic reticulum ATPase revisited: (II) equilibrium and kinetic evidence for a two-route mechanism. J. Biol. Chem. 268, 10961–10968.

Froehlich, J. P., & Heller, P. F. (1985). Transient-state kinetics of the ADP-insensitive phosphoenzyme in sarcoplasmic reticulum. Biochemistry 24, 126–136.

Fujimori, T., & Jencks, W. P. (1990). Lanthanum inhibits steady-state turnover of the sarcoplasmic reticulum calcium ATPase by replacing magnesium as the catalytic ion. J. Biol. Chem. 265, 16262–16270.

Fujimori, T., & Jencks, W. P. (1992a). The kinetics for the phosphoryl transfer steps of the sarcoplasmic reticulum calcium ATPase are the same with strontium and with calcium bound to the transport sites. J. Biol. Chem. 267, 18466–18474.

Fujimori, T., & Jencks, W. P. (1992b). Binding of two Sr^{2+} ions changes the chemical specificities for phosphorylation of the sarcoplasmic reticulum calcium ATPase through a stepwise mechanism. J. Biol. Chem. 267, 18475–18487.

Gafni, A., & Boyer, P. D. (1984). Characterization of sarcoplasmic reticulum ATPase purified by selective column adsorption. Biochemistry 23, 4362–4367.

Gould, G. W., McWhirter, J. M., East, J. M., & Lee, A. G. (1987). A fast passive Ca^{2+} efflux mediated by the (Ca^{2+} + Mg^{2+})-ATPase in reconstituted vesicles. Biochim. Biophys. Acta 904, 45–54.

Guillain, F., Champeil, P., Lacapère, J. J., & Gingold, M. P. (1981). Stopped flow and rapid quenching measurement of the transient steps induced by calcium binding to sarcoplasmic reticulum adenosine triphosphatase: Competition with Ca^{2+}-independent phosphorylation. J. Biol. Chem. 256, 6140–6147.

Guimaraes-Motta, H., Sande-Lemos, M. P., & de Meis, L. (1984). Energy interconversion in sarcoplasmic reticulum vesicles in the presence of Ca^{2+} and Sr^{2+} gradients. J. Biol. Chem. 259, 8699–8705.

Hanel, A. M., & Jencks, W. P. (1991). Dissociation of calcium from the phosphorylated calcium-transporting adenosine triphosphatase of sarcoplasmic reticulum: Kinetic equivalence of the calcium ions bound to the phosphorylated enzyme. Biochemistry 30, 11320–11330.

Hara, H., & Kanazawa, T. (1986). Selective inhibition by ionophore A23187 of the enzyme isomerization in the catalytic cycle of sarcoplasmic reticulum Ca^{2+}-ATPase. J. Biol. Chem. 261, 16584–16590.

Hasselbach, W., & Makinose, M. (1961). The calcium pump of muscle "relaxing factor" and its dependence on ATP by hydrolysis. [*Die Calciumpumpe der "Erschlaffungsgrana" des Muskels und ihre Abhängigkeit von der ATP-spaltung*]. Biochemische Zeitschrift 333, 518–528.

Hasselbach, W., & Makinose, M. (1963). Mechanism of calcium transport through sarcoplasmic reticulum membranes. [*Über den Mechanismus des Calciumtransportes durch die Membranen des sarkoplasmatischen Reticulums*]. Biochemische Zeitschrift 339, 94–111.

Hasselbach, W., & Migala, A. (1985). Invariance of stoichiometry of the sarcoplasmic reticulum calcium pump at physiological calcium concentrations — a reevaluation. Z. Naturforsch. 40c, 571–575.

Heegard, C. W., le Maire, M., Gulik-Krzywicki, T., & Møller, J. V. (1990). Monomeric state and Ca^{2+} transport by sarcoplasmic reticulum Ca^{2+}-ATPase, reconstituted with an excess of phospholipid. J. Biol. Chem. 265, 12020–12028.

Henao, F., de Foresta, B., Orlowski, S., Cuenda, A., Gutiérrez-Merino, C., & Champeil, P. (1991). Kinetic characterization of the normal and procaine-perturbed reaction cycles of the sarcoplasmic reticulum calcium pump. Eur. J. Biochem. 202, 559–567.

Henderson, I. M. J., Khan, Y. M., East, J. M., & Lee, A. G. (1994a). Binding of Ca^{2+} to the (Ca^{2+}–Mg^{2+})-ATPase of sarcoplasmic reticulum: Equilibrium studies. Biochem. J. 297, 615–624.

Henderson, I. M. J., Starling, A. P., Wictome, M., East, J. M., & Lee, A. G. (1994b). Binding of Ca^{2+} to the (Ca^{2+}–Mg^{2+})-ATPase of sarcoplasmic reticulum: Kinetic studies. Biochem. J. 297, 625–636.

Hidalgo, C. (1987). Lipid–protein interactions and the function of the Ca^{2+}-ATPase of sarcoplasmic reticulum. CRC Crit. Rev. Biochem. 21, 319–347.

Hobbs, A. S., Albers, R. W., Froehlich, J. P., & Heller, P. F. (1985). ADP stimulates hydrolysis of the "ADP-insensitive" phosphoenzyme in Na^{+},K^{+}-ATPase and Ca^{2+}-ATPase. J. Biol. Chem. 260, 2035–2037.

Hussain, A., Lewis, D., Sumbilla, C., Lai, L. C., Melera, P. W., & Inesi, G. (1992). Coupled expression of Ca^{2+} transport ATPase and a dihydrofolate reductase selectable marker in a mammalian cell system. Arch. Biochem. Biophys. 296, 539–546.

Ikemoto, N. (1975). Transport and inhibitory Ca^{2+} binding sites on the ATPase enzyme isolated from the sarcoplasmic reticulum. J. Biol. Chem. 250, 7219–7224.

Inao, S., & Kanazawa, T. (1986). Characterization of the phosphoenzyme that is involved in the Ca^{2+}–Ca^{2+} exchange catalyzed by the Ca^{2+}-ATPase of sarcoplasmic reticulum vesicles. Biochim. Biophys. Acta 857, 28–37.

Inesi, G. (1987). Sequential mechanism of calcium binding and translocation in sarcoplasmic reticulum adenosine triphosphatase. J. Biol. Chem. 262, 16338–16342.

Inesi, G., & Sagara, Y. (1992). Thapsigargin, a high affinity and global inhibitor of intracellular Ca^{2+} transport ATPases. Arch. Biochem. Biophys. 298, 313–317.

Inesi, G., Maring, E., Murphy, A. J., & McFarland, B. H. (1970). A study of the phosphorylated intermediate of sarcoplasmic reticulum ATPase. Arch. Biochem. Biophys. 138, 285–294.

Inesi, G., Kurzmack, M., Coan, C., & Lewis, D. E. (1980). Cooperative calcium binding and ATPase activation in sarcoplasmic reticulum vesicles. J. Biol. Chem. 255, 3025–3031.

Inesi, G., Kurzmack, M., Kosk-Kosicka, D., Lewis, D., Scofano, H., & Guimaraes-Motta, H. (1982). Equilibrium and kinetic studies of calcium transport and ATPase activity in sarcoplasmic reticulum. Z. Naturforsch. 37c, 685–691.

Inesi, G., Kurzmack, M., & Lewis, D. (1988). Kinetic and equilibrium characterization of an energy-transducing enzyme and its partial reactions. In: Methods in Enzymology, Vol. 157 (Fleischer, S., & Fleischer, B., eds.) pp. 154–189. Acad. Press, London.

Inesi, G., Sumbilla, C., & Kirtley, M. E. (1990). Relationships of molecular structure and function in Ca^{2+}-transport ATPase. Physiol. Rev. 70, 749–760.

Inesi, G., Lewis, D., Nikic, D., Hussain, A., & Kirtley, M. E. (1992). Long-range intramolecular linked functions in the calcium transport ATPase. Advances in Enzymology and Related Areas in Molecular Biology 65, 185–215.

James, P., Inui, M., Tada, M., Chiesi, M., & Carafoli, E. (1989). Nature and site of phospholamban regulation of the Ca^{2+} pump of sarcoplasmic reticulum. Nature 342, 90–92.

Jencks, W. P. (1989). How does a calcium pump pump calcium? J. Biol. Chem. 264, 18855–18858.

Jencks, W. P., Yang, T., Peisach, D., & Myung, J. (1994). Calcium ATPase of sarcoplasmic reticulum has four binding sites for calcium. Biochemistry 32, 7030–7034.

Kanazawa, T., & Boyer, P. D. (1973). Occurrence and characteristics of a rapid exchange of phosphate oxygens catalyzed by sarcoplasmic reticulum vesicles. J. Biol. Chem. 248, 3163–3172.

Kass, G. E. N., Duddy, S. K., Moore, G. A., & Orrenius, S. (1989). 2,5-di(*tert*-butyl)-1,4-benzohydroquinone rapidly elevates cytosolic Ca^{2+} concentration by mobilizing the inositol 1,4,5-triphosphate sensitive Ca^{2+} pool. J. Biol. Chem. 264, 15192–15198.

Kawashima, T., Hara, H., & Kanazawa, T. (1990). Selective inhibition by lasolacid of hydrolysis of the ADP-insensitive phosphoenzyme in the catalytic cycle of sarcoplasmic reticulum Ca^{2+}-ATPase. J. Biol. Chem. 265, 10993–10999.

Kijima, Y., Ogunbunmi, E., & Fleischer, S. (1991). Drug action of thapsigargin on the Ca^{2+} pump protein of sarcoplasmic reticulum. J. Biol. Chem. 266, 22912–22918.

Knowles, A. F., & Racker, E. (1975). Formation of adenosine triphosphate from P_i and adenosine diphosphate by purified Ca^{2+} ATPase. J. Biol. Chem. 250, 1949–1951.

Kolassa, N., Punzengruber, C., Suko, J., & Makinose, M. (1979). Mechanism of calcium-independent phosphorylation of sarcoplasmic reticulum ATPase by orthophosphate: Evidence of magnesium–phosphoprotein formation. FEBS Lett. 108, 495–500.

Kosk-Kosicka, D., Kurzmack, M., & Inesi, G. (1983). Kinetic characterization of detergent-solubilized sarcoplasmic reticulum adenosinetriphosphatase. Biochemistry 22, 2559–2567.

Lacapère, J. J., & Guillain, F. (1993). The reaction mechanism of Ca^{2+}-ATPase of sarcoplasmic reticulum: Direct measurements of the Mg-ATP dissociation constant gives similar values in the presence or absence of calcium. Eur. J. Biochem. 211, 117–126.

Lacapère, J. J., Gingold, M. P., Champeil, P., & Guillain, F. (1981). Sarcoplasmic reticulum ATPase phosphorylation from inorganic phosphate in the absence of a calcium gradient: Steady-state and kinetic fluorescence studies. J. Biol. Chem. 256, 2302–2306.

le Maire, M., & Møller, J. V. (1986). Protein–protein and protein–lipid interactions of the sarcoplasmic reticulum Ca^{2+}-ATPase. In: Sarcoplasmic Reticulum in Muscle Physiology, (Entman, M. L., & Van Winkle, W. B., eds.) pp. 101–126. CRC Press Inc., Boca Raton, FL.

le Maire, M., Lind, K. E., Jørgensen, K. E., Røigaard, H., & Møller, J. V. (1978). Enzymatically active Ca^{2+}-ATPase from sarcoplasmic reticulum membranes, solubilized by nonionic detergents: Role of lipid for aggregation of the protein. J. Biol. Chem. 253, 7051–7060.

Lentz, B. R., Clubb, K. W., Alford, D. R., Höchli, M., & Meissner, G. (1985). Phase behavior of membranes reconstituted from dipentadecanoylphosphatidylcholine and the Mg^{2+}-dependent, Ca^{2+}-stimulated adenosinetriphosphatase of sarcoplasmic reticulum: Evidence for disrupted lipid domain surrounding protein. Biochemistry 24, 433–442.

Lévy, D., Seigneuret, M., Bluzat, A., & Rigaud, J. L. (1990). Evidence for proton countertransport by the sarcoplasmic reticulum Ca^{2+}-ATPase during calcium transport in reconstituted proteoliposomes with low ionic permeability. J. Biol. Chem. 265, 19524–19534.

Loomis, C. R., Martin, D. W., & Tanford, C. (1982). Phosphorylation of calcium adenosine triphosphatase by inorganic phosphate: Reversible inhibition at high magnesium ion concentrations. Biochemistry 21, 151–156.

Luckie, D. B., Lemas, V., Boyd, K. L., Fambrough, D. M., Inesi, G., & Takeyasu, K. (1992). Molecular dissection of functional domains of the E_1E_2-ATPase using sodium and calcium pump chimeric molecules. Biophys. J. 62, 220–227.

Lund, S., & Møller, J. V. (1988). Biphasic kinetics of sarcoplasmic reticulum Ca^{2+}-ATPase and the detergent-solubilized monomer. J. Biol. Chem. 263, 1654–1664.

Lund, S., Orlowski, S., de Foresta, B., Champeil, P., le Maire, M., & Møller, J. V. (1989). Detergent structure and associated lipid as determinants in the stabilization of solubilized Ca^{2+}-ATPase from sarcoplasmic reticulum. J. Biol. Chem. 264, 4907–4915.

Lytton, J., & Nigam, S. K. (1992). Intracellular calcium: Molecules and pools. Current Opinion in Cell Biology 4, 220–226.

Lytton, J., Westlin, M., & Hanley, M. R. (1991). Thapsigargin inhibits the sarcoplasmic or endoplasmic reticulum Ca^{2+}-ATPase family of calcium pumps. J. Biol. Chem. 266, 17067–17071.

Lytton, J., Westlin, M., Burk, S. E., Shull, G. E., & MacLennan, D. H. (1992). Functional comparisons between isoforms of the sarcoplasmic or endoplasmic reticulum family of calcium pumps. J. Biol. Chem. 267, 14483–14489.

MacLennan, D. H., Brandl, C. J., Korczak, B., & Green, N. M. (1985). Amino-acid sequence of a Ca^{2+} + Mg^{2+}-dependent ATPase from rabbit muscle sarcoplasmic reticulum, deduced from its complementary DNA sequence. Nature 316, 696–700.

MacLennan, D. H., Clarke, D. M., Loo, T. W., & Skerjanc, I. S. (1992). Site-directed mutagenesis of the Ca^{2+}-ATPase of sarcoplasmic reticulum. Acta Physiol. Scand. 146, 141–150.

Madeira, V. M. C. (1984). State of translocated Ca^{2+} by sarcoplasmic reticulum inferred from kinetic analysis of calcium oxalate precipitation. Biochim. Biophys. Acta 769, 284–290.

Mahaney, J. E., Kleinschmidt, J., Marsh, D., & Thomas, D. D. (1992). Effects of melittin on lipid–protein interactions in sarcoplasmic reticulum membranes. Biophys. J. 63, 1513–1522.

Makinose, M. (1969). The phosphorylation of the membranal protein of the sarcoplasmic vesicles during active calcium transport. Eur. J. Biochem. 10, 74–82.

Makinose, M. (1971). Calcium efflux-dependent formation of ATP from ADP and orthophosphate by the membranes of the sarcoplasmic vesicles. FEBS Lett. 12, 269–270.

Makinose, M. (1972). Phosphoprotein formation during osmo-chemical energy conversion in the membrane of the sarcoplasmic reticulum. FEBS Lett. 25, 113–115.

Makinose, M. (1973). Possible functional states of the enzyme of the sarcoplasmic Ca^{2+} pump. FEBS Lett. 37, 140–143.

Makinose, M., & Hasselbach, W. (1971). ATP synthesis by the reverse of the sarcoplasmic calcium pump. FEBS Lett. 12, 271–272.

Makinose, M., & Boll, W. (1979). The role of magnesium on the sarcoplasmic calcium pump. In: Cation Flux across Membranes (Mukohata, Y., ed.) pp. 89–100, Academic Press, New York.

Martin, D. W. (1983). Active unit of solubilized sarcoplasmic reticulum calcium adenosinetriphosphatase: An active enzyme centrifugation analysis. Biochemistry 22, 2276–2282.

Martonosi, A. (1969). Properties of a phosphoprotein intermediate implicated in calcium transport. J. Biol. Chem. 244, 613–620.

Martonosi, A., & Feretos, R. (1964). Correlation between adenosinetriphosphatase activity and Ca^{2+} uptake. J. Biol. Chem. 239, 648–658.

Maruyama, K., & MacLennan, D. H. (1988). Mutation of aspartic acid-351, lysine-352, and lysine-515 alters the Ca^{2+} transport activity of the Ca^{2+}-ATPase expressed in COS-1 cells. Proc. Natl. Acad. Sci. USA 85, 3314–3318.

Masuda, H., & de Meis, L. (1973). Phosphorylation of the sarcoplasmic reticulum membrane by orthophosphate. Inhibition by calcium ions. Biochemistry 12, 4581–4585.

McIntosh, D. B., & Boyer, P. D. (1983). Adenosine 5′-triphosphate modulation of catalytic intermediates of calcium ion activated adenosinetriphosphatase of sarcoplasmic reticulum subsequent to enzyme phosphorylation. Biochemistry 22, 2867–2875.

McIntosh, D. B., & Davidson, G. A. (1984). Effects of nonsolubilizing and solubilizing concentrations of Triton X-100 on Ca^{2+} binding and Ca^{2+}-ATPase activity of sarcoplasmic reticulum. Biochemistry 23, 1959–1965.

McIntosh, D. B., & Ross, D. C. (1985). Role of phospholipid and protein–protein associations in activation and stabilization of soluble Ca^{2+}-ATPase of sarcoplasmic reticulum. Biochemistry 24, 1244–1251.

McIntosh, D. B., & Ross, D. C. (1988). Reaction cycle of solubilized monomeric Ca^{2+}-ATPase of sarcoplasmic reticulum is the same as that of the membrane form. J. Biol. Chem. 263, 12220–12223.

McIntosh, D. B., Ross, D. C., Champeil, P., & Guillain, F. (1991). Crosslinking the active site of sarcoplasmic reticulum Ca^{2+}-ATPase completely blocks Ca^{2+} release to the vesicle lumen. Proc. Natl. Acad. Sci. USA 88, 6437–6441.

McIntosh, D. B., Wooley, D. G., & Berman, M. C. (1992). 2′3′-O-(2,4,6-trinitrophenyl)-8-azido-AMP and -ATP photolabel Lys-492 at the active site of sarcoplasmic reticulum Ca^{2+}-ATPase. J. Biol. Chem. 267, 5301–5309.

Meissner, G. (1973). ATP and Ca^{2+} binding by the Ca^{2+} pump protein of sarcoplasmic reticulum. Biochim. Biophys. Acta 298, 906–926.

Meissner, G., Conner, G. E., & Fleischer, S. (1973). Isolation of sarcoplasmic reticulum by zonal centrifugation and purification of Ca^{2+}-pump and Ca^{2+}-binding proteins. Biochim. Biophys. Acta 298, 246–269.

Mintz, E., Lacapère, J. J., & Guillain, F. (1990). Reversal of the sarcoplasmic reticulum ATPase cycle by substituting various cations for Mg^{2+}: Phosphorylation and ATP synthesis when Ca^{2+} replaces Mg^{2+}. J. Biol. Chem. 265, 18762–18768.

Møller, J. V., le Maire, M., & Andersen, J. P. (1986). Uses of non-ionic and bile salt detergents in the study of membrane proteins. In: Progress in Protein–Lipid Interactions, Vol. 2 pp. 147–196. Elsevier Science Publisher BV.

Murphy, A. J. (1976). Sulfhydryl group modification of sarcoplasmic reticulum membranes. Biochemistry 15, 4492–4496.

Myung, J., & Jencks, W. P. (1991). The vectorial specificity for calcium binding to the Ca^{2+}-ATPase of sarcoplasmic reticulum is controlled by phosphorylation, not by an E–E^* conformational change. FEBS Lett. 278, 35–37.

Nakamura, Y. (1984). Two alternate kinetic routes for the decomposition of the phosphorylated intermediate of sarcoplasmic reticulum Ca^{2+}-ATPase. J. Biol. Chem. 259, 8183–8189.

Nakamura, J. (1989). pH and temperature resolve the kinetics of two pools of calcium bound to the sarcoplasmic reticulum Ca^{2+}-ATPase. J. Biol. Chem. 264, 17029–17031.

Nakamura, Y., & Tonomura, Y. (1982). Changes in affinity for calcium ions with the formation of two kinds of phosphoenzyme in the Ca^{2+},Mg^{2+}-dependent ATPase of sarcoplasmic reticulum. J. Biochem. (Tokyo) 91, 449–461.

Navarro, J., Toivio-Kinnucan, M., & Racker, E. (1984). Effect of lipid composition on the calcium/adenosine 5′-triphosphate coupling ratio of the Ca^{2+}-ATPase of sarcoplasmic reticulum. Biochemistry 23, 130–135.

Nishie, I., Anzai, K., Yamamoto, T., & Kirino, Y. (1990). Measurements of steady-state Ca^{2+} pump current caused by purified Ca^{2+}-ATPase of sarcoplasmic reticulum incorporated into a planar bilayer membrane. J. Biol. Chem. 265, 2488–2491.

Ogurusu, T., Wakabayashi, S., & Shigekawa, M. (1991). Activation of sarcoplasmic reticulum Ca^{2+}-ATPase by Mn^{2+}: A Mn^{2+} binding study. J. Biochem. (Tokyo) 109, 472–476.

Orlowski, S., & Champeil, P. (1991a). Kinetics of calcium dissociation from its high affinity transport sites on sarcoplasmic reticulum ATPase. Biochemistry 30, 352–361.

Orlowski, S., & Champeil, P. (1991b). The two calcium ions initially bound to nonphosphorylated sarcoplasmic reticulum Ca^{2+}-ATPase can no longer be kinetically distinguished when they dissociate from phosphorylated ATPase toward the lumen. Biochemistry 30, 11331–11342.

Orlowski, S., & Champeil, P. (1993). Strontium binding to sarcoplasmic reticulum ATPase. FEBS Lett. 328, 296–300.

Orlowski, S., Lund, S., Møller, J. V., & Champeil, P. (1988). Phosphoenzymes formed from Mg-ATP and Ca-ATP during presteady-state kinetics of sarcoplasmic reticulum ATPase. J. Biol. Chem. 263, 17576–17583.

Pedersen, P. L., & Carafoli, E. (1987). Ion motive ATPases. I. Ubiquity, properties, and significance to cell function. TIBS 12, 143–147.

Petithory, J. R., & Jencks, W. P. (1986). Phosphorylation of the Ca^{2+}-ATPase of sarcoplasmic reticulum: Rate-limiting conformational change followed by rapid phosphoryl transfer. Biochemistry 25, 4494–4497.

Petithory, J. R., & Jencks, W. P. (1988a). Sequential dissociation of Ca^{2+} from the calcium adenosine triphosphatase of sarcoplasmic reticulum and the calcium requirement for its phosphorylation by ATP. Biochemistry 27, 5553–5564.

Petithory, J. R., & Jencks, W. P. (1988b). Binding of Ca^{2+} to the calcium adenosine triphosphatase of sarcoplasmic reticulum. Biochemistry 27, 8626–8635.

Pick, U., & Karlish, S. J. (1982). Regulation of the conformational transition in the Ca^{2+}-ATPase from sarcoplasmic reticulum by pH, temperature, and calcium ions. J. Biol. Chem. 257, 6120–6126.

Pickart, C. M., & Jencks, W. P. (1982). Slow dissociation of ATP from the calcium ATPase. J. Biol. Chem. 257, 5319–5322.

Rauch, B., von Chak, D., & Hasselbach, W. (1978). An estimate of the kinetics of calcium binding and dissociation of the sarcoplasmic reticulum transport ATPase. FEBS Lett. 93, 65–68.

Ronzani, N., Migala, A., & Hasselbach, W. (1979). Comparison between ATP-supported and GTP-supported phosphate turnover of the calcium-transporting sarcoplasmic reticulum membranes. Eur. J. Biochem. 101, 593–606.

Ross, D. C., Davidson, G. A., & McIntosh, D. B. (1991). Mechanism of inhibition of sarcoplasmic reticulum Ca^{2+}-ATPase by active site cross-linking: Impairment of nucleotide binding slows nucleotide-dependent phosphoryl transfer, and loss of active site flexibility stabilizes occluded forms and blocks E_2–P formation. J. Biol. Chem. 266, 4613–4621.

Sagara, Y., & Inesi, G. (1991). Inhibition of the sarcoplasmic reticulum Ca^{2+} transport ATPase by thapsigargin at subnanomolar concentrations. J. Biol. Chem. 266, 13503–13506.

Scofano, H., Vieyra, A., & de Meis, L. (1979). Substrate regulation of the sarcoplasmic reticulum ATPase. Transient kinetic studies. J. Biol. Chem. 254, 10227–10231.

Seebregts, C., & McIntosh, D. B. (1989). 2′3′-*O*-(2,4,6-trinitrophenyl)-8-azido-adenosine mono-, di-, and triphosphates as photoaffinity probes of the Ca^{2+}-ATPase of sarcoplasmic reticulum: Regulatory/superfluorescent nucleotides label the catalytic site with high efficiency. J. Biol. Chem. 264, 2043–2052.

Seidler, N. W., Jona, I., Vegh, M., & Martonosi, A. (1989). Cyclopiazonic acid is a specific inhibitor of the Ca^{2+}-ATPase of sarcoplasmic reticulum. J. Biol. Chem. 264, 17816–17823.

Serpersu, E., Kirch, U., & Schoner, W. (1982). Demonstration of a stable occluded form of Ca^{2+} by the use of the chromium complex of ATP in the Ca^{2+}-ATPase of sarcoplasmic reticulum. Eur. J. Biochem. 122, 347–354.

Shigekawa, M., & Dougherty, J. P. (1978). Reaction mechanism of Ca^{2+}-dependent ATP hydrolysis by skeletal muscle sarcoplasmic reticulum in the absence of added alkali metal salts: (III) sequential occurrence of ADP-sensitive and ADP-insensitive phosphoenzymes. J. Biol. Chem. 253, 1458–1464.

Shigekawa, M., & Akowitz, A. (1979). On the mechanism of Ca^{2+}-dependent adenosine triphosphatase of sarcoplasmic reticulum: Occurrence of two types of phosphoenzyme intermediates in the presence of KCl. J. Biol. Chem. 254, 4726–4730.

Shigekawa, M., & Kanazawa, T. (1982). Phosphoenzyme formation from ATP in the ATPase of sarcoplasmic reticulum: Effect of KCl or ATP and slow dissociation of ATP from precursor enzyme–ATP complex. J. Biol. Chem. 257, 7657–7665.

Shigekawa, M., Wakabayashi, S., & Nakamura, H. (1983). Effect of divalent cation bound to the ATPase of sarcoplasmic reticulum. J. Biol. Chem. 258, 14157–14161.

Skerjanc, I. S., Toyofuku, T., Richardson, C., & MacLennan, D. H. (1993). Mutation of glutamate 309 to glutamine alters one Ca^{2+}-binding site in the Ca^{2+}-ATPase of sarcoplasmic reticulum expressed in Sf9 cells. J. Biol. Chem. 268, 15944–15950.

Soler, F., Teruel, J. A., Fernandez-Belda, F., & Gomez-Fernandez, J. C. (1990). Characterization of the steady-state calcium fluxes in skeletal sarcoplasmic reticulum vesicles: Role of the Ca^{2+} pump. Eur. J. Biochem. 192, 347–354.

Sorensen, M. M. (1983). Calcium control of passive permeability to calcium in sarcoplasmic reticulum vesicles. J. Biol. Chem. 258, 7684–7690.

Souza, D. O. G., & de Meis, L. (1976). Calcium and magnesium regulation of phosphorylation by ATP and ITP in sarcoplasmic reticulum vesicles. J. Biol. Chem. 251, 6355–6359.

Squier, T. C., Bigelow, D. J., & Thomas, D. D. (1988). Lipid fluidity directly modulates the overall protein rotational mobility of the Ca^{2+}-ATPase in sarcoplasmic reticulum. J. Biol. Chem. 263, 9178–9186.

Stahl, N., & Jencks, W. P. (1987). Reactions of the sarcoplasmic reticulum calcium adenosine triphosphatase with adenosine 5′-triphosphate and Ca^{2+} that are not satisfactorily described by an E_1–E_2 model. Biochemistry 26, 7654–7667.

Starling, A. P., East, J. M., & Lee, A. G. (1993). Effects of phosphatidylcholine fatty acyl chain length on calcium binding and other functions of the (Ca^{2+}–Mg^{2+})-ATPase. Biochemistry 32, 1593–1600.

Stefanova, H., Napier, R. M., East, J. M., & Lee, A. G. (1987). Effects of Mg^{2+}, anions and cations on the Ca^{2+} + Mg^{2+}-activated ATPase of sarcoplasmic reticulum. Biochem. J. 245, 723–730.

Suko, J., Plank, B., Preis, P., Kolassa, N., Hellmann, G., & Conca, W. (1981). Formation of magnesium-phosphoenzyme and magnesium–calcium-phosphoenzyme in the phosphorylation of adenosine triphosphatase by orthophosphate in sarcoplasmic reticulum: Models of a reaction sequence. Eur. J. Biochem. 119, 225–236.

Sumbilla, C., Cantilina, T., Collins, J. H., Malak, H., Lakowicz, J. R., & Inesi, G. (1991). Structural perturbation of the transmembrane region interferes with calcium binding by the Ca^{2+} transport ATPase. J. Biol. Chem. 266, 12682–12689.

Sumbilla, C., Lu, L., Lewis, D. E., Inesi, G., Ishii, T., Takeyasu, K., Feng, Y., & Fambrough, D. M. (1993). Ca^{2+}-dependent and thapsigargin-inhibited phosphorylation of Na^+,K^+-ATPase catalytic domain following chimeric recombination with Ca^{2+}-ATPase. J. Biol. Chem. 268, 21185–21192.

Sumida, M., Wang, T., Mandel, F., Froehlich, J. P., & Schwartz, A. (1978). Transient kinetics of Ca^{2+} transport of sarcoplasmic reticulum: A comparison of cardiac and skeletal muscle. J. Biol. Chem. 253, 8772–8777.

Suzuki, H., Kubota, K., Kubo, K., & Kanazawa, T. (1990). Existence of low-affinity ATP-binding site in the unphosphorylated Ca^{2+}-ATPase of sarcoplasmic reticulum vesicles: Evidence from binding of 2′,3′-*O*-(2,4,6-trinitrocyclohexadienylidene)-[^{3}H]AMP and -[^{3}H]ATP. Biochemistry 29, 7040–7045.

Takenaka, H., Adler, P. N., & Katz, A. M. (1982). Calcium fluxes across the membrane of sarcoplasmic reticulum vesicles. J. Biol. Chem. 257, 12649–12656.

Takisawa, H., & Makinose, M. (1983). Occlusion of calcium in the ADP-sensitive phosphoenzyme of the adenosine triphosphatase of sarcoplasmic reticulum. J. Biol. Chem. 258, 2986–2992.

Toyofuku, T., Kurzyldlowski, K., Lytton, J., & MacLennan, D. H. (1992). The nucleotide binding/hinge domain plays a crucial role in determining isoform-specific Ca^{2+} dependence of organellar Ca^{2+}-ATPase. J. Biol. Chem. 267, 14490–14496.

Toyofuku, T., Kurzyldlowski, K., Tada, M., & MacLennan, D. H. (1993). Identification of regions in the Ca^{2+}-ATPase of sarcoplasmic reticulum that affect functional association with phospholamban. J. Biol. Chem. 268, 2809–2815.

Verjovski-Almeida, S., Kurzmack, M., & Inesi, G. (1978). Partial reactions in the catalytic and transport cycle of sarcoplasmic reticulum ATPase. Biochemistry 17, 5006–5013.

Vianna, A. L. (1975). Interaction of calcium and magnesium in activating and inhibiting the nucleoside triphosphatase of sarcoplasmic reticulum vesicles. Biochim. Biophys. Acta 410, 389–406.

Vilsen, B., & Andersen, J. P. (1986). Occlusion of Ca^{2+} in soluble monomeric sarcoplasmic reticulum ATPase. Biochim. Biophys. Acta 855, 429–431.

Vilsen, B., & Andersen, J. P. (1987). Effect of phospholipid, detergent and protein–protein interaction on stability and phosphoenzyme isomerization of soluble sarcoplasmic reticulum ATPase. Eur. J. Biochem. 170, 421–429.

Vilsen, B., & Andersen, J. P. (1992a). Interdependence of Ca^{2+} occlusion sites in the unphosphorylated sarcoplasmic reticulum Ca^{2+}-ATPase complexes with CrATP. J. Biol. Chem. 267, 3539–3550.

Vilsen, B., & Andersen, J. P. (1992b). CrATP-induced Ca^{2+} occlusion in mutants of the Ca^{2+}-ATPase of sarcoplasmic reticulum. J. Biol. Chem. 267, 25739–25743.

Vilsen, B., Andersen, J. P., & MacLennan, D. H. (1991). Functional consequences of alterations to aminoacids located in the hinge domain of the Ca^{2+}-ATPase of sarcoplasmic reticulum. J. Biol. Chem. 266, 16157–16164.

Waas, W., & Hasselbach, W. (1981). Interference of nucleoside diphosphates and inorganic phosphate with nucleoside–triphosphate-dependent calcium fluxes and calcium-dependent nucleoside–triphosphate hydrolysis in membranes of sarcoplasmic reticulum vesicles. Eur. J. Biochem. 116, 601–608.

Wakabayashi, S., & Shigekawa, M. (1987). Effect of metal bound to the substrate site on calcium release from the phosphoenzyme intermediate of sarcoplasmic reticulum ATPase. J. Biol. Chem. 262, 11524–11531.

Wakabayashi, S., & Shigekawa, M. (1990). Mechanism for activation of the 4-nitrobenzo-2-oxa-1,3-diazole-labeled sarcoplasmic reticulum ATPase by Ca^{2+} and its modulation by nucleotides. Biochemistry 29, 7309–7318.

Wakabayashi, S., Ogurusu, T., & Shigekawa, M. (1986). Factors influencing calcium release from the ADP-sensitive phosphoenzyme intermediate of the sarcoplasmic reticulum ATPase. J. Biol. Chem. 261, 9762–9769.

Wakabayashi, S., Ogurusu, T., & Shigekawa, M. (1988). Mechanism for 3,3′,4′,5-tetrachloresalicylanilide-induced activation of sarcoplasmic reticulum ATPase. J. Biol. Chem. 263, 15304–15312.

Wang, T. (1986). Transient-state kinetics of phosphoenzyme transformation in the rabbit skeletal sarcoplasmic reticulum calcium-dependent adenosine triphosphatase reaction: Two distinct modes of ADP and K^+ regulation. J. Biol. Chem. 261, 6307–6316.

Watanabe, T., Lewis, D., Nakamoto, R., Kurzmack, M., Fronticelli, C., & Inesi, G. (1981). Modulation of calcium binding in sarcoplasmic reticulum adenosinetriphosphatase. Biochemistry 20, 6617–6625.

Wictome, M., Michelangeli, F., Lee, A. G., & East, J. M. (1992). The inhibitors thapsigargin and 2,5-di(*tert*-butyl)-1,4-benzohydroquinone favour the E_2 form of the Ca^{2+},Mg^{2+}-ATPase. FEBS Lett. 304, 109–113.

Yamada, S., & Ikemoto, N. (1980). Reaction mechanism of Ca^{2+}-ATPase of sarcoplasmic reticulum: Substrates for phosphorylation reaction and back reaction, and further resolution of phosphorylated intermediates. J. Biol. Chem. 255, 3108–3119.

Yamada, S., Fujui, J., & Katayama, H. (1986). Sarcoplasmic reticulum Ca^{2+}-ATPase: Distinction of phosphoenzymes formed from MgATP and CaATP as substrates and interconversion of the phosphoenzymes by Mg^{2+} and Ca^{2+}. J. Biochem. (Tokyo) 100, 1329–1342.

Yamaguchi, M., & Kanazawa, T. (1984). Protonation of the sarcoplasmic reticulum Ca^{2+}-ATPase during ATP hydrolysis. J. Biol. Chem. 259, 9526–9531.

Yamaguchi, M., & Kanazawa, T. (1985). Coincidence of H^+ binding and Ca^{2+} dissociation in the sarcoplasmic reticulum Ca^{2+}-ATPase during ATP hydrolysis. J. Biol. Chem. 260, 4896–4900.

Yamamoto, T., & Tonomura, Y. (1968). Intermediate formation of phosphoryl protein. J. Biochem. (Tokyo) 64, 137–145.

CARDIAC Ca^{2+}-ATPASE AND PHOSPHOLAMBAN

A. G. Lee

I. CONTROL OF CARDIAC Ca^{2+}-ATPASE BY INTERACTION WITH PHOSPHOLAMBAN

Contraction and relaxation of heart muscle is controlled by the level of free Ca^{2+} in the myoplasm. The Ca^{2+} concentration is regulated by the sarcoplasmic reticulum (SR) and by the sarcolemma and its invaginations, the transverse tubules. Depolarization of the sarcolemma and transverse tubules leads to opening of Ca^{2+} channels in the SR membrane and a rise in cytoplasmic Ca^{2+} concentration; this

Biomembranes
Volume 5, pages 77–100.

ISBN: 1-55938-662-2.

induces contraction of the muscle. Re-accumulation of Ca^{2+} in the SR mediated by a Ca^{2+}-ATPase results in relaxation.

As described in Chapter 1, distinct isoforms of the Ca^{2+}-ATPase exist in the various muscle types. Three different genes for the SERCA family of sarcoplasmic/endoplasmic reticulum Ca^{2+}-ATPases have been found in mammalian tissue: SERCA1, SERCA2, and SERCA3. SERCA1 is found in fast-twitch skeletal muscle. Alternate splicing of the SERCA2 gene transcript gives rise to two distinct isoforms. SERCA2a is found in slow-skeletal muscle and heart (and also at low levels in smooth muscle). SERCA2b is expressed in smooth muscle and in non-muscle tissue. The two isoforms differ in that the last four amino acids of SERCA2a are replaced by an extended tail of 49 amino acids in SERCA2b. SERCA3 is expressed at high levels in non-muscle tissues.

The function of the Ca^{2+}-ATPase in cardiac muscle is modified by β-adrenergic agonists which increase contractile force and increase the rate of relaxation of cardiac muscle. Effects on the Ca^{2+}-ATPase are mediated by phosphorylation of a small protein, phospholamban (PLN), present in the SR membrane (Tada et al., 1974, 1975; Luo et al., 1994). The molar ratio of PLN to Ca^{2+}-ATPase in the cardiac SR has been estimated to be ca. 2:1 (Colyer and Wang, 1991) and the Ca^{2+}-ATPase makes up to ca. 40% of the total protein of cardiac SR (Tada and Katz, 1982).

Phosphorylation of PLN by various kinases leads to activation of the Ca^{2+}-ATPase (Tada et al., 1974, 1975). The C-terminal portion of PLN is highly hydrophobic and is assumed to form a single α-helix spanning the membrane once. The N-terminal portion of PLN is amphiphilic and contains the target sequences for the kinases (Wegener et al., 1986; Fujii et al., 1987). Binding of the nonphosphorylated PLN to the Ca^{2+}-ATPase leads to inhibition of the ATPase, and phosphorylation of PLN causes dissociation of PLN from the Ca^{2+}-ATPase, relieving the inhibition (Kirchberger et al., 1986; Suzuki and Wang, 1986).

PLN is found in slow-twitch skeletal muscle and smooth muscle as well as in cardiac muscle, but not in fast-twitch skeletal muscle (Jorgensen and Jones, 1986; Raeymaekers and Jones, 1986). Whereas phosphorylation of PLN in cardiac muscle leads to ca. 200% stimulation of Ca^{2+} uptake by SR, phosphorylation of PLN in slow-twitch muscle SR leads to only a ca. 30% increase in the rate of Ca^{2+} accumulation (Kirchberger and Tada, 1976). Small effects of PLN are also seen when fast-twitch skeletal muscle is changed to slow-twitch muscle by chronic electrostimulation. On electrostimulation, the protein profile of the muscle changes so that the SERCA1 isoform of the Ca^{2+}-ATPase is replaced by the slow-twitch isoform, SERCA2a, and PLN appears in the SR membrane (Leberer et al., 1989; Briggs et al., 1992). Again, however, interaction of PLN has minimal effect on the activity of the Ca^{2+}-ATPase (Briggs et al., 1992). The reason for the poor coupling between PLN and the Ca^{2+}-ATPase in slow-twitch muscle is unclear.

PLN is phosphorylated by cAMP-dependent protein kinase (PK-A) and by an SR-associated Ca^{2+}/calmodulin-dependent kinase, on Ser-16 and Thr-17 respectively (Kirchberger and Antonetz, 1982; Tada and Katz, 1982; Simmerman et al.,

1986). The increase in phosphorylation of PLN in response to β-adrenergic agonists *in vivo* is six-fold within 60s. The residue initially phosphorylated is Ser-16, followed by Thr-17 (Raeymaekers et al., 1988). It has been suggested that PK-A, which is activated in response to β-agonists, phosphorylates Ser-16 directly and that an increase in intracellular Ca^{2+}, resulting from adrenergic activation of the slow inward Ca^{2+} channel of the sarcolemma, subsequently activates the Ca^{2+}/calmodulin-dependent protein kinase, leading to phosphorylation of Thr-17 (MacDougall et al., 1991; Figure 1). PLN is also phosphorylated by a cGMP-activated protein kinase, probably at Ser-16, the site of phosphorylation by PK-A (Raeymaekers et al., 1988) and by protein kinase C, at a site(s) other than at Ser-16 or Thr-17 (Iwasa and Hosey, 1984; Movsesian et al., 1984). The Ca^{2+}/calmodulin-dependent kinase has been purified and shown to be composed of 10 identical subunits, each of M_r 55,000; it has been characterized as a multifunctional Ca^{2+}/calmodulin-dependent kinase II (Gupta and Kranias, 1989). The kinase has also been reported to phosphorylate the Ca^{2+}-ATPase itself in cardiac SR (Xu et al., 1993). Cardiac Ca^{2+}-ATPase contains three residues in minimal consensus sequences R–X–X–S/T for phosphorylation by the kinase; Ser-38, Ser-167, and Thr-531. The Ca^{2+}-ATPase of fast

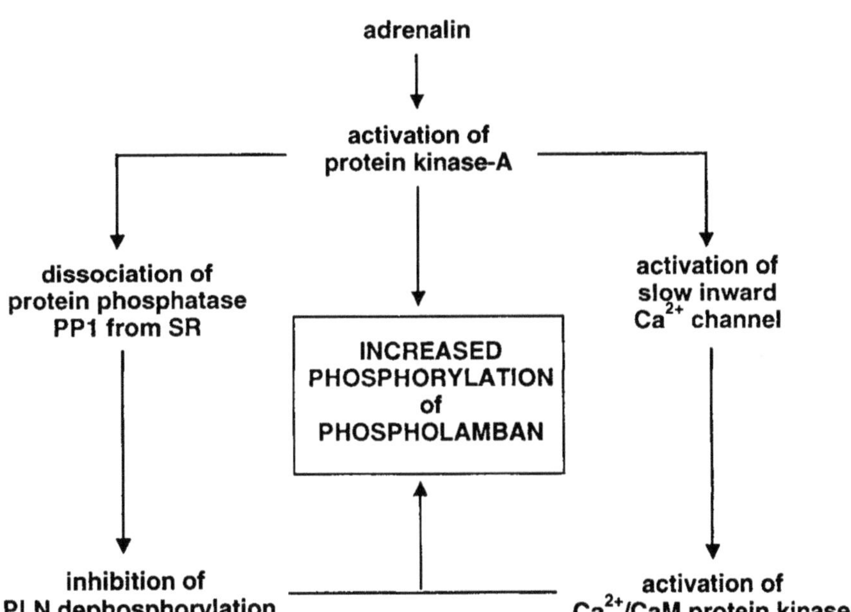

Figure 1. Effects of β-adrenergic agonists on phospholamban. Adrenaline activation of protein kinase A leads to direct phosphorylation of PLN and activation of the slow inward Ca^{2+} channel of sarcolemma leads to activation of Ca^{2+}/calmodulin-dependent protein kinase. Inactivation of protein phosphatase PPI can also lead to an increase in phosphorylation of PLN. Adapted from MacDougall et al. (1991).

skeletal muscle is not phosphorylated by the kinase, and contains two of these residues, Ser-167 and Thr-531. It is, therefore, likely that phosphorylation of cardiac Ca^{2+}-ATPase by the kinase occurs on Ser-38 (Xu et al., 1993). The role of this phosphorylation is unclear; the plasma membrane Ca^{2+}-ATPase of red blood cells has also been reported to be phosphorylated by a cAMP-dependent protein kinase (James et al., 1989b).

Protein phosphatase activity, capable of dephosphorylating Ser-16 and Thr-17, has been found associated with cardiac SR. The protein primarily responsible, a form of protein phosphatase 1, termed $PP1_G$, has also been found associated with skeletal muscle SR (Hubbard et al., 1990; MacDougall et al., 1991; Steenaart et al., 1992). The protein is very similar, if not identical, to that associated with glycogen particles. It is composed of a 37 kD catalytic subunit, complexed to a much larger (ca. 160 kD) subunit, termed the G-subunit, which is responsible for association of the phosphatase to the SR membrane or to the glycogen particle. The carboxyl-terminus of the G-subunit contains a stretch of 32 hydrophobic amino acids which is likely to constitute the anchor to the SR membrane. Phosphorylation of the G-subunit by cAMP-dependent protein kinase triggers dissociation of the catalytic subunit from the membrane or glycogen particle. The released catalytic subunit has much reduced activity (MacDougall et al., 1991). Thus, phosphorylation of the phosphatase may provide an additional mechanism for increasing the level of phosphorylation of PLN in response to β-adrenergic agonists (Figure 1).

II. THE STRUCTURE OF PHOSPHOLAMBAN

As shown in Figure 2, PLN contains 52 amino acids, corresponding to a molecular weight of 6,080 Da. The NH_2-terminal of PLN is acetylated (Fujii et al., 1987). The sequence is well conserved between human, dog, rabbit, pig, and chicken (Figure 3).

The pattern of proteolytic digestion of PLN suggests the presence of two distinct domains; a cytosolic, hydrophilic domain (Met-1 to Arg-25) containing the two sites of phosphorylation, and a membrane-associated, predominantly hydrophobic, domain (Gln-26 to Leu-52; Wegener et al., 1986). These experiments are consistent with the sequence of PLN (Tada, 1992; Figure 3). This shows the presence of a C-terminal domain (Domain II; Leu-31 to Leu-52) which is hydrophobic and likely to be a transmembrane α-helix. The hydrophilic domain contains two regions, an N-terminal region from, in human, Met-1 to Pro-21 (Domain IA) and a second, smaller region from Asn-22 to Asn-30 (Domain IB) linking Domain 1A to the transmembrane Domain II. It has been suggested that Domain IA forms an amphipathic α-helix, broken at Pro-21, whereas Domain IB is a less structured region (Simmerman et al., 1989). The conservation of a proline residue at position 21 or 22 (in chicken) argues in favor of the importance of a helix breaker at this position. It has been suggested that Domain IB could act as a hinge between the cytoplasmic and membrane-spanning domains, allowing a change in angle between these two domains on phosphorylation of PLN (Simmerman et al., 1989).

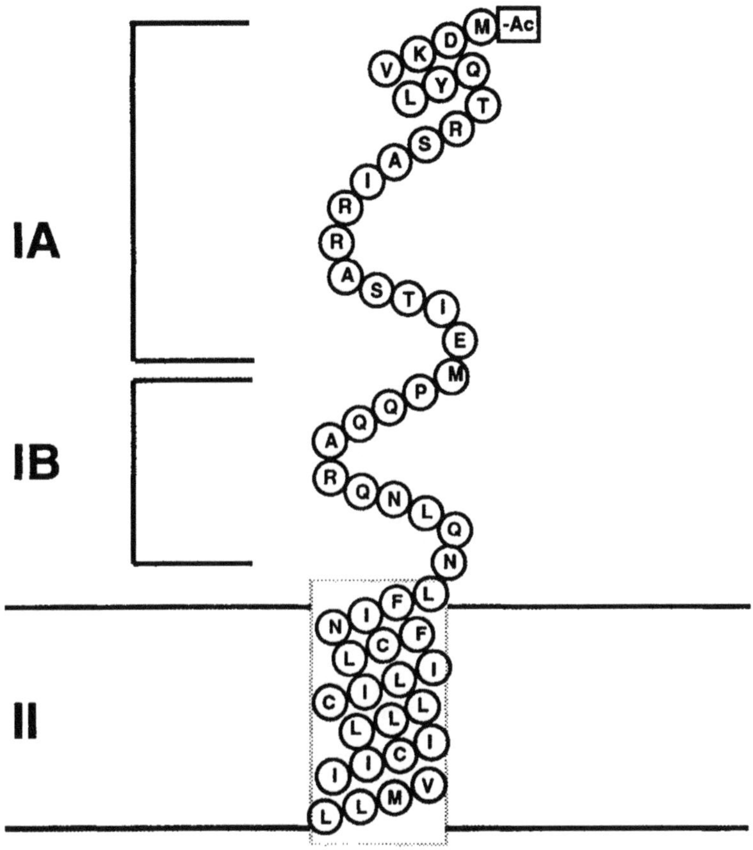

Figure 2. The structure of phospholamban. The hydrophobic Domain II is shown as an α-helix, and domain 1A as a mixture of α-helix and extended structure, and the linking domain 1B as an extended structure. Adapted from Fujii et al. (1987) and Toyofuku et al. (1994a).

```
Human     MEKVQYLTRSAIRRASTIEMPQQARQKLQNLFINFCLILICLLLICIIVMLL
Rabbit    *E****L****I*******MPQ****N**N**I*******************
Dog       *D****L****I*******MPQ****N**N**I*******************
Pig       *D****L****I*******MPQ****N**N**I*******************
Chicken   *E****I****L*******VNP****R**E**V*******************
```

Figure 3. Comparison of the sequences of PLN from human, dog, rabbit, pig, and chicken (Fujii et al., 1987, 1988, 1991; Verboomen et al., 1989; Toyofuku and Zak, 1991). Residues marked * are identical in all the species.

CD spectra of PLN in the detergent $C_{12}E_8$ suggest a structure which is 78% α-helix, and 22% β-sheet (Simmerman et al., 1989). However, the CD spectrum of a peptide corresponding to residues 2–33 of PLN showed no well-defined, ordered, structure in buffer at pH 7.0, although trifluoroethanol was able to induce α-helix formation (Terzi et al., 1992). A peptide corresponding to residues 1–25 of PLN has also been shown to adopt no well-defined structure in buffer (Hughes and Lee, unpublished observations). Evidence arguing against an amphipathic α-helical structure for Domain IA also comes from consideration of the structural requirements for phosphorylation of PLN by kinases. The phosphorylation sequence $RRAST^{17}$ in PLN must be exposed to cAMP-dependent protein kinase A and to Ca^{2+}/calmodulin-dependent kinase. Studies of a high-affinity peptide inhibitor of protein kinase A have shown that the phosphorylation sequence (RRAS) of the peptide adopts an extended conformation (Knighton et al., 1991), and thus, the corresponding region of PLN is unlikely to be α-helical.

PLN shows a strong tendency to aggregate into pentamers when extracted from the membrane; such pentamers are formed even in the presence of sodium dodecyl sulphate (Fujii et al., 1986). It has been suggested that the formation of the pentamers is driven by interactions between the transmembrane α-helices (Figure 4). Related structures have been suggested by Tatulian et al. (1995), Mortishire-Smith et al. (1995), Arkin et al. (1995), and Simmerman et al. (1996). On SDS-PAGE, purified PLN migrates with an apparent M_r of 25,000. Boiling the protein in SDS prior to PAGE decreases the apparent M_r to 5,000–6,000. On heating to intermediate temperatures, pentameric, tetrameric, trimeric, dimeric, and monomeric forms can be observed (Fujii et al., 1986). Phosphorylation of PLN by PK-A on Ser-16 gives pentameric species containing up to five phosphate groups, indicating that all the PLN molecules in the pentamer can be phosphorylated (Jakab and Kranias, 1988; Li et al., 1990; Colyer and Wang, 1991). The time course of phosphorylation was consistent with a non-cooperative, random pathway for phosphorylation of the monomeric units in the pentamer (Li et al., 1990).

Oligomerization of PLN is driven by the hydrophobic Domain II since the tryptic-digest product of PLN, which has lost the phosphorylation sites, is still pentameric (Wegener et al., 1986). It has been suggested that the pentameric structure is stabilized by the cysteine residues situated at every fifth residue in the putative transmembrane region (Cys-36, Cys-41, and Cys-46). Replacement of one or more of these residues by site-directed mutagenesis gave mutants which dissociated from pentamer to monomer at lower temperatures than the wild-type pentamer (Fujii et al., 1989). Of these cysteine residues, Cys-41 is particularly important (Fujii et al., 1989). The cysteine residues exist as free SH groups and not as disulphides (Simmerman et al., 1986); it has been suggested that stabilization follows from hydrogen bonding between Cys residues in the hydrophobic interior of the membrane (Fujii et al., 1989). The importance of the cysteine residues is also supported by chemical labeling studies. Thus, reaction of cysteine with N-ethylmaleimide or iodoacetic acid led to the disappearance of the pentameric species in

SDS-PAGE (Young et al., 1989). However, modification of the cysteine residues with the hydrophobic p-azidophenyl bromide under conditions where cross-linking did not occur, did not lead to the abolition of pentameric species in SDS-PAGE (Young et al., 1989).

The state of PLN in the SR membrane itself is unclear. Young et al. (1989) investigated the state of oligomerization of PLN in cardiac SR membranes by cross-linking with hydrophilic and hydrophobic cross-linkers. They observed only monomeric and dimeric species, with no evidence for pentameric species.

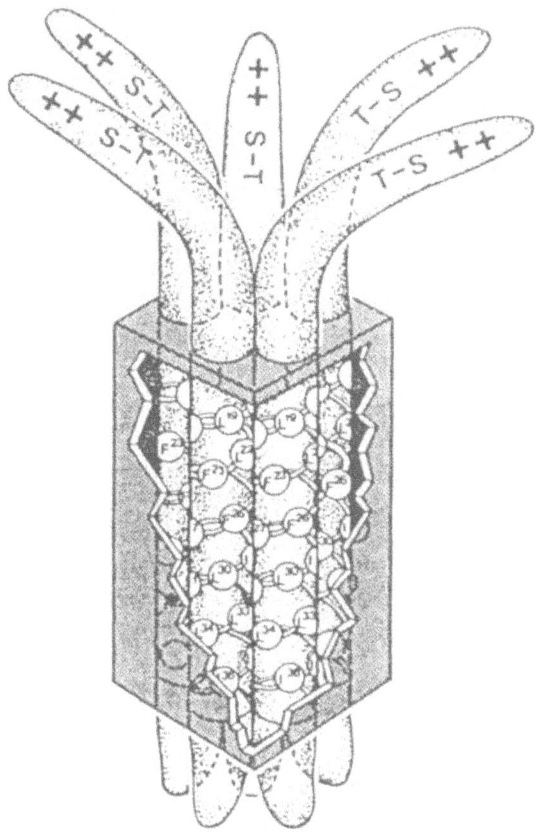

Figure 4. A model for the PLN pentamer. The pentameric structure involves aggregation of the hydrophobic domains in the SR membrane. The transmembrane helices are oriented with polar faces inward and leucine, isoleucine, and phenylalanine residues facing the lipid bilayer. S and T show the serine and threonine residues phosphorylated on PLN. (Reproduced, with permission, from Simmerman et al., 1986.)

It has been reported that incorporation of PLN into lipid bilayers leads to the formation of cation-selective ion channels, with a higher conductance for Ca^{2+} than for K^+ (Kovacs et al., 1988). It is likely that this is associated with aggregation of PLN into oligomers in the membrane. However, the formation of Ca^{2+} channels in cardiac SR would be surprising since the high concentration of PLN in the cardiac SR membrane would then make the membrane relatively permeable to Ca^{2+}.

Evidence for changes in the structure of PLN on phosphorylation have come from studies using CD and proteolytic digestion. Although, as described above, Terzi et al. (1992) reported that a peptide corresponding to residues 2 to 33 of PLN had no well defined structure in buffer, either in the phosphorylated or nonphosphorylated state, they observed that in the presence of trifluoroethanol the proportion of α-helix was greater for the phosphorylated peptide than for the nonphosphorylated peptide (Terzi et al., 1992). In contrast, Simmerman et al. (1989) found that phosphorylation of PLN had no significant effect on the CD spectrum in $C_{12}E_8$. Huggins and England (1987) found that whereas PLN in the nonphosphorylated state was digested by trypsin, in the phosphorylated state it was trypsin-resistant; digestion by papain was unaffected by phosphorylation (Huggins and England, 1987). This would be consistent with a conformational change in the linker region around Arg-25 on phosphorylation. Quirk et al. (1996) reported that phosphorylation results in increased structural constraints in the region around the phosphorylation site.

III. CARDIAC Ca^{2+}-ATPASE

Comparison of the sequences of fast skeletal (SERCA1) and cardiac muscle (SERCA2a) Ca^{2+}-ATPases show a similar overall structure (Brandl et al., 1986; Komuro et al., 1989). The predicted 10-transmembrane α-helices are highly conserved, including the residues believed to be involved in Ca^{2+} binding, as are the hinge- and stalk-regions (see Chapter 1). The most variation occurs in the region of the phosphorylation and nucleotide-binding domains, where antibody studies have shown considerable surface exposure in SERCA1 (Mata et al., 1992). The C- and N-terminal sequences also differ markedly, and again antibody studies have shown these regions to be surface exposed (Mata et al., 1992).

The mechanisms of the cardiac and fast skeletal muscle Ca^{2+}-ATPases are very similar. Both involve a hydroxylamine-labile acylphosphate intermediate (Shigekawa et al., 1976), and both ATPases are inhibited by vanadate, an analog of phosphate (Wang et al., 1979), and by the specific inhibitor, thapsigargin (Kirby et al., 1992). For both, the stoichiometry of Ca^{2+} uptake appears to be two Ca^{2+} ions per ATP molecule hydrolyzed. True coupling ratios between Ca^{2+} uptake and ATP hydrolysis are difficult to determine for the cardiac Ca^{2+}-ATPase because of rapid efflux of Ca^{2+} from cardiac SR vesicles. Thus, values close to 1 Ca^{2+} ion transported per ATP molecule hydrolyzed have often been reported (Kirchberger and Antonetz, 1982; Chamberlain et al., 1983; Kimura et al., 1991). However, if the ryanodine-sensitive Ca^{2+} channel in the cardiac SR is blocked with ruthenium red or ryanodine, then the coupling ratio increases towards two (Chamberlain et al., 1984).

The rates of dephosphorylation of the phosphorylated forms of both ATPases have been observed to decrease with decreasing pH (Mandel et al., 1982) and to increase on addition of K+ (Jones, 1979). High concentrations of Ca2+ inhibit both, possibly due to the formation of CaATP, a less good substrate than MgATP (Jones, 1979).

The mechanism of the Ca^{2+}-ATPase can be interpreted in terms of the E2–E1 scheme shown in Figure 5 and discussed in Chapters 1 and 2. The scheme proposes that, in the absence of any ligands, the Ca^{2+}-ATPase exists as a mixture of two conformations, E1 and E2. In the E1 conformation, the ATPase possesses two high-affinity binding sites for Ca^{2+}, exposed to the cytoplasm. Following binding of Ca^{2+} and MgATP, the ATPase is phosphorylated on Asp-351. A conformation change $E1PCa_2 \rightarrow E2PCa_2$ then leads to exposure of the Ca^{2+}-binding sites to the lumen of the SR, and to a conversion from high affinity in $E1PCa_2$ to low affinity in $E2PCa_2$. Ca^{2+} can then be lost from $E2PCa_2$ to give E2P, which then dephosphorylates to E2. The existence of $E2PCa_2$ as a distinct intermediate in the $E1PCa_2 \rightarrow$ E2P now seems unlikely.

Comparison of the turnover numbers of cardiac and fast-skeletal muscle Ca^{2+}-ATPase is difficult because of problems in purification of the cardiac Ca^{2+}-ATPase, but activities of the pure proteins appear to be rather similar, with the turnover number of the cardiac Ca^{2+}-ATPase being perhaps a factor of 2 or so less than for the fast-skeletal muscle Ca^{2+}-ATPase. The rates of phosphoenzyme formation observed on addition of ATP to the Ca^{2+}-ATPase incubated in the presence of Ca^{2+} have been reported to be identical for cardiac and fast-skeletal muscle (Sumida et al., 1978). However, the rate of phosphoenzyme formation observed on simultaneous addition of Ca^{2+} and ATP to the ATPase incubated in EGTA was slower for cardiac than for fast-skeletal muscle ATPase (Sumida et al., 1978). This would imply that either the E2 $\rightarrow$ E1 transition or some step in the Ca^{2+}-binding process from E1 to $E1Ca_2$ was slower for cardiac than for fast-skeletal muscle. Although PLN had no effect on the rate of phosphorylation of the ATPase observed when the Ca^{2+}-bound ATPase was mixed with ATP, binding of PLN to the ATPase decreased the rate of phosphorylation when the Ca^{2+}-free ATPase was mixed simultaneously with Ca^{2+} and ATP (Tada et al., 1980).

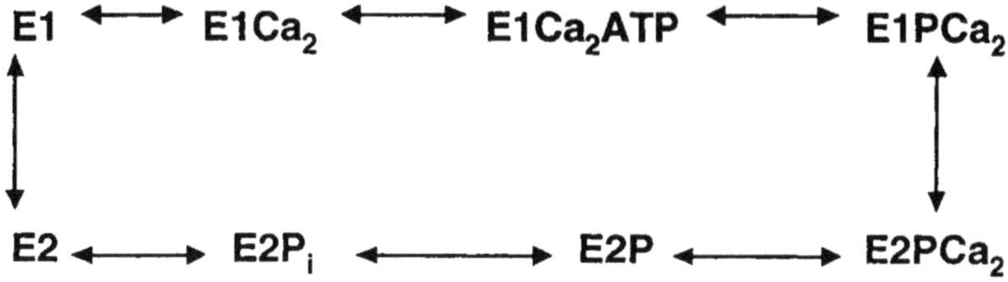

Figure 5. A scheme for the mechanism of the Ca^{2+}-ATPase.

Experiments in which the ATPase incubated in the presence of Ca^{2+} was mixed with EGTA and ATP suggested that the rate of dissociation of Ca^{2+} from $E1Ca_2$ could be slower for cardiac than for fast skeletal muscle (Sumida et al., 1978). Since affinities for Ca^{2+} measured from $^{45}Ca^{2+}$ binding were identical for cardiac and fast-skeletal muscle (Cantilina et al., 1993) this would also imply slower Ca^{2+} binding for cardiac than for fast-skeletal muscle.

The rate of dephosphorylation observed on mixing the phosphorylated ATPase with excess unlabeled ATP was faster for fast-skeletal muscle than for cardiac muscle (Sumida et al., 1978). Binding of PLN to the cardiac Ca^{2+}-ATPase was observed to decrease the rate of dephosphorylation (Tada et al., 1980). Addition of ADP to the phosphorylated ATPase led to a rapid loss of ca. 40% of phosphoenzyme for both cardiac and fast-skeletal muscle Ca^{2+}-ATPase (Wang et al., 1981).

A major difference between the functions of the cardiac and fast-skeletal muscle Ca^{2+}-ATPases might be in their reaction with GTP. It has been suggested that the GTPase activity of cardiac SR is not Ca^{2+} dependent, does not involve an acyl-phosphate intermediate, and that hydrolysis of GTP does not lead to accumulation of Ca^{2+} (Van Winkle et al., 1981; Bick et al., 1983; Tate et al., 1991). In contrast, fast-skeletal muscle Ca^{2+}-ATPase hydrolyzes GTP by the same pathway as ATP, albeit at a lower rate, leading to accumulation of Ca^{2+}, as for ATP (Van Winkle et al., 1981). It has, therefore, been suggested that GTP is hydrolyzed by cardiac Ca^{2+}-ATPase by a pathway different to that for ATP (Tate et al., 1991). However, Ogurusu et al. (1989) reported that an acyl-phosphate intermediate was observed with GTP, and that GTP hydrolysis did lead to accumulation of Ca^{2+} in the normal way. The rate of formation of phosphoenzyme from GTP was, however, reported to be very slow compared to that from ATP (Ogurusu et al., 1989).

The reported phosphorylation of the cardiac Ca^{2+}-ATPase by Ca^{2+}/calmodulin-dependent kinase, possibly on Ser-38, with a doubling of activity is also of interest (Xu et al., 1993). In the SERCA1 isoform, Ser-38 is replaced by a histidine residue. This region of the ATPase occurs immediately before the first stalk region (see Chapter 1) and, as yet, has no known functional role.

IV. EFFECTS OF PHOSPHORYLATION OF PHOSPHOLAMBAN

Phosphorylation of PLN by cAMP-dependent protein kinase leads to an increase in the rate of Ca^{2+} uptake by cardiac SR vesicles and to an increase in the rate of ATP hydrolysis (Figure 6), but with the stoichiometry of Ca^{2+} ions transported per ATP molecule hydrolyzed remaining at 2 (Tada et al., 1979). The apparent affinity of the ATPase for Ca^{2+} was observed to increase by a factor of ca. 0.3 (Figure 7) and the rate of ATP hydrolysis at maximally stimulating Ca^{2+} and high ATP was observed to double (Figure 6) (Tada et al., 1979). The same effects were seen on mild proteolysis of cardiac SR, which leads to removal of the cytoplasmic domain I of PLN (Kirchberger et al., 1986; Figure 7) or on binding monoclonal antibodies

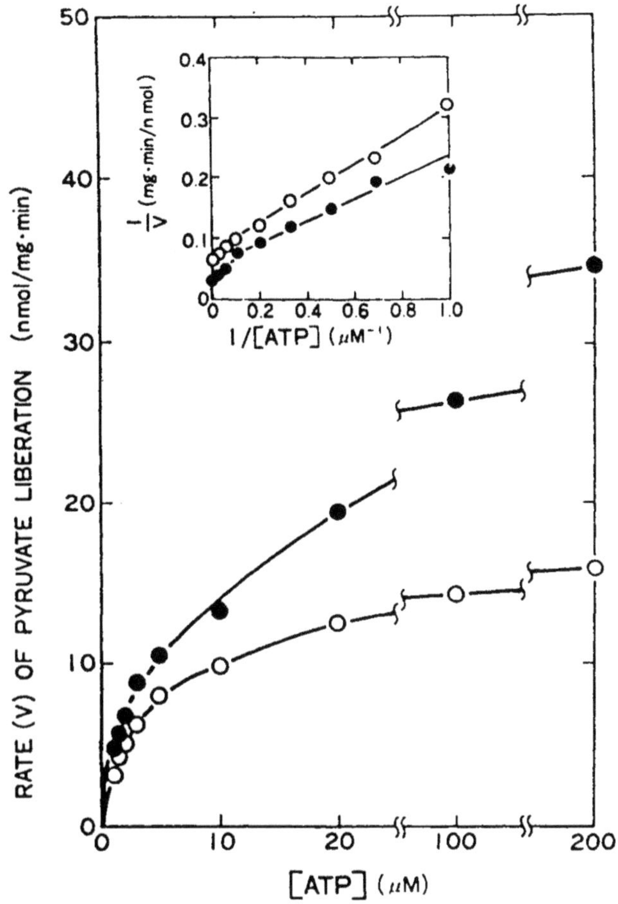

Figure 6. Effect of phosphorylation of PLN by cAMP-dependent protein kinase on the rate of ATP hydrolysis by the Ca^{2+}-ATPase of cardiac SR. ATPase activities were assayed at the given concentrations of ATP in the presence of an ATP regenerating system (phosphoenolpyruvate and pyruvate kinase) in order to maintain the concentrations of ATP constant. The rate of production of pyruvate is then equivalent to the rate of formation of ADP by the Ca^{2+}-ATPase. Assays were performed at pH 7.0, 1 mM Mg^{2+}, 100 mM KCl, 10 μM Ca^{2+}, 25 °C. (●) phosphorylated and (○) unphosphorylated PLN, respectively. The insert shows the data at low ATP concentrations as a Lineweaver–Burk plot. (Reproduced, with permission, from Tada et al., 1979.)

against PLN (Kimura et al., 1991; Morris et al., 1991; Cantilina et al., 1993). The apparent affinity of the ATPase for Ca^{2+}, determined from studies of the rate of ATP hydrolysis, was also observed to increase by a factor of 3 when the ATPase was purified from cardiac SR vesicles (Kim et al., 1990). All these observations suggest that interaction of PLN with the Ca^{2+}-ATPase leads to inhibition of activity, which

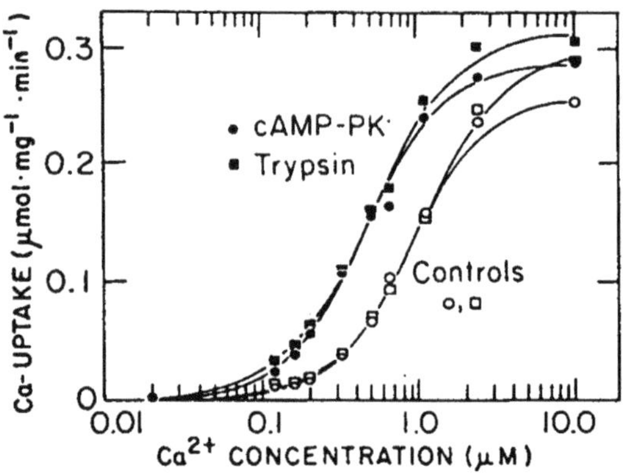

Figure 7. The effect of PLN on the apparent affinity of the Ca^{2+}-ATPase in cardiac SR for Ca^{2+}. Shown are the rates of Ca^{2+} accumulation by cardiac SR vesicles either untreated (○,□) or following phosphorylation with cAMP dependent protein kinase (●) or treatment with trypsin (■). Uptake was measured at pH 6.8, 120 mM KCl, 1 mM Mg^{2+}, 1 mM ATP, and 2.5 mM Tris/oxalate, at 25 °C. (Reproduced, with permission, from Kirchberger et al., 1986.)

is reversed when PLN is displaced from the ATPase, either by phosphorylation or by antibody binding, or by removal of the cytoplasmic domain of PLN by proteolysis. The increase in rate of Ca^{2+} uptake by SR vesicles in the presence of oxalate resulting from trypsin digestion of PLN (Lu et al., 1993) was much less dependent on the concentration of ATP than was the rate of ATP hydrolysis shown in Figure 6. This would follow if the rate of Ca^{2+} uptake in the presence of oxalate was dependent on the rate of transport of oxalate ions across the membrane, as suggested for skeletal muscle SR (Stefanova et al., 1991).

The relationship between ATPase activity and the level of phosphorylation of PLN has been determined in order to test whether monomeric or pentameric PLN is involved in interaction with the ATPase (Colyer and Wang, 1991). The observed data were consistent with each phosphorylation event contributing equally to stimulation of the Ca^{2+}-ATPase, or to a model in which the introduction of one phosphate group into a PLN pentamer had no effect on the interaction with the Ca^{2+}-ATPase, but that formation of successively higher phosphorylated forms of the pentamer did modify its interaction with the ATPase (Colyer and Wang, 1991). Thus, although the data are consistent with interaction between the Ca^{2+}-ATPase and a monomer of PLN leading to inhibition, more complex scenarios involving interaction of the Ca^{2+}-ATPase with the pentamer cannot be ruled out. Evidence that PLN monomers cause inhibition comes from studies using site-directed mutagenesis. Replacement of Cys-41 in the putative transmembrane α-helix with Phe was found to disrupt pentamer formation, but had no effect on the interaction between PLN and the ATPase (Fujii et al., 1989).

Phosphorylation of PLN at both Ser-16 and Thr-17 had no greater effect than phosphorylation at just one of these residues alone (Colyer and Wang, 1991). The effect of phosphorylation of PLN on the rate of ATP hydrolysis was observed to vary with ATP concentration, relatively small effects being observed at low concentrations of ATP, but with a doubling in rate at high concentrations of ATP (Figure 6). This suggests that interaction of PLN with the Ca^{2+}-ATPase has no effect on the affinity of the ATPase for ATP at the catalytic site, but that the interaction decreases the rate of a step(s) stimulated by binding of ATP.

Phosphorylation of PLN had no effect on the rate of phosphorylation of the Ca^{2+}-ATPase by ATP at saturating concentrations of Ca^{2+} (Kranias et al., 1980). However, at less than saturating concentrations of Ca^{2+}, phosphorylation of PLN did result in an increase in the rate of phosphorylation of the Ca^{2+}-ATPase, but this can be attributed to the decreased apparent affinity for Ca^{2+} of the PLN-bound Ca^{2+}-ATPase (Kranias et al., 1980).

The concentration of Ca^{2+} causing half-maximal stimulation of ATP hydrolysis by cardiac Ca^{2+}-ATPase is higher in the presence of PLN than in its absence (Figure 7), as described above. Surprisingly, however, Cantilina et al. (1993) showed that binding of a monoclonal antibody against PLN to cardiac SR had no effect on the affinity of the ATPase for Ca^{2+} when measured directly using $^{45}Ca^{2+}$. The measured K_d for Ca^{2+} binding to the cardiac Ca^{2+}-ATPase was 0.32 μM in the presence or absence of PLN (pH 7.0, 5 mM Mg^{2+}, and 80 mM KCl); an identical value was measured for fast-skeletal muscle Ca^{2+}-ATPase (Cantilina et al., 1993). The shift in apparent K_d for Ca^{2+} observed in kinetic studies must therefore follow from a PLN-induced change in the rate of some step in the Ca^{2+}-binding process. Cantilina et al. (1993) showed that if Ca^{2+} binding follows the sequence, $E1 \rightarrow E1Ca \rightarrow E1'Ca \rightarrow E1'Ca_2$, with the rate of the $E1Ca \rightarrow E1'Ca$ step being slow, then inhibition of this step by PLN would result in an apparent increase in K_d for Ca^{2+} in kinetic studies.

The effect of PLN on the rate of dephosphorylation of the ATPase from $E1Ca_2P \rightarrow E2$ has been determined by first phosphorylating the ATPase in the presence of Ca^{2+} and ATP, and then initiating dephosphorylation by the addition of EGTA to remove Ca^{2+}. The rate of loss of phosphoenzyme was found to double on phosphorylation of PLN (Tada et al., 1979; Kranias et al., 1980).

V. INTERACTION OF PHOSPHOLAMBAN WITH THE Ca^{2+}-ATPASE

There is strong evidence that charge–charge interactions are involved in binding PLN to the ATPase. The hydrophilic segment of phospholamban is highly basic, and high ionic strength eliminates regulation of the Ca^{2+}-ATPase by PLN (Chiesi and Schwaller, 1989). It is possible that release of PLN from the ATPase on phosphorylation follows from reduction in the net positive charge on PLN. A number of polycationic compounds have been observed to inhibit Ca^{2+} uptake by cardiac SR, including poly-L-arginine, poly-L-lysine, spermine, spermidine, histone, and polymyxin B (Xu and Kirchberger, 1989). However, effects show some selectivity since ruthenium red with six positive charges has no effect on either Ca^{2+} accumulation by SR vesicles or on the rate of Ca^{2+} binding even though it does bind to the Ca^{2+}-ATPase (Corbalan-Garcia et al., 1992; Moutin et al., 1992).

The region of the cytoplasmic domain of PLN critical for its interaction with the Ca^{2+}-ATPase lies between residues 1 and 18 (Morris et al., 1991; Toyofuku et al., 1994a). Site-directed mutagenesis has identified the charged residues Glu-2, Lys-3, Arg-9, Arg-13, and Arg-14, the hydrophobic residues Val-4, Leu-7, Ile-12, and Ile-18, and the phosphorylatable residues Ser-16 and Thr-17 as important (Toyofuku et al., 1994a). The importance of the overall charge is clear (Toyofuku et al., 1994a). Thus, the net charge for the first 20 residues is +2 (2 Glu^{-}, 3Arg^{+}, 1 Lys^{+}). Function of PLN was found to be retained if the net positive charge was maintained at +2 or +1, but was lost if the net positive charge was increased to +3 or decreased to 0 or negative (Toyofuku et al., 1994a). The effect of phosphorylation of PLN will be, of course, to decrease the net positive charge. Antibodies binding to PLN around residues 7 to 16 block the effect of PLN on the Ca^{2+}-ATPase, also suggesting that this region of PLN is important in the interaction with the Ca^{2+}-ATPase (Kimura et al., 1991; Morris et al., 1991).

Cross-linking experiments have suggested that phospholamban interacts with the Ca^{2+}-ATPases at a region just C-terminal of the residue (Asp-351) phosphorylated by ATP, and, using ^{125}I-labeled PLN, radioactivity was recovered in Lys-397 and Lys-400 (James et al., 1989a; Vorherr et al., 1992). These two residues occur in a region conserved in the SERCA1 and SERCA2 isoforms of the Ca^{2+}-ATPase, but not in the SERCA3 isoform (Burk et al., 1989) or in more distantly related Ca^{2+}-ATPases such as those in *Plasmodium* (Kimura et al., 1993), *Artemia* (Palmero and Sastre, 1989), or plants (Wimmers et al., 1992).

It has been observed that the activity of the SERCA1 isoform of the Ca^{2+}-ATPase in fast-skeletal muscle SR is modulated by interaction with PLN if both the ATPase and PLN are co-expressed in COS cells (Toyofuku et al., 1993), even though PLN is not found in the fast-skeletal muscle SR. Similarly, if the purified Ca^{2+}-ATPase from skeletal muscle SR is reconstituted with PLN, then inhibition is observed as for the Ca^{2+}-ATPase from cardiac SR (Szymanska et al., 1990; Vorherr et al., 1992). However PLN has been shown not to affect the apparent Ca^{2+} affinity of the SERCA3 isoform of the Ca^{2+}-ATPase and the construction of chimeric Ca^{2+}-ATPases between SERCA2 and SERCA3 has shown that the region between residues 336 and 412 is essential for interaction with PLN (Toyofuku et al., 1993). Presumably, binding of the positively charged PLN would be to the negatively charged residues found in this region of the ATPase (e.g., Glu-392, Glu-394, and Asp-399). Site-directed mutagenesis has confirmed the importance of the sequence Lys–Asp–Asp–Lys–400 in the cardiac Ca^{2+}-ATPase (Toyofuku et al., 1994b). Although removal of any one of these charged residues had no effect on interaction with PLN, removal of both Lys residues or both Asp residues prevented binding (Toyofuku et al., 1994b). This is consistent with the observation that fast-skeletal muscle Ca^{2+}-ATPase, in which Asp-398 is replaced by Asn, is also inhibited by PLN. An antipeptide antibody raised to residues 381–400 of the Ca^{2+}-ATPase bound to the native ATPase, suggesting surface exposure for this region (Matthews et al., 1989; Mata et al., 1992).

Toyofuku et al. (1993) have shown that the nucleotide-binding/hinge region between Arg-467 and Arg-743 is also involved in the interaction with PLN. Antipeptide antibody binding studies have also shown considerable surface exposure in this region of the ATPase (Mata et al., 1992).

Since PLN has similar effects on the Ca^{2+} affinity of the SERCA2a and SERCA2b isoforms of the Ca^{2+}-ATPase when expressed in COS cells, the extended C-terminal tail present in the SERCA2b isoform can have no effect on the interaction with PLN (Verboomen et al., 1992).

The various domains of PLN appear to have distinct effects on the activity of the ATPase. Binding of a peptide corresponding to residues 1–31 of PLN to cardiac Ca^{2+}-ATPase led to a reduction in V_{max} for the ATPase with no effect on the apparent affinity for Ca^{2+}, whereas binding of a peptide corresponding to residues 28–47 led to a reduction in the apparent affinity of the ATPase for Ca^{2+} without any effect on V_{max} (Sasaki et al., 1992). The observation shown in Figure 7 that trypsin treatment of cardiac SR leads to the same increase in apparent affinity for Ca^{2+} as phosphorylation of PLN (Kirchberger et al., 1986), suggests that the hydrophobic domain of PLN, at the concentrations found in the native membrane, can bind only weakly to the Ca^{2+}-ATPase.

Binding of the peptide corresponding to residues 1 to 25 of PLN (PLN(1–25)) to the Ca^{2+}-ATPase of fast-skeletal muscle was found to result in a maximum inhibition of ATPase activity of 44%, with half-maximal inhibition at 5 μM PLN(1–25) (Hughes et al., 1994a). Sasaki et al. (1992) obtained up to 37% inhibition of the ATPase activity of cardiac Ca^{2+}-ATPase with a peptide corresponding to residues 1–32 of phospholamban, maximal inhibition being observed with 110 μM peptide. Kim et al. (1990) and Vorherr et al. (1992) observed much smaller effects of similar peptides on Ca^{2+} uptake into sealed SR vesicles measured in the presence of oxalate, but Ca^{2+} uptake in the presence of oxalate is known to be at least partly dependent on the rate of oxalate transport across the membrane (Stefanova et al., 1991) and the oxalate transporter is unlikely to be affected by PLN (Starling et al., 1995). Although the molar ratio of peptide:ATPase required for inhibition is relatively high in these experiments (half-maximal inhibition at a molar ratio of PLN/ATPase of 136:1 in Hughes et al. (1994a)), this presumably reflects the relatively high water solubility of the polar peptide.

The peptide PLN(1–25) could be cross-linked to either the fast-skeletal muscle or cardiac muscle isoforms of the Ca^{2+}-ATPase, but only when the peptide was unphosphorylated (Szymanska et al., 1990). It was also found that the presence of 100 μM Ca^{2+} prevented cross-linking of the peptide to the ATPase (Szymanska et al., 1990). The meaning of this observation is, however, unclear since the observed effects of peptides on the activity of the ATPase described below imply that they must be able to bind to the Ca^{2+}-ATPase in the presence of Ca^{2+}.

PLN(1–32) had no significant effect on the Ca^{2+} dependence of ATPase activity of cardiac Ca^{2+}-ATPase (Sasaki et al., 1992), and PLN(1–25) had no significant effect on the Ca^{2+} dependence of ATPase activity of fast-skeletal muscle Ca^{2+}-ATPase (Hughes et al., 1994a). Similarly, inhibitory polycationic compounds such as spermine were found to have no effect on Ca^{2+} affinity (Hughes et al., 1994b). Inhibition of ATPase activity by spermine was found to follow from a decrease in the rate of the $E1PCa_2 \rightarrow E2P$ transition, with no effect on the rate of phosphorylation or dephosphorylation (Hughes et al., 1994b). As shown in Figure 8, 25 μM PLN(1–26) also reduced the rate of the $E1PCa_2 \rightarrow E2P$ transition, by a factor of 2.4, accounting for the observed inhibition of ATPase activity. Poly-L-Arg, which decreased ATPase activity by up to 41%, caused a decrease in the rate of this transition by a factor of 2.2 at a concentration of 10 μM (Hughes et al., 1994a).

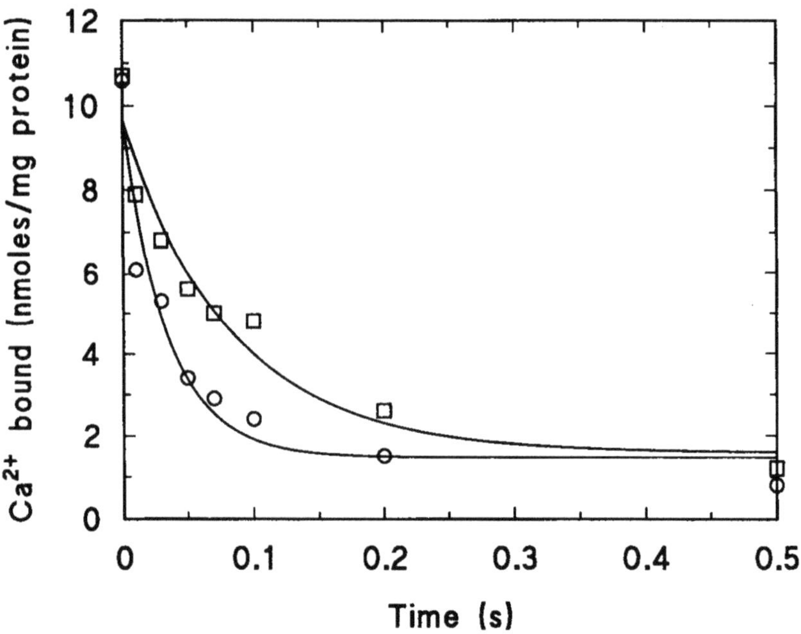

Figure 8. Effect of peptide PLN(1–25) on the ATP-induced rate of release of $^{45}Ca^{2+}$ from the ATPase. SR vesicles (1.2 μM ATPase) were first equilibrated in pH 6.0 buffer (150 mM Mes/Tris, 20 mM Mg^{2+}) containing 100 μM $^{45}Ca^{2+}$ and 0.5 mM [^{3}H] sucrose in the presence of 4% (w/w protein) A23187 either in the absence (○) of peptide or in the presence (□) of 25 μM PL(1–25). Vesicles were then adsorbed onto Millipore filters and the filters perfused for the given times with the same buffer containing 100 μM unlabeled Ca^{2+}, 0.5 mM [^{3}H] sucrose and 2 mM ATP; in the absence (○) or presence (□) of PLN(1–25) (Hughes et al., 1994a).

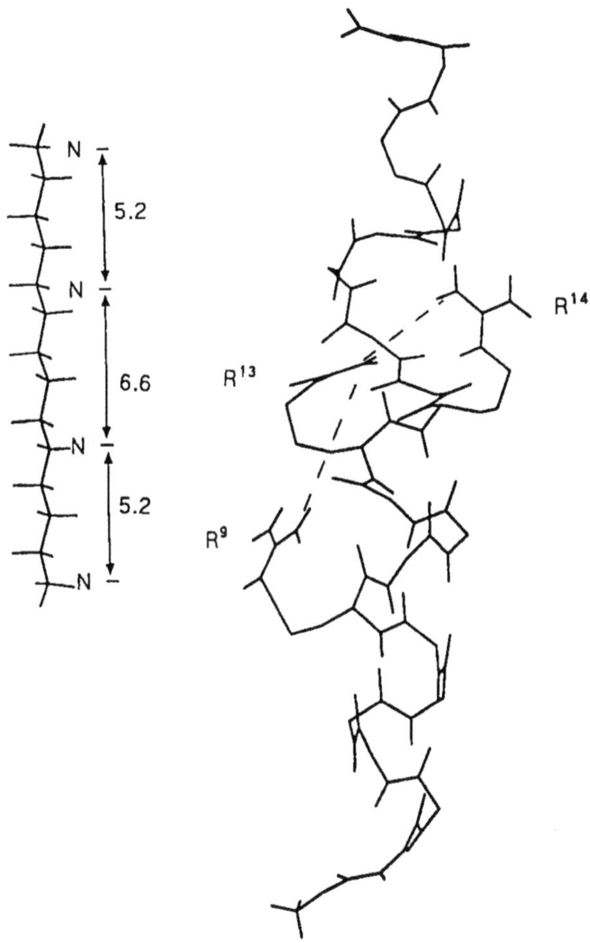

Figure 9. The structures of spermine and the hydrophilic region of phospholamban. Shown is a comparison of the structures of spermine (left) and residues 2–23 of phospholamban (right). Spermine is shown in a fully extended conformation, with distances between nitrogen atoms given in Å. For phospholamban, just the α-carbon backbone and residues Arg-9, Arg-13, and Arg-14 are shown for clarity. Residues 2–23 of phospholamban have been modeled as a slightly distorted α-helix, in an energy-minimized conformation with separations between nitrogens of residues Arg-9 and Arg-13, and Arg-13 and Arg-14 (as shown by the dotted lines) of 6.8 and 5.2 Å, respectively (Hughes et al., 1994a).

PLN(1–25) was also found to shift the E2–E1 equilibrium for the ATPase towards E1 (Hughes et al., 1994a). The observed shift was consistent with stronger binding of PLN(1–25) to the E1 conformation of the ATPase than to the E2 conformation: spermine was also found to bind more strongly to the E1 conformation of the ATPase (Hughes et al., 1994b).

The inhibition of the $E1PCa_2 \rightarrow E2P$ step by spermine, poly-L-Arg, and PLN(1–25) suggests that cationic sites on phospholamban are involved in the interaction (Hughes et al., 1994a, b). Since the peptides PLN(13–25) (Hughes et al., 1994a) and PLN(8–47) (Sasaki et al., 1992) had no detectable effects on ATPase activity, cationic residues in the region 1–12 are presumably important. Since spermine and spermidine inhibit ATPase activity with similar potencies, but N^1-acetylspermidine causes no inhibition (Hughes et al., 1994b), it is likely that binding of spermine to the ATPase involves three of the four nitrogen centers (Figure 9). As discussed above, it has been suggested, that the hydrophilic domain of PLN between residues 1 and 21 can adopt an α-helical conformation (Tada, 1992). As shown in Figure 9, this region of phospholamban can be modeled as a slightly distorted α-helix with a separation between the nitrogens of Arg-9, Arg-13, and Arg-14 equal to that between the terminal and central nitrogens of spermine. If the less ordered region between Gln-22 and Asn-30 is fully extended, it would have an end-to-end length of 30 Å. The separation between Pro-21 and Arg-9 in the structure shown in Figure 9 is 17 Å. Thus, the maximum height above the membrane surface of the binding site region proposed in Figure 9 would be ca. 47 Å. Heights of Cys-344, Glu-439, and Cys-670/Cys-674 above the membrane surface have been estimated to be 45-, 70-, and 54-Å, respectively (Mata et al., 1993; Stefanova et al., 1993; Baker et al., 1994). Thus, the proposed structure for the binding-site region on PLN is compatible with the suggested sites of interaction on the ATPase.

VI. EFFECTS OF PHOSPHOLIPIDS

The predominant phospholipids in cardiac SR are the zwitterionic phosphatidylcholine and phosphatidylethanolamine, with relatively small amounts of the negatively charged phosphatidylserine and phosphatidylinositol (Owens et al., 1973; Jakab and Kranias, 1988; Table 1). A membrane fraction enriched in PLN prepared by detergent treatment of SR followed by sulfhydryl-group affinity chromatography was found to contain a much higher fraction of negatively charged phospholipids (Table 1). This suggests preferential interaction between negatively charged phospholipids and PLN. The fatty acid composition of the PLN-enriched fraction was fairly typical of most biological membranes, being ca. 40% saturated (predominantly palmitic acid (C16:0), and stearic acid (C18:0)) with the major unsaturated fatty acids being oleic acid (C18:1) and linoleic acid (C18:2), but with a higher than normal (25%) fraction of linolenic acid (C18:3) (Jakab and Kranias, 1988). Stimulation of cAMP-dependent protein kinases led to phosphorylation of

Table 1. Phospholipid Composition of Cardiac SR and Purified PLN and Ca^{2+}-ATPase Fractions[a]

	% Composition		
Phospholipid	*Cardiac SR*	*PLN-Enriched Fraction*	*Purified Ca^{2+}-ATPase*
Phosphatidylcholine	61	22	48
Phosphatidylethanolamine	23	9	32
Phosphatidylinositol	9	13	13
Phosphatidylserine	3	34	—
Sphingomyelin	2	17	3

Note: [a]Jakab and Kranias, 1988; Kim et al., 1990

the phosphatidylinositiol to phosphatidylinositiol 4-monophosphate and phosphatidylinositiol 4,5-bisphosphate (Jakab and Kranias, 1988).

A Ca^{2+}-ATPase preparation purified from cardiac SR was also reported to have a phospholipid composition distinct from that of the native membrane with a lower content of phosphatidylcholine and a higher content of phosphatidylethanolamine (Kim et al., 1990; Table 1). The level of phosphatidylinositol associated with the purified Ca^{2+}-ATPase preparation was also higher than that present in the SR membrane. However, this phosphatidylinositol was not phosphorylated by cAMP-dependent protein kinase (Kim et al., 1990).

Possible effects of phospholipid composition on ATPase activity are controversial. Kinsella's group reported that the lipids of mouse cardiac SR could be enriched in ω-3 and ω-6 polyunsaturated fatty acids by dietary means (Croset et al., 1989; Swanson et al., 1989). It was reported that these changes resulted in no change in ATPase activity of cardiac SR vesicles, but did lead to a decrease in Ca^{2+} uptake, attributed to oxidative damage of membranes containing a high proportion of polyunsaturated fatty-acyl chains (Croset et al., 1989). However, it was also reported that both ATPase activity and Ca^{2+} accumulation were decreased (Swanson et al., 1989). Taffet et al. (1993) found that the lipids of rat cardiac SR could also be enriched in ω-3 and ω-6 polyunsaturated fatty acids by dietary means and reported decreases in both Ca^{2+} accumulation and ATPase activity.

REFERENCES

Arkin, I. T., Rothman, M., Ludlam, C. F. C., Aimoto, S., Engelman, D. M., Rothschild, K. J., & Smith, S. O. (1995). Structural model of the phospholamban ion channel complex in phospholipid membranes. J. Mol. Biol. 248, 824–834.

Baker, K. J., East, J. M., & Lee, A. G. (1994). Localization of the hinge region of the Ca^{2+}-ATPase of sarcoplasmic reticulum using resonance energy transfer. Biochim. Biophys. Acta 1192, 53–60.

Baltas, L. G., Karczewski, P., & Krause, E. G. (1995). The cardiac sarcoplasmic reticulum phospholamban kinase is a distinct delta-CaM kinase isozyme. FEBS. Lett. 373, 71–75.

Bick, R. J., Van Winkle, W. B., Tate, C. A., & Entman, M. L. (1983). Nucleotide triphosphate utilization by cardiac and skeletal muscle sarcoplasmic reticulum. Further evidence for an alternative substrate hydrolysis cycle and the effect of calcium NTPase purification. J. Biol. Chem. 258, 4447–4452.

Brandl, C. J., Green, N. M., Korczak, B., & MacLennan, D. H. (1986). Two Ca^{2+}-ATPase genes: Homologies and mechanistic implications of deduced amino acid sequences. Cell 44, 597–607.

Briggs, F. N., Lee, K. F., Wechsler, A. W., & Jones, L. R. (1992). Phospholamban expressed in slow-twitch and chronically stimulated fast-twitch muscles minimally affects calcium affinity of sarcoplasmic reticulum Ca^{2+}-ATPase. J. Biol. Chem. 267, 26056–26061.

Burk, S. E., Lytton, J., MacLennan, D. H., & Shull, G. E. (1989). cDNA cloning, functional expression, and mRNA tissue distribution of a third organellar Ca^{2+} pump. J. Biol. Chem. 264, 18561–18568.

Cantilina, T., Sagara, Y., Inesi, G., & Jones, L. R. (1993). Comparative studies of cardiac and skeletal sarcoplasmic reticulum ATPases — effect of a phospholamban antibody on enzyme activation by Ca^{2+}. J. Biol. Chem. 268, 17018–17025.

Chamberlain, B. K., Levitsky, D. O., & Fleischer, S. (1983). Isolation and characterization of canine cardiac sarcoplasmic reticulum with improved Ca^{2+} transport properties. J. Biol. Chem. 258, 6602–6609.

Chamberlain, B. K., Volpe, P., & Fleischer, S. (1984). Inhibition of calcium-induced calcium release from purified cardiac sarcoplasmic reticulum vesicles. J. Biol. Chem. 259, 7547–7553.

Chiesi, M., & Schwaller, R. (1989). Involvement of electrostatic phenomena in phospholamban-induced stimulation of Ca^{2+} uptake into cardiac sarcoplasmic reticulum. FEBS Lett. 244, 241–244.

Colyer, J., & Wang, J. H. (1991). Dependence of cardiac sarcoplasmic reticulum calcium pump activity on the phosphorylation status of phospholamban. J. Biol. Chem. 266, 17486–17493.

Corbalan-Garcia, S., Teruel, J. A., & Gomez-Fernandez, J. C. (1992). Characterization of ruthenium red-binding sites of the Ca^{2+}-ATPase from sarcoplasmic reticulum and their interaction with Ca^{2+} binding sites. Biochem. J. 287, 767–774.

Croset, M., Black, J. M., Swanson, J. E., & Kinsella, J. E. (1989). Effects of dietary n-3 polyunsaturated fatty acids on phospholipid composition and calcium transport in mouse cardiac sarcoplasmic reticulum. Lipids 24, 278–285.

Fujii, J., Kadoma, M., Tada, M., Toda, H., & Sakiyama, F. (1986). Characterisation of structural unit of phospholamban by amino acid sequencing and electrophoretic analysis. Biochem. Biophys. Res. Commun. 138, 1044–1050.

Fujii, J., Ueno, A., Kitano, K., Tanaka, S., Kadoma, M., & Tada, M. (1987). Complete complementary DNA-derived amino acid sequence of canine cardiac phospholamban. J. Clin. Invest. 79, 301–304.

Fujii, J., Lytton, J., Tada, M., & MacLennan, D. H. (1988). Rabbit cardiac and slow-twitch muscle express the same phospholamban gene. FEBS Lett. 227, 51–55.

Fujii, J., Maruyama, K., Tada, M., & MacLennan, D. H. (1989). Expression and site-specific mutagenesis of phospholamban: Studies of residues involved in phosphorylation and pentamer formation. J. Biol. Chem. 264, 12950–12955.

Fujii, J., Zarainherzberg, A., Willard, H. F., Tada, M., & MacLennan, D. H. (1991). Structure of the rabbit phospholamban gene, cloning of the human cDNA, and assignment of the gene to human chromosome- 6. J. Biol. Chem. 266, 11669–11675.

Gupta, R. C., & Kranias, E. G. (1989). Purification and characterization of a calcium–calmodulin-dependent phospholamban kinase from canine myocardium. Biochemistry 28, 5909–5916.

Hubbard, M. J., Dent, P., Smythe, C., & Cohen, P. (1990). Targetting of protein phosphatase 1 to the sarcoplasmic reticulum of rabbit skeletal muscle by a protein that is very similar or identical to the G subunit that directs the enzyme to glycogen. Eur. J. Biochem. 189, 243–249.

Huggins, J. P., & England, P. J. (1987). Evidence for a phosphorylation-induced conformational change in phospholamban from effects of three proteases. FEBS Lett. 217, 32–36.

Hughes, G., East, J. M., & Lee, A. G. (1994a). The hydrophilic domain of phospholamban inhibits the Ca^{2+}-ATPase transport step of the Ca^{2+}-ATPase. Biochem. J. 303, 511–516.

Hughes, G., Starling, A. P., East, J. M., & Lee, A. G. (1994b). The mechanism of inhibition of the Ca^{2+}-ATPase by spermine and other polycationic compounds. Biochemistry.33, 4745–4754.

Iwasa, Y., & Hosey, M. M. (1984). Phosphorylation of cardiac sarcolemma proteins by the calcium-activated phospholipid-dependent protein kinase. J. Biol. Chem. 259, 534–540.

Jakab, G., & Kranias, E. G. (1988). Phosphorylation and dephosphorylation of purified phospholamban and associated phosphoinositides. Biochemistry 27, 3799–3806.

James, P., Inui, M., Tada, M., Chiesi, M., & Carafoli, E. (1989a). Nature and site of phospholamban regulation of the Ca^{2+} pump of sarcoplasmic reticulum. Nature 342, 90–92.

James, P. H., Pruschy, M., Vorherr, T. E., Penniston, J. T., & Carafoli, E. (1989b). Primary structure of the cAMP-dependent phosphorylation site of the plasma-membrane calcium pump. Biochemistry 28, 4253–4258.

Jones, L. R. (1979). Mg^{2+} and ATP effects on K^+ activation of the Ca^{2+}-transport ATPase of cardiac sarcoplasmic reticulum. Biochim. Biophys. Acta 557, 230–242.

Jorgensen, A. O., & Jones, L. R. (1986). Localization of phospholamban in slow but not fast canine skeletal muscle fibers. An immunological and biochemical study. J. Biol. Chem. 261, 3775–3781.

Kim, H. W., Steenaart, N. A., Ferguson, D. G., & Kranias, E. G. (1990). Functional reconstitution of the cardiac sarcoplasmic reticulum Ca^{2+}-ATPase with phospholamban in phospholipid vesicles. J. Biol. Chem. 265, 1702–1709.

Kimura, Y., Inui, M., Kadoma, M., Kijima, Y., Sasaki, T., & Tada, M. (1991). Effects of monoclonal antibody against phospholamban on calcium pump ATPase of cardiac sarcoplasmic reticulum. J. Mol. Cell Cardiol. 23, 1223–1230.

Kimura, M., Yamaguchi, Y., Takada, S., & Tanabe, K. (1993). Cloning of a Ca^{2+}-ATPase gene of *Plasmodium falciparum* and comparison with vertebrate Ca^{2+}-ATPases. J. Cell Sci. 104, 1129–1136.

Kirby, M. S., Sagara, Y., Gaa, S., Inesi, G., Lederer, W. J., & Rogers, T. B. (1992). Thapsigargin inhibits contraction and Ca^{2+} transient in cardiac cells by specific inhibition of the sarcoplasmic reticulum Ca^{2+} pump. J. Biol. Chem. 267, 12545–12551.

Kirchberger, M. A., & Tada, M. (1976). Effects of adenosine 3′:5′-monophosphate-dependent protein kinase on sarcoplasmic reticulum isolated from cardiac and slow and fast contracting skeletal muscles. J. Biol. Chem. 251, 729–735.

Kirchberger, M. A., & Antonetz, T. (1982). Calmodulin-mediated regulation of calcium transport and (Ca–Mg)-activated ATPase activity in isolated cardiac sarcoplasmic reticulum. J. Biol. Chem. 257, 5685–5691.

Kirchberger, M. A., Borchman, D., & Kasinathan, C. (1986). Proteolytic activation of the canine cardiac sarcoplasmic reticulum calcium pump. Biochemistry 25, 5484–5492.

Knighton, D. R., Zheng, J., Ten Eyck, L. F., Xuong, N.-H., Taylor, S., & Sowadski, J. M. (1991). Structure of a peptide inhibitor bound to the catalytic subunit of cyclic adenosine-monophosphate dependent protein kinase. Science 253, 414–420.

Komuro, I., Kurabayashi, M., Shibazaki, Y., Takaku, F., & Yazaki, Y. (1989). Molecular cloning and characterization of a Ca^{2+} + Mg^{2+}-dependent adenosine triphosphatase from rat cardiac sarcoplasmic reticulum. Regulation of its expression by pressure overload and developmental stage. J. Clin. Invest. 83, 1102–1108.

Kovacs, R. J., Nelson, M. T., Simmerman, H. K. B., & Jones, L. R. (1988). Phospholamban forms Ca^{2+} selective channels in lipid bilayers. J. Biol. Chem. 263, 18364–18368.

Kranias, E. G., Mandel, F., Wang, T., & Schwartz, A. (1980). Mechanism of the stimulation of calcium ion dependent adenosine triphosphatase of cardiac sarcoplasmic reticulum by adenosine 3′,5′-monophosphate dependent protein kinase. Biochemistry 19, 5434–5439.

Leberer, E., Hartner, K. T., Brandl, C. J., Fujii, J., Tada, M., MacLennan, D. H., & Pette, D. (1989). Slow/cardiac sarcoplasmic reticulum Ca^{2+}-ATPase and phospholamban mRNAs are expressed in chronically stimulated rabbit fast-twich muscle. Eur. J. Biochem. 185, 51–54.

Li, C. F., Wang, J. H., & Colyer, J. (1990). Immunological detection of phospholamban phosphorylation states facilitates the description of the mechanism of phosphorylation and dephosphorylation. Biochemistry 29, 4535–4540.

Lu, Y. Z., Xu, Z. C., & Kirchberger, M. A. (1993). Evidence for an effect of phospholamban on the regulatory role of ATP in calcium uptake by the calcium pump of the cardiac sarcoplasmic reticulum. Biochemistry 32, 3105–3111.

Luo, W. S., Grupp, I. L., Harrer, J., Ponniah, S., Grupp, G., Duffy, J. J., Doetschman, T., & Kranias, E. G. (1994). Targeted ablation of the phospholamban gene is associated with markedly enhanced myocardial contractility and loss of beta-agonist stimulation. Circ. Res. 75, 401–409.

MacDougall, L. K., Jones, L. R., & Cohen, P. (1991). Identification of the major protein phosphatases in mammalian cardiac muscle which dephosphorylate phospholamban. Eur. J. Biochem. 196, 725–734.

Mandel, F., Kranias, E. G., de Grassi, G. E., Sumida, M., & Schwartz, A. (1982). The effect of pH on the transient-state kinetics of Ca^{2+}–Mg^{2+}-ATPase of cardiac sarcoplasmic reticulum. A comparison with skeletal sarcoplasmic reticulum. Circ. Res. 50, 310–317.

Mata, A. M., Matthews, I., Tunwell, R. E. A., Sharma, R. P., Lee, A. G., & East, J. M. (1992). Definition of surface exposed and trans-membranous regions of the (Ca^{2+}–Mg^{2+})-ATPase of sarcoplasmic reticulum using anti-peptide antibodies. Biochem. J. 286, 567–580.

Mata, A. M., Stefanova, H. I., Gore, M. G., East, J. M., & Lee, A. G. (1993). Localization of Cys-344 on the (Ca^{2+}–Mg^{2+})-ATPase of sarcoplasmic reticulum using fluorescence energy transfer. Biochim. Biophys. Acta 1147, 6–12.

Matthews, I., Colyer, J., Mata, A. M., Green, N. M., Sharma, R. P., Lee, A. G., & East, J. M. (1989). Evidence for the cytoplasmic location of the N- and C-terminal segments of sarcoplasmic reticulum (Ca^{2+}–Mg^{2+})-ATPase. Biochem. Biophys. Res. Commun. 161, 683–688.

Morris, G. L., Cheng, H. C., Colyer, J., & Wang, J. H. (1991). Phospholamban regulation of cardiac sarcoplasmic reticulum (Ca^{2+}–Mg^{2+})-ATPase. Mechanism of regulation and site of monoclonal antibody interaction. J. Biol. Chem. 266, 11270–11275.

Mortishire-Smith, R. J., Pitzenberger, S. M., Burke, C. J., Middaugh, C. R., Garsky, V. M., & Johnson, R. G. (1995). Solution structure of the cytoplasmic domain of phospholamban: Phosphorylation leads to a local perturbation in secondary structure. Biochemistry 34, 7603–7613.

Moutin, M. J., Rapin, C., & Dupont, Y. (1992). Ruthenium red affects the intrinsic fluorescence of the calcium-ATPase of skeletal muscle sarcoplasmic reticulum. Biochim. Biophys. Acta 1100, 321–328.

Movsesian, M. A., Nishikawa, M., & Adelstein, R. S. (1984). Phosphorylation of phospholamban by calcium-activated, phospholipid-dependent protein kinase. J. Biol. Chem. 259, 8029–8032.

Ogurusu, T., Wakabayashi, S., Watanabe, T., & Shigekawa, M. (1989). Guanosine triphosphate utilization by canine cardiac muscle sarcoplasmic reticulum. J. Biochem. 106, 599–605.

Owens, K., Weglicki, W. B., Ruth, R. C., Stam, A. C., & Sonnenblick, E. H. (1973). Lipid composition, Ca^{2+} uptake and Ca^{2+} stimulated ATPase activity of sarcoplasmic reticulum of the cardiomyopathic hamster. Biochim. Biophys. Acta 296, 71–78.

Palmero, I., & Sastre, L. (1989). Complementary DNA cloning of a protein highly homologous to mammalian sarcoplasmic reticulum Ca^{2+}-ATPase from the crustacean *Artemia*. J. Mol. Biol. 210, 737–748.

Quirk, P. G., Patchell, V. B., Colyer, J., Drago, G. A., & Gao, Y. (1996). Conformational effects of serine phosphorylation in phospholamban peptides. Eur. J. Biochem. 236, 85–91.

Raeymaekers, L., & Jones, L. R. (1986). Evidence for the presence of phospholamban in the endoplasmic reticulum of smooth muscle. Biochim. Biophys. Acta 882, 258–265.

Raeymaekers, L., Hofmann, F., & Casteels, R. (1988). Cyclic GMP-dependent protein kinase phosphorylates phospholamban in isolated sarcoplasmic reticulum from cardiac and smooth muscle. Biochem. J. 252, 269–273.

Sasaki, T., Inui, M., Kimura, Y., Kuzuya, T., & Tada, M. (1992). Molecular mechanism of regulation of Ca^{2+} pump ATPase by phospholamban in cardiac sarcoplasmic reticulum. Effects of synthetic phospholamban peptides on Ca^{2+} pump ATPase. J. Biol. Chem. 267, 1674–1679.

Shigekawa, M., Finegan, J. A., & Katz, A. M. (1976). Calcium transport ATPase of canine cardiac sarcoplasmic reticulum. A comparison with that of rabbit fast skeletal muscle sarcoplasmic reticulum. J. Biol. Chem. 251, 6894–6900.

Simmerman, H. K. B., Collins, J. H., Theibert, J. L., Wegner, A. D., & Jones, L. R. (1986). Sequence analysis of phospholamban. J. Biol. Chem. 261, 13333–13341.

Simmerman, H. K. B., Lovelace, D. E., & Jones, L. R. (1989). Secondary structure of detergent-solubilized phospholamban, a phosphorylatable, oligomeric protein of cardiac sarcoplasmic reticulum. Biochim. Biophys. Acta 997, 322–329.

Simmerman, H. K. B., Kobayashi, Y. M., Autry, J. M., & Jones, L. R. (1996). A leucine zipper stabilizes the pentameric membrane domain of phospholamban and forms a coiled-coil pore structure. J. Biol. Chem. 271, 5941–5946.

Starling, A. P., Hughes, G., Sharma, R. P., East, J. M., & Lee, A. G. (1995). The hydrophilic domain of phospholamban inhibits the Ca^{2+}-ATPase—the importance of the method of assay. Biochem. Biophys. Res. Commun. 215, 1067–1079.

Steenaart, N. A., Ganim, J. R., Di-Salvo, J., & Kranias, E. G. (1992). The phospholamban phosphatase associated with cardiac sarcoplasmic reticulum is a type 1 enzyme. Arch. Biochem. Biophys. 293, 17–24.

Stefanova, H. I., East, J. M., & Lee, A. G. (1991). Covalent and non-covalent inhibitors of the phosphate transporter of sarcoplasmic reticulum. Biochim. Biophys. Acta 1064, 321–328.

Stefanova, H. I., Mata, A. M., Gore, M. G., East, J. M., & Lee, A. G. (1993). Labeling the (Ca^{2+}–Mg^{2+})-ATPase of sarcoplasmic reticulum at Glu-439 with 5-(bromomethyl)-fluorescein. Biochemistry 32, 6095–6103.

Sumida, M., Wang, T., Mandel, F., Froehlich, J. P., & Schwartz, A. (1978). Transient kinetics of Ca^{2+} transport of sarcoplasmic reticulum. A comparison of cardiac and skeletal muscle. J. Biol. Chem. 253, 8772–8777.

Suzuki, T., & Wang, J. H. (1986). Stimulation of bovine cardiac sarcoplasmic reticulum Ca^{2+}-pump and blocking of phospholamban phosphorylation and dephosphorylation by a phospholamban monoclonal antibody. J. Biol. Chem. 261, 7018–7023.

Swanson, J. E., Lokesh, B. R., & Kinsella, J. E. (1989). Ca^{2+}–Mg^{2+} ATPase of mouse cardiac sarcoplasmic reticulum is affected by membrane n-6 and n-3 polyunsaturated fatty acid content. J. Nutr. 119, 364–372.

Szymanska, G., Kim, H. W., Cuppoletti, J., & Kranias, E. G. (1990). Regulation of the skeletal sarcoplasmic reticulum Ca^{2+} pump by phospholamban in reconstituted phospholipid vesicles. Membr. Biochem. 9, 191–202.

Tada, M. (1992). Molecular structure and function of phospholamban in regulating the calcium pump from sarcoplasmic reticulum. Ann. N. Y. Acad. Sci. 671, 92–103.

Tada, M., & Katz, A. M. (1982). Phosphorylation of the sarcoplasmic reticulum and sarcolemma. Annu. Rev. Physiol. 44, 401–423.

Tada, M., Kirchberger, M. A., Repke, D. I., & Katz, A. M. (1974). The stimulation of calcium transport in cardiac sarcoplasmic reticulum by adenosine 3′:5′-monophosphate-dependent protein kinase. J. Biol. Chem. 249, 6174–6180.

Tada, M., Kirchberger, M. A., & Katz, A. M. (1975). Phosphorylation of a 22,000-dalton component of the cardiac sarcoplasmic reticulum by adenosine 3′:5′-monophosphate-dependent protein kinase. J. Biol. Chem. 250, 2640–2647.

Tada, M., Ohmori, F., Yamada, M., & Abe, H. (1979). Mechanism of the stimulation of Ca dependent ATPase of cardiac sarcoplasmic reticulum by adenosine 3′:5′-monophosphate-dependent protein kinase. J. Biol. Chem. 254, 319–326.

Tada, M., Yamada, M., Ohmori, F., Kuzuya, T., Inui, M., & Abe, H. (1980). Transient state kinetic studies of Ca^{2+}-dependent ATPase and calcium transport by cardiac sarcoplasmic reticulum. J. Biol. Chem. 255, 1985–1992.

Taffet, G. E., Pham, T. T., Bick, D. L. M., Entman, M. L., Pownall, H. J., & Bick, R. J. (1993). The calcium uptake of the rat heart sarcoplasmic reticulum is altered by dietary lipid. J. Membrane. Biol. 131, 35–42.

Tate, C. A., Shin, G., Walseth, T. F., Taffet, G. E., Bick, R. J., & Entman, M. L. (1991). Nucleotide specificity of cardiac sarcoplasmic reticulum. J. Biol. Chem. 276, 16165–16170.

Tatulian, S. A., Jones, L. R., Reddy, L. G., Stokes, D. L., & Tamm, L. K. (1995). Secondary structure and orientation of phospholamban reconstituted in supported bilayers from polarized attenuated total reflection FTIR spectroscopy. Biochemistry 34, 4448–4456.

Terzi, E., Poteur, L., & Trifilieff, E. (1992). Evidence for a phosphorylation-induced conformational change in phospholamban cytoplasmic domain by CD analysis. FEBS Lett. 309, 413–416.

Toyofuku, T., & Zak, R. (1991). Characterization of cDNA and genomic sequences encoding a chicken phospholamban. J. Biol. Chem. 266, 5375–5383.

Toyofuku, T., Kurzydlowski, K., Tada, M., & MacLennan, D. H. (1993). Identification of regions in the Ca^{2+}-ATPase of sarcoplasmic reticulum that affect functional association with phospholamban. J. Biol. Chem. 268, 2809–2815.

Toyofuku, T., Kurzydlowski, K., Tada, M., & MacLennan, D. H. (1994a). Amino acids Glu2 to Ile18 in the cytoplasmic domain of phospholamban are essential for functional association with the Ca^{2+}-ATPase of sarcoplasmic reticulum. J. Biol. Chem. 269, 3088–3094.

Toyofuku, T., Tada, M., & MacLennan, D. H. (1994b). Amino acids Lys–Asp–Asp–Lys-400 in the cardiac Ca^{2+}-ATPase (SERCA2) are critical or functional association with phospholamban. Biophys. J. 66, A121.

Van Winkle, W. B., Tate, C. A., Bick, R. J., & Entman, M. L. (1981). Nucleotide triphosphate utilization by cardiac and skeletal muscle sarcoplasmic reticulum. Evidence for a hydrolysis cycle not coupled to intermediate acyl phosphate formation and calcium translocation. J. Biol. Chem. 256, 2268–2274.

Verboomen, H., Wuytack, F., Eggermont, J. A., De Jaegere, S., Missiaen, L., Raeymaekers, L., & Casteels, R. (1989). cDNA cloning and sequencing of phospholamban from pig stomach smooth muscle. Biochem. J. 262, 353–356.

Verboomen, H., Wuytack, F., Desmedt, H., Himpens, B., & Casteels, R. (1992). Functional difference between SERCA2a and SERCA2b Ca^{2+} pumps and their modulation by phospholamban. Biochem. J. 286, 591–595.

Vorherr, T., Chiesi, M., Schwaller, R., & Carafoli, E. (1992). Regulation of the calcium ion pump of sarcoplasmic reticulum: Reversible inhibition by phospholamban and by the calmodulin binding domain of the plasma membrane calcium ion pump. Biochemistry 31, 371–376.

Wang, T., Tsai, L. I., Solaro, R. J., de Grassi, G. E., & Schwartz, A. (1979). Effects of potassium on vanadate inhibition of sarcoplasmic reticulum Ca^{2+}-ATPase from dog cardiac and rabbit skeletal muscle. Biochem. Biophys. Res. Commun. 91, 356–361.

Wang, T., de Grassi, G. E., Tsai, L. I., & Schwartz, A. (1981). Influence of monovalent cations on the Ca^{2+}-ATPase of sarcoplasmic reticulum isolated from rabbit skeletal and dog cardiac muscles. An interpretation of transient-state kinetic data. Biochim. Biophys. Acta 637, 523–529.

Wegener, A. D., Simmerman, H. K., Liepnieks, J., & Jones, L. R. (1986). Proteolytic cleavage of phospholamban purified from canine cardiac sarcoplasmic reticulum vesicles. Generation of a low resolution model of phospholamban structure. J. Biol. Chem. 261, 5154–5159.

Wimmers, L. E., Ewing, N. N., & Bennett, A. B. (1992). Higher plant Ca^{2+}-ATPase : primary structure and regulation of mRNA abundance by salt. Proc. Natl. Acad. Sci. USA 89, 9205–9209.

Xu, Z. C., & Kirchberger, M. A. (1989). Modulation by polyelectrolytes of canine cardiac microsomal calcium uptake and the possible relationship to phospholamban. J. Biol. Chem. 264, 16644–16651.

Xu, A., Hawkins, C., & Narayanan, N. (1993). Phosphorylation and activation of the Ca^{2+}-pumping ATPase of cardiac sarcoplasmic reticulum by Ca^{2+}/calmodulin-dependent protein kinase. J. Biol. Chem. 268, 8394–8397.

Young, E. F., McKee, M. J., Ferguson, D. G., & Kranias, E. G. (1989). Structural characterization of phospholamban in cardiac sarcoplasmic reticulum membranes by cross-linking. Membr. Biochem. 8, 95–106.

THE CALCIUM PUMP OF PLASMA MEMBRANES

Joachim Krebs and Danilo Guerini

Biomembranes
Volume 5, pages 101–131.

ISBN: 1-55938-662-2.

I. INTRODUCTION

Calcium plays a pivotal role in biological systems. Inside cells it is one of the second messengers which requires the maintenance of a very low cytosolic Ca^{2+}-concentration, that is, 100–200 nM in a resting cell. Since in the extracellular space or intracellularly in the reticular systems the Ca^{2+}-concentration is in the mM range and the permeability of these membranes for Ca^{2+} is carefully controlled, it results in a very steep ion-gradient over the plasma or reticular membrane. Therefore, several transmembrane Ca^{2+}-transporting systems participate in controlling the free Ca^{2+} concentration in the cell including Ca^{2+}-channels, a Ca^{2+}-pump, and a Na^{+}/Ca^{2+}-exchanger in the plasma membrane; a Ca^{2+}-pump and Ca^{2+}-release channels in the sarco(endo)plasmic reticulum, and Ca^{2+}-uptake and -release systems in mitochondria. Recently, evidence has accumulated that Ca^{2+}-transport systems also exist in the nuclear envelope, but the characterization of the latter system is still fragmentary.

In this review, we will focus on the Ca^{2+}-pump of the plasma membrane and will concentrate on the molecular characterization of the structural and functional aspects of this enzyme. We will give brief accounts of the mechanistic and regulatory properties and will discuss, in detail, the structure and distribution of the different isoforms resulting from different genes or due to alternative splicing. A number of comprehensive reviews dealing with various aspects of the plasma membrane Ca^{2+}-pump have appeared recently (Schatzmann, 1982; Penniston, 1983; Carafoli, 1991, 1992; Carafoli and Guerini, 1993).

II. HISTORICAL ASPECTS AND GENERAL PROPERTIES

In 1961, Dunham and Glynn first described a Ca^{2+}-dependent ATPase in erythrocyte membranes (Dunham and Glynn, 1961), but it was Schatzmann in 1966 who provided evidence that Ca^{2+} is pumped out of the cell on the expense of ATP against a Ca^{2+}-gradient across the membrane (Schatzmann, 1966). The general reaction mechanism of the enzyme was first described in 1969 by Schatzmann and Vincenzi (1969). Like most of the ion pumps, the enzyme belongs to the P-type pump family as classified by Pedersen and Carafoli (1987a, b), that is, these enzymes are characterized by forming a phosphorylated high-energy intermediate. The ion-mediated high-affinity binding of ATP by these enzymes results in the formation of an acyl-phosphate, usually an aspartylphosphate, which provides the enzyme with sufficient energy to pump the ion across the membrane against the ion gradient. Therefore, the enzyme is thought to exist in at least two different conformational states, E_1 and E_2. As can be seen from Figure 1 a general reaction cycle can characterize the P-class ATPases, that is, after the ion is bound to the enzyme in its high-affinity E_1-form and the phosphorylated intermediate has been formed as indicated by E_1~P, the ion—here Ca^{2+}—is transported across the body of the ATPase to the other side of the membrane during the E_1~P to E_2~P transition. In

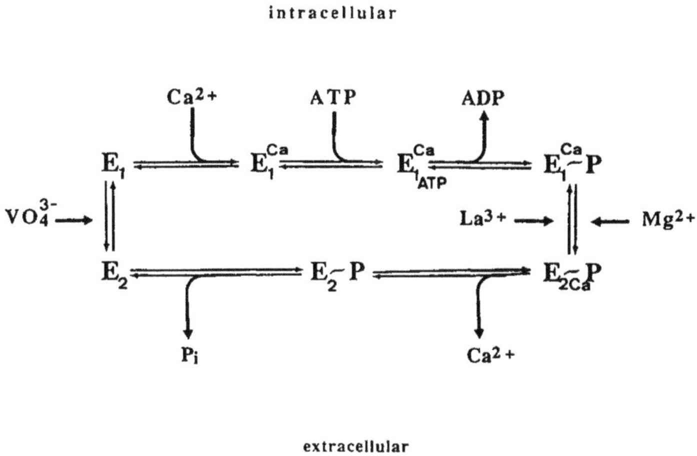

Figure 1. Schematic diagram of the reaction cycle of the Ca^{2+}-pump.

the E_2~P form, the enzyme is in a low-affinity state for the ion, that is, the ion can be released to the environment and after dephosphorylation the enzyme reverts back to the E_1 state. The difference between the two conformational states, E_1 and E_2, is reflected by changes in the secondary structure as demonstrated by circular dichroism and fluorescence spectroscopy (Krebs et al., 1987). It is interesting to note that the Ca^{2+}-bound form can exist in two different states depending on whether the phosphorylated intermediate has been formed or not (Krebs et al., 1987).

An important aspect for the validity of this reaction cycle is its reversibility. As it has been first shown for the homologous Ca^{2+}-pump of the sarcoplasmic reticulum by Makinose and Hasselbach (1971), the plasma membrane Ca^{2+}-pump in closed erythrocyte vesicles could also make use of a Ca^{2+}-gradient to produce ATP (Rossi et al., 1978; Wüthrich et al., 1979). According to a series of experiments performed by De Meis and co-workers (1980, 1982), the critical step to gain energy to form ATP comes from the solvation energy of the reactants P_i, ADP, and Ca^{2+}, that is, a transmembrane ion gradient is not necessary provided the water activity at the active site of the enzyme is reduced to a minimum. This was achieved by performing the reaction in the presence of an excess of dimethylsulfoxide. These conditions greatly facilitate the formation of E_2~P due to the reduction of the free energy of hydrolysis of the phosphorylated intermediate as compared in the presence of water. Conversion of E_2~P into the high-energy form E_1~P was then achieved by rehydration of the catalytic site in the presence of Ca^{2+}. Similar experiments following essentially the protocol of De Meis and co-workers succeeded in the formation of ATP by using the detergent-solubilized form of the purified Ca^{2+}-pump from erythrocyte membranes (Chiesi et al., 1984).

The Ca^{2+}-pump of plasma membranes has some interesting features which can help to identify the enzyme even in crude membranes and in the presence of the ER/SR Ca^{2+}-pump. Using ^{32}P γ-labeled ATP, the enzyme can be conveniently radioactively labeled which gives rise to a phosphoenzyme band on SDS-PAGE with an apparent molecular weight of 140,000 Da. The conversion of $E_1{\sim}P$ to $E_2{\sim}P$ is promoted by Mg^{2+}, that is, high concentrations of Mg^{2+} accelerate the dephosphorylation of the phosphoenzyme. On the other hand, low concentrations of the well known inhibitor, La^{3+} (Quist and Roufogalis, 1975), increases the steady-state level of phosphoenzyme formation of the Ca^{2+}-pump of plasma membranes. This is in contrast to the Ca^{2+}-pump of the sarco- or endoplasmic reticulum and this difference can be used as a convenient way to distinguish between the two enzymes from the reticular or plasma membranes on polyacrylamide gels applying crude membrane preparations (Wuytack et al., 1982).

Orthovanadate, a pentacoordinate analog of phosphate, is a common inhibitor of all P-type ion pumps. At the high-affinity site, it acts as a noncompetitive inhibitor of ATP, whereas at the low-affinity site, vanadate is a mixed, partly competitive inhibitor (Barrabin et al., 1980). In the case of the plasma membrane Ca^{2+}-pump the inhibitor constant K_i can be as low as 2–3 μM depending on the conditions; the ion composition of the medium, for example, concentrations of Na^+, K^+, and Mg^{2+}, are especially important. Concerning the reaction cycle (see Figure 1), it is generally assumed that vanadate interacts with the E_2 conformational state of the pump, thereby blocking the $E_2 \rightarrow E_1$ transition.

One important feature which is still debated is the stoichiometry between transported Ca^{2+} and hydrolyzed ATP. Thermodynamically, it would be conceivable that a stoichiometry of 2 also exists for the plasma membrane Ca^{2+}-pump as it has been shown for the enzyme of the sarcoplasmic reticulum (Hasselbach and Wass, 1982), but most experimental data indicate that the Ca^{2+}/ATP stoichiometry approaches 1 in the case of the enzyme of plasma membranes *in situ* as well as in a reconstituted system (Schatzmann, 1973; Niggli et al., 1981a; Clark and Carafoli, 1983). Results obtained with this enzyme reconstituted into liposomes indicate an electro-neutral Ca^{2+}/H^+ exchanger (Niggli et al., 1982). This view was recently challenged by Inesi and co-workers (Hao et al., 1994) who confirmed the Ca^{2+}/ATP stoichiometry to be 1, but reported the Ca^{2+}/H^+ exchange to be electrogenic. However, Hao et al. could not exclude the possibility that the Ca^{2+}/H^+ stoichiometry, that is, whether the exchange is electro-neutral or electrogenic, could critically depend on the method used to reconstitute the enzyme, that is, which detergent, which phospholipid composition, and which protein/lipid ratio was used to obtain optimal results. In other words, it could not be demonstrated unambiguously whether the Ca^{2+}/H^+ ratio leading to electrogenicity of the pump is an intrinsic property of the enzyme or whether it is influenced by the method of reconstitution.

III. ACTIVATION BY CALMODULIN OR BY OTHER MEANS

In 1973, Bond and Clough reported that a soluble activator extracted from erythrocytes could activate the Ca^{2+}-pump of these cells. This observation was corroborated in 1977 by Gopinath and Vincenzi and, independently, by Jarrett and Penniston reporting that the Ca^{2+}-pump was stimulated by the same protein originally described by Cheung and, independently, by Kakiuchi (1970) as the activator of the cyclic nucleotide phosphodiesterase. This activator later became known as the Ca^{2+}-binding protein, calmodulin (CaM), responsible for the regulation of many different enzymes and ubiquitous in all eucaryotic cells (for a recent review see Cohen and Klee, 1988). The interaction between CaM and the Ca^{2+}-pump of plasma membranes is direct and has some important consequences for the structural and functional properties of the enzyme as will be described in detail below. By contrast, the reticular Ca^{2+}-pump can be regulated by CaM only indirectly.

The influence of CaM on the functional properties of the Ca^{2+}-pump of plasma membranes is two-fold (see Table 1): it decreases the K_M-value for Ca^{2+} from about 20 μM to ca. 0.5 μM and increases the V_{Max} up to 10-fold in line with an increase of the turnover of the pump. These findings suggest that the plasma membrane Ca^{2+}-pump is involved in the fine and rapid tuning of the free Ca^{2+} concentration in resting cells due to its high Ca^{2+} affinity. On the other hand, this enzyme has a low capacity and is, therefore, not able to transport bulk quantities across membranes. This is in contrast to the Na^+/Ca^{2+}-exchanger, a transport system of plasma membranes which has a low affinity, but high transport capacity for Ca^{2+}.

The direct, Ca^{2+}-dependent interaction between CaM and the Ca^{2+}-pump of plasma membranes has been successfully exploited to isolate the enzyme in pure form from erythrocytes by affinity chromatography (Niggli et al., 1979). The

Table 1. General Properties of the Plasma Membrane Ca^{2+}-pump

Moleclar mass	ca. 135,000
Pump type	P-class
	(formation of a phosphorylated intermediate, i.e., aspartyl phosphate)
Ca^{2+}-affinity	>10 μM in the low-affinity, resting state
	<0.5 μM in the activated, high-affinity state
Activators	CaM (K_D ca. 1 nM)
	polyunsaturated fatty acids, negatively charged phospholipids, phosphorylation by PKA, PKC in at least 1 isoform, oligomerization
Inhibitors	Vanadate (K_i ca. 3–5 μM)
	La^{3+} (K_i ca. 1 μM)
Stoichiometry	Ca/ATP 1:1
Charge balance	Ca^{2+}/H^+ electroneutral?
Distribution	All eucaryotic cells

enzyme is a single polypeptide of apparent molecular mass of 138,000 Da on sodium dodecyl sulfate polyacrylamide gels (SDS-PAGE). The purified enzyme could be reconstituted into liposomes providing Ca^{2+}-transport activity which was only dependent on CaM if the phospholipids used for the reconstitution were neutral, for example, phosphatidylcholine and not acidic like phosphatidylserine (Niggli et al., 1981b; see following).

Next to CaM, the Ca^{2+}-pump can be activated by a variety of compounds or treatments which could well have a physiological meaning. On the other hand, some of these treatments helped to characterize functional domains of the enzyme in more detail as it is discussed below.

A. Phospholipids

Early experiments on the Ca^{2+}-pump of erythrocytes indicated the importance of the phospholipid environment for the activity of the enzyme, since treatment of the erythrocyte membrane with phospholipases inactivated the enzyme (Roelofsen and Schatzmann, 1977). In a detailed study by Ronner et al. (1977), evidence was provided that the Ca^{2+}-pump demonstrated a marked preference for negatively charged phospholipids, such as phosphatidylserine or phosphatidylinositol. In addition, polyunsaturated fatty acids could also reactivate the Ca^{2+}-ATPase indicating that the polar properties of the lipid environment could become important for the activity of the enzyme.

B. Proteolysis

In 1980, Taverna and Hanahan first showed that in isolated erythrocyte membranes the Ca^{2+}-pump could be activated by controlled proteolysis using trypsin or chymotrypsin as the proteolytic enzymes. In a more detailed study by Sarkadi and co-workers using inverted erythrocyte vesicles, it was established that treatment of those vesicles with trypsin could mimic the effect of CaM on the Ca^{2+}-pump and that the enzyme thus activated could no longer be stimulated by CaM (Enyedi et al., 1980). In view of later studies using controlled proteolysis on the purified Ca^{2+}-pump (see following), it was of interest to note that a ca. 30 kD fragment was removed which was thought to comprise a regulatory unit of the enzyme (Enyedi et al., 1980).

Trypsin or chymotrypsin have been instrumental in elucidating functional domains of the Ca^{2+}-pump, but obviously have no physiological meaning. However, application of the Ca^{2+}-dependent cellular protease calpain could be of physiological relevance. As has been shown for trypsin or chymotrypsin, calpain could increase the basal activity of the Ca^{2+}-pump, thereby removing a ca. 12 kD fragment (Au, 1986; Wang et al., 1988). Depending on the duration of the proteolysis, the high molecular fragment of ca. 124 kD could still bind CaM as demonstrated in gel-overlay experiments using ^{125}I-labeled CaM. This property gradually vanished with prolonged time of proteolysis concomitant with approaching the level of

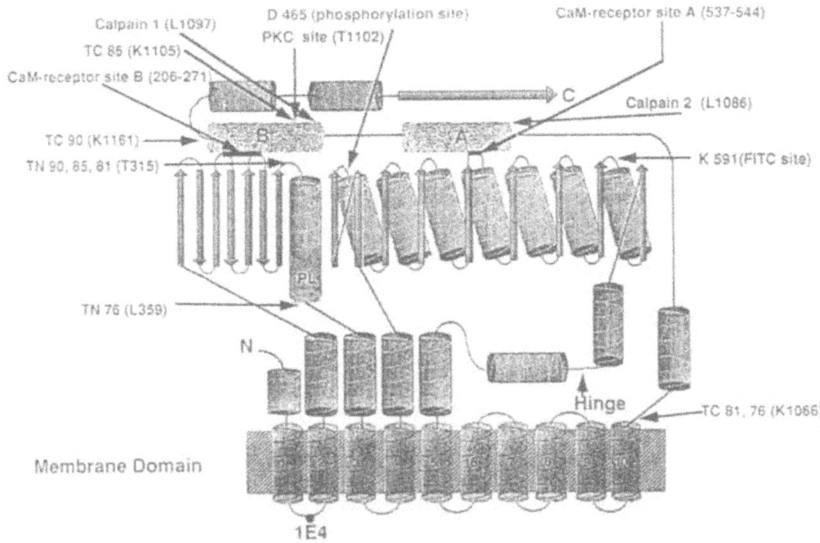

Figure 2. Topographical model of the Ca^{2+}-pump. The secondary structure elements (cylinders = α-helices, arrows = β-sheets) are indicated according to secondary-structure prediction algorithms as discussed in the text. The numbering of the amino acids in the Figure is based on the sequence of hPMCA 4 (Strehler et al., 1990). 1E4 = binding site of the monoclonal antibody 1E4; PL = suggested phospholipid responsive domain; TC = C-terminal amino acid of different fragments (90-, 85-, 81-, and 76-kD) obtained by trypsin cleavage; TN = N-terminal amino acid of the same fragments obtained by trypsin cleavage; A, B = Different parts of the CaM-binding domain defining their internal receptor sites as discussed in the text; PKC site = identification of the amino acid (T1102) phosphorylated by protein kinase C; Calpain 1 and 2 = identification of the two calpain cleavage sites.

activity attained with CaM as demonstrated by James et al. (1989a). These authors also provided evidence that calpain digests the CaM-binding domain of the pump in two successive steps: first by cleaving the domain in the middle, thereby retaining its CaM-binding property and second, by removing the binding domain completely (see Figure 2).

C. Phosphorylation

Caroni and Carafoli (1981a) first reported the influence of phosphorylation by protein kinase A (PKA) on the activity and transport property of the Ca^{2+}-pump of plasma membranes. This observation was made by characterizing the Ca^{2+}-pump in heart cell plasma membranes (sarcolemma) and could only be observed after dephosphorylation of the membranes. Later experiments provided evidence for the direct phosphorylation of the Ca^{2+}-pump by PKA (Neyses et al., 1985), the site being downstream of the CaM-binding domain (James et al., 1989b; see Figure 2). It is of interest that a detailed analysis of the Ca^{2+}-pump from erythrocyte membranes provided evidence for the existence of two different isoforms of the enzyme stemming from two different genes (Strehler et al., 1990) only one of which contained the PKA phosphorylation site. Smallwood et al. (1988) reported that the Ca^{2+}-ATPase activity as well as its Ca^{2+}-transport rate could also be increased due to phosphorylation by PKC. This phosphorylation site was later established by Wang et al. (1991) as being located within the CaM-binding domain (see Figure 2). These authors also provided evidence that phosphorylation by PKC can be prevented by the presence of CaM.

D. Dimerization

On SDS-PAGE, sometimes a protein band can be observed which corresponds roughly to twice the molecular weight of the monomeric Ca^{2+}-pump. Incubation of the enzyme with γ-[^{32}P]-ATP under conditions where a phosphorylated intermediate can be formed indicated that this high molecular weight band indeed represented a functional ATPase since it was phosphorylated only in the presence of Ca^{2+}, that is, possibly indicating the formation of a dimer. Such dimerization as a second active form of the enzyme was studied in detail by Kosk-Kosicka and her coworkers (1988, 1990). It could be shown that oligomerization/dimerization activated the enzyme in a similar way to CaM, that is, deriving a CaM-independent form which as the authors claim should also play a role *in vivo*. On the other hand, due to the low abundance of the enzyme (0.1% or less of total membrane protein) such a function is difficult to reconcile. Recent reports indicated that such a dimerization probably occurs via the CaM-binding domain (Vorherr et al., 1991). Since, as will be shown later, the latter can interact in the enzyme with its own receptor site in the absence of CaM, such an association could occur via *inter*molecular interaction between the CaM-binding site of one molecule and its receptor located in a neighboring molecule.

IV. DISTRIBUTION OF THE Ca^{2+}-PUMP OF PLASMA MEMBRANES

For many years, the Ca^{2+}-pump of erythrocyte membranes was the sole object of investigation for this enzyme. This was not only due to the experimental advantage dealing with blood cells but also due to the general belief that in most cells, especially excitable ones, extrusion of Ca^{2+} occurred only via the Na^+/Ca^{2+} exchanger. Thus, the finding that an ATP-dependent, Na^+-independent system existed in plasma membranes of squid axons (Di Polo, 1979) or in mammalian heart cells (Caroni and Carafoli, 1981b) to extrude Ca^{2+} came as a surprise. Since then, this enzyme has been found in the plasma membranes of many different cells and organisms (including plants) leading to the conclusion that this enzyme is ubiquitous and essential for the fine tuning of Ca^{2+}-homeostasis of eucaryotic cells. As has been documented by many different laboratories, the general properties of this enzyme in a variety of systems seem to be very similar: it is a monomeric protein of an apparent molecular weight of ca. 140,000 depending on the isoform, with high Ca^{2+}-sensitivity and CaM-dependence in its activation.

The only organ which seems to possess a Ca^{2+}-pump with peculiar properties is the liver. This enzyme differs in many respects from the normal enzyme of plasma membranes including the requirement for exceedingly low Mg^{2+} concentrations, extremely high affinity for Ca^{2+}, that is, below 0.5 μM Ca^{2+}, a lack in specificity for the nucleotide requirement (usually ATP) and the insensitivity to CaM. In addition, the apparent molecular weight seemed to be lower, that is, 105,000 (Chan and Junger, 1983), and the enzyme seemed to demonstrate a peculiar sensitivity to hormones (Lotersztajn et al., 1984). Due to the insensitivity to CaM, purification of the enzyme turned out to be difficult, but recent results on this enzyme shed some new light on the properties of the Ca^{2+}-pump of hepatic plasma membranes. Monoclonal antibodies raised against the human erythrocyte Ca^{2+}-pump cross-reacted with a protein of ca. 150 kD enriched in the sinusoidal domain of the rat liver plasma membrane (Kessler et al., 1990). It was of great interest to note that the same protein also demonstrated CaM-binding in gel-overlay experiments. The same antibodies were used in an affinity chromatography step to isolate the protein from liver plasma membranes (Kessler et al., 1990) permitting the analysis of the differences in properties compared to the enzyme from erythrocytes, for example, the reason for the difference in molecular weight or in the different CaM-binding properties.

V. PROPERTIES OF THE ISOLATED Ca^{2+}-PUMP

It was indicated above that the crucial step in the purification of this enzyme was the application of an affinity chromatography step using CaM as a ligand (Niggli et al., 1979). Previous attempts to purify this enzyme with conventional chromatography methods (e.g., Peterson et al., 1978) have been unsuccessful mainly due to the fact that this enzyme is of low abundance in plasma membranes (i.e., 0.1–0.2% of the total membrane protein). The crucial observation for the successful application of CaM-affinity chromatography was the report that the enzyme solubilized by Triton-X100 still could be stimulated by CaM (Lynch and Cheung, 1979; Niggli et al., 1979). In the original isolation protocol, erythrocyte plasma membranes were extensively washed with an EDTA-containing hypotonic buffer to remove all bound CaM from the membranes and then solubilized using Triton-X100 in the presence of asolectin as the phospholipid component. After removing unsolubilized membrane components by centrifugation, the required amount of Ca^{2+} was added to the supernatant containing the Ca^{2+}-pump, and the solution was applied to a CaM-Sepharose column equilibrated with a Ca^{2+}-containing buffer. After extensive washing using the same Ca^{2+}-containing buffer, the enzyme was eluted from the column using an EGTA-containing buffer to chelate Ca^{2+}. The enzyme eluted from the column corresponded to a protein of an approximate molecular weight of 138,000 on SDS-PAGE, with a minor component of roughly twice the molecular weight indicating a dimer formation as mentioned above.

In the original report by Niggli et al. (1979), the purified Ca^{2+}-ATPase was no longer CaM-sensitive. This was a surprising finding, but later could be explained due to a systematic study (Niggli et al., 1981b): the stimulation of the enzyme by CaM could only be observed in the absence of negatively charged phospholipids. In addition, reconstitution experiments with varying ratios of acidic to neutral phospholipids provided evidence that Ca^{2+}-transport became up to 50% CaM-insensitive depending on the amount of acidic phospholipids present. This observation could be of physiological importance implying that in native membranes, that is, in erythrocytes, containing sufficient amounts of negatively charged phospholipids a substantial amount of the enzyme could be CaM insensitive.

VI. PRIMARY STRUCTURE OF THE Ca^{2+}-PUMP AND CHARACTERIZATION OF FUNCTIONAL DOMAINS

Due to its size, its low abundance, and its hydrophobic properties as a membrane protein, direct sequence information was difficult to obtain. Nevertheless, several tryptic fragments yielded partial sequences containing the ATP-binding site as characterized by fluorescein-isothiocyanate (FITC; Filoteo et al., 1987), the site of acylphosphate formation (James et al., 1987), and the CaM-binding domain (James et al., 1988). The identification and sequencing of the latter was of special interest since it was the first direct identification of such a binding domain of any CaM-dependent enzyme. The characterization was achieved by using a radioactively labeled, heterobifunctional, cleavable photoreactive cross-linker (James et al., 1988). The latter was linked to CaM in the dark, incubated with the purified enzyme in the presence of Ca^{2+}, cross-linked to the pump by photoactivation, and finally cleaved to remove CaM from the target. In this way, the radioactive marker was introduced into the target domain which, after exhaustive proteolytic cleavage of the protein, was separated and purified from the other fragments by HPLC and sequenced (James et al., 1988). This method also proved to be very useful in identifying other target domains.

Based on partial protein sequences as indicated above, it was possible to construct degenerate oligonucleotides for the cloning of DNA coding for the plasma membrane Ca^{2+}-ATPase (PMCA). This led to the first complete sequence of a Ca^{2+}-pump from human plasma membranes (Verma et al., 1988). At the same time, two other independent approaches led to the isolation of PMCA-specific cDNA clones from other species. A partial clone was obtained by screening an expression library with antibodies against the PMCA (Brandt et al., 1988) and two full-length clones coding for two rat PMCA isoforms were isolated by low-stringency screening with oligonucleotides derived from the conserved catalytic site of the P-type pumps

(Shull and Greeb, 1988). This last approach was first successfully employed for the isolation of clones of the H^+/K^+-ATPase (Shull and Lingrel, 1986). From these results, it soon emerged that the plasma membrane Ca^{2+}-pump belongs to a multigene family with additional isoforms due to alternative splicing as will be detailed below.

According to the primary structure of the enzyme and based on sequence analysis software used to predict secondary structure, a topographical scheme was suggested as shown in Figure 2. The model is based on that of the SR Ca^{2+}-pump as published by MacLennan et al. (1985) which seemed reasonable due to the high degree of structural and functional similarity between the two proteins. As indicated by the model, 10-transmembrane domains could be identified. The N- and C-termini are both located on the cytosolic side together with about 80% of the protein mass. Extracellularly, short loops are predicted to connect the different putative transmembrane domains. To test the proposed membrane topography, antibodies against peptides corresponding to sequences of the extracellular loops have been raised, but either due to the shortness of these sequences or due to their conserved nature, most of these attempts have failed so far. Nevertheless, it is interesting to note that the location for at least one of the proposed extracellular loops, that is, between transmembrane domains 1 and 2, could be recognized by a monoclonal antibody in intact erythrocytes and thus confirmed the correctness of the proposed localization (Feschenko et al., 1992).

Alignment with sequences of other P-type ATPases provided evidence for the conservation of the domain of the phosphorylated intermediate (Asp 465, hPMCA4; see Strehler et al., 1990), the ATP-binding site around the residue labeled by FITC (Lys 591), and the so-called "hinge-region," probably responsible for bringing the phosphorylation site close to the ATP-binding site. These functional domains are contained within the second and largest protruding unit between transmembrane domains 4 and 5. The first protruding unit located according to Figure 2 between transmembrane domain 2 and 3 contains the so-called "transduction unit" which, according to MacLennan et al. (1985), is responsible for the coupling between ATP hydrolysis and Ca^{2+}-translocation. It also contains a domain which, at least in part, is responsible for the phospholipid sensitivity of the enzyme (Brodin et al., 1992).

The part C-terminal to the last transmembrane domain contains a number of different regulatory domains, that is, the CaM-binding domain, the cAMP-dependent phosphorylation site (hPMCA1), and at least 2 rather acidic domains which may be involved in the regulation of Ca^{2+}-binding to the enzyme. In this context, it is of interest that part of the C-terminal domain, that is, E1079–P1180, expressed in a bacterial system (Hofmann et al., 1993) contained three ions of Ca^{2+} bound/mole of peptide even after SDS-PAGE, indicating a high affinity to these sites (Hofmann et al., 1993). The latter probably have a regulatory function, whereas the high-affinity site(s) responsible for the transport of Ca^{2+} through the membrane are probably formed within the transmembrane domains as suggested by Clarke et al.

(1989) for the SR Ca^{2+}-pump. Site-directed mutagenesis experiments performed on polar residues located within transmembrane domains 4, 5, 6, and 8 of the SR Ca^{2+}-pump (most of which are also conserved in the plasma membrane enzyme) indicated the importance of these residues for the Ca^{2+}-transport.

One aspect which developed from experiments to activate the Ca^{2+}-pump by controlled proteolysis (Zurini et al., 1984) was the suggestion that the CaM-binding domain could have autoinhibitory functions. Similar suggestions have also been made for other CaM-dependent enzymes (Buschmeier et al., 1987; Kennelly et al., 1987; Payne et al., 1988). Using similar cross-linking techniques as have been applied to identify the CaM-binding domain of the pump, a photoactivatable derivative of a radioactively-labeled peptide corresponding to the CaM-binding domain (C28W) has been prepared to identify the site(s) of interaction within the pump. Toward this end, phenylalanine residues in positions 9 and 25 of the peptide have been replaced by analogs containing a photoactivatable diazirine group and have reacted with the calpain truncated form of the ATPase which lacks the CaM-binding domain (Falchetto et al., 1991). After photoactivation and cross-linking to the receptor site(s), the enzyme has been fragmented using different proteases and the cross-linked, radioactively labeled peptides have been purified and sequenced. If the reactive group was placed near the N-terminus of peptide C28W, that is, Phe 9, a peptide corresponding to sequence 537–544 of the Ca^{2+}-pump was labeled, which is located between the phosphorylation site (Asp 465) and the ATP-binding site as labeled by FITC (Lys 591), a receptor site with obvious inhibitory properties (see Figure 2). On the other hand, if the diazirine analog of Phe was placed near the C-terminus of C28W, that is, Phe 25, a sequence corresponding to residues 206–277 of the ATPase was labeled (Falchetto et al., 1992), which is located within the so-called transduction domain. These results suggest very strongly that the CaM-binding domain exhibits its inhibitory function within the Ca^{2+}-pump by "bridging" two functionally important domains of the enzyme in the absence of CaM and thereby, possibly limiting the access of the substrate Ca^{2+} to the catalytic site.

As indicated before, the activation of the Ca^{2+}-pump by CaM could be mimicked by controlled proteolysis. In a detailed study (Benaim et al., 1984; Zurini et al., 1984), a number of fragments were obtained. Those which still displayed ATPase activity (90-, 85-, 81-, and 76-kD) differed in their CaM sensitivity. The largest fragment of 90 kD could be fully activated by CaM, whereas the fragment of 85 kD could still be isolated by a CaM-column, but could no longer be activated by CaM. However, the fragments of masses 81- and 76-kD have lost all CaM sensitivity. These results indicated that the pump progressively lost its CaM-sensitivity and became irreversibly activated by proteolysis to an extent corresponding to CaM stimulation, that is, the CaM-binding domain played an inhibitory function as discussed above. In a later study, the N- and C-terminal sequences of those fragments have been determined to understand the progressive change of their activity properties (Zvaritch et al., 1990). In these investigations, it could be determined by C-terminal sequencing that the fragment of 90 kD was cleaved C-terminal to the CaM-binding domain (Lys 1161), in contrast to the 85 kD fragment which was cleaved in the middle of the CaM-binding domain (Lys 1105). C-terminal cleavage of fragments 81- and 76-kD occurred at the interface between transmembrane 10 and the C-terminal part of the pump (Lys 1066), removing the entire regulatory region (see Figure 2). These results explained the difference in the activity properties of these fragments. On the other hand, at the N-terminal cleavage sites, one interesting difference could be observed. As can be seen from Figure 2 fragments 90-, 85-, and 81-kD all have the same N-terminus (Thr 315), whereas fragment 76 kD was cleaved some 50 amino acid residues downstream of the former (Leu 359; Zvaritch et al., 1990). This resulted in an interesting difference in sensitivity to acidic phospholipids (Enyedi et al., 1987). Fragments 90-, 85-, and 81-kD were all still sensitive to acidic phospholipids, whereas fragment 76 kD had lost this sensitivity, indicating that the sequence responsible for this difference is important for the interaction with acidic phospholipids. This view has later been corroborated by a study of Brodin et al. (1992), identifying this region as a possible phospholipid binding domain.

VII. INTERACTION BETWEEN CALMODULIN AND THE Ca^{2+}-PUMP

Early studies undertaken to understand the interaction between CaM and the Ca^{2+}-pump of plasma membranes demonstrated that the C-terminal half of CaM was sufficient to fully stimulate the Ca^{2+}-pump, whereas the N-terminal half did not show this ability (Guerini et al., 1984). As originally shown by Drabikowski and his co-workers (Drabikowski et al., 1977; Walsh et al., 1977), CaM can be cleaved into several fragments depending on whether Ca^{2+} is present or not, that is, in the presence of Ca^{2+}, CaM is cleaved by trypsin in the middle of the central helix whereas in its absence, the first cleavage point is downstream of the Ca^{2+}-binding loop III (Arg 106) yielding two fragments of different length, that is, 1–106 and 107–148. After purification of all these fragments and studying their activation properties with respect to the Ca^{2+}-pump (Guerini et al., 1984), it became apparent that not only the primary contact site between CaM and the pump is located within the C-terminal half of CaM as indicated above, but also that the intact third Ca^{2+}-binding loop of CaM, as counted from the N-terminus, is of special importance for this interaction (Guerini et al., 1984). These results were corroborated by additional studies using chemically (Guerini et al., 1987) and genetically modified CaM (Krebs et al., 1990), but were in contrast to other studies (e.g., with the CaM-dependent enzymes phosphodiesterase or calcineurin, see Newton et al., 1984) indicating differences within the interphases of the protein–protein contact sites between CaM and its different targets. This aspect was investigated recently from a different angle by studying the activation of four target enzymes by two series of CaM mutants in which the Ca^{2+}-binding site properties have been modified systematically (Gao et al., 1993). In each mutant, the conserved bidentate glutamate in the last position of the individual Ca^{2+}-binding sites were mutated to glutamine or lysine and studied with four different enzymes, including the Ca^{2+}-pump. For three of these enzymes, that is, MLCK from smooth and skeletal muscle and the adenylate cyclase, mutations in Ca^{2+}-binding site IV had the most severe effect followed by mutations in sites II, III, and then I. By contrast, mutants of sites III and IV were much poorer activators of the Ca^{2+}-pump than those of sites II and I, again demonstrating the importance of an intact C-terminal half of CaM for the activation of the Ca^{2+}-pump of plasma membranes (Gao et al., 1993).

To gain a more complete view on the structural details of the interaction between CaM and its targets at the molecular level, studies have been initiated on the interaction between CaM and peptides corresponding to binding domains of different target enzymes. The first data obtained by using small angle X-ray scattering (SAXS) techniques provided the surprising result that some of these peptides could bend the extended, dumbbell-like structure of CaM to a more globular one (e.g., Heidorn et al., 1989; Kataoka et al., 1989). This observation gained full support by the detailed three-dimensional structure of the complex between CaM and the CaM-binding domain peptide of MLCK as obtained either by NMR (Ikura et al., 1992) or by X-ray crystallography (Meador et al., 1992). However, a somewhat different result was obtained with peptides of the CaM-binding domain of the Ca^{2+}-pump. Depending on the length of the peptides, that is, corresponding either to the N-terminal 20- or to the C-terminal 24-amino acids of the domain which is 28 amino acid residues in length (see James et al., 1988), the complex between these peptides and CaM showed either an extended dumbbell-like or a more compact globular structure according to SAXS measurements (Kataoka et al., 1991). These and other results (e.g., see Vorherr et al., 1990, 1992; Yazawa et al., 1992) underlined the importance of the C-terminal half of CaM as the primary recognition site for the Ca^{2+}-pump and the N-terminal half being needed for high-affinity binding and full cooperativity of the system.

VIII. ISOLATION OF cDNA CODING FOR PLASMA MEMBRANE Ca^{2+}-ATPase (PMCA) ISOFORMS

Comparison of the primary structure of isoforms from rat and human tissues indicated that one of the rat isoforms, that is, rPMCA1, was almost completely identical to the human isoform hPMCA1, whereas hPMCA2 showed only around 82% homology on the protein level to rPMCA1, and these differences were spread over the total length of the sequence. From these results, it became clear that PMCA were coded by more than one gene. The search continued and in a short time two additional sequences became available, PMCA3 in rat (Greeb and Shull, 1989) and PMCA4 in human (Strehler et al., 1990). Due to an extensive search in different laboratories, a great variety of different isoforms have been characterized (see also Table 2), and it is now generally accepted that at least four different genes for PMCA

Table 2. Cloned cDNAs for PMCA

Isogene	*nt*[a]	*MW*	*Chromosomal Location (human) References: g,o,q*[b]	*Sequence Organism*	*References*[b]
PMCA1	1220	134,700	12q21–23	human, rat, pig, rabbit	a,b,c,d,e
PMCA2	1198	132,600	3p25–26	human, rat	b,f,g,h
PMCA3	1206	132,900	X	rat, human (partial)	i,l
PMCA4	1205	133,900	1q25–32	human, rat (partial), cow (partial)	m,n,p

Notes: [a]nt and MW for the major forms are given: PMCA1b; PMCA2b,z; PMCA3b,x; PMCA4b,x.

[b]References: (a) Verma et al., 1988; (b) Shull and Greeb, 1988; (c) De Jaegere et al., 1990; (d) Kahn and Grover, 1991; (e) Kumar et al., 1993; (f) Heim et al., 1992; (g) Latif et al., 1993a; (h) Brandt et al., 1992a; (i) Greeb and Shull, 1989; (l) Stauffer et al., 1993; (m) Strehler et al., 1990; (n) Keeton et al., 1993; (o) Olson et al., 1991; (p) Brandt et al., 1992b; (q) Wang et al., 1994.

are present in the rat and in the human genome. Full length PMCA1 cDNA was found in rat (Shull and Greeb, 1988), in human (Verma et al., 1988; Kumar et al., 1993), in pig (De Jaegere et al., 1990), and in rabbit (Kahn and Grover, 1991); PMCA2 was found in rat (Shull and Greeb, 1988) and in human (Brandt et al., 1992b; Heim et al., 1992a; Latif et al., 1993a); PMCA3 was found in rat (Greeb and Shull, 1989); and PMCA4 was found in human (Strehler et al., 1990). Although the complete sequence of rat and human PMCA3 are not yet available, their presence has been confirmed by the isolation of partial cDNA clones (see Table 2). Partial sequence of a cow PMCA4 isoform has also been described (Brandt et al., 1988, 1992a). So far, the search for additional genes coding for the PMCA have failed, although it was proposed that one other type of Ca^{2+}-ATPase, a hybrid between SERCA and PMCA, may exist (Gunteski-Hamblin et al., 1992). This type of pump has been characterized by cloning, and it shows homology to the yeast PMR1 gene (Schlesser et al., 1988; Rudolf et al., 1989). Preliminary results indicated that this pump may be localized in the vacuoles of the yeast, *Saccharomices cerevisiae* (Antebi and Fink, 1992), but little is known about the corresponding protein in higher organisms. Other plasma membrane P-type pumps were detected by different approaches and they could belong to a different gene family (Pavoine et al., 1987). A different type of Ca^{2+}-ATPase, called PMC1, was found in the yeast, *Saccharomices cerevisiae.* The sequence demonstrates a much higher degree of homology to PMCA than to SERCA protein sequences, but does not seem to contain a calmodulin-binding domain (Cunningham and Fink, 1994). PMC1 in conjunction with PMR1 seems to confer Ca^{2+}-tolerance to *Saccharomices cerevisiae* (Cunningham and Fink, 1994).

IX. STRUCTURE OF THE PMCA GENES

The genes for the rat PMCA3 and the human PMCA1 have been isolated and characterized in detail (Burk and Shull, 1992; Hilfiker et al., 1993). A simplified scheme of the structure of the two genes and a topological model of the pump are shown in Figure 3. The gene for the rat PMCA3 is composed of 24 exons, the gene for human PMCA1 of 22 exons (eventually 23 exons, see below), with the coding sequence distributed over 22 exons in the case of rPMCA3 and 21 exons in the case of hPMCA1. Exon 0, the first exon in the rPMCA3 gene, is not present in hPMCA1

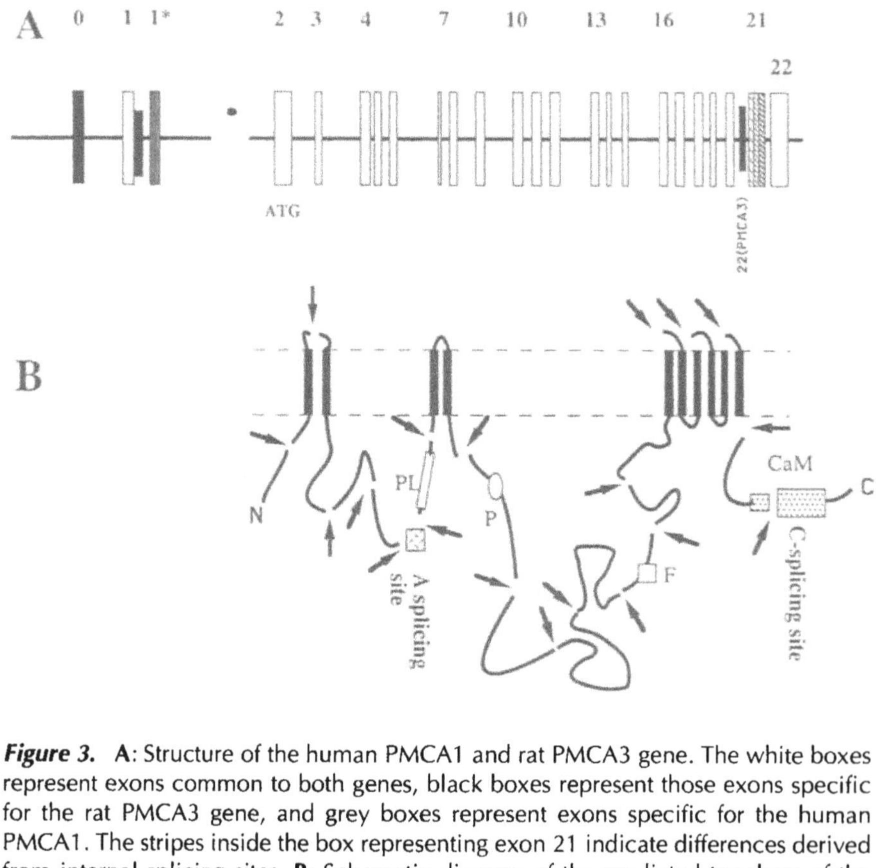

Figure 3. **A**: Structure of the human PMCA1 and rat PMCA3 gene. The white boxes represent exons common to both genes, black boxes represent those exons specific for the rat PMCA3 gene, and grey boxes represent exons specific for the human PMCA1. The stripes inside the box representing exon 21 indicate differences derived from internal splicing sites. **B**: Schematic diagram of the predicted topology of the plasma membrane calcium ATPase (PMCA), see also Figure 2. Intron/exon boundaries are indicated by an arrow. The two major splicing sites, A and C, are shown in the Figure. The black boxes represent the 10-transmembrane domains. PL = putative phospholipid responsive domain; P = phosphoenzyme intermediate site; F = FITC binding site; CaM = calmodulin binding domain.

(Figure 3a). By contrast, one additional exon 1 (exon 1^*, Figure 3a) was found in the hPMCA1 gene, after comparing sequences from the 5′ untranslated regions from pig and human cDNAs. Since no human mRNA containing exon 1^* was detected, its functional role is not yet clear (Hilfiker et al., 1993).

The sizes of the exons coding for the protein sequence are highly conserved (although the homology of hPMCA1 and rPMCA3 on the level of nucleotide sequence is less than 80%) in contrast to the differences observed in the 5′-untranslated coding exons. This difference is also reflected by the fact that the cDNA sequences of the untranslated-region from both genes are completely unrelated. The exact starting point of the transcription of the rat PMCA3 has not been determined and it is not known whether the foremost 5′ exon (exon 0 in Figure 3a) is the first one of the gene (Burk and Shull, 1992). By contrast, evidence for the mRNA transcription starting point was collected for the human PMCA1. Reverse-priming experiments revealed the presence of different possible initiation sites which mapped just ahead of exon 1. Among these, the one providing the most prominent signal was taken as the possible mRNA start. No TATA box has been found in the region preceding it, but a CAAT-box and SP1 sites were detected (Hilfiker et al., 1993). These results, together with the fact that the same region is highly enriched in GC basepairs, would support the idea that this is the promoter region for a housekeeping gene.

It is worth emphasizing how well the size of exons 2 to 20 (the exons containing coding sequences) is conserved, bearing in mind that these two genes code for two different isoforms and that they have been isolated from different organisms. Four out of nineteen exons showed a difference of three nucleotides, the others were identical in size. Exons 4, 7, and 8 in the rat PMCA3 contained 3 additional nucleotides at the 3′ end, just before the splice donor site; exon 12 contained 3 nucleotides less at its 3′ end than the corresponding one in the human PMCA1. A relevant difference between the two genes is the presence of an additional exon in rat PMCA3 (Figure 3a, 22(PMCA3)). This exon was demonstrated to be alternatively spliced (Burk and Shull, 1992) and together with the exon 21 (Figure 3a) is responsible for the very complex alternative splicing pattern at the C-site (Figure 4, see below). In both genes, the next-to-last exon (Figure 3a, exon 21) contains potential internal alternative splicing sites. Exon 7 of the rat PMCA3 gene can be inserted alternatively, the corresponding one of PMCA1 is always inserted in the mature mRNA (Hilfiker et al., 1993; Burk and Shull, 1992).

Both rat and human genes are very large, around 90 kbp for the rat PMCA3 and more than 100 kbp for the human PMCA1. In fact, in the human gene there is a gap between exons 1 and 2 (● in Figure 3a), which has not been filled to date. It is not known whether this gap is due to a sequence refractory to cloning or whether the distance is too large to be covered by conventional cosmid cloning. The use of YAC-libraries will possibly allow this question to be answered.

The four human PMCA genes have been mapped. PMCA1 was localized to chromosome 12q21–23 and PMCA4 to chromosome 1q25–q32 (Olson et al., 1991,

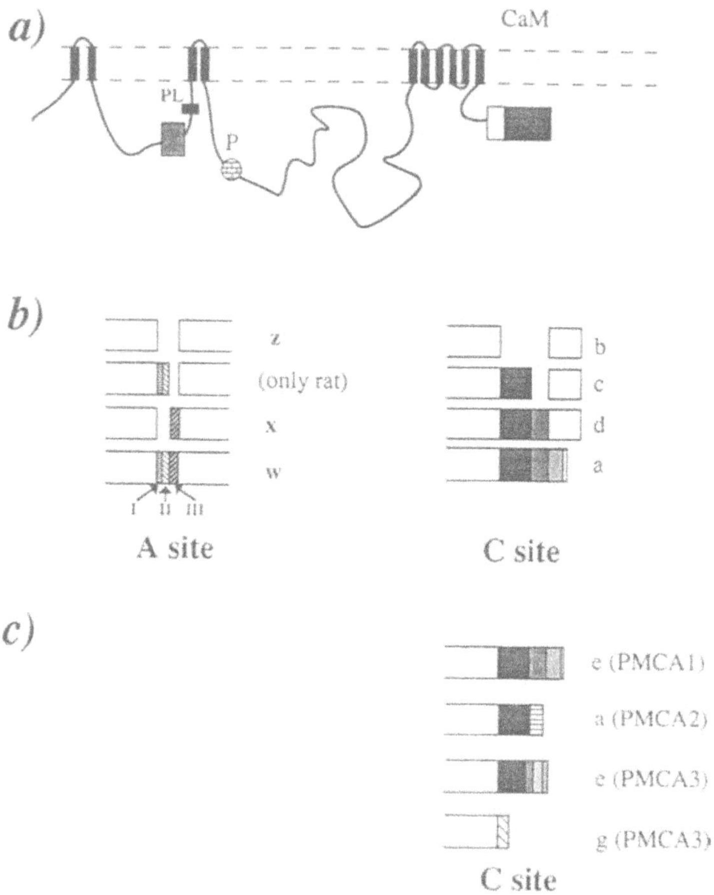

Figure 4. **a**: Representation of the putative topology of the pump (see also Figures 2 and 3). P = phosphoenzyme site; PL = phospholipid responsive domain; CaM = calmodulin binding domain. **b**: A general scheme for the patterns of alternative splicing at sites A and C. Different patterns of shadows indicate variable usage of an internal splicing site (indicated by a vertical bar). The integration of additional independent exons is indicated by striped boxes. **c**: Specific patterns of alternative splicing for individual isoforms. Names in parenthesis indicate the corresponding isoform.

see Table 2). PMCA2 was mapped to chromosome 3 (Brandt et al., 1992a) and more precisely to 3p25–26 (Latif et al., 1993a). The locus of the PMCA2 co-localized with the Van Hippel–Lindau syndrome, and in an early study it was proposed that the PMCA2 could be involved in the onset of this disease (Latif et al., 1993a). Further evidence unequivocally demonstrated that the gene responsible for the Van

Hippel–Lindau disease was not PMCA2, but another gene, 200,000–300,000 bp telomeric to the same location (Latif et al., 1993b). PMCA3 was mapped to chromosome Xq28 (Wang et al., 1994).

X. ALTERNATIVE SPLICING OF THE PMCA ISOFORMS

As mentioned before, the genes for PMCA undergo a very complex pattern of alternative splicing. Up to four sites were proposed to be involved in this process. For sites B and D (Strehler, 1991; Carafoli and Guerini, 1993), little experimental evidence has been collected and they were suggested to be products of cloning artifacts (Keeton et al., 1993; Stauffer et al., 1993). Nevertheless, one point should be made concerning the B-site. This alternative splicing, resulting in an open reading frame for a PMCA lacking the 10th transmembrane domain, has been found during the cloning of cDNA for the human PMCA4 (Strehler et al., 1990). By comparing the gene structure of hPMCA1 and rPMCA3, this alternative splicing would correspond exactly to the deletion of exon 19 (see Figure 3). Although the structure of the PMCA4 gene is not yet available, due to the high level of conservation between the human PMCA1 and rat PMCA3 genes, the PMCA4 gene structure is expected to be very similar. It is possible, therefore, that this exon would be spliced out under special conditions as some new data may indicate (Howard et al., 1993). The effect of such an alternative splicing on protein function is difficult to predict: it would be important to know if such a protein retained activity, since deletion of transmembrane domains of PMCA or SERCA caused a total loss of activity (Heim et al., 1992b; Skerjanc et al., 1993).

However, alternative splicing at sites A and C (see Figure 4) have been supported by a wealth of experimental data. The A-splice site occurs just N-terminal to a region which was proposed to interact with acidic phospholipids (Zvaritch et al., 1990; Brodin et al., 1992) and just C-terminal to a sequence interacting with the CaM-binding domain of the PMCA (Falchetto et al., 1992; e.g., see Figure 2). In addition, the domain encompassing the A-site has only been found in PMCA and not in other P-type pumps. The most complex splicing pattern has been found for PMCA2 (Adamo and Penniston, 1992; Heim et al., 1992a; Stauffer et al., 1993). In fact, up to four different mRNA could be generated in rat and three in human (see Figure 4). In the human gene, exons I+II (30- and 63-nucleotides, respectively) were invariably inserted with exon III (42 nucleotides) leading to an insertion of a total of 135 nucleotides. By contrast, in the case of rat PMCA2, these exons can apparently be spliced out independently of exon III (see Figure 4). Alternative splicing at the A-site has also been detected for PMCA3 and 4 (Stauffer et al., 1993). However, for these two genes only two alternatively spliced products could be detected, corresponding to the z and x types (Figure 4). In fact, only one exon of 42 nucleotides in the case of PMCA3 (exon 7 in Figure 3a, corresponding to exon III of PMCA2, Figure 4), and 36 nucleotides in the case of PMCA4 are responsible for the splicing. No splicing at site A was found for the PMCA1 gene, i.e., exon 7 is always present (type-x splicing) in the mature mRNA (Hilfiker et al., 1993; Stauffer et al., 1993).

Alternative splicing at site C is also rather complex. Differences could be observed by comparing human and rat isoforms (Keeton et al., 1993; Stauffer et al., 1993). The highest number of alternatively spliced isoforms was described for the rPMCA3 gene, giving rise to up to six different spliced mRNA as detected by RT-PCR (Keeton et al., 1993), whereas in human tissues, only four of them could be detected (Stauffer et al., 1993). This complexity is due to the fact that the PMCA3 gene contains an additional exon (exon 22 (PMCA3) in Figure 3). In human, as well as in rat, splicing types b, a, e(3), and g were found (Figure 4). Type a is derived from the insertion of a part of exon 21 (154 nucleotides; see Figure 3 and Figure 4) and because of a frame-shift associated with this insertion (154 nucleotides is not a multiple of 3), type a results in a protein with a shorter C-terminus. Type e is derived from the insertion of the complete, 222-nucleotide exon 21, whereas in type b, this exon is missing completely. Type g is characterized by the presence of an exon, 68 nucleotides in length, (exon 22 (PMCA3)) in the mature mRNA (see Figures 3 and 4) resulting in a shorter protein as in splicing form a. In fact, this last alternative splicing form gave rise to the protein with the smallest predicted molecular mass of all PMCA. The possibility of other derivatives due to internal splicing sites in exon 21 was reported for the rat PMCA3 mRNA (Keeton et al., 1993) and is responsible for two additional mRNA products (type c and d). The corresponding spliced mRNA have not been found in the case of human PMCA3 isoforms (Stauffer et al., 1993). In addition, the use of alternative polyadenylation sites present in exon 23 of PMCA3 resulted in a smaller mRNA in rat skeletal muscle and also partially in rat brain tissues (Burk and Shull, 1992).

For PMCA4 and PMCA2, two different alternatively spliced forms have been detected, type a and type b (Figure 4). In both cases, type a resulted in a protein shorter than type b, as reported for PMCA3. In PMCA4, an exon of 178 nucleotides (similar to the "21st" of PMCA1) (see Figure 3), is either inserted or deleted. In the case of rPMCA2, the inserted sequence of 227 nucleotides was demonstrated to be distributed within two exons of 172- and 54-nucleotides, respectively (Keeton et al., 1993). The sequence coded by the first exon corresponded to the sequence present in exon 21 of PMCA1. Since the size of the insertion of the human PMCA2 was identical to the one of the rat, it was proposed that in human PMCA2, two exons would also be responsible for the alternative splicing (Stauffer et al., 1993).

The alternative splicing forms of the PMCA1 gene were the first to be characterized in detail (Strehler et al., 1989). These resulted as a consequence of multiple acceptor/donor sites present in exon 21 (Figure 3). Splicing types a, b, c, and d have been well characterized in a variety of tissues (Strehler et al., 1989; Stauffer et al., 1993; Keeton et al., 1993) and recent work demonstrated an additional small variation of type a: type e(1) (Stauffer et al., 1993; see Figure 4). This alternative splicing was derived from a frame-shift of two nucleotides at the splice donor site of exon 21, resulting in an insertion of 152 nucleotides. Type a corresponded to the

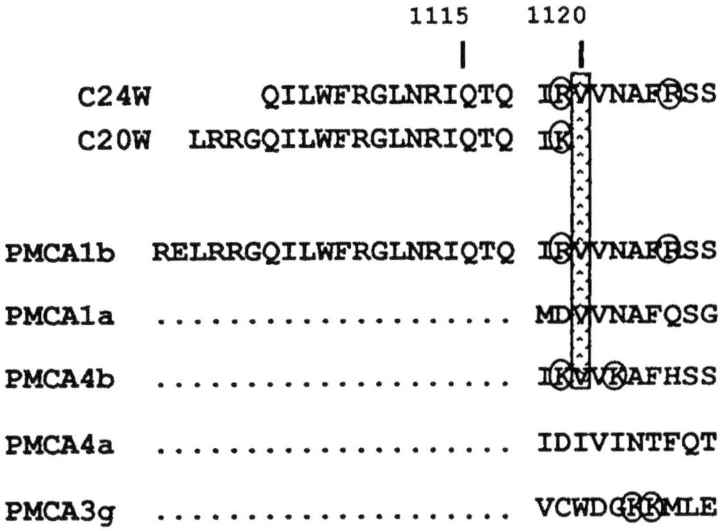

Figure 5. Alignment of the calmodulin-binding regions of some representative PMCA isoforms and comparison with two synthetic calmodulin binding peptides. The conserved Val at position 1120 is boxed (see text). Positively charged amino acids are encircled. The sequences were taken from the following references: Strehler et al., 1989: PMCA1a and b; Strehler, 1991: PMCA4a,b; Stauffer et al., 1993: PMCA3g. The peptides were described by Vorherr et al., 1990.

complete insertion of the 154 bp of exon 21, type c and type d resulted in the insertion of only a partial exon 21, that is, 87- and 114-nucleotides, respectively. Type b lacked the exon 21 completely.

Alternative splicing at the C-site creates a PMCA protein with amino acid differences at the C-terminal region of the CaM-binding domain (see Figure 5). It is known that the part of the CaM-binding peptide encoded by exon 20, possesses high affinity binding to CaM (Vorherr et al., 1990). The alternative splicing occurs after the QTQ-sequence (1115–1117 of PMCA1b), corresponding to the C-terminus minus two amino acids of the C20W peptide, a peptide known to bind CaM with nM affinity (Vorherr et al., 1990). From the structure of the CaM–MLCK peptide complex, it became evident that the hydrophobic residue after the QTQ sequence (i.e., V1120 of the PMCA1b, see Figure 5) is important for the binding of the peptide to CaM as discussed above in Section VII (see Ikura et al., 1992). Therefore, differences in binding to CaM could be foreseen as a result of the alternative splicing at site C, as will be shown later (Section XII).

XI. TISSUE DISTRIBUTION OF THE PMCA ISOFORMS

The tissue distribution of the four different genes was studied at the level of mRNA by PCR, northern blot, and *in situ* hybridization. An overview is given in Table 3.

Table 3. Tissue Distribution of the Different Genes

Tissue	*PMCA1*	*PMCA2*	*PMCA3*	*PMCA4*
Brain	+	+	+	+
Skeletal muscle	+	*a*	*b*	+
Heart muscle	+	*a,b*	–	+
Smooth muscle	+	*a*	–	+
Other tissues	+	–	–	+

Notes: [a]Only detected by PCR.
[b]Very small amount detected by Northern blot.

mRNA for genes 1 and 4 (PMCA1 and PMCA4) was found in all analyzed human and rat tissues (Table 3) (Greeb and Shull, 1989; Keeton et al., 1993; Stauffer et al., 1993). The amount of mRNA for PMCA1 was 40 to 60% and those for PMCA4 25 to 50% of the total PMCA mRNA (Stauffer et al., 1993). These results were confirmed by protein sequencing of purified PMCA from red blood cells (Strehler et al., 1990): all peptides isolated and sequenced were specific for either PMCA1 or PMCA4 gene products. PMCA1 and PMCA4 therefore represent the housekeeping pumps. By contrast, PMCA2 and PMCA3 genes are expressed only in a limited number of tissues. Northern blots of rat mRNA (Greeb and Shull, 1989) and quantification of human mRNA (Stauffer et al., 1993) indicated that the mRNA for these two genes are present in detectable amounts only in the brain. The presence of mRNA for these isoforms is at least 10 to 100 times lower in other tissues than in brain. *In situ* hybridization on rat brain slices indicated that PMCA1, 2, and 3 are expressed to a different extent in different subcellular populations (Stahl et al., 1992). For example, a high level of PMCA2 mRNA was detected in the cerebellum and, more specifically, in Purkinje cells, whereas a high level of mRNA specific for PMCA1 was detected in the hippocampus. PMCA3 generally showed a 5–10 times lower signal, but a very strong hybridization was found in the habenula and the choroid plexus (Stahl et al., 1992).

Most of the data collected on the tissue distribution of the PMCA gene products were obtained by studying mRNA. Even if the presence of the PMCA1 and PMCA4 pump protein had been demonstrated in human red blood cells by direct sequencing, only the availability of isoform specific antibodies permitted a more systematic analysis of the distribution of the corresponding proteins (Stauffer et al., 1995). PMCA2 and PMCA3 proteins could only be detected in the human brain. In all other tissues analyzed to date, PMCA1 was more abundant than PMCA4, the only exception being the human red blood cells. Interestingly, PMCA2 protein was most abundant in the cerebellum (Stauffer et al., 1995). For the PMCA3 isoform protein, only faint bands could be visualized, in line with the observation reported by Stahl et al. (1992) using *in situ* hybridization.

Table 4. Tissue Distribution of Alternative Splicing Products at Site C

Tissue	*PMCA1*	*PMCA2*	*PMCA3*	*PMCA4*
Brain	1b 1c 1a 1e	2b 2c	3a 3b 3g 3e	4b 4a
Skeletal muscle	1b 1c	—	—	4b
Heart muscle	1b 1c	—	—	4b 4a
Smooth muscle	1b	—	—	4b
Other tissues	1b	—	—	4b

These results indicated that not only the different gene products, but also their alternatively spliced isoforms demonstrated differences in tissue distribution (Strehler et al., 1989; Brandt et al., 1992a; Keeton et al., 1993; Stauffer et al., 1993). Data for the two major alternative splicing sites (A and C) have been collected (Table 4). Type b isoforms of PMCA1 and PMCA4 have been found in all tissues tested, whereas type a isoforms were restricted to brain (PMCA1 and PMCA4) and to heart muscle (PMCA4, where ca. 50% of the PMCA4 mRNA is of type a; Stauffer et al., 1993). PMCA1c isoform was detected in brain, skeletal, and heart muscle (Stauffer et al., 1993; Table 4). Differences in induction of alternatively spliced PMCA isoforms were observed during muscle and nerve cells differentiation (Hammes et al., 1994). In general, the brain and muscle were those tissues where the highest amount of different spliced isoforms have been detected. Similar experiments have been carried out for splicing site A, and also in this case, a tissue-specific expression for PMCA4 isoforms were observed: in heart two possible spliced isoforms were found, whereas in other tissues, only the PMCA4x could be detected (Table 4) (Stauffer et al., 1993).

XII. OVEREXPRESSION OF PMCA

Cloning of the different PMCA genes raised the question of their function and regulation of expression. As indicated before, conclusions concerning the possible isoforms were based on the analysis of mRNA. To analyze the influence of differently expressed isoforms, efforts have been made to develop expression methods using DNA-recombinant technology.

One of the results determining splicing site C within the CaM-binding domain was the identification of four different isoforms, that is, PMCA1 a, b, c, and d. In order to study their influence on the affinity to CaM, C-terminal peptides starting right after the tenth transmembrane domain have been expressed in bacteria and were purified (Kessler et al., 1992). The results indicated a difference in affinity to CaM. These findings have been corroborated by expression of the full-length PMCA4 in COS cells: type a spliced isoform demonstrated a weaker affinity to CaM than type b (Enyedi et al., 1994).

Expression of full-length PMCA4 cDNA has been achieved by transient expression in COS cells (Adamo et al., 1992; Heim et al., 1992b; Enyedi et al., 1994) and by using the baculovirus expression system (Heim et al., 1992b). The latter system permitted the expression of the pump in amounts amenable to purification in an active state by CaM-affinity chromatography (Heim et al., 1992b). It was also demonstrated that an N-terminally truncated pump (lacking the region encompassing transmembranes 1 and 2) was not active, even if it was targeted correctly to the plasma membrane (Heim et al., 1992b). This result underlined the importance of this region for a functional, competent pump.

An interesting observation was made concerning a difference in calmodulin-affinity between the isoforms PMCA2 and PMCA4. The overexpression of these isoforms in insect cells permitted a direct comparison of their biochemical properties: the PMCA2b protein bound 5–10 times stronger to calmodulin than the PMCA4b protein (Hilfiker et al., 1994). In addition, the two isoforms also demonstrated differences in their affinity for ATP (Hilfiker et al., 1994).

XIII. CONCLUSIONS

In summary, the following conclusions can be made:

1. The Ca^{2+}-pump of plasma membranes (PMCA) belongs to the P-type ATPases as classified by the formation of a high-energy acylphosphate intermediate. The enzyme is responsible for the fine tuning of Ca^{2+} homeostasis in eucaryotic cells by pumping Ca^{2+} out of cells against its ion gradient with a Ca^{2+}/ATP stoichiometry of 1.
2. The PMCA is a protein with a molecular weight of ca. 140 kD depending on its isoform. The enzyme can be activated by calmodulin due to a direct, Ca^{2+}-dependent protein–protein interaction thereby lowering the K_M value for calcium to the nanomolar range and increasing the V_{Max} value up to 10-fold.
3. Next to calmodulin, PMCA can also be activated by negatively charged phospholipids, by polyunsaturated fatty acids, by phosphorylation, by controlled proteolysis, or by oligomerization.
4. A topographical model of the primary structure of PMCA indicates 10-transmembrane domains with both the N- and the C-terminus located on the cytosolic side of the plasma membrane. 80% of the protein mass protrudes into the cytosol.
5. In mammalian organisms, at least four different genes exist, each giving rise to additional isoforms due to alternative splicings. Two main splice sites have been identified, one located near the N-terminus, the other in the C-terminal region within the calmodulin-binding domain. The latter is responsible for possible differences in the calmodulin affinity of the enzyme.
6. In human, the chromosomal localization of the 4 different genes have been mapped: PMCA 1 is located in 12q21–23, PMCA 2 in 3p25–26, PMCA 3 in the Xq28 chromosome, and PMCA 4 in 1q25–32. PMCA 1 and 4 are

"housekeeping" genes, whereas 2 and 3 seem to be more specific in their tissue distribution.

7. The structure of 2 genes have been characterized in detail, that is, PMCA1 for human and PMCA 3 for rat. Both genes are very large, that is, rPMCA 3 contains ca. 90 kbp and hPMCA 1 more than 100 kbp. hPMCA 1 is composed of 22 exons, rPMCA 3 contains 24 exons. By comparing the 2 different genes, the conserved size of the exons is striking.

ACKNOWLEDGMENTS

The authors are indebted to Dr. Ernesto Carafoli for his continuous interest and support. Communication of results by Drs. Inesi and McBride prior to publication is gratefully acknowledged. Rolf Moser is thanked for his help in designing the figures.

REFERENCES

Adamo, H. P., & Penniston, J. T. (1992). New Ca^{2+} pump isoforms generated by alternative splicing of rPMCA2 mRNA. Biochem. J. 283, 355–359.

Adamo, H. P., Verma, A. K., Sanders, M. A., Heim, R., Salisbury, J. L., Wieben, E. D., & Pennistan, J. T. (1992). Overexpression of the erythrocyte plasma membrane Ca^{2+} pump in COS-1 cells. Biochem. J. 285, 791–797.

Antebi, A., & Fink, G. R. (1992). The yeast Ca^{2+} ATPase homologue, PMR1, is required for normal Golgi function and localizes in a novel Golgi-like distribution. Mol. Biol. Cell. 3, 633–654.

Au, K. S. (1986). Activation of erythrocyte membrane Ca^{2+}-ATPase by calpain. Biochim. Biophys. Acta 905, 273–278.

Barrabin, H., Garrahan, P. J., & Rega, A. F. (1980). Vanadate inhibition of the Ca^{2+}-ATPase from human red cell membranes. Biochim. Biophys. Acta 600, 796–804.

Benaim, G., Zurini, M., & Carafoli, E. (1984). Different conformational states of the purified Ca^{2+}-ATPase of the erythrocyte plasma membrane revealed by controlled proteolysis. J. Biol. Chem. 259, 8471–8477.

Bond, G. H., & Clough, D. L. (1973). A soluble protein activator of (Mg^{2+} + Ca^{2+})-dependent ATPase in human red cell membranes. Biochim. Biophys. Acta 323, 592–599.

Brandt, P., Zurini, M., Neve, R. L., Rhoads, R. E., & Vanaman, T. C. (1988). A C-terminal, calmodulin-like regulatory domain of the plasma membrane Ca^{2+}-pumping ATPase. Proc. Natl. Acad. Sci. USA 85, 2914–2918.

Brandt, P., Neve, R. L., Kammesheidt, A., Rhoads, R. E., & Vanaman, T. C. (1992a). Analysis of the tissue-specific distribution of mRNAs encoding the plasma membrane calcium-pumping ATPase and characterisation of an alternately spliced form of PMCA4 at the cDNA and genomic levels. J. Biol. Chem. 267, 4376–4385.

Brandt, P., Ibrahim, E., Bruns, G. A. P., & Neve, R. L. (1992b). Determination of the nucleotide sequence and chromosomal localization of the ATP2B2 gene encoding the human Ca^{2+}-pumping ATPase isoform PMCA2. Genomics 14, 484–487.

Brodin, P., Falchetto, R., Vorherr, T., & Carafoli, E. (1992). Identification of two domains which mediate the binding of activating phospholipids to the plasma membrane Ca^{2+} pump. Eur. J. Biochem. 204, 939–946.

Burk, S. E., & Shull, G. E. (1992). Structure of the rat membrane Ca^{2+}-ATPase isoform 3 gene and characterisation of alternative splicing and transcription products. J. Biol. Chem. 267, 19683–19690.

Buschmeier, B., Meyer, H. E., & Mayr, G. W. (1987). Characterization of the calmodulin-binding sites of muscle phosphofructokinase and comparison with known calmodulin-binding domains. J. Biol. Chem. 262, 9454–9462.

Carafoli, E. (1991). Calcium pump of the plasma membrane. Physiol. Rev. 71, 129–153.

Carafoli, E. (1992). The Ca^{2+}-pump of the plasma membrane. J. Biol. Chem. 267, 2115–2118.

Carafoli, E., & Guerini, D. (1993). Molecular and cellular biology of plasma membrane calcium ATPase. Trends Cardiovasc. Med. 3, 177–184.

Caroni, P., & Carafoli, E. (1981a). The Ca^{2+} pumping ATPase of heart sarcolemma is activated by a phosphorylation–dephosphyralation process. J. Biol. Chem. 256, 9371–9373.

Caroni, P., & Carafoli, E. (1981b). The Ca^{2+}-pumping ATPase of heart sarcolemma. Characterization, calmodulin dependence, and partial purification. J. Biol. Chem. 256, 3263–3270.

Chan, K. M., & Junger, K. D. (1983). Calcium transport and phosphorylated intermediate of (Ca^{2+} + Mg^{2+})-ATPase in plasma membranes of rat liver. J. Biol. Chem. 253, 4404–4410.

Cheung, W. Y. (1970). Cyclic 3′,5′-nucleotide phosphodiesterase. Demonstration of an activator. Biochem. Biophys. Res. Commun. 38, 533–538.

Chiesi, M., Zurini, M., & Carafoli, E. (1984). ATP synthesis catalyzed by the purified erythrocyte Ca-ATPase in the absence of calcium gradients. Biochemistry 23, 2595–2600.

Clark, A., & Carafoli, E. (1983). The stoichiometry of the Ca^{2+}-pumping ATPase of erythrocytes. Cell Calcium 4, 83–88.

Clarke, D. M., Loo, T. W., Inesi, G., & MacLennan, D. H. (1989). Location of high affinity Ca^{2+}-binding sites within the predicted transmembrane domain of the sarcoplasmic reticulum Ca^{2+}-ATPase. Nature (London) 239, 476–478.

Cohen, P., & Klee, C. B. (1988). Calmodulin. In: Molecular Aspects of Cellular Regulation, Vol. 5, (Cohen, P., & Klee, C. B., eds.) Elsevier, Amsterdam.

Cunningham, K. W., & Fink, G. R. (1994). Calcineurin-dependent growth control in *Saccharomyces cerevisiae* mutants lacking PMC1, a homologue of plasma membrane Ca^{2+} ATPases. J. Cell. Biol. 124, 351–362.

De Jaegere, S., Wuytack, F., Eggermont, J. A., Verboomen, H., & Casteel, R. (1990). Molecular cloning and sequencing of the plasma-membrane Ca^{2+} pump of pig smooth muscle. Biochem. J. 271, 655–660.

De Meis, L., Martins, O. B., & Alves, E. W. (1980). Role of water, hydrogen ion, and temperature on the synthesis of adenosinetriphosphate by the sarcoplasmic reticulum adenosinetriphosphatase in the absence of calcium ion gradient. Biochemistry 19, 4252–4261.

De Meis, L., Souza-Otero, L. A., Martins, O. B., Alves, E. W., Inesi, G., & Nakamoto, R. (1982). Phosphorylation of sarcoplasmic reticulum ATPase by orthophosphate in the absence of Ca^{2+}-gradient. J. Biol. Chem. 257, 4993–4998.

Di Polo, R. (1979). Ca pump driven by ATP in squid axon. Nature (London) 274, 390–392.

Drabikowski, W., Kuznicki, J., & Grabarek, Z. (1977). Similarity in Ca^{2+}-induced changes between troponin C and protein activator of 3′:5′-cyclic nucleotide phosphodiesterase and their tryptic fragments. Biochim. Biophys. Acta 485, 124–133.

Dunham, E. T., & Glynn, I. M. (1961). Adenosine-triphosphatase activity and the active movements of alkali metal ions. J. Physiol. (London). 156, 274–293.

Enyedi, A., Sarkadi, B., Szasz, I., Bot, B., & Gardos, G. (1980). Molecular properties of the red cell calcium pump. II. Effects of proteolysis, proteolytic digestion and drugs on the calcium-induced phosphorylation by ATP in inside/out red cell membrane vesicles. Cell Calcium 1, 299–310.

Enyedi, A., Flura, M., Sarkadi, B., Gardos, G., & Carafoli, E. (1987). The maximal velocity and the calcium affinity of the red cell calcium pump may be regulated independently. J. Biol. Chem. 262, 6425–6430.

Enyedi, A., Verma, A. K., Heim, R., Adamo, H. P., Filoteo, A. G., Strehler, E. E., & Penniston, J. P. (1994). The Ca^{2+} affinity of the plasma membrane Ca^{2+} pump is controlled by alternative splicing. J. Biol. Chem. 269, 41–43.

Falchetto, R., Vorherr, T., Brunner, J., & Carafoli, E. (1991). The plasma membrane Ca^{2+}-pump contains a site that interacts with its calmodulin-binding domain. J. Biol. Chem. 266, 2930–2936.

Falchetto, R., Vorherr, T., & Carafoli, E. (1992). The calmodulin binding site of the plasma membrane Ca^{2+} pump interacts with the transduction domain of the enzyme. Protein Sci. 1, 1613–1621.

Feschenko, M. S., Zvaritch, E. I., Hofmann, F., Shakhparonov, M. I., Modyanov, N. N., Vorherr, T., & Carafoli, E. (1992). A monoclonal antibody recognizes an epitope in the first extracellular loop of the plasma membrane Ca^{2+} pump. J. Biol. Chem. 267, 4097–4101.

Filoteo, A. G., Gorski, J. P., & Penniston, J. T. (1987). The ATP-binding site of the erythrocyte membrane Ca^{2+} pump. J. Biol. Chem. 262, 6526–6530.

Gao, Z. H., Krebs, J., VanBerkum, M. F. A., Tang, W-J., Maune, J. F., Means, A. R., Stull, J. T., & Beckingham, K. (1993). Activation of four enzymes by two series of calmodulin mutants with point mutations in individual Ca^{2+} binding sites. J. Biol. Chem. 268, 20096–20104.

Gopinath, R. M., & Vincenzi, F. F. (1977). Phosphodiesterase protein activator mimics red blood cell cytoplasmatic activator of the (Ca^{2+} + Mg^{2+}) ATPase. Biochem. Biophys. Res. Commun. 77, 1203–1209.

Greeb, J., & Shull, G. E. (1989). Molecular cloning of a third isoform of the calmodulin-sensitive plasma membrane Ca^{2+} transporting ATPase that is expressed predominantly in brain and skeletal muscle. J. Biol. Chem. 264, 18569–18576.

Guerini, D., Krebs, J., & Carafoli, E. (1984). Stimulation of the purified erythrocyte Ca^{2+}-ATPase by tryptic fragments of calmodulin. J. Biol. Chem. 259, 15172–15177.

Guerini, D., Krebs, J., & Carafoli, E. (1987). Stimulation of the erythrocyte Ca^{2+}-ATPase and of bovine brain cyclic nucleotide phosphodiesterase by chemically modified calmodulin. Eur. J. Biochem. 170, 35–42.

Gunteski-Hamblin, A.-M., Clark, D. M., & Shull, G. E. (1992). Molecular cloning and tissue distribution of alternatively spliced mRNAs encoding possible mammalian homologues of the yeast secretory pathway calcium pump. Biochemistry 31, 7600–7608.

Hammes, A., Oberdorf, S., Strehler, E. E., Stauffer, T., Carafoli, E., Vetter, H., & Neyses, L. (1994). Differentiation-specific isoform mRNA expression of the calmodulin-dependent plasma membrane Ca^{2+}-ATPase. FASEB J. 8, 428–435.

Hao, L., Rigaud, J-L., & Inesi, G. (1994). Ca2+/H^+ countertransport and electrogenicity in proteoliposomes containing erythrocyte plasma membrane Ca-ATPase and exogenous lipids. J. Biol. Chem. 269, 16656–16661.

Hasselbach, W., & Wass, W. (1982). Energy coupling in sarcoplasmic reticulum Ca^{2+} transport. An overview. Ann. N. Y. Acad. Sci. USA 402, 459–469.

Heidorn, D. B., Seeger, P. A., Rokop, S. E., Blumenthal, D. K., Means, A. R., Crespi, H., & Trewhella, J. (1989). Changes in the structure of calmodulin induced by a peptide based on the calmodulin-binding domain of myosin light chain kinase. Biochemistry 28, 6757–6764.

Heim, R., Hug, M., Iwata, T., Strehler, E. E., & Carafoli, E. (1992a). Microdiversity of human-plasma-membrane calcium-pump isoform 2 generated by alternative RNA splicing in the N-terminal coding region. Eur. J. Biochem. 205, 333–340.

Heim, R., Iwata, T., Zvaritch, E., Adamo, H. P., Rutishauser, B., Strehler, E. E., Guerini, D., & Carafoli, E. (1992b). Expression, purification, and properties of the plasma membrane Ca^{2+} pump and its N-terminally truncated 105-kDa fragment. J. Biol. Chem. 267, 24476–24484.

Hilfiker, E., Strehler, M-A., Carafoli, E., & Strehler, E. E. (1993). Structure of the gene encoding the human plasma membrane calcium isoform 1. J. Biol. Chem. 268, 19717–19725.

Hilfiker, H., Guerini, D., & Carafoli, E. (1994). Cloning and expression of isoform 2 of the human plasma membrane Ca^{2+} ATPase. J. Biol. Chem. 269, 26178–26183.

Hofmann, F., James, P., Vorherr, T., & Carafoli, E. (1993). The C-terminal domain of the plasma membrane Ca^{2+} pump contains three high affinity Ca^{2+} binding sites. J. Biol. Chem. 268, 10252–10259.

Howard, A., Legon, S., & Walters, J. R. (1993). Human and rat intestinal plasma membrane pump isoforms. Am. J. Physiol. 265, P:G917–25.

Ikura, M., Clore, G. M., Groneborn, A. M., Zhu, G., Klee, C. B., & Bax, A. (1992). Solution structure of a calmodulin-target peptide complex by multidimensional NMR. Science 256, 632–638.

James, P., Zvaritch, E., Hakhparanov, M. S., Penniston, J. T., & Carafoli, E. (1987). The amino acid sequence of the phosphorylation domain of the erythrocyte Ca^{2+}-ATPase. Biochem. Biophys. Res. Commun. 149, 7–12.

James, P., Maeda, M., Fischer, R., Verma, A. K., Krebs, J., Penniston, J. T., & Carafoli, E. (1988). Identification and primary structure of a calmodulin binding domain of the Ca^{2+}-pump of human erythrocytes. J. Biol. Chem. 263, 2905–2910.

James, P., Vorherr, T., Krebs, J., Morelli, A., Castello, G., McCormick, D. J., Penniston, J. T., De Flora, A., & Carafoli, E. (1989a). Modulation of erythrocyte Ca^{2+}-ATPase by selective calpain cleavage of the calmodulin binding domain. J. Biol. Chem. 264, 8289–8296.

James, P. H., Pruschy, M., Vorherr, T., Penniston, J. T., & Carafoli, E. (1989b). Primary structure of the cAMP-dependent phosphorylation site of the plasma membrane calcium pump. Biochemistry 28, 4253–4258.

Jarrett, H. W., & Penniston, J. T. (1977). Partial purification of the (Ca^{2+} + Mg^{2+}) ATPase activator from human erythrocytes: Its similarity to the activator of 3′-5′-cyclic nucleotide phosphodiesterase. Biochem. Biophys. Res. Commun. 77, 1210–1216.

Kahn, I., & Grover, A. K. (1991). Expression of cyclic-nucleotide-sensitive and -insensitive isoforms of the plasma membrane Ca^{2+} pump in smooth muscle and other tissues. Biochem. J. 277, 345–349.

Kakiuchi, S., & Yamasuki, R. (1970). Calcium dependent phosphodiesterase activity and its activating factor (PAF) from brain. Biochem. Biophys. Res. Commun. 41, 1104–1110.

Kataoka, M., Head, J. F., Seaton, B. A., & Engelman, D. M. (1989). Melittin binding causes a large calcium-dependent conformational change in calmodulin. Proc. Natl. Acad. Sci. USA 86, 6944–6948.

Kataoka, M., Head, J. F., Vorherr, T., Krebs, J., & Carafoli, E. (1991). Small-angle X-ray scattering study of calmodulin bound to two peptides corresponding to parts of the calmodulin-binding domain of the plasma membrane Ca^{2+} pump. Biochemistry 30, 6247–6251.

Keeton, T. P., Burk, S. E., & Shull, G. E. (1993). Alternative splicing of exons encoding the calmodulin-binding domains and C-termini of the plasma membrane Ca^{2+}-ATPase isoforms 1, 2, 3 and 4. J. Biol. Chem. 268, 2740–2748.

Kennelly, P. J., Edelman, A. M., Blumenthal, D. K., & Krebs, E. G. (1987). Rabbit skeletal muscle myosin light chain kinase. The calmodulin binding domain as a potential active site-directed inhibitor domain. J. Biol. Chem. 262, 11958–11963.

Kessler, F., Bennardini, F., Bachs, O., Serratosa, J., James, P., Caride, A. J., Gazzotti, P., Penniston, J. T., & Carafoli, E. (1990). Partial purification and characterization of the pumping ATPase of the liver plasma membrane. J. Biol. Chem. 265, 16012–16019.

Kessler, F., Falchetto, R., Heim, R., Meili, R., Vorherr, T., Strehler, E. E., & Carafoli, E. (1992). Study of calmodulin binding to the alternatively spliced C-terminal domain of the plasma membrane Ca^{2+} pump. Biochemistry 31, 11785–11792.

Kosk-Kosicka, D., & Bzdega, T. (1988). Activation of the erythrocyte Ca^{2+}-ATPase by either self-association or interaction with calmodulin. J. Biol. Chem. 263, 18184–18189.

Kosk-Kosicka, D., Bzdega, T., & Johnson, J. D. (1990). Fluorescence studies on calmodulin binding to erythrocyte Ca^{2+}-ATPase different oligomerization states. Biochemistry 29, 1875–1879.

Krebs, J., Vasak, M., Scarpa, A., & Carafoli, E. (1987). Conformational differences between the E1 and E2 states of the calcium adenosinetriphosphatase of the erythrocyte plasma membrane as revealed by circular dichroism and fluorescence spectroscopy. Biochemistry 26, 3921–3926.

Krebs, J., Vorherr, T., James, P., Carafoli, E., Craig, T. A., & Watterson, D. M. (1990). Structural details of the interaction of calmodulin with the plasma membrane Ca^{2+}-ATPase. Adv. Exp. Med. Biol. 269, 163–167.

Kumar, R., Haugen, J. D., & Penniston, J. T. (1993). Molecular cloning of the plasma membrane calcium pump from normal human osteoblasts. J. Bone Miner. Res. 8, 505–513.

Latif, F., Duh, F.-M., Gnarra, J., Kalman, T., Kuzmin, I., Yao, M., Stackhouse, T., Modi, W., Geil, L., Schmidt, L., Hua, L., Orcutt, M. L., Maher, E., Richards, F., Phipps, M., Ferguson-Smith, M., Le Paslier, D., Linhan, W. M., Zbar, B., & Lerman, M. I. (1993a). Von Hippel–Lindau syndrome: Cloning and identification of the plasma membrane Ca^{++}-transporting ATPase isoform 2 gene that resides in the van Hippel–Lindau gene region. Cancer Res. 53, 861–867.

Latif, F., Tory, K., Gnarra, J., Yao, M., Duh, F.-M., Orcutt, M. L., Stackhouse, T., Kuzmin, I., Modi, W., Geil, L., Schmidt, L., Zhou, F., Li, H., Wei, M. H., Chen, F., Glenn, G., Choyke, P., Walther, Mc. C. M., Weng, Y., Duan, D.-SR., Dean, M., Glavac, D., Richards, F. M., Crossey, P. A., Ferguson-Smith, M. A., Le Paslier, D., Chumakov, I., Cohen, D., Chinault, C., Maher, E. R., Linhan, W. M., Zbar, B., & Lerman, M. (1993b). Identification of the von Hippel–Lindau disease tumor suppressor gene. Science 260, 1317–1320.

Lotersztajn, S., Epand, R. M., Malat, A., & Pecker, F. (1984). Inhibition by glucagon of the Ca^{2+}-pump in liver plasma membranes. J. Biol. Chem. 259, 8195–8201.

Lynch, T. J., & Cheung, W. Y. (1979). Human erythrocyte (Ca^{2+} Mg^{2+})-ATPase: Mechanism of stimulation by Ca^{2+}. Arch. Biochem. Biophys. 194, 165–170.

MacLennan, D. H., Brandl, C. J., Korczak, B., & Green, N. M. (1985). Amino-acid sequence of a Ca^{2+} + Mg^{2+}-dependent ATPase from rabbit muscle sarcoplasmic reticulum, deduced from its complementary DNA sequence. Nature (London) 316, 696–700.

Makinose, M., & Hasselbach, W. (1971). ATP synthesis by the reverse of the sarcoplasmic reticulum calcium pump. FEBS Lett. 12, 271–272.

Meador, W. E., Means, A. R., & Quiocho, F. A. (1992). Target enzyme recognition by calmodulin: 2.4 Å structure of a calmodulin–peptide complex. Science 257, 1251–1255.

Newton, D. L., Oldewurtel, M. D., Krinks, M. H., Shiloach, J., & Klee, C. B. (1984). Agonist and antagonist properties of calmodulin fragments. J. Biol. Chem. 259, 4419–4426.

Neyses, L., Reinlib, L., & Carafoli, E. (1985). Phosphorylation of the Ca^{2+}-pumping ATPase of heart sarcolemma and erythrocyte plasma membrane by the cAMP-dependent protein kinase. J. Biol. Chem. 260, 10283–10287.

Niggli, V., Penniston, J. T., & Carafoli, E. (1979). Purification of the (Ca^{2+}-Mg^{2+})-ATPase from human erythrocytes membranes using a calmodulin affinity column. J. Biol. Chem. 254, 9955–9958.

Niggli, V., Adunyah, E. S., Penniston, J. T., & Carafoli, E. (1981a). Purified (Ca^{2+} + Mg^{2+}) ATPase of the erythrocyte membrane: Reconstitution and effect of calmodulin and phospholipids. J. Biol. Chem. 256, 395–401.

Niggli, V., Adunyah, E. S., & Carafoli, E. (1981b). Acidic phospholipids, unsaturated fatty acids, and limited proteolysis mimic the effect of calmodulin on the purified erythrocyte Ca^{2+}-ATPase. J. Biol. Chem. 256, 8588–8592.

Niggli, V., Sigel, E., & Carafoli, E. (1982). The purified Ca^{2+}-pump of human erythrocyte membranes catalyzes an electroneutral Ca^{2+}-H^{+} exchange in reconstituted liposomal systems. J. Biol. Chem. 257, 2350–2356.

Olson, S., Wang, M. G., Carafoli, E., Strehler, E. E., & McBride, O. W. (1991). Localisation of two genes encoding plasma membrane Ca^{2+}-transporting ATPases to human chromosomes 1q25–32 and 12q21–23. Genomics 9, 629–641.

Pavoine, C., Lotersztayn, S., Mallat, A., & Pecker, F. (1987). The high affinity (Ca^{2+}-Mg^{2+})-ATPase in liver plasma membranes is a Ca^{2+} pump. J. Biol. Chem. 262, 5113–5117.

Payne, M. E., Fong, Y.-L., Ono, T., Colbran, R. J., Kemp, B. E., Soderling, T. R., & Means, A. R. (1988). Calcium/calmodulin-dependent protein kinase II. Characterization of distinct calmodulin binding and inhibitory domains. J. Biol. Chem. 263, 7190–7195.

Pedersen, P. L., & Carafoli, E. (1987a). Ion motive ATPases. I. Ubiquity, properties, and significance for cell function. Trends Biochem. Sci. 12, 146–150.

Pedersen, P. L., & Carafoli, E. (1987b). Ion motive ATPases. II. Energy coupling and work output. Trends Biochem. Sci. 12, 186–189.

Penniston, J. T. (1983). Plasma membrane Ca^{2+}-ATPases as active Ca^{2+} pumps. In: Calcium and Cell Function, Vol. 4, (Cheung, W. Y., ed.) pp. 100–149. Academic Press, N. Y.

Peterson, S. W., Ronner, P., & Carafoli, E. (1978). Partial purification and reconstitution of the (Ca^{2+} + Mg^{2+})-ATPase of erythrocyte membranes. Arch. Biochem. Biophys. 186, 202–210.

Quist, E. E., & Roufogalis, B. D. (1975). Determination of the stoichiometry of the calcium pump in human erythrocytes using lanthanum as a selective inhibitor. FEBS Lett. 50, 135–139.

Roelofsen, B., & Schatzmann, H. J. (1977). The lipid requirement of the (Ca^{2+} Mg^{2+}) ATPase in the erythrocyte membrane, as studied by various highly purified phospholipases. Biochim. Biophys. Acta 464, 17–36.

Ronner, P., Gazzotti, P., & Carafoli, E. (1977). A lipid requirement of the (Ca^{2+} Mg^{2+})-activated ATPase of erythrocyte membranes. Arch. Biochem. Biophys. 179, 578–583.

Rossi, J. P. F. C., Garrahan, P. J., & Rega, A. F. (1978). Reversal of the calcium pump in human red cells. J. Membr. Biol. 44, 37–46.

Rudolph, H., Antebi, A., Fink, G. R., Buckley, C. M., Dorman, T. E., LeVitre, J.-A., Davidow, L. S., Mao, J.-I., & Moir, D. T. (1989). The yeast secretory pathway is perturbed by mutations in PMR1, a member of a Ca^{2+} ATPase family. Cell 58, 133–145.

Schatzmann, H. J. (1966). ATP-dependent Ca^{2+} extrusion from human red cells. Experientia (Basel) 22, 364–368.

Schatzmann, H. J. (1973). Dependence on calcium concentration and stoichiometry of the calcium pump in human red cells. J. Physiol. (London). 235, 551–569.

Schatzmann, H. J. (1982). The calcium pump of erythrocytes and other animal cells. In: Membrane Transport of Calcium (Carafoli, E., ed.) pp. 41–108. Academic Press, London.

Schatzmann, H. J., & Vincenzi, F. F. (1969). Calcium movements across the membrane of human red cells. J. Physiol. 201, 369–395.

Schlesser, R. B., Ulaszewski, S., Ghislain, M., & Goffeau, A. (1988). A second transport ATPase gene in *Saccharomyces cerevisiae*. J. Biol. Chem. 263, 19480–19487.

Shull, G. E., & Lingrel, J. B. (1986). Molecular cloning of the rat stomach (H^{+} + K^{+})-ATPase. J. Biol. Chem. 261, 16788–16791.

Shull, G. E., & Greeb, J. (1988). Molecular cloning of two isoforms of the plasma membrane Ca^{2+}-transporting ATPase from rat brain. J. Biol. Chem. 263, 8646–8657.

Skerjanc, I. S., Clarke, D. M., Loo, T. W., & MacLennan, D. H. (1993). Deletion of the NH_2- and COOH-terminal sequences destroys function of the Ca^{2+}ATPase of rabbit fast-twitch skeletal muscle sarcoplasmic reticulum. FEBS Lett. 336, 168–170.

Smallwood, J. I., Gügi, B., & Rasmussen, H. (1988). Modulation of erythrocyte Ca^{2+}pump activity by protein kinase C. J. Biol. Chem. 263, 2195–2202.

Stahl, W. L., Eakin, T. J., Owens, J. W. M., Breininger, J. F., Filuk, P. E., & Anderson, W. R. (1992). Plasma membrane Ca^{2+}-ATPase isoforms: Distribution of mRNAs in rat brain by in situ hybridisation. Mol. Brain Res. 16, 223–231.

Stauffer, T. P., Hilfiker, H., Carafoli, E., & Strehler, E. E. (1993). Quantitative analysis of alternative splicing options of human plasma membrane calcium pump. J. Biol. Chem. 268, 25993–26003.

Stauffer, T., Guerini, D., & Carafoli, E. (1995). Tissue distribution of the four gene products of the plasma membrane Ca^{2+} pump. J. Biol. Chem. 270, 12184–12190.

Strehler, E. E. (1991). Recent advances in the molecular characterisation of plasma membrane Ca^{2+} pumps. J. Membrane Biol. 120, 1–15.

Strehler, E. E., Strehler-Page, M.-A., Vogel, G., & Carafoli, E. (1989). mRNAs for plasma membrane calcium pump isoforms differing in their regulatory domain are generated by alternative splicing that involves two internal donor sites in a single exon. Proc. Natl. Acad. Sci. USA 86, 6908–6912.

Strehler, E. E., James, P., Fischer, R., Heim, R., Vorherr, T., Filoteo, A., Penniston, J. T., & Carafoli, E. (1990). Peptide sequence analysis and molecular cloning reveal two calcium isoforms in the human erythrocyte. Membrane. J. Biol. Chem. 265, 2835–2842.

Taverna, R. D., & Hanahan, D. H. (1980). Modulation of human erythrocyte Ca^{2+}/Mg^{2+} ATPase activity by phospholipase A_2 and proteases. A comparison with calmodulin. Biochem. Biophys. Res. Commun. 94, 652–659.

Verma, A. K., Filoteo, A. G., Stanford, D. R., Wieben, E. D., Penniston, J. T., Strehler, E. E., Fischer, R., Heim, R., Vogel, G., Mathews, S., Strehler-Page, M.-A., James, P., Vorherr, T., Krebs, J., & Carafoli, E. (1988). Complete primary structure of a human plasma membrane Ca^{2+} pump. J. Biol. Chem. 263, 14152–14159.

Vorherr, T., James, P., Krebs, J., Enyedi, A., McCormick, D. J., Penniston, J. T., & Carafoli, E. (1990). The interaction of calmodulin with the calmodulin binding domain of the plasma membrane Ca^{2+} pump. Biochemistry 29, 355–365.

Vorherr, T., Kessler, T., Hofmann, F., & Carafoli, E. (1991). The calmodulin-binding domain mediates the self-association of the plasma membrane Ca^{2+} pump. J. Biol. Chem. 266, 22–27.

Vorherr, T., Quadroni, M., Krebs, J., & Carafoli, E. (1992). Photoaffinity labeling study of the interaction of calmodulin with the plasma membrane Ca^{2+} pump. Biochemistry 31, 8245–8251.

Walsh, M., Stevens, F. C., Kuznicki, J., & Drabikowski, W. (1977). Characterization of tryptic fragments obtained from bovine brain protein modulator of cyclic nucleotide phosphodiesterase. J. Biol. Chem. 252, 7440–7443.

Wang, K. W. K., Villalobo, A., & Roufogalis, B. D. (1988). Activation of the Ca^{2+}-ATPase of human erythrocyte membrane by an endogenous Ca^{2+}-dependent neutral protease. Arch. Biochem. Biophys. 260, 696–704.

Wang, K. W. K., Wright, L. C., Machan, C. L., Allen, B. G., Conigrave, A. D., & Roufogalis, B. D. (1991). Protein kinase C phosphorylates the carboxyl terminus of the plasma membrane Ca^{2+}-ATPase from human erythrocytes. J. Biol. Chem. 266, 9078–9085.

Wang, M. G., Yi, H.-F., Hilfiker, H., Carafoli, E., Strehler, E. E., & McBride, O. W. (1994). Localization of two genes encoding plasma membrane Ca^{2+} ATPases isoforms 2 (ATP2B2) and 3 (ATP2B3) to human chromosomes 3p26→p25 and Xq28, respectively. Cytogenet. Cell Genet. 67, 41–45.

Wüthrich, A., Schatzmann, H. J., & Romero, P. (1979). Net ATP synthesis by running the red cell calcium pump backwards. Experientia (Basel). 35, 1589–1590.

Wuytack, F., Raeymaekers, L., Deschutter, G., & Casteels, R. (1982). Demonstration of the phosphorylated intermediates of the Ca^{2+} transport ATPase in a microsomal fraction and in a (Ca^{2+} + Mg^{2+})-ATPase purified from smooth muscle by means of calmodulin affinity chromatography. Biochim. Biophys. Acta 693, 45–52.

Yazawa, M., Vorherr, T., James, P., Carafoli, E., & Yagi, K. (1992). Binding of calcium by calmodulin: Influence of the calmodulin binding domain of the plasma membrane calcium pump. Biochemistry 31, 3171–3176.

Zurini, M., Krebs, J., Penniston, J. T., & Carafoli, E. (1984). Controlled proteolysis of the Ca^{2+} ATPase of the erythrocyte membrane. A correlation between structure and function of the enzyme. J. Biol. Chem. 259, 618–627.

Zvaritch, E., James, P., Vorherr, T., Falchetto, R., Modyanov, N., & Carafoli, E. (1990). Mapping of functional domains in the plasma membrane Ca^{2+} pump using trypsin proteolysis. Biochemistry 29, 8070–8076.

THE SODIUM PUMP

Flemming Cornelius

Biomembranes
Volume 5, pages 133–184.

ISBN: 1-55938-662-2.

I. INTRODUCTION

The sodium pump is an integral membrane bound protein which is found in all higher animal cells. It was named the sodium pump (Dean, 1941) because it was believed to primarily expel sodium from the cell. Actually, it couples the sodium transport out of the cell to potassium transport into the cell. The direction of both transports is against the prevailing electrochemical ion gradients, and the pump derives the necessary energy from the splitting of ATP: the protein is at the same time an ion-activated ATPase, the Na^+,K^+-ATPase (EC 3.6.1.37) and an ATP hydrolyzing ion-transporter, the sodium pump.

The sodium pump belongs to a family of membrane bound ion-motive ATPases which has been named the P-type ATPases, as opposed to both the vacuolar H^+-ATPases (V-type) and the F_0F_1 H^+-ATPases (F-type; Pedersen and Carafoli, 1987). The P-type ATPases are characterized by undergoing a phosphorylation during their reaction cycle, in which the γ-phosphoryl group from ATP is transferred to an aspartyl (Asp) residue in the enzyme (E), which forms a covalent phosphorylated intermediate (EP). Other members of the P-type ATPases are the Ca^{2+}-ATPases from reticulum, lysosomes, and the Golgi apparatus; the Ca^{2+}-ATPase from plasma membranes; the gastric H^+,K^+-ATPase; some H^+-ATPases of lower and higher eucaryotes; and the K^+-ATPase from some bacteria. Apart from their similarity in forming a covalent phosphorylated intermediate, they may appear quite different functionally, since their specificity to cation are very different, and structurally, since only the Na^+,K^+-ATPase and the H^+,K^+-ATPase have a β-subunit (see following).

The Na^+,K^+-ATPase was first identified in an isolated form in 1957 by J.C. Skou, who demonstrated the coupled activation by Na^+ and K^+ of ATPase activity in a preparation from crab nerves. Later, its specific inhibition by the digitaloid from Foxglove, ouabain (Schatzmann, 1953), was demonstrated (Skou, 1960; Dunham and Glynn, 1961). The Na^+,K^+-ATPase was isolated and purified for the first time from the outer medulla of the kidney (Jørgensen and Skou, 1971; Jørgensen, 1974, 1988) using ultracentrifugation or zonal centrifugation followed by activation by the detergents, DOC or SDS, to open up vesicular fractions. The best preparations are more than 90% pure and have a specific hydrolytic activity of 50 μmol P_i/mg protein·min.

The general aspects of the structure and function of the sodium pump have recently been extensively reviewed (Jørgensen and Andersen, 1988; Skou, 1990; Läuger, 1991b; Jørgensen, 1992; Glynn, 1993), including aspects of its enzyme kinetics (Cornelius, 1991; Glynn and Karlish, 1990; Repke and Schön, 1992; Robinson and Pratap, 1993) and electrogenecity (De Weer et al., 1988; Apell, 1989), physiological regulation and expression (Sweadner, 1989; Clausen and Everts, 1989; Geering, 1990, 1991; Ewart and Klip, 1995), chemical modification (Pedemonte and Kaplan, 1990), as well as its molecular biology and evolution employing the rapidly growing field of molecular genetics (Lingrel, 1991; Horisberger et al.,

1991). Since 1973, the proceedings from The International Conference on Na^+,K^+-ATPase held every third or fourth year contain the latest developments in this field. The latest publications are from 1991 (Kaplan and De Weer, 1991a, 1991b) and 1994 (Bamberg and Schoner, 1994).

II. PHYSIOLOGICAL FUNCTIONS

The importance of the sodium pump in cell physiology cannot be overemphasized. Some of the known direct and indirect impacts of the sodium pump on cellular physiology are briefly described following.

A. Osmoregulation

At an early stage in evolution, when a compliant cell membrane was found profitable at the expense of the constant volume keeping cell wall, the sodium pump as a dynamic volume regulatory device could have been developed, maybe by gene recombination of exons controlling cation specificity from existing proton pumps of the P-types in certain bacteria, fungal, and plant cells. A lucid perspective of the history of pumping has recently been given by Glynn (1993).

The animal cell membrane is a flexible lipid bilayer containing a mosaic of dissolved and attached proteins, with different specialized functions. The lipid bilayer will not withstand much stress, and the cell is accordingly dependent on controlling its volume within quite narrow limits. On the other hand, the cell contains impermeable, charged macromolecules which imposes an uneven passive distribution of the permeable ions across the cell membrane. This is the well known Donnan system, which if uncompensated leads to cell swelling and eventually lysis, since the higher osmotic concentrations due to macromolecules and permeable ions causes water to flow into the cell. The system can only attain equilibrium if a hydrostatic pressure develops due to the volume increase, imposing a fatally high wall tension in the cell membrane. In plant cells and bacteria, the problem is solved by the rigid cellulose cell walls. Animal cells, however, regulate their volume by membrane transport processes (for references see Macknight, 1987) in which the sodium pump is essentially involved (Conway, 1957), by actively creating an effectively low intracellular Na^+ concentration, thereby compensating for the surplus of macromolecules in the cell. A necessary condition for the sodium pump to work as an osmoregulatory device is the low ground permeability of the cell membrane to Na^+, otherwise the maintenance of a low intracellular steady-state concentration of Na^+ would be energetically too costly for the cell. This pump-and-leak theory (Tosteson and Hoffman, 1960) forms the basis for the volume regulatory action of the sodium pump.

The importance of the sodium pump in maintaining cell volume is demonstrated by its inhibition, which eventually induces swelling. The time that it takes, however, depends on the balance of passive permeabilities primarily for Na^+, K^+, and Cl^-, as well as the activation of cotransport systems with a volume regulatory function (Hofmann and Simonsen, 1989, and see next paragraph).

B. Energy Coupling

The primary active transport of the sodium pump maintains steady-state electrochemical gradients for both Na^+ (inwardly directed) and K^+ (outwardly directed) across the cell membrane. In most cells, a 30-fold concentration gradient for K^+ and a 10-fold concentration gradient for Na^+ is established. The liberation of free energy in the dissipation of these gradients can be coupled to the otherwise energetically unfavorable (uphill) transports of, e.g., nutrients or waste products across the cell membrane. Such transports are performed by specialized proteins which, by the passive transport of, for example, Na^+ (inward) or K^+ (outward) and the binding of a co-substrate or co-ligand, are fueled either for co-transport or counter-transport of the latter substance. These transport systems are named secondary active, as opposed to the primary active systems like the sodium pump itself, which derive their energy directly from the splitting of ATP or other energy rich compounds.

A large number of symporters and antiporters are known today and many of them share a common structural motif of 12 transmembrane helices with a central cytoplasmic loop containing an ATP-binding domain. Being the device that primarily makes the energy available to these symporters/antiporters, the sodium pump is of great, though indirect, importance for such vital cell functions as the uptake of nutrients (sugars and amino acids), the expulsion of metabolic end products (e.g., H^+), the regulation of internal pH via the Na^+/H^+ antiporter (which may be of importance in hypertension), the internal signaling by Ca^{2+} via the Na^+/Ca^{2+} antiporter (important also for the inotropic effect on heart muscle), the thyroid uptake of iodine via the Na^+/I^+ exchanger, and the volume regulatory cotransport systems like the $Na^+/K^+/2Cl^-$, the Na^+/Cl^-, and the Na^+/H^+ (probably involved in regulatory volume increase), and the K^+/Cl^-, and K^+/H^+ exchangers involved in regulatory volume decrease (for references see Hoffmann and Simonsen, 1989).

C. Heat Production

The contribution of ATP-hydrolyzing proteins is significant in metabolic energy turnover and important for the heat production: as much as 50% of the body weight of ATP per day is in the form of ATP hydrolysis at rest, increasing up to 25 times during exercise. In brain and kidney, the main sources of sodium pumps, more than 45% of the energy turnover is utilized for active Na^+,K^+-transport by the sodium pumps (Ismail-Beigi and Edelman, 1971); in skeletal muscle, heart, and liver somewhat less, about 10% (Clausen et al., 1991).

D. Water and Salt Balance

Transport epithelial cells are polarized with the sodium pumps restricted to the basolateral plasma membranes, essential for their function in transepithelial water and salt transport.

In the kidney tubulus cells, the sodium pump plays a key role for water resorption through CHIP28 water channels (Denker et al., 1988; Preston and Agre, 1991; Smith and Agre, 1991) and for the salt balance, by Na^+ reabsorption from the ultrafiltrate. In the transcellular transport of solutes, the spatial separation of sodium pumps and K^+-channels is important. The latter are localized exclusively to the basolateral plasma membrane, whereas sodium channels and co-transporters are present in the apical membrane domains. The CHIP28 water channels are present in both the apical and the basolateral membranes (Nielsen et al., 1993). The uptake of Na^+ and water through the channels in the apical membranes facing the lumen of the tubules is driven by the sodium pumps which expel the Na^+ and an equivalent amount of water through the water channels, into the blood supply. Paracellular transport is blocked to various degree by tight junctions, and the uptake of K^+ by the sodium pumps is recycled by the basolateral K^+-channels. Incorrect targeting of the sodium pumps, leading to their incorporation also in apical membrane domains, is associated with severe kidney diseases and ischemic damage.

E. Membrane Potential

The gradients for Na^+ and K^+ maintained by the sodium pump are the basis for the cell membrane potential, an inside-negative diffusion potential which results from the much lower ground permeability of Na^+ compared to that of K^+: the cell behaves much like a K^+-electrode.

Under physiological conditions, the sodium pump is electrogenic (see Section IVb) and generates current, since it per turnover transports $3Na^+$ for only $2K^+$. The net-positive outward current adds to the negative membrane diffusion potential, depending on the dissipative ion permeabilities (mainly for Cl^-). Inhibition of the sodium pump initially depolarizes the cell membrane potential by a few mV, corresponding to its electrogenic contribution.

The membrane potential established by the sodium pump forms the basis for the excitability of nerves and muscle cells, and the sodium pump restores the steady state Na^+/K^+-gradients after the action potentials, where Na^+ flows in and K^+ out by the opening of potential dependent Na^+- and K^+-channels.

F. K^+-Homeostasis

The skeletal muscle sodium pumps are essential in controlling the K^+ concentration in plasma. The K^+ pool of the skeletal muscle tissue constitutes about 75% of the total body pool in man. During muscle exercise, K^+ is lost from the muscle cells at a rate which can substantially influence the plasma K^+ concentration leading to hyperkalemia with severe impact on excitation–contraction coupling and metabolism. The K^+ lost during muscle exercise, which ends up in plasma, is removed by the sodium pumps, controlling the K^+ homeostasis. Regulation of the sodium pumping capacity according to muscle activity is therefore observed, and essentially established by (1) feed-back regulation of catalytic activity by Na^+ and K^+ concentration dependent ion activation, (2) acute hormonal regulation of pump activity by catecholamines (epinephrine/norepinephrine, dopamine) and possibly insulin, and (3) long-term regulation of pump density in the membranes depending on cell status and, for example, thyroid level (see reviews by Clausen, 1986; Clausen and Everts, 1990).

The K^+-homeostasis controlled by the sodium pump is also of direct importance for the activity of many cellular enzymes dependent on a high K^+-concentration, for example, the enzymes responsible for protein synthesis.

III. MOLECULAR STRUCTURE

The sodium pump is a heterodimer consisting of two non-covalently linked subunits, the α-subunit and the glycosylated β-subunit (Figure 1). The physiologically functional unit may be an oligomeric combination of the αβ-units, for example, consisting of two heterodimers. The α-unit contains all known catalytic properties of the enzyme, whereas the function of the β-unit is less clear (see Section V).

A. Primary Structure

The amino acid sequence of both the α-unit and the β-unit is known from several species. The α-unit consists of ~1016 amino acyl residues corresponding to an M_r 112,000; the β-unit of ~302 amino acyl residues of M_r 35,000, or 55,000 including the sugars.

The α-subunit from sheep (Shull et al., 1985) and eel (Kawakami et al., 1985) was the first to be sequenced, soon after followed the α and β subunits from several

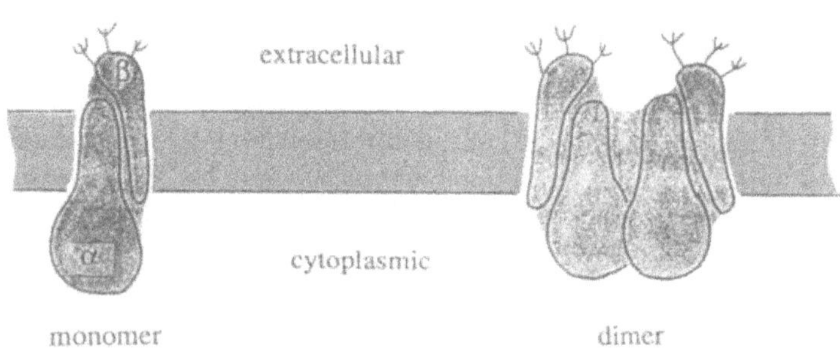

Figure 1. Schematic of αβ-heterodimer coarse topography in the membrane, assembled either as a monomer or as a dimer. Three sugar moieties are indicated on the β-subunit.

mammals (Shull et al., 1985, 1986a, b; Mercer et al., 1986; Ovchinnikov et al., 1986; Brown et al., 1987; Hara et al., 1987; Herrera et al., 1987; Young et al., 1987), including man (Kawakami et al., 1986a, b; Chehab et al., 1987), from birds (Takeyasu et al., 1987, 1988), fish (Kawakami et al., 1985; Noguchi et al., 1986), amphibians (Verrey et al., 1989), and insects (Lebovitz et al., 1989). The homology in primary sequence of the α-unit is very high between different animal species: in birds and mammals >90%—higher than between the isoforms within each individual species (see following), where about 80% sequence identity is found (Shull et al., 1986; Herrera et al. 1987; Takeyasu et al., 1990).

B. Membrane Topography

The deduction of the secondary structure from the primary amino acid sequences has been pursued by topographical approaches involving labeling with hydrophobic/hydrophilic probes (Nicholas, 1984; Kyte et al., 1987; Modyanov et al., 1991), the patterns of sided proteolytic digestion (Jørgensen and Collins, 1986; Kyte et al., 1987; Bayer, 1990; Karlish et al., 1993), chemical modifications (for references see Pedemonte and Kaplan, 1990), or antigenic probes raised against specific sequences (Ball, 1984; Kyte et al., 1987; Ovchinnikov et al., 1988; Abbott and Ball, 1993). Recently, molecular genetics methods producing chimeric cDNA which encodes regions of Na^+,K^+-ATPase and Ca^{2+}-ATPase to deduce membrane topology have also been employed (Lemas et al., 1992).

Transmembrane Domains

The folding pattern of the subunits has mainly been inferred from hydropathy plots indicating the distribution of hydrophobic residues (Kyte and Doolittle, 1982), thereby identifying putative membrane-spanning stretches from their highly hydrophobic nature, and the length of membrane-spanning domains (20–25 residues for an α-helix, 9–12 residues for a β-sheet).

Agreement exists as to the membrane topography of the first ~325 N-terminal amino acids of the α-subunit, which form four transmembrane domains (two hairpins). However, it is at present unresolved if the last ~300 residues in the C-terminal end form three-, four-, or six-transmembrane domains (Shull et al., 1985, 1986a; Ovchinnikov et al., 1986; Takeyasu et al., 1990; Capasso et al., 1992; Lemas et al., 1992), giving a total of either 7-, 8-, or 10-transmembrane domains (M1–M10) for the sodium pump α-subunit (Figure 2). At present the most favored models are the ones with 8- or 10-transmembrane segments (M8- or M10-models), since it seems now fairly certain that the C-terminus is on the cytoplasmic side (Thibault, 1993). The disagreement about the two latter models mainly concerns the putative M6 and M8 in the M8-model, which in the 10-transmembrane model are each split into two. It is also possible that M9/M10 do not span, but are only partially embedded into the membrane. In all models, a large cytoplasmic loop containing some 440 amino acid residues is placed between M4 and M5.

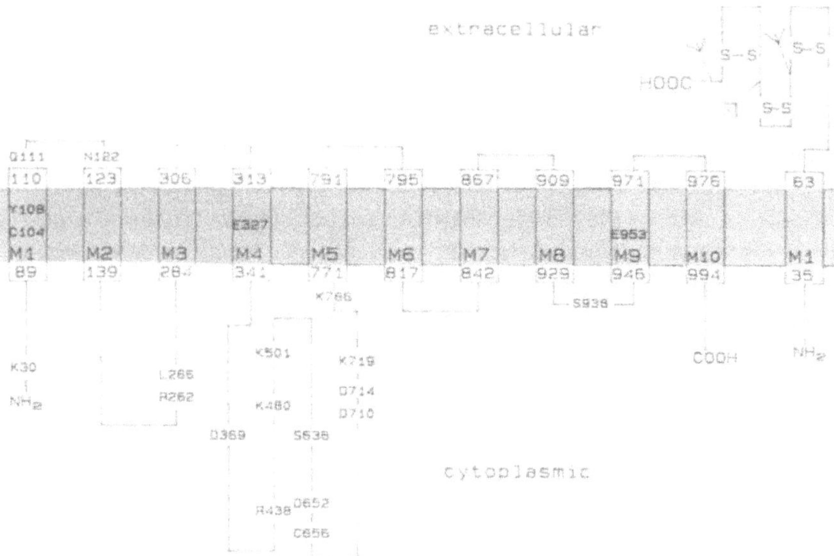

Figure 2. Unfolded model for the possible disposition of transmembrane domains (M1–M10) of the sodium pump αβ-unit. Ten transmembrane segments are assumed (M10-model) for the α-subunit and one for the β-subunit. The following identified amino acid residues are shown for the α-subunit: K30, R262, L266, and R438, the tryptic and chymotryptic splits; C104, Y108, Q111, and N122, amino acid residues important for ouabain binding and sensitivity; E327, P328, and L332 are residues important in cation binding; D369 is the phosphorylation site; K480 is the AP_2PL, PLP, and 8AA site; K501 and K766 are the FITC-, NIPI-, and SITS-binding sites; the stretch S636–D652 is the epitope involved in M7-PB-E9 antibody binding; C656 and K719 are the FSBA sites; D710 and D714 are the C1R-ATP sites; E953 and E954 are DCCD labeling sites important for cation binding; S938 is the cAMP dependent protein kinase phosphorylation site. For the β-subunit three disulfide bonds are indicated at C127–C150, C159–C176, and C214–C277, and three *N*-linked glucosylation sites are indicated: N158, N193, and N265.

The β-subunit has one transmembrane domain (Figure 2), a small cytoplasmic N-terminal and a large extracellular C-terminal domain containing three disulfide bonds (S-S bridges; Ohta et al., 1986b; Kirley, 1989), and three glycosylation sites (Sweadner and Gilkeson, 1985; Miller and Farley, 1988; Fambrough, 1983).

Mass Distribution

The distribution of protein mass of the αβ-unit between the hydrophobic membrane compartment and the hydrophilic external and cytoplasmic media can be inferred from reconstructed three-dimensional (3D) models of two-dimensional (2D) crystals (Hebert et al., 1985, 1988), from sided chemical labeling of the sodium pump (O'Connell, 1982; Sharkey, 1983; Dzhandzhugazyan and Jørgensen, 1985), and from extensive proteolytic digestion experiments (Karlish et al., 1977). Such calculations indicate that 30–40% of the protein mass is located in close connection with the membrane. This is somewhat more than predicted even from the M10 model of the α-subunit together with the β-subunit, where only about 21% is in the membrane. It is possible that some hydrophilic stretches of the protein are in close connection with, or even embedded in, the membrane.

Crystallization in two-dimensional arrays of purified, membrane bound Na^+,K^+-ATPase is possible by incubation in media favoring the E_2 conformation of the enzyme: vanadate and Mg^{2+} (Skriver et al., 1981, 1988), in phosphate and Mg^{2+} (Skriver et al., 1981; Zamphigi et al., 1984), with phospholipase A_2 (Mohraz et al., 1987), or with cobalt–tetramine-ATP (Skriver et al., 1989). It is also possible to crystallize in the E_1 conformation using oligomycin and high Na^+-concentration (Skriver et al., 1985). In acidic media of sodium citrate, crystallization may also be in the E_1 conformation (Tahara et al., 1993). The latter method and the phospholipase treatment give crystallization in yields high enough to allow electron cryomicroscopy (Tahara et al., 1993).

The 2D crystals have different crystal symmetry with space groups P1, P2, $P2_1$, or P4 and have unit cells that contain either one protomer (P1), one dimer (P2 and $P2_1$), or four dimers (P4). The monomers protrude 6 nm and 2 nm on the two sides of the 4 nm bilayer, whereas the protrusion of the dimers is slightly lower, about 4 nm (Figure 3). The 3D reconstruction model of the dimers shows a deep, probably cytoplasmic, and a shallow, probably extracellular, cleft and a connection at the level of the bilayer. The resolution obtained with the present crystals is still only 2–2.5 nm, a better resolution awaiting more regular crystals, the ultimate goal being 3D crystals which can be analyzed by X-ray diffraction.

C. Functional Significance of Domains

The sodium pump can be envisaged as segregated laterally and vertically into functional domains. The vertical separation by the lipid bilayer into an extracellular and a cytoplasmic part with well known kinetic characteristics will be dealt with under the kinetics of the enzyme (see The Albers–Post model). Less known are the functional details of the lateral segregation on both the extracellular and the cytoplasmic side, although, it is evident that properties like ATP-binding, phosphorylation, cation binding, occlusion, and translocation, energy transduction, and inhibitor binding are probably attributed to special domains within the protein topography (see Table 1).

The ATP-binding Domain

The ATP-binding region and the phosphorylation site have been unambiguously related to the major cytoplasmic loop between M4 and M5, that is, between residues 341–771 in an M10-model. This part of the protein contains highly conserved

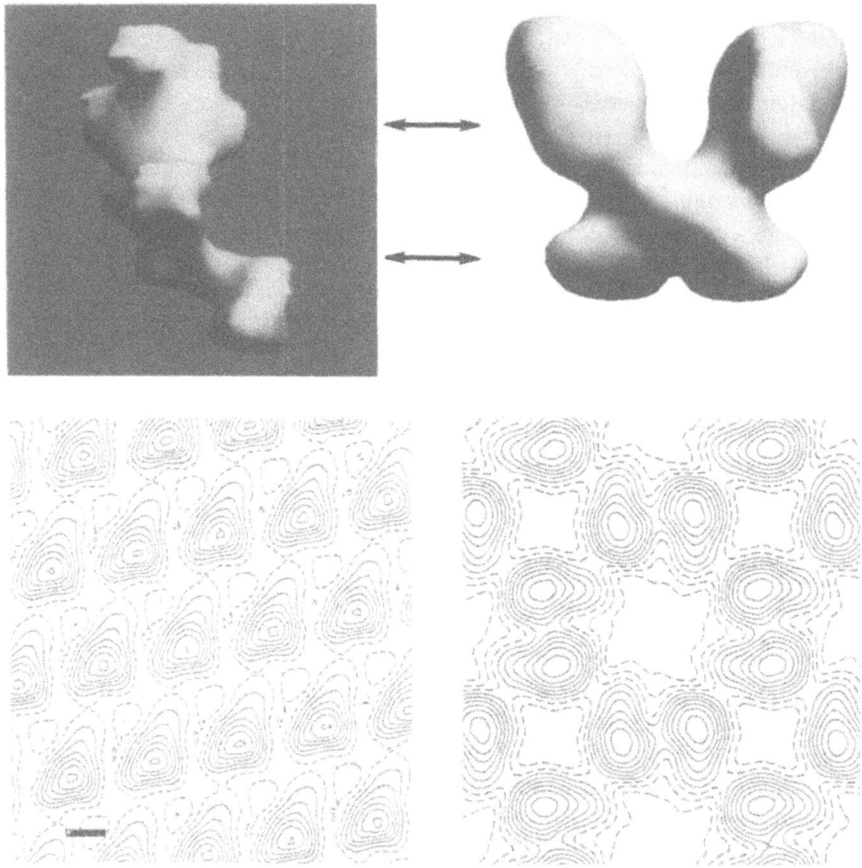

Figure 3. Two types of three-dimensional models of Na^+,K^+-ATPase reconstructed from two-dimensional membrane crystals together with their projection maps (below). To the left, the monomeric vanadate-induced P1 crystal (for details see Hebert et al., 1988) and to the right, the dimer from the cobalt–tetramine-ATP induced P4 tetrameric crystal (see Skriver et al., 1989, 1992). In the projection maps protein-rich regions are drawn with full lines and negative stain regions with dashed lines. For the monomer, the unit cell dimensions are 69 × 53 Å, and for the tetrameric crystal 141 × 141 Å. The suggested position of the membrane is indicated by arrows and the intracellular side is facing upward. The bar: 2.5 nm. (Kindly provided by E. Skriver, A. B. Maunsbach, and H. Hebert.)

sequences among P-type ATPases and also shows some homologies with other ATP binding proteins (Taylor and Green, 1989).

The residue which is phosphorylated by ATP is aspartate-369 (Post and Kume, 1973; Shull et al., 1985), indicating that this residue is close to the γ-phosphate of the bound ATP. Mutagenesis of this aspartate residue completely blocks enzyme

Table 1. Suggested Location in the α-subunit Topography of Specific Functions

Function	*Suggested Location[a] (M10-model)*
ATP-binding/phosphorylation	CD3 (342–770)
Cation-binding/occlusion	M4, M5, M6, M9 (313–341), (771–191), (795–817), (946–971)
E_1/E_2 transformation and energy transduction	CD1, CD2, CD3 (1–88), (140–283), (342–770)
Gating	CD1 (1–88)
Channel	M4, M5, M6, M9 (331–341), (771–791), (795–817), (946–971)
Ouabain-binding	M1, ED1, M2 (89–110), (111–122), (123–139)
α/β assembly	ED3, ED4 (868–908), (972–975)
Regulation	CD1 (1–88) CD5 (930–945)

Note: [a]ED is extracellular domain, CD is cytoplasmic domain, and M is transmembrane domain.

activity (Ohtsubo et al., 1990). The ATP derivative, γ-[4-(N-2-chloroethyl-N-methylamino)]benzylamide (C1R-ATP) that contains an alkylating moiety bound to the γ-phosphate, modifies Asp-710 and 714 (Dzhandzhugazyan and Modyanov, 1985; Ovchinnikov et al., 1987) indicating that these residues may be close to Asp-369 in the tertiary structure. Other amino acids involved in ATP binding and identified by chemical modifications are spread over the primary sequence, indicating a complicated folding to form the ATP-binding pocket. Lys-501 is the major fluorescein isothiocyanate (FITC) labeling site (Farley et al., 1984; Kirley et al., 1984), and recent observations also involving Lys-480 and Lys-766 (Xu, 1989) indicate, in fact, that a cluster of lysines may be involved in FITC-binding. 4-acetamido-4′-isothyocyanostilbene-2,2′-disulphonic acid (SITS) and N-isothiocyanophenylimidazola (NIPI) inhibition of Na^+,K^+-ATPase activity was recently demonstrated to be caused by binding to the same Lys-501 residue (Ellis-Davis and Kaplan, 1990; Pedemonte et al., 1992). Although ATP protects against inhibition and labeling by FITC, SITS, and NIPI, antibody raised against the segment including Lys-501 fails to inhibit Na^+,K^+-ATPase (Ball and Friedman, 1987), and site-directed mutagenesis of the homologous region in the Ca^{2+}-ATPase shows modest inhibition in Ca^{2+}-transport (Maruyama and MacLennan, 1988; Maruyama et al., 1989). Labeling with another adenosine analog, 5′-p-flourosulfonylbenzoyladenosine (FSBA), implies Cys-656 and Lys-719 to be near the ATP-binding domain (Ohta et al., 1986a). Binding of the ATP analogs, pyridoxal-5′-diphospho-5′-adenosine (AP_2PL) and pyridoxal phosphate (PLP), have been shown to involve Lys-480 (Hinz and Kirley, 1990), within a conserved sequence also labeled by 8-azido-ATP (8-AA). It is currently believed that Lys-480 is not essential for ATP binding or hydrolysis, but is likely to be located in the vicinity of the ATP-binding site, since its replacement by site-directed mutagenesis with an acidic amino acid (alanine, arginine, or glutamine) decreases the affinity of mutant Na^+,K^+-ATPase for ATP and P_i (Wang and Farley, 1992).

Biochemical modification studies, as well as studies using site-directed mutagenesis are prone to ambiguities as previously discussed (Pedemonte and Kaplan, 1990; Williams, 1993), since modifications apart from being direct can induce disturbances in the protein conformation in neighboring loci that are important to functional integrity. Site-directed mutagenesis of the reactive amino acids in this putative ATP-binding pocket is still scarce for the Na^+,K^+-ATPase and a complete model for the kinase-like ATP-binding pocket is not yet available, however the spatial organization of important residues are more or less fixed: the Asp-369 and Lys-710 are in the same vicinity at one end of the pocket interacting with the γ-phosphate position of ATP. Opposite is a stretch including Lys-501 to Lys-480 which either interacts with the adenine-ring structure of ATP, or provides space for it, and finally, the pocket is flanked at its two borders by Cys-656 and Lys-719 respectively, possibly interacting with the ribose ring of ATP (Figure 4).

It is a well known property of the Na^+,K^+-ATPase that it has a high ATP binding-affinity ($K_D \sim 0.1$ μM) in the presence of Na^+ (E_1-conformation), whereas the de-occlusion of K^+ in the E_2-conformation is accelerated by ATP interaction with a 10^3-times lower affinity. This implies that either the gross conformational change of the protein during the E_1/E_2 transition is followed by a rearrangement in the ATP-binding pocket giving rise to a different ATP-affinity (Moczydlowski and Fortes, 1981), or that the αβ-subunit contains two ATP-binding sites with different affinities (Sheiner-Bobis et al., 1987). The first possibility is currently favored.

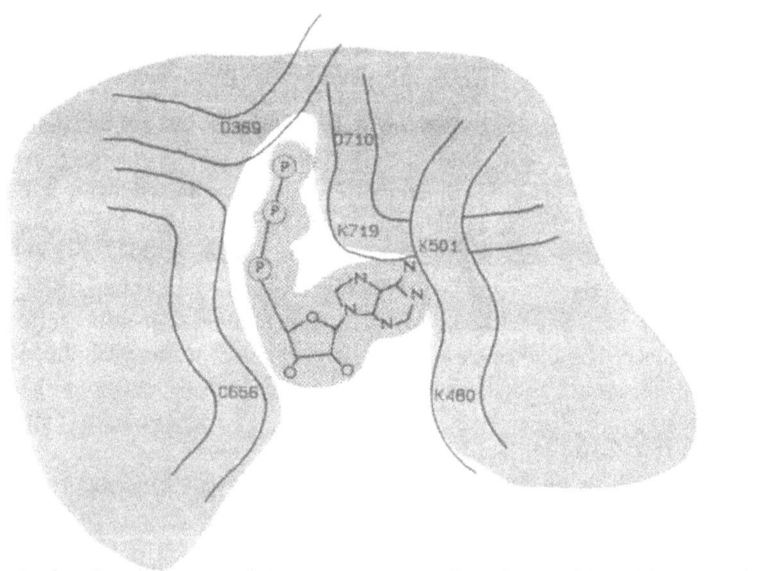

Figure 4. A sketch of the possible arrangement of amino acid residues in the ATP-binding pocket of the α-subunit.

Cation Binding Domain

The alkali cation binding- and occlusion-domain of the sodium pump is less clearly defined than the ATP-binding domain. Strategies for the identification of amino acid residues as forming putative cation-binding sites have been largely based on the assumption of local charge interaction, where negatively charged amino acids side-groups are supposed to interact with the transport cations. Carboxyl residues of, for example, aspartic or glutamic acid are obvious choices, and modification of these by carbodiimide (DCCD) as originally initiated by Robinson (1974), has been extensively used as probes for potential cation sites, even though their specificity may be dubious due to secondary cross-linking within the protein (Gorga, 1985; Pedemonte and Kaplan, 1986a, b).

Sequence homology with other P-type ATPases, which is of great help for the identification of the ATP-binding domain, may be of little help in screening for the cation-binding sites, since the various P-type ATPases have different cation specificities, probably reflecting different sequences at exactly their cation sites.

The general picture that emerges is that the cation-binding sites are presumably membrane-embedded, since removal of up to 70% of the protein by proteolytic digestion in the absence of divalent metal ions, but in the presence of Rb^+, preserves the Rb^+ occlusion (Capasso et al., 1992).

Recent evidence suggests that there is a single cation-binding site common to alkali cations and DCCD located within a 19 kD membrane fragment resisting extensive tryptic digestion of purified pig kidney Na^+,K^+-ATPase (Karlish et al., 1990; Goldshleger et al., 1992). The 19 kD fragment can be demonstrated to occlude Na^+ and Rb^+ (K^+) and with a capacity similar to the control enzyme. Although the hydrolytic capacity of the fragment is lost, as well as other ATP-dependent functions, it retained passive Rb/Rb exchange after reconstitution into liposomes, suggesting that not only is the occlusion cage preserved, but also the complete transport path (Karlish et al., 1990). The 19 kD membrane-embedded fragment from pig kidney Na^+,K^+-ATPase spans Asn-831 to the C-terminus, corresponding either to transmembrane fragments M7–M10 in an M10-model, or to M7 and M8 in an M8-model, in either case the N-terminal Asn-831 is located on the cytoplasmic side (Karlish et al., 1993). Comparable results have been obtained for shark Na^+,K^+-ATPase where a 19 kD segment with retained occlusion capacity has a similar sequence of 40 N-terminal residues as kidney enzyme (Esmann and Sottrup-Jensen, 1992). Sequencing of the 19 kD membrane fragment from purified pig kidney showed that the DCCD label was located to Glu-953 and, possibly, the adjacent Glu-954 (Goldshleger et al., 1992). Rb^+ protects against DCCD labeling; however, site-directed mutagenesis with single- or even double-substitutions of this amino acid affects cation activation only slightly, which seems to contradict that this residue is essential for cation binding and occlusion (Van Huysse et al., 1993), although it may still be present and form part of a cation-binding groove.

Another putative cation-binding site apart from Glu-953 in the pig kidney enzyme was suggested by DCCD labeling of undigested enzyme which showed a stoichiometry of about 2 DCCD per enzyme molecule (Goldshleger et al., 1992), whereas the ratio for the 19 kD membranes was 1. The putative second binding site for cations has not been identified, however, the DCCD label is present in a small intramembraneous tryptic peptide fragment, which is probably not derived from the 19 kD fragment itself. In the Ca^{2+}-ATPase, 4 carboxylic residues are proposed as Ca^{2+} ligands (Clarke et al., 1989; Vilsen and Andersen, 1992). Glu-327, in the highly conserved PEGLL motif of predicted transmembrane segment M4 of sheep, is homologous to one of the four residues in the Ca^{2+}-ATPase and has been suggested to form the second cation site in Na^+,K^+-ATPase. Site-directed mutagenesis of residues in this motif, Pro-328 → Ala (Vilsen, 1992), but not the Glu-329 (Glu-327 in sheep), reduces the apparent cation affinity and indicates a role for Pro-328 in high-affinity Na^+-binding, either as an electron donor to ion binding, or to bend the helix in order to place the adjacent Glu-329 in optimal position for Na^+ binding. Conserved residues important for Ca^{2+} binding and occlusion in the Ca^{2+}-ATPase (Vilsen and Andersen, 1993) are also present in the Na^+,K^+-ATPase M5 and M6, such as Asp-804 and Asp-808 in M6 (see Table 2), but their functional significance has not yet been probed in the Na^+,K^+-ATPase.

In both Ca^{2+}-ATPase and Na^+,K^+-ATPase, the suggested locations of cation-binding sites indicate that they are far apart on each side of the major cytoplasmic loop in the unfolded models, namely in Na^+,K^+-ATPase in M4 and in either M7, or M9, depending on the model (Karlish et al., 1993; see Figure 2). While this may seem paradoxical, it may reflect the proximity of C-terminal and N-terminal transmembrane helices in the three-dimensional packing of the peptides in the membrane. Such a three-dimensional model has been constructed by Modyanov et al. (1991) using an M7 model in which the putative cation-binding sites formed by Glu-327, Glu-953, and Glu-954 are located in close proximity inside the protein. Transmembrane helices M1–M4, which are connected by short hydrophobic sequences, are associated in pairs, and hydrophobic residues are facing the lipid environment (Figures 5 and 6).

E_1/E_2 Conformational Change

The Na^+,K^+-ATPase reaction mechanism, and those of other P-type ATPases as well, have usually been described by an E_1/E_2 model, with two major conformational states (see Subsection "How the Na^+,K^+-ATPase Pumps"). Many more conformational states are likely to exist, and there may well be a conformational change for each step in the reaction of the catalytic cycle. However, the E_1/E_2 (and E_1P/E_2P) conformations can be identified by the tryptic and chymotryptic cleavage pattern (Jørgensen, 1975, 1977). Lys-30 (T_2) and Arg-262 (T_3) are sites exposed to trypsin, and Leu-266 (C_3) is the chymotrypsin site cleaved, when the enzyme is in the E_1-conformation. In the E_2-conformation, the chymotrypsin site is protected

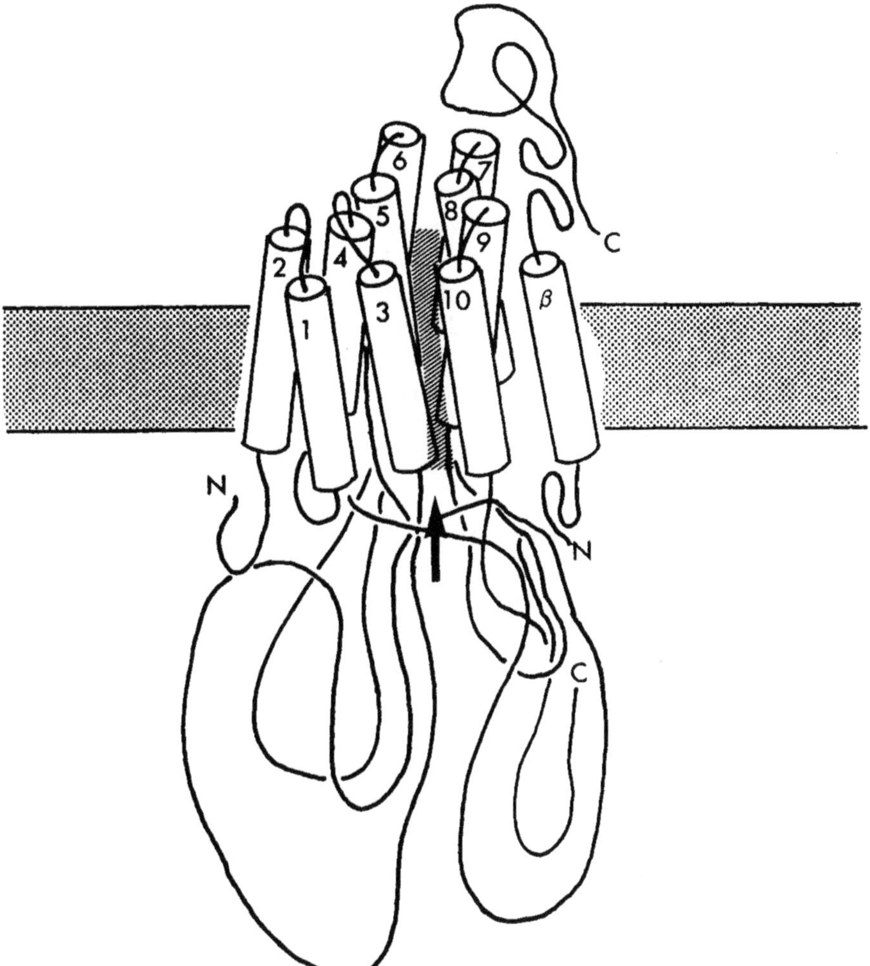

Figure 5. Drawing of the folded tertiary structure of the αβ-unit of the sodium pump. Ten α-helical transmembrane segments are clustered for the α-subunit, whereas only one is supposed for the β-subunit. In the center of the helices, a gated channel-like structure is assumed (hatched) through which ion translocation takes place (arrow). The major protein mass of the α-subunit is cytoplasmic, whereas for the β-subunit it is extracellular.

and trypsin cleaves first at Arg-438 (T_1) and then at T_2. Tryptic cleavage results in a change in the steady-state ratios of E_1/E_2 and E_1P/E_2P favoring the E_1-conformations (Jørgensen and Karlish, 1980; Jørgensen and Andersen, 1988). Cleavage at T_2 probably increases E_1/E_2 by accelerating the E_2 to E_1 transition, whereas cleavage at T_3 may cause a blockage in the E_1P to E_2P transition (Wierzbiki and Blostein, 1993). During the conformational change from E_1 to E_2, rearrangement of the protein structure in the second cytoplasmic domain (CD2) between M2 and M3 takes place, involving protection of T_3 and C_3 bonds. At the same time, T_1 is exposed indicating that structural changes in CD3, the major cytoplasmic domain between M4 and M5, also occur. This has also been inferred from binding studies using the antibody, M7-PB-E9, that binds to an epitope containing Ser-636 to Asp-652 (Abbott and Ball, 1993) and which enhances the E_1 to E_2 transition (Ball, 1984). Predominant labeling of the E_2 conformation with hydrophobic probes like iodonaphthylazide (INA) (Karlish et al., 1977) or 3-(trifluoromethyl)-3-(m-iodophenyl)diazarin (TID) (Modyanov et al., 1991) suggests that the E_1/E_2 transition increases the amount of residues buried in or in close contact to the lipid bilayer by 10–30%. These conformational changes probably reflect gross structural rearrangements in connection with the ion translocation processes including energy transduction, ion occlusion, and translocation.

Gating Mechanism

The first cytoplasmic domain (CD1, residues 1–88) contains clusters of highly charged residues: 44 charged residues are contained in this stretch in the α1 from sheep. The rating is especially high among the first 30 residues (i.e., up to tryptic split T_2), that contain 9 lysines. It has been suggested that this segment serves as an ion-selective gate (Shull et al., 1985, 1986a). Experimental evidence indicates that this lysine-rich stretch modulates the energy barrier for K^+ de-occlusion (Vasilets et al., 1991; Wierzbiki and Blostein, 1993), apparently supporting this notion. However, it may seem difficult to explain the high sequence diversity among species and isoenzymes of this region (Shull et al., 1986a) for such a general functional role. Moreover, in the Ca^{2+}-ATPase, the gating mechanism is apparently an integral part of the channel, composed of a group of charged residues in the middle of the helices (Green, 1989; Clarke et al., 1989).

If the N-terminal cytoplasmic domain functions as a gate in ion occlusion, which structure could then be envisaged as the corollary extracellular gate? No obvious candidate seems to be present, unless the β-subunit is implemented. Such a function could have evolved to allow counter-transport of two cations, which demands the presence of two gates, as appear specific to the Na^+,K^+-ATPase and the H^+,K^+-ATPase, both of which are the only members of the P-type ATPases that have the β-subunit. Potentially, extracellular motifs could be the highly conserved and charged sequence Lys-216, Glu-219, Asp-220 (see Section IIID). Actually, Rb^+ or Na^+ were found by Capasso et al. (1992) to protect against proteolysis of the β-subunit, as expected for the β-subunit to play such a role.

Ouabain Binding Domain

The sodium pump is the target for cardiac glycosides like ouabain (Schatzmann, 1953) and other glycosylated steroids, which inhibit the enzyme from the extracellular aspect when phosphorylated (Charnock and Post, 1963; Matsui and Schwartz,

1968; Schwartz et al., 1968), and with the highest affinity to the E_2P-conformation (Post et al., 1969; Yoda and Yoda, 1982b).

The epitope that is the receptor for cardiac glycosides has been identified as located to border residues of the first extracellular domain (ED1; Price and Lingrel, 1988; Price et al., 1989, 1990) where a low ouabain sensitivity seems to be connected with the presence of two charged amino acid residues: substitution of Arg-111 with Glu, and Asp-122 with Asn, confers ouabain sensitivity to the insensitive rat α1-subunit (Price and Lingrel, 1988). Likewise, the insensitivity of the Monarch butterfly seems to rests on a charged histidine residue at position 122, where ouabain sensitive insects have asparagine (Holzinger et al., 1992). In addition, at least two residues in the M1 transmembrane domain, Cys-104 and Tyr-108, are important in determining ouabain sensitivity (Canessa et al., 1992; Schultheis and Lingrel, 1993).

The highly conserved ouabain-binding site in Na^+,K^+-ATPases from different species and among isoenzymes has prompted speculation as to the existence of an endogenous, ouabain-like factor or hormone (Hamlyn et al., 1982), to which the sodium pump served as a receptor (Anner, 1985). Such an ouabain-like substance could act extracellularly on the mature Na^+,K^+-ATPase regulating Na^+-homeostasis, or cytoplasmic to regulate its intracellular maturation processes (Kelly and Smith, 1989). It remains controversial whether a synthetic pathway for ouabain exists in mammals, and whether it acts as a regulator of sodium homeostasis. However, endogenous ouabain has recently been identified in human plasma and in the plasma from several mammals (Hamlyn et al., 1991; Ludens et al., 1991), and the adrenal gland has been found to be its major source in dog (Boulanger et al., 1993).

αβ-assembly Domain

The β-subunit has no known function in the catalytic reaction cycle of the Na^+,K^+-ATPase, but its presence is necessary for expression of the α-subunit (Noguchi et al., 1987), probably since subunit assembly is required for exit from the endoplasmic reticulum (ER; Jaunin et al., 1992), and to increase Na^+,K^+-ATPase activity and ouabain binding in transfected yeast cells (Horowitz et al., 1990). It is believed to be involved in the folding and maturation of the α-subunit (Geering et al., 1987), and subsequent transport to the plasma membrane (Noguchi et al., 1987; Fambrough, 1988; Takeyasu et al., 1989). Deletions of more than a few C-terminal amino acid residues of the β-subunit results in failure to assemble with the α-subunit, whereas the cytosolic N-terminal domain is unnecessary for this assembly (Renaud and Fambrough, 1991).

By constructing chimeric cDNA that encodes different regions of Na^+,K^+-ATPase and Ca^{2+}-ATPase and expressing them together with cDNA for Na^+,K^+-ATPase β-subunit, Lemas et al. (1992) find an assembly region confined to the one or two C-terminal extracellular domains of the α-subunit (see Figures 2 and 6). The region comprising the C-terminal 161 or less residues is apparently common among Na^+,K^+-ATPase isoforms from one and the same species (Kone et al., 1990; Jaunin et al., 1992), but also among different species (Horowitc et al., 1990a, 1990b; Noguchi et al., 1990; Takeyasu et al., 1990), and among α-subunit from Na^+,K^+-ATPase and β-subunit from H^+,K^+-ATPase (Horisberger et al., 1992), since their α and β subunits are able to assemble into functional pumps.

Channel-forming Segments

Little is known at present about the three-dimensional folding of helices and sheets of the sodium pump in the plasma membrane and predictions of channel-forming domains are purely speculative, even though the existence of such channel domains may be demonstrated under special conditions (Teissie and Tsong, 1980; Last et al., 1983; Mironova et al., 1986; Halperin and Cornelius, 1991). The voltage dependence of the apparent Na^+ affinity during electroneutral Na^+:Na^+ exchange indicates that more than half of the membrane potential drops between the Na^+-binding site and the extracellular aspect, indicating an access channel functionally analogous to an ion channel (Gadsby et al., 1993).

In ligand-gated receptor proteins, transmembrane α-helices are known to cluster, thereby forming a hydrophilic channel for ions (Greningloh et al., 1987; Schofield et al., 1987; Tanabe et al., 1987). If, by analogy, transmembrane segments in the sodium pump form a channel structure, helices containing the presumptive cation-binding sites are likely to contribute. In the M10-model, this would be M4 and M9, and in the M8-model, M4 and M7. The clustering of four α-helices seems enough to form a narrow channel (Inesi and Kitley, 1990), which could involve transmembrane segments M5 and M6, too. In agreement with this, the homologous segments in the Ca^{2+}-ATPase have been demonstrated by site-directed mutagenesis to include high-affinity Ca^{2+}-binding sites (Clarke et al., 1989), some of which are essential for Ca^{2+} transport and occlusion (Vilsen and Andersen, 1993). The four presumptive channel forming helices have amphiphilic characters and can be arranged with charged and polar residues lining the channel interior, and with apolar residues on the opposite side, stabilizing contact with the lipid phase (Figure 6). At present, models for an actually aqueous channel are not supported, due to the many available oxygen ligands in the involved helical segments. As seen from Table 2, the putative channel forming helices in the sodium pump do contain one or several proline residues which are absent in all other transmembrane segments. It has been suggested that these prolines are obligatory components of some ion channels (Woolfson et al., 1991), inducing kinks into the α-helical arrangement that may play a functional role for dynamical channel operation (Williams and Deber, 1991; von Heijne, 1991). However, the validity of such predictions may be limited since in channel proteins presumptive extramembraneous residues may form pores from β-barrel structures that do not contain charged or hydrophilic residues (Yool and Schwarz, 1991).

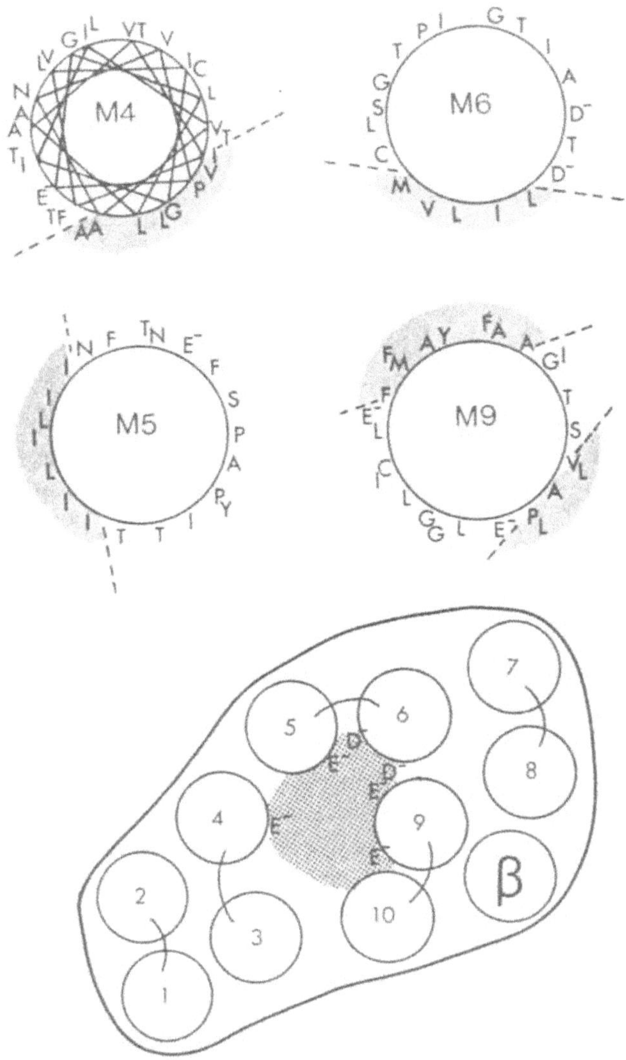

Figure 6. Helical-wheel model (indicated only in M4) of suggested channel-forming transmembrane segments M4, M5, M6, and M9. The helices are viewed from the cytoplasmic side, the first residue being located at 9 o'clock. Shaded area indicate predominately hydrophobic stretches.

Below is shown one of several possible arrangement of transmembrane segments viewed from the extracellular aspect of the Na^+,K^+-ATPase. When the putative channel-forming helices are arranged with individual stretches of polar residues in mutual contact a hydrophilic path including six charged residues, four glutamic acids, and two aspartic acids, is formed (shaded area), which could comprise a channel-like structure.

Table 2. Transmembrane Segments Suggested To Be Included in the Putative Ion-Channel Formation

Segment	*Amino Acid Residues*[a]
M4	AVIFLIGIIVANVPEGLLATVTVCLTLTA
	313 — 341
M5–M6	YTLTSNIPEITPFLIFIIANI -PLP- LGTVTILCIDLGTDMVPAIS
	771 — 791 795 — — 817
M9–M10	ILIFGLFEETALAAFLSYCPGMGVAL -RMYP- LKPTWWFCAFPYLLIFVY
	946 -- 971 976+ 994

Note: [a]Polar and charged residues are indicated with bold-faced types and prolines are underlined. Amino acid residue numbers are given below the one letter amino acid code.

D. Isoenzymes

The α-subunit of Na^+,K^+-ATPase exists in multiple isoforms encoded by different genes. At present, three isoforms of the α-subunit have been described and named α1, α2, and α3 (see reviews by Sweadner, 1989, 1991; Lingrel et al., 1990). Screening of cDNA clones from rat revealed three classes, and sequence analysis demonstrated they encode three polypeptides of 1023, 1020, and 1013 amino acids, respectively (Shull et al., 1986a), corresponding to the three isoforms. The homology between isoforms is pronounced, ~80% at the amino acid level (Shull et al., 1986a; Herrera et al., 1987) even between different species, suggesting an early divergence and a high degree of conservation in evolution (Takeyasu et al., 1990).

There seems to be a clear tissue-specific as well as developmental expression of the isoform subunits. Specific tissues predominantly contain more of one isoform than of the others, however most tissues express more than one isozyme. This is revealed by mRNA blot analysis showing α1-mRNA to be expressed in almost any tissue, albeit to various degree, α2- and α3-mRNAs are both expressed predominantly in neural tissue, the α2-mRNA also in muscles (Young and Lingrel, 1987), and the α3-mRNA specifically in brain (Gick et al., 1993). The α-subunit gene distribution in various cultured cell lines follows an identical pattern with α1-mRNA present in all investigated cell lines, while α2- and α3-mRNAs are restricted to neural (α3) or/and muscle cells (α2). The distribution suggests a function in fundamental cell physiology for the α1, while α2 and α3 may be expressed to fulfill more specialized tasks in cell functions (Lingrel et al., 1990).

As with the α-subunit, three isoforms of the β-subunit have been identified, called β1, β2, and β3 (see reviews by Sweadner, 1989, 1991; Lingrel et al., 1990). Although different sized cDNA clones for the β-subunit have been found, their coding sequences are identical, indicating a single gene for encoding the β-subunit by alternative RNA splicing (Mercer et al., 1986; Young et al., 1987).

The six cysteine residues involved in disulfide bridging and important for folding are conserved in all β-subunits (Kirley, 1989; Miller and Farley, 1990), whereas the number of potential N-linked glycosylation sites differs from three in β1 and β3, to seven in β2. Two extracellular regions contain conserved motifs in all β-subunits: Tyr-242 to Tyr-246, with one conservative change in β2 that has Phe-243 instead of Tyr-243; and a charged motif (Lys-216, Glu-219, Asp-220).

The β1 is ubiquitously expressed in several vertebrate species, whereas the β2 is more restricted to neural tissue and identical to the adhesion molecule (AMOG) of glia (Pagliusi et al., 1989; Gloor et al., 1990). The β3 is brain-specific and isolated from *Xenopus* (Good et al., 1990).

E. Cytoskeleton

In polarized transport epithelia cells, a nonrandom distribution of sodium pumps to the basolateral plasma membranes is essential for their function as transcellular mediators of salt and solute. This uneven distribution of sodium pumps is maintained by the membrane cytoskeleton (Simon and Fuller, 1985; Rodriguez-Boulan and Nelson, 1989; Luna and Hitt, 1992) through contact to an ankyrin–fodrin tetramer with the sodium pump (Morrow et al., 1989; Nelson and Hammerton, 1989). It has been shown that purified, membrane bound Na^+,K^+-ATPase from canine kidney epithelial cells has a high-affinity binding site (K_D ~10^{-8} M) for ankyrin (Nelson and Vesnock, 1987). The cell adhesion molecule, uvomorulin, has been found to induce this assembly of the cytoskeleton with the sodium pump: in an elegant series of experiments, McNeil et al. (1990) expressed uvomorulin in nonpolarized mouse L cells (fibroblasts), which express ankyrin, fodrin, and Na^+,K^+-ATPase, but not uvomorulin, by transfection with cDNA encoding either full-length or truncated uvomorulin. Only in the former did the cells form tight connections, as found in epithelia cells, and a nonrandom distribution of uvomorulin, fodrin, and Na^+,K^+-ATPase localized to cell–cell contacts. Once the polarization of the cells has been established by uvomorulin mediated cell–cell contacts, the newly synthesized Na^+,K^+-ATPase must be targeted to the basolateral membrane domain from the Golgi apparatus.

IV. MOLECULAR FUNCTION

A. The Reaction Cycle

The Na^+,K^+-ATPase catalyzes the fully reversible coupled vectorial reaction:

$$3Na^+_{cyt} + 2K^+_{ext} + ATP \rightleftharpoons 3Na^+_{ext} + 2K^+_{cyt} + ADP + P_i \quad (1)$$

which couples the energetically uphill vectorial transport of both Na^+ and K^+ to the energetically downhill chemical breakdown of ATP. In the cell, the transport of Na^+ as well as K^+ is uphill, against both their electrochemical gradients, the steepest being that for Na^+. The concentration gradients are about 30 (out/in) for K^+ and 10 (in/out) for Na^+ in many animal cells. This gives equilibrium potentials of roughly –90 mV for K^+ and +60 mV for Na^+ according to the Nernst equation:

$$E_{eq} = -(RT/zF) \cdot \ln([X_i]/[X_o]) \quad (2)$$

where X is the ion species, z is the charge number, R the gas constant, T the absolute temperature, and F the Faraday number. With a membrane potential of $E_m \sim -60$ mV, the electrochemical potential difference ($\Delta\tilde{\mu}$) is therefore ~ 11.5 kJ/mol for Na^+ and ~ 3 kJ/mol for K^+ according to:

$$\Delta\tilde{\mu} = zF(E_m - E_{eq}) \quad (3)$$

Assuming the 3:2 Na^+:K^+ stoichiometry, one turn-over of the pump must consume ~ 40.5 kJ. The free energy of ATP hydrolysis is estimated to be about 50 kJ/mol (Veech et al., 1979), which means that about 70–85% of the free energy for hydrolysis of ATP is used in pumping, which is ideally efficient, demonstrating the tight coupling of vectorial ion transport with the chemical reaction of ATP hydrolysis.

How the Na^+,K^+-ATPase Pumps

The fundamental kinetic reaction steps in this mechanism of the sodium pump were clarified by investigations by Post and his colleagues (Post et al., 1969) and Albers and coworkers (Fahn et al., 1966) during the 1960s (for additional references see Glynn, 1993). The general reaction mechanism emerging from this work is now known as the Albers–Post model. The general features contained in this reaction are depicted in Figure 7 below, and consist of: (1) a major conformational change where the enzyme can reversibly alternate between two major conformations: an E_1-form, which is the high Na^+-affinity "sodium form" and an E_2-form which is the low Na^+-affinity "potassium-form," (2) a phosphorylation/dephosphorylation sequence in which both enzyme conformations can exist as either phosphorylated, (E_1P, E_2P), or unphosphorylated, (E_1, E_2), forms, (3) an occlusion/de-occlusion cycle, in which the cation-binding sites on the sodium pump, alternatingly accessible to either the extracellular (*cis*) or the cytoplasmic (*trans*) medium, are transferred into the membrane phase, where their dissociation rate is slow, shielded from the medium by (4), a gating mechanism, where two gates or energy barriers alternatingly control the access from the *cis*- and *trans*-side of the pump. The experimental evidence for these basic steps are described below.

What makes the model function as a vectorial ion-pump mechanism is the mutual intervention of chemical and vectorial specificities (Jencks, 1989): the E_1 form of the enzyme reacts reversibly with ATP to form a high-energy E_1~P phospho-intermediate, whereas E_2 reacts reversibly with P_i to form a low-energy E_2P phospho-intermediate. Simultaneously, the unphosphorylated conformation of the enzyme preferentially binds and dissociates Na^+, whereas the phosphorylated enzyme species preferentially binds and dissociates K^+.

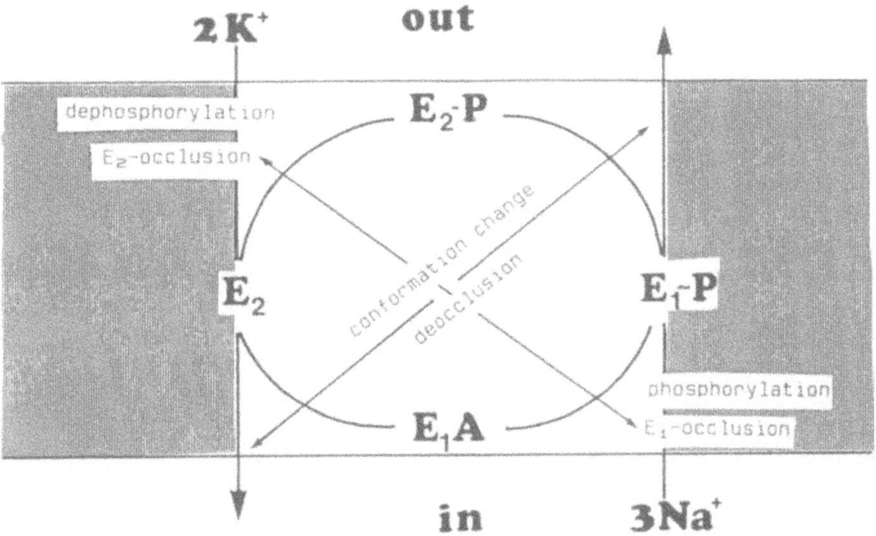

Figure 7. Principal scheme for the catalytic cycle of the sodium pump, illustrating how occlusion/de-occlusion and ion-translocation are linked to phosphorylation/dephosphorylation processes and change in protein conformation. Two phospho-intermediates are depicted: the 'high-energy' $E_1{\sim}P$ form and the 'low-energy' E_2-P form. Two dephospho-conformations are also shown: the E_2, which contain a low-affinity ATP-binding site, and the E_1A, which has ATP (A) bound at a high-affinity site.

The Na^+,K^+-ATPase can hydrolyze various phosphoric anhydrides (e.g., *p*-nitrophenyl phosphate, acetyl phosphate, and others), although, in the case of *p*-nitrophenyl phosphate, it is not fully clarified if these reactions represent an uncoupling between catalytic reactions and cation transport, or if they are linked to cation transports (Berberián and Beaugé, 1992). It is also uncertain if the hydrolysis of this substance produces comparable phospho-intermediates, as when ATP is the substrate, transferring their phosphoryl to the enzyme. Acetyl phosphate has been shown to substitute at the catalytic site, but not at the regulatory site (see following), promoting de-occlusion of K^+ (Campos et al., 1988).

In the Albers–Post model as described here, the translocations of Na^+ as well as of K^+ are associated with a spontaneous change in protein conformation, from E_1 to E_2 in the case of cytoplasmic Na^+-translocation, and from E_2P to E_1P in the case of extracellular K^+-translocation. In other versions of the reaction cycle, the translocation of ions and the change in conformations are not simultaneous (Nørby et al., 1983; Cornelius and Skou, 1985).

The Albers–Post Model

A more elaborated kinetic scheme of the Albers–Post model is shown in Figure 8, and the measured or calculated kinetic parameters are given in Table 3. The

Table 3. Kinetic Parameters (20 °C) for the Albers–Post Scheme (Figure 7)

Reaction	*Value*	*Key References*
$E_1A + 3Na^+ \leftrightarrows E_1ANa_3$	20; 4; 3 mM	Cornelius and Skou, 1988
$E_2P + 3Na^+ \leftrightarrows E_2PNa_3$	1; 60; 80 mM	Cornelius and Skou, 1988
$E_1A + 2K^+ \leftrightarrows E_1AK_2$	10; 40 mM	Läuger, 1991b
$E_2P + 2K^+ \leftrightarrows E_2PK_2$	0.5; 2 mM	Läuger, 1991b
$E_1ANa_3 \rightarrow E_1P(Na_3)$	180 s^{-1}	Mårdh and Zetterquist, 1974
$E_1ANa_3 \leftarrow E_1P(Na_3)$	$2{\cdot}10^4$ $M^{-1}s^{-1}$	Mårdh, 1975
$E_1P(Na_3) \rightarrow E_2PNa_3$	20 s^{-1}	Mårdh and Post, 1977
$E_1P(Na_3) \leftarrow E_2PNa_3$	2 s^{-1}	Stürmer et al., 1989
$E_2PK_2 \rightarrow E_2(K_2)$	>>100 s^{-1}	Forbush, 1988
$E_2PK_2 \leftarrow E_2(K_2)$	$>6{\cdot}10^4$ $M^{-1}s^{-1}$	Läuger, 1991b
$E_2(K_2) + ATP \rightarrow E_2A(K_2)$	$>1{\cdot}10^6$ $M^{-1}s^{-1}$	Läuger, 1991b
$E_2(K_2) + ATP \leftarrow E_2A(K_2)$	>100 s^{-1}	Moczydlowski and Fortes, 1981
$E_2A(K_2) \rightarrow E_1AK_2$	50 s^{-1}	Karlish and Yates, 1978
$E2A(K2) \leftarrow E1AK2$	30 s^{-1}	Läuger, 1991b

scheme depicts some additional intermediates predicted from model analysis or inferred from kinetic experiments. In Na^+:K^+ exchange, ATP is bound and hydrolyzed by acting at a high-affinity (apparent $K_D \sim 0.1$ μM) intracellular catalytic site (Glynn and Karlish, 1976; Cantley et al., 1978; Modczydlowski and Fortes, 1981). Three Na^+ are bound with positive cooperativity ($K_1 = 20$ mM, K_2 and $K_3 = 3$ mM) to high-affinity sites on E_1A at the cytoplasmic aspect (Cornelius and Skou, 1988, 1991). A β-aspartyl carboxyl group in the enzyme becomes phosphorylated (Post and Kume, 1973) and the three Na^+ become occluded in a conformation that only very slowly releases them. A spontaneous change in conformation from $E_1P \rightarrow E^*P$ reduces the number of binding sites to two, with the release of one Na^+ to the extracellular side, which reduces the affinity for Na^+ at the two remaining sites and makes these sites accessible to the extracellular side, where the remaining two Na^+ are released concomitantly with a change in enzyme conformation to an E_2P-form (Yoda and Yoda, 1987a, b). The E_2P-conformation, with high K^+-affinity, binds two extracellular K^+ and translocates the K^+ ions into the membrane phase in an occluded $E_2(K_2)$-form. By a low-affinity binding ($K_D \sim 0.1$ mM) of ATP (Robinson and Flashner, 1979; Moczydlowski and Fortes, 1981), or nonhydrolyzable analogs including ADP (Simons, 1975) to a regulatory site of this form, K^+-release to the cell interior is accelerated due to a lowered binding affinity caused by the ATP-induced change in conformation to an E_1A-form, which closes the cycle. A selection of experimentally determined kinetic parameters can be found in Läuger (1991b). It is apparent that during physiological conditions, the rate-limiting step in the over-all reaction scheme is the occlusion/de-occlusion of Na^+, whereas during Na^+:Na^+ exchange, it is probably the dephosphorylation step.

It should be emphasized that several experimental observations cannot easily be accommodated by the Albers–Post model in its simple form. For example, it is still debated whether the mechanism is *consecutive* as in the Albers–Post scheme, where Na^+ is bound at the cytoplasmic side and released to the extracellular side before K^+ is bound and translocated, or if Na^+ and K^+ are *simultaneously* bound in some part of the reaction cycle (see Sachs, 1991). Certain *trans*-allosteric effects observed in Na^+:K^+ exchange (Karlish and Stein, 1985) and in ATP-driven Na^+:Na^+ exchange (Cornelius and Skou, 1988) which indicate the presence of regulatory sites separate from the transport sites are not incorporated into the Albers–Post model. The presence of a third phospho-intermediate, the $E_2P(Na_2)$ or E^*P in Figure 8, was inferred from experiments indicating that the sum of ADP-sensitive (E_1P) and K^+-sensitive (E_2P) phosphoforms in some cases became greater than 100% (Yoda and Yoda, 1982a, 1986; Nørby et al., 1983; Klodos and Nørby, 1987), and from a stoichiometry of $1Na^+$:ATP found under some conditions (Yoda and Yoda, 1987a).

One persisting problem with the consecutive-reaction mechanism is the experimental findings that dephosphorylation from steady-state EP-level in Na^+ by the addition of a chase solution (e.g., cold ATP) is bi-exponential with a delayed phase

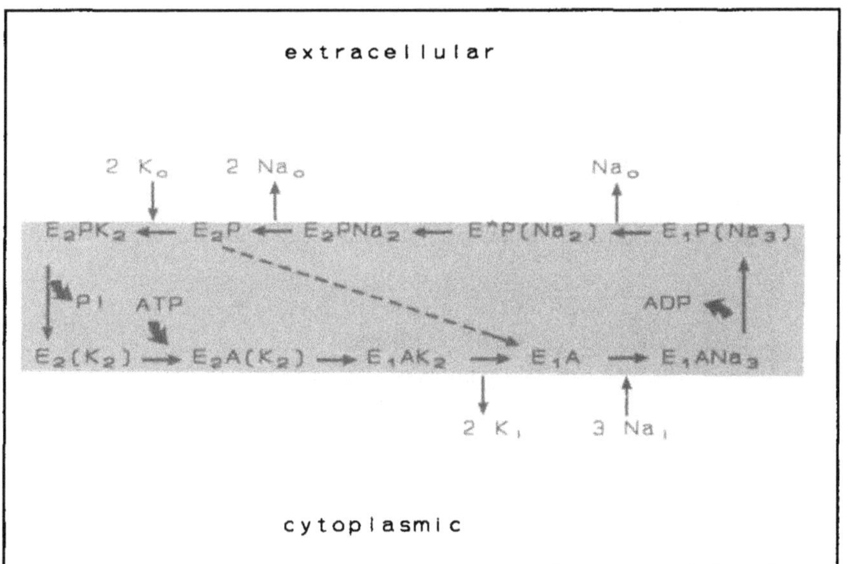

Figure 8. The catalytic cycle or the Albers–Post kinetic scheme of the Na^+,K^+-ATPase as modified from Karlish et al., 1978. Na_i and K_i, and Na_o, and K_o indicate cations inside and outside, respectively. Occluded ions are depicted in brackets. The sequential release of Na^+ to the outside leads to formation of a E^*-conformation; however, this intermediate could possibly also represent an E_2-subconformation with only two occluded Na^+ ions.

which is too slow to accommodate the rate of the over-all reaction both in the absence (Klodos et al., 1981; Plesner and Plesner, 1981a, b; Plesner et al., 1981) and in the presence of K^+ (J.C. Skou, personal communication). Parallel dephosphorylation of phospho-intermediates in a heterogeneous population has been suggested to account for this result (Martin and Sachs, 1991), alternatively, interference from the chase could cause the problem.

Occlusion and Conformations

From a mechanistic point of view, efficient active transmembrane ion-transport by a fixed carrier may demand two gates or energy barriers that alternatingly allow access via a channel to and from the two sides of the membrane (Jardetzky, 1966; Läuger, 1979; Klingenberg, 1981; Tanford, 1983). The two alternating-access states may correspond to the two major conformations the sodium pump can attain (Figure 9). If the molecular design is such that an intermediate state is present that is simultaneously shielded by energy barriers at both the cytoplasmic and the extracellular aspect of the pump, occlusion is present.

Although the molecular nature of occlusion sites is essentially unknown, they can be imagined, by analogy with ionophores or crown ethers, as a crevice with the protein backbone wrapped around the occluded ions. About 4–6 coordinating groups including backbone-peptide carbonyls, hydroxyls, and carboxyls, are usually involved in high-affinity sites, replacing some of the water molecules of the ion-hydration shell. Therefore, occluded ions can be imagined to be wholly or partially dehydrated in the occluded state. Tanford (1982) has described a model in which a slight twist of one of the transmembrane helices alternatingly exposes the binding pocket to either side of the membrane and simultaneously changes the number of peptide groups that constitutes the coordinating groups in the binding site. An increasing number of coordinating groups, say from four to six, would then

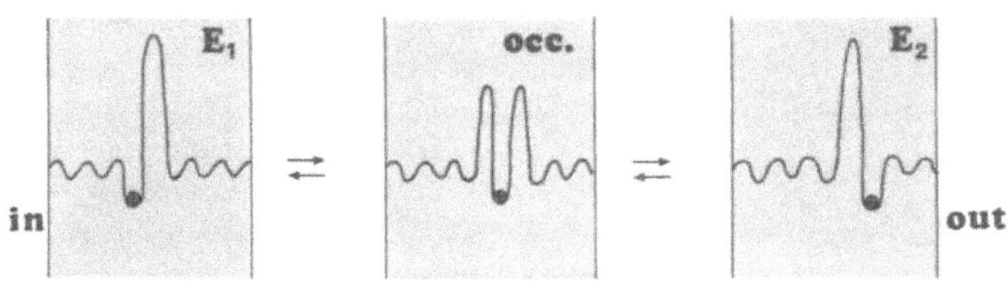

Figure 9. The moving barrier model of E_1–E_2 conformational change. In the E_1-conformation, the energy barrier is to the right of the cation-binding site and prevents the release of the ion to the outside (the extracellular gate). In the E_2-conformation, the energy-barrier has moved and obstructs ion-release to the cytoplasmic aspect (the cytoplasmic gate). An intermediate state with the ion enclosed by barriers on either side represents the occluded state (occ.).

result in an increased binding affinity of the cation by replacing water molecules in the hydration sphere, much like the binding of K^+ to valinomycin (Grell et al., 1987). Since it has been demonstrated that monomerization by detergents of the membrane-bound enzyme apparently preserves occlusion capacity for both K^+ and Na^+, it is probable that the occlusion cavity is within the αβ-unit itself (Vilsen et al., 1987, but cf. Esmann, 1985).

The occluded state may be too short-lived to be detected experimentally, however, this has not been the case with the sodium pump, and recently Rb^+-occlusion within the H^+,K^+-ATPase has been demonstrated, too (Rabon et al., 1993). Although some of the intermediate species in many versions of the Albers–Post model seems speculative, and kinetic models tend to take on a life of their own, there is ample evidence for the existence of occluded forms both with Na^+ and K^+. The first suggestion that an intermediate with occluded K^+-ions existed came from experiments carried out by Post et al. (1972) in order to explain that the rate of rephosphorylation depended on the cation present during the preceding dephosphorylation. In a study by Glynn and Hoffman (1971) of Na^+:Na^+ exchange in red cells, an intermediate state, comparable to a state with occluded Na^+, had to be assumed in order to explain the stimulatory effect of oligomycin of ATP:ADP exchange and its inhibitory effect on Na^+:Na^+ exchange.

K^+-occlusion/De-occlusion

Direct experimental evidence for K^+-occlusion followed after the demonstration of the slow transition of the unphosphorylated enzyme when changing from a K^+-medium to a Na^+-medium (Karlish and Yates, 1978; Karlish et al., 1978). This permitted measurements of K^+-release by ion-exchange chromatography that finally proved the existence of the K^+-occluded enzyme species (Beaugé and Glynn, 1979). The experiments demonstrated that, in the absence of ATP, approximately two K^+ ions are occluded per enzyme phosphorylation site (see Figure 7). K^+ can also be occluded after phosphorylation in the presence of a small concentration of ATP, corresponding to the forward running in the Albers–Post scheme, and with the same stoichiometry of two per phosphate binding-site (Beaugé and Glynn, 1979; Glynn and Richards, 1982; Forbush, 1987b; Shani-Sekler et al., 1988). The ATP concentration must be small enough to ensure phosphorylation at the high-affinity site, but too low to act appreciably on the low-affinity regulatory site. These findings are in concert with the basic Albers–Post model in which K^+ can be occluded by either (1) acting from the extracellular side by high-affinity binding to an E_2P-form followed by a dephosphorylation, or (2) acting from the cytoplasmic side by low-affinity binding to the E_1-form (the "direct route"). Glynn, Hara, Richards, and Steinberg (1987) later showed that the rate of K^+-release and the rate of conformational change are associated and that they are both accelerated by low-affinity ATP binding at the regulatory site. The left hand part of the Albers–Post model constituting its "K^+-limb" of the cycle seems, therefore, thoroughly supported by experiments.

De-occlusion of K^+ to the extracellular side by addition of Mg^{2+} and P_i after occlusion by the direct route, leads to an ordered release of the two trapped K^+ ions, suggesting heterogeneous extracellular leaving-sites with either a fast or a slow sequential release of ions (Glynn et al., 1985; Forbush, 1987a; Glynn and Richards, 1989). This indicates a single-file release of K^+ to the extracellular side through a narrow channel in the protein. However, spatial restriction on the release, with obligatory dissociation from a "fast-site" before release from a "slow-site," apparently cannot account alone for the kinetics of P_i-induced de-occlusion, and additional restriction on extracellular K^+-release, supposed to be rate-limiting, had to be assumed in the form of a "flickering gate" with intermittently opening (Forbush, 1987a). In this model, the extracellular gate opens infrequently and briefly, only long enough to allow the release of K^+ from the fast-site, whereas K^+ bound at the slow-site must first jump to the fast-site before being released. In contrast, the release of occluded K^+ after phosphorylation from ATP is found to be random (Forbush, 1987b), immediately following the conformational change from E_2P to E_1P.

Na^+-occlusion/De-occlusion

Occlusion of Na^+ within E_1P has been experimentally more difficult to detect since it is difficult to prevent either its dephosphorylation, or the transformation into E_2P, both of which lead to de-occlusion of Na^+. By pretreatment of the enzyme with chymotrypsin or N-ethylmaleimide, stabilizing the E_1P conformation, or oligomycin, stabilizing the E_1 conformation, it is possible to demonstrate Na^+-occlusion with a stoichiometry of close to three Na^+ per phosphorylation site within either E_1P (Glynn et al., 1984), or E_1 (Esmann and Skou, 1985).

The de-occlusion of Na^+ is probably a sequential reaction in which one Na^+ ion is first released in conjunction with a spontaneous conformational change from E_1P to a sub-conformation, E^*P, followed by the de-occlusion of the remaining two Na^+ ions, concomitant with the change in conformation from E^*P to E_2P (Yoda and Yoda, 1987a, 1987b, and see following).

Phosphorylation/Dephosphorylation

As previously mentioned, the P-type ATPases are characterized by having a phosphorylation event as part of their ion-translocation mechanism. The phosphorylation, in which the terminal phosphoryl group is transferred from ATP to an Asp-369 carboxyl group, together with movements within the major cytoplasmic domain, is somehow transmitted to the cytoplasmic gate which closes, thereby preventing access to the binding sites where 3 Na^+ ions are bound from the cytoplasmic medium. This phosphorylation step constitutes the major transference of free energy from the ATP energy source to the sodium pump, where it is transiently stored in the conformation $E_1{\sim}P$. The free energy increase resulting from phosphorylation is released stepwise in conjunction with a transformation in protein conformation, associated with a re-orientation of the ion-binding sites, to the more stable form E_2-P. Two other energetically uphill steps are comprised by the cation-releasing steps, either of Na^+ to the extracellular side, or of K^+ to the cytoplasmic side. A calculation of the free-energy levels of various enzyme species comprising the Albers–Post scheme has been given by Stein (1990).

The two major phosphorylated conformations differ in the reactivity of their *trans*-phosphorylation reactions. The E_1P conformation with three bound Na^+ ions has its acyl phosphate shielded from water and can return the covalently bound phosphate only to ADP, whereas E_2P, with two K^+ ions bound can donate its acyl phosphate to water, but not to ADP. Therefore, in dephosphorylation studies the fraction of total phospho-intermediates found to be ADP-sensitive would correspond to E_1P, whereas the fraction found to be K^+-sensitive apparently corresponds to the E_2P-fraction. It has become clear, however, that this notion is probably an over-simplification, since it has been demonstrated that the sum of phospho-enzyme in some experimental conditions exceeded the amount of total EP (Yoda and Yoda, 1982a, 1986; Nørby et al., 1983). Furthermore, the kinetics of dephosphorylation do not conform to such a simple two-pool model (Plesner and Plesner, 1981a,b; Plesner et al., 1981; Klodos et al., 1991). Therefore, it must be assumed that at least three phospho-enzyme intermediates exist, the former two and a third that is reactive towards both ADP and K^+, with a proportion that depends on the Na^+-concentration (Yoda and Yoda, 1986). Whether the intermediate phosphoform should be considered a separate conformation, E^*P, or is assigned a sub-conformation of E_2P with the two binding sites occupied by Na^+ as opposed to $E_2P(K_2)$, is controversial. However, the proteolytic digestion pattern is apparently independent on whether or not two K^+ or two Na^+ are occluded, indicating similar gross E_2P-conformations for both (Jørgensen, 1992). One could, therefore, speculate if it is the occupancy of three or two cation sites which determines the variable exposition of the acyl-phosphate to water as reflected in the E_1P/E_2P conformations.

B. Electrogenicity

In each turnover of the sodium pump, 3 Na^+ are expelled and 2 K^+ are taken up for each molecule of ATP split. The 3:2:1 stoichiometry which is attained during physiological conditions can probably vary according to the experimental conditions (De Weer et al., 1988; Cornelius, 1990).

The electrogenicity of the pump causes a contribution to the resting potential of the cell. In steady-state, one can show this contribution to be at most ~10 mV (Mullins and Noda, 1963) due to the passive permeabilities, especially for Cl^-. By artificially lowering the membrane conductance, much higher electrogenic contributions can be obtained. In proteo-liposomes with low Cl^--permeability or with

SO_4^{2-} replacing Cl^-, the electrogenic contribution can be about 200 mV (Cornelius, 1989, 1990).

The converse, that the transmembrane potential affects the pumping rate, must therefore also be true. Hyperpolarization in cells to about –200 mV should arrest the pump, and further hyperpolarization reverse it. The possible effect of transmembrane potential on specific rate constants in the kinetic scheme (Figure 8) has recently drawn much attention. Such effects have been studied using reconstituted vesicles, and sodium pump containing membranes adsorbed to planar bilayers.

In the Albers–Post kinetic scheme, it is the "sodium-limb" of the cycle that is probably associated with the transfer of the electrical net charge (Fendler et al., 1985; Nakao and Gadsby, 1986; Rephaeli et al., 1986; Borlinghaus et al., 1987).

In a simple model for ion-translocating pumps, it is proposed that the binding domain is located inside a channel structure so that a part of the membrane potential drops across this channel and only a part over the membrane dielectric (Läuger, 1979; Tanford, 1983). This means that the ion-binding sites only have to move over a fraction of the total membrane dielectric, presumably in association with a conformational change. If a part of the transmembrane potential drops across the narrow access channel, a high-field access channel or ion well is present (Mitchell and Moyle, 1974). There is strong evidence that the electrogenicity of the sodium-limb in the Na^+,K^+-ATPase reaction cycle can be accounted for by the de-occlusion and release of Na^+ through such an extracellular narrow access channel (Läuger and Apell, 1986, 1988; Nakao and Gadsby, 1986; Borlinghaus et al., 1987; Stürmer et al., 1989; Rakowski et al., 1991; Stürmer et al., 1991; Gadsby et al., 1993; Rakowski, 1993).

The pathway by which extracellular K^+ is translocated during physiological conditions, the "potassium-limb," is generally thought to be electroneutral (Fendler et al., 1985; Nakao and Gadsby, 1986; Goldshleger et al., 1987, 1990; Bahinski et al., 1988; Rakowski and Paxson, 1988; Stürmer et al., 1989), and only during conditions of low extracellular K^+ can a voltage dependency be shown (Rakowski, 1991; Rakowski et al., 1991). These observations seem to provide strong evidence against conformational changes occurring in the K^+-limb to be electrogenic themselves, involving charge translocation through the membrane field, whereas the binding of extracellular K^+ ions and their subsequent occlusion are believed to be electrogenic, probably due to their passage through an ion well to arrive at their binding sites (Läuger, 1991b; Stürmer et al., 1991).

In order to account for the presumably electroneutral K^+ translocation in the physiological Na^+:K^+ exchange, and for the lack of a negative slope in the I–V curves (Nakao and Gadsby, 1986), a simple electrostatic model for the charge translocation has been suggested, in which the cytoplasmic ligand-binding domain, which normally binds Na^+, encloses two negative charges (Nakao and Gadsby, 1986; Goldschleger et al., 1987) that comigrate with the ligands during turnover. Whether or not the K^+-translocating pathway includes charge movement is still controversial, and to observe it may depend on the establishment of suitable experimental conditions, especially nonsaturating extracellular K^+-concentrations (Rakowski, 1991; Rakowski et al., 1991). It is still being debated if a shallow ion well is involved in the cytoplasmic Na^+ binding, at least for one of the three Na^+ ions, which could explain an apparent weak electrogenic effect on cytoplasmic Na^+ affinity (Goldschleger et al., 1987).

C. Coupling of the Catalytic Reaction with Fluxes

Taken together, the E_1–E_2 model has been successful as a general frame of reference in the investigation of the sodium pump. Several transport modes or partial reactions of the sodium pump have been important in elucidating the overall reaction mechanism: the Albers–Post model has been remarkably successful in that it explains all of the experimentally observed modes of ion-exchange that the sodium pump can accommodate (Figures 10a–10f).

Na^+:K^+ Exchange and Its Offshoots

The physiological mode (Figure 10a), where 3 internal Na^+ are exchanged for 2 external K^+, at the expense of energy derived from the splitting of 1 ATP (Garrahan and Glynn, 1967a; Glynn et al., 1971), can be varied by changing the cation ligands at both the Na^+-sites and the K^+-sites. A number of cations can function as K^+-congeners: Rb^+, Cs^+, Tl^+, NH_3^+, Li^+, and even Na^+. In the absence of extracellular K^+, an exchange of internal Na^+ for external Na^+ takes place as though external Na^+ were acting as K^+ in promoting dephosphorylation of E_2P, becoming occluded and translocated. Na^+ acts as a "poor K^+-substitute" in an otherwise apparently identical reaction cycle as the normal Na^+:K^+ exchange. The effectiveness of extracellular Na^+ in substituting for K^+ is low, however, only 5–10% (Cornelius and Skou, 1985) of maximal Na^+:K^+ turnover can be obtained, and other details concerning the de-occlusion of extracellular Na^+ and its acceleration by ATP may be different in the two reaction modes (see Cornelius and Skou, 1987, 1991).

The ATP-driven Na^+:Na^+ exchange mode was first observed in red cell ghosts by Lee and Blostein (1980) and further investigated by Blostein (1983). In reconstituted liposomes, this mode has been demonstrated to be electrogenic with a stoichiometry close to $3Na_{cyt}$:$2Na_{ext}$:1 ATP (Forgac and Chin, 1981; Cornelius and Skou, 1985; Karlish and Stein, 1985; Apell et al., 1990).

The number of Na^+-congeners is much smaller, and apparently only Li^+, and H^+ can replace Na^+. Recently, it has been shown that H^+ has Na^+-like effects (Hara and Nakao, 1981; Blostein, 1985) and can substitute (poorly) for Na^+ in an ATP-hydrolysis dependent exchange for K^+ (Polvani and Blostein, 1988). The H^+:K^+ exchange has been demonstrated in reconstituted systems by Hara and Nakao (1986). The electrogenicity of the exchange reaction has not been investigated. Cytoplasmic K^+ competes with Na^+ for binding to E_1A (Sachs, 1986; Garray and Garrahan, 1973), but apparently the formed species containing K^+ cannot translo-

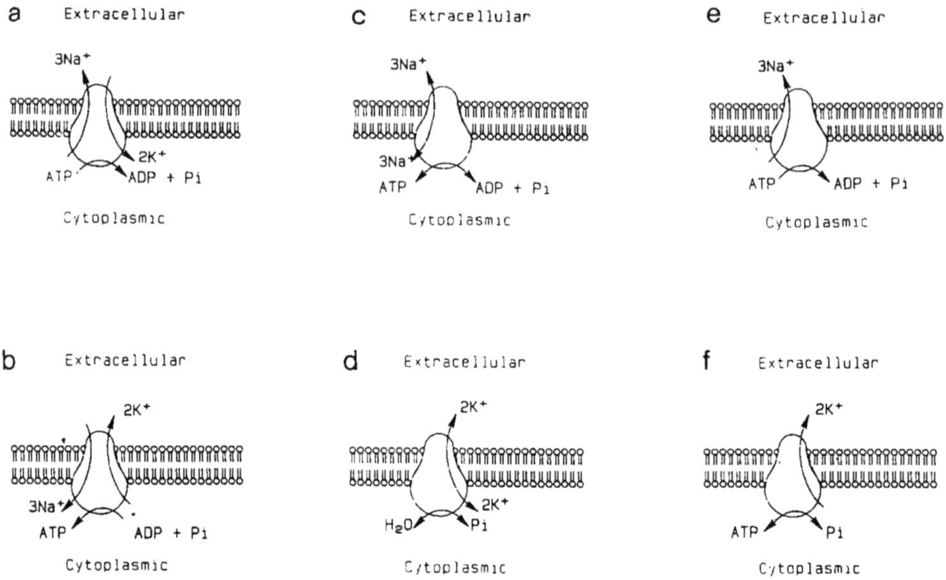

Figure 10. The 6-flux modes the sodium pump can accommodate by varying experimental arrangements. (**a**) The physiological Na^+:K^+ exchange in which 3 Na^+_{cyt} are exchanged for 2 K^+_{ext} for each molecule of ATP split. This mode, in fact, represents several exchange submodes in which different alkali cations substitute for either extracellular K^+, Li^+, or H^+ and act as possible cytoplasmic Na^+ congeners. (**b**) The reversed Na^+:K^+ exchange. (**c**) and (**d**), the shuttling of Na^+ or K^+ through the right-hand part or the left-hand part of the Albers–Post scheme, respectively. (**e**) and (**f**), the uncoupled effluxes of Na^+ or K^+, respectively.

cate, and/or occlude the K^+, and inhibition results (Cornelius, 1992; Matsui and Homereda, 1982).

ADP Stimulated Na^+:Na^+ Exchange

In the absence of extracellular K^+ and with a high Na^+-concentration, a one for one exchange of internal and external Na^+ takes place, Figure 10c (Garrahan and Glynn, 1967b, c). The exchange is electroneutral and requires both ATP and ADP (De Weer, 1970; Glynn and Hoffman, 1971; Cavieres and Glynn, 1979; Kennedy et al., 1986), but proceeds without net hydrolysis (Glynn and Hoffman, 1971).

The exchange is assumed to represent a shuttling back and forth through the right-hand part of the Albers–Post scheme in accordance with the observation that it is accompanied by an ATP:ADP exchange (cf. Kaplan, 1982). The ADP-stimulated Na^+:Na^+ exchange has been studied in a variety of cells, such as red cells, squid axons, and muscle cells, but has been poorly characterized in reconstituted systems (Anner and Moosmayer, 1982; Karlish and Stein, 1982a, 1982b; Karlish et al., 1988).

K^+:K^+-exchange

In the absence of Na^+ and in the presence of K^+, P_i, Mg^{2+}, and nucleotide, a shuttling through the left-hand part of the Albers–Post scheme gives rise to the observed K^+:K^+ exchange, Figure 10d (Simons, 1974). The mode is an electroneutral one—for one exchange of internal and external K^+ accompanied by a H_2O:P_i exchange of oxygen. The exchange has been mostly studied in red cells and their ghosts (Glynn et al., 1970, 1971; Sachs, 1980, 1981), but has also been investigated in detail in reconstituted systems (Karlish and Stein, 1982a, 1982b; Karlish et al., 1982).

An ouabain-sensitive uncoupled Rb^+-efflux observed in reconstituted renal Na^+,K^+-ATPase (Karlish and Stein, 1982a, 1982b), which is independent of ATP and P_i, is different from the above uncoupled K^+-efflux, and probably represents a reversible slippage of Rb^+ from the occluded $E_2(Rb_2)$-form without phosphorylation from P_i.

Uncoupled Na^+-efflux

In the absence of Na^+ and K^+ on the extracellular side, a small efflux of Na^+ is observed, Figure 10e (Garrahan and Glynn, 1967b, c). The stoichiometry is 2–3 Na^+ ions expelled for one ATP hydrolyzed (Glynn and Karlish, 1976). The electrogenicity, however, does not apparently result in a measurable transmembrane voltage in red cells (Dissing and Hoffman, 1990), and the experiments suggest that it is accompanied by a concomitant transport of anions, which makes the transport electrically silent.

An investigation of this transport in a reconstituted system was recently carried out, which demonstrated its electrogenicity and measured the positive net charge stoichiometry to be three positive charges per ATP split, in accordance with three Na^+ expelled per ATP hydrolyzed (Cornelius, 1989); however this stoichiometry varied according to pH (Cornelius, 1990).

A variation of this mode with a stoichiometry of 1 Na_{cyt}:1 ATP is probably explained by dephosphorylation of E^*P with the release of only one Na^+ (Yoda and Yoda, 1987a).

Uncoupled K^+-efflux

According to the scheme in Figure 7, the uncoupled Na^+ efflux proceeds via a dephosphorylation of "empty" E_2P to E_2 (via the dotted line in the scheme). Since the release of K^+ to the outside via a back-reaction of the left-hand part of the scheme also results in the formation of E_2P, an uncoupled K^+-efflux is feasible (Figure 10f). This is shown to be true by Sachs (1986), but not noted by Glynn et al. (1970) due to the presence of external Na^+ which inhibits this reaction.

Reversed Na^+:K^+ Exchange

The sodium-pump is reversible and by proper arrangement of the ion-gradients and phosphorylation potential ((ATP)/(ADP)·(P_i)), it is possible to reverse the pump and synthesize ATP at the expenditure of energy derived from the run-down of the ion-gradient, Figure 10b (Garrahan and Glynn, 1967d).

Reversal of the pump has only been demonstrated in red cells, red cell ghosts, and in squid axons; in the latter, the reversed pump retains the 3:2 Na^+:K^+ stoichiometry (De Weer and Rakowski, 1984).

V. BIOSYNTHESIS

The polypeptide chain synthesis of the α and β subunits of the sodium pump is produced from distinct mRNAs and not necessarily correlated: in some tissues the α-subunit is overexpressed, whereas in others, the synthesis of the β-subunit is favored (Geering et al., 1989; McDonough et al., 1990; Taormino and Fambrough, 1990; Lescale-Matys et al., 1993). The α and β subunits are inserted independently into the endoplasmic reticulum (ER; Geering et al., 1985; Cayanis et al., 1990), where they are immediately assembled (Fambrough, 1983; Tamkun and Fambrough, 1986; Ackermann and Geering, 1990). The assembly is not necessarily co-translatory, since newly synthesized α-subunits can associate with preexisting β-subunits, synthesized from injected cDNA before expression of the α-subunit (Noguchi et al., 1990). This indicates that it is the number of β-subunits that limits in the formation of functional αβ-units.

The co-translatory insertion into the ER takes place via interaction with the signal recognition particle and probably the signal recognition particle receptor (Kawakami and Nagano, 1988; Cayanis et al., 1990). Neither of the subunits contain a cleavable amino-terminal leader peptide, that would function as a signal sequence for insertion, and probably both contain an internal signal sequence (Homareda et al., 1988; Kawakami and Nagano, 1988). Soon after ER insertion, the β-subunit acquires core sugars by N-linked glycosylation (Geering et al., 1985).

The assembly of subunits seems to be required in order for the complex to leave the ER (Jaunin et al., 1992) and appear in the Golgi apparatus. During the transfer and in the Golgi apparatus, the core sugars are modified and in the *trans*-Golgi, a further addition of complex sugars takes place. A possible functional significance of glycosylation for the correct folding of newly synthesized β-subunit and subsequent efficient assembly with the α-subunit *in vivo* has been proposed (Zamofing et al., 1988, 1989); however, inhibition of glycosylation has no effect upon either the assembly or the incorporation of αβ-subunits into plasma membranes (Tamkun and Fambrough, 1986). Nor does the inhibition of correct glycosylation interfere with the catalytic or transport functions of the sodium pump (Zamofing et al., 1988; Caplan et al., 1990).

In the ER, and during its intracellular routing, the α-subunit undergoes a structural maturation process indicated, for example, by its susceptibility to trypsinolysis in parallel with the ability to undergo conformational changes (Geering et al., 1987) and for ouabain binding (Caplan et al., 1990). Following modifications in the *trans*-Golgi apparatus, the sodium pump is transported to the plasma membrane mediated by general vesicular trafficking (Bennett and Scheller, 1993), and is eventually targeted to the basolateral membranes in polarized cells (Caplan et al., 1986).

Degradation of the sodium pumps probably takes place by proteolysis in the lysosomes. The removal from the plasma membrane can be rapid (Takeyasu et al., 1989; Lescale-Matys et al., 1993). In a pig kidney cell line, the degradation of β-subunits is initially faster than that of α-subunits, presumably to take care of the excess nascent β (Lescale-Matys et al., 1993), and then the degradation of α- and β-units becomes identical, indicating that the target is then the αβ-unit (Wolitzky and Fambrough, 1986).

VI. REGULATION

It is inconceivable that a membrane protein like the sodium pump, which is of such vital importance in cell housekeeping, should not be under regulatory control. Nonetheless, it is only recently that a picture of how that might be achieved at the molecular level has begun to emerge. Both the activity and the concentration of the sodium pumps are regulated hormonally as well as nonhormonally. The regulation takes place at various stages of maturation, ranging from the level of transcription to direct control of the pump at the plasma membrane (Figure 11).

Investigations of the transcriptional regulation of the three genes coding for the three α-isoforms, explaining their tissue-specific and developmental expression patterns (Lingrel et al., 1990), are in progress in several laboratories. It is probable that each isoform gene has its own set of regulatory elements for potential transcription factors and hormones sites: by preparing transgenic mice, Lingrel et al. (1991) were able to locate a 1.6 kb 5′-flanking sequence of the α3 gene sufficient to confer tissue-specific expression of this isoform primarily in brain.

Apart from the tissue-specific distribution of isoforms (Young and Lingrel, 1987; Gick et al., 1993), many cells also exhibit a developmental pattern: in cardiac tissue, e.g., the α1 mRNA is the predominant isoform gene transcript at all developmental stages, whereas the α3 mRNA transcript is primarily expressed in fetal and neonatal stages, then declines concomitant with a replacement by α2 mRNA isoform (Orlowski and Lingrel, 1988). The switch in this isoform expression is, in part, induced by the thyroid hormone, which increases both α2 and α3 isoform mRNA, and, in part, by glucocorticoids, which in the presence of thyronine repress the induction of the α3 isoform mRNA (Lingrel et al., 1991). The importance of the two hormones in the regulation of gene expression is supported by *in vivo* studies using hypothyroid animals, which express primarily α1 mRNA and α2 mRNA upon thyronine treatment (Gick et al., 1990; Horowitz et al., 1990a, 1990b).

Transcriptional regulation of the β-gene is also apparently possible: by increased intracellular Na^+-concentration in myoblasts, induced by the opening of the Na^+-channels, an up-regulation of Na^+,K^+-ATPase takes place driven by an increased β-subunit biosynthesis, resulting in the assembly of more αβ-units (Wolitzky and

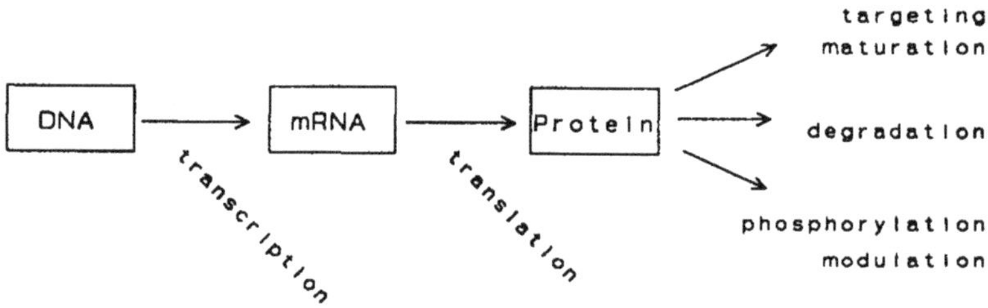

Figure 11. Diagram of various levels of regulation of Na^+,K^+-ATPase involving the expression of α- and β-genes (transcriptional regulation), the biosynthesis and assembly of α- and β-subunits (translational regulation), and post-translational processing. The latter includes targeting to and maturation in the plasma membrane, the rate of degradation, and the modulation of activity by ligands and phosphorylation by protein kinases.

Fambrough, 1986; Taormino and Fambrough, 1990). The up-regulation of Na^+,K^+-ATPase involves an increase in β1 mRNA which accounts for the increased biosynthesis of β1-subunit, indicating increased transcriptional activity on the β1-gene induced by increased cytosolic Na^+ (Fambrough et al., 1991). The results are consistent with the functional role of the β-subunit in the intracellular transport and maturation of the sodium pump (Geering, 1990, 1991). If the Na^+ entry is blocked in the myotubes, the Na^+,K^+-ATPase is rapidly removed from the plasma membrane and internalized (Takeyasu et al., 1989) without increasing the rate of degradation. A direct role for intracellular Na^+ as a major determinant of the sodium pump recruitment factor has also been suggested for rabbit kidney (Blot-Chabaud et al., 1990; Coutry et al., 1992).

Translational regulation of subunit production is indicated by unequal changes in subunit mRNA levels and enzyme activity induced by hormones such as thyronine. In rat hepatic tissue, thyroid hormone induces a seven-fold increase in α-subunit mRNA, whereas the β-subunit mRNA is unchanged, and the Na^+,K^+-ATPase activity increases by 30% (Gick et al., 1988). Recent studies indicate that a different mRNA translation efficiency mediated by the 5′-UT region of α mRNA is important in regulation of subunit gene expression.

Apart from the long-term hormonal regulation described above, acute hormonal regulation of the sodium pump has also been observed. Catecholamines, insulin, vasopressin, glucagon, and EGF act on the sodium pump at the plasma membrane. The mechanisms by which these agents regulate sodium pump activity have not yet been fully explored, but recent lines of evidence indicate that the regulatory mechanism of catecholamines may be mediated by a covalent phosphorylation of the α-subunit by protein kinases: elevated levels of protein kinase C (PKC) or cAMP-activated protein kinase A (PKA) induce a change in Na^+,K^+-ATPase activity (Marver et al., 1986; Satoh and Endou, 1990; Navran et al., 1991).

Stimulation of PKA and PKC mediate phosphorylation of the Na^+,K^+-ATPase both *in vitro* and in intact cells (Ling and Cantley, 1984; Lowndes et al., 1990; Chibalin et al., 1991, 1992; Beguin et al., 1994, 1996; Feschenko and Sweadner, 1994, 1995; and cf. Ewart and Klip, 1995). The location of the PKA phosphorylation site is Ser-938 (Feschenko and Sweadner, 1994; Fisone et al., 1994) which is conserved among all three α-subunits of the Na^+,K^+-ATPase and the homologues residue Lys-953 is one of two potential sites of phosphorylation in the H^+,K^+-ATPase (Maeda et al., 1988). The PKC phosphorylation sites are serine and threonine residues in the N-terminal part of the α-subunit (Bequin et al., 1994; Feschenko and Sweadner, 1995). It is still under dispute, however, whether the protein kinase phosphorylations form the molecular basis for the observed acute hormonal regulation: conflicting results on the physiological effects of PKA and PKC mediated phosphorylations of Na^+,K^+-ATPase have been obtained in different *in vivo* systems and in different preparations *in vitro* (Bertorello et al., 1991; Vasilits and Schwarz, 1992; Feschenko and Sweadner, 1994; Fisone et al., 1994; Féraille et al., 1995; Cornelius and Logvinenko, 1996). This may be caused in part by methodological problems and in part by protein kinase phosphorylation being both isoform- and species-specific as shown by Feschenko and Sweadner (1994) and by Cornelius and Logvinenko (1996).

An acute feedback regulation of the Na^+,K^+-ATPase activity and associated pumping is also effective via the activation/deactivation exerted from the ligands of the catalytic reaction itself. During steady-state the sodium pump has a large reserve capacity: at a physiological intracellular/extracellular Na^+-concentration of 10–20 mM/140 mM, and K^+-concentration of about 120 mM/4 mM the sodium pump turnover is only 10–20% of maximum, which is about 10^4 min^{-1}.

ACKNOWLEDGMENTS

Tom Blucher is gratefully acknowledged for scrutinizing the manuscript, and Jesper V. Møller for comments and suggestions.

REFERENCES

Abbott, A., & Ball, W. J. (1993). The epitope for the inhibitory antibody M7-PB-E9 contains Ser-646 and Asp-652 of the sheep Na^+,K^+-ATPase α-subunit. Biochemistry 32, 3511–3518.

Ackermann, U., & Geering, K. (1990). Mutual dependence of Na^+,K^+-ATPase α- and β-subunits for correct posttranslational processing and intracellular transport. FEBS Lett. 269, 105–108.

Anner, B. M. (1985). The receptor function of the Na^+,K^+-activated adenosine triphosphatase system. Biochem. J. 227, 1–11.

Anner, M. B., & Moosmayer, M. (1982). On the kinetics of the Na^+:K^+ exchange in the initial and final phase of sodium pump activity in liposomes. J. Memb. Sc. 11, 27–37.

Apell, H.-J. (1989). Electrogenic properties of the Na^+,K^+-pump. J. Memb. Biol. 110, 103–114.

Apell, H.-J., Häring, V., & Roudna, M. (1990). Na^+,K^+-ATPase in artificial lipid vesicles. Comparison of Na^+,K^+ and Na^+-only pumping mode. Biochim. Biophys. Acta 1023, 81–90.

Bahinski, A., Nakao, M., & Gadsby, D. C. (1988). Potassium translocation by the Na^+/K^+ pump is voltage insensitive. Proc. Natl. Acad. Sci. USA 85, 3412–3416.

Ball, W. J. (1984). Immunochemical characterization of a functional site of $(Na^{+}+K^{+})$-ATPase. Biochemistry 23, 2275–2281.

Ball, W. J., & Friedman, M. L. (1987). Immunochemical evidence that the FITC-labeling site on Na^{+},K^{+}-ATPase is not the ATP binding site. Biochem. Biophys. Res. Commun. 148, 246–253.

Bamberg, E., & Schoner, W. (1994). The solium pump. Steinkopff Darmstadt and Springer, New York.

Bayer, R. (1990). Topological disposition of the sequence -QRKIVE- and -KETYY in native $(Na^{+}+K^{+})$-ATPase. Biochemistry 29, 2251–2256.

Beaugé, L. A., & Glynn, I. M. (1979). Occlusion of K^{+} ions in the unphosphorylated sodium pump. Nature 280, 510–512.

Beguin, P., Beggah, A. T., Chibalin, A. V., Burgener-Kairuz, P., Jaisser, F., Mathews, P. M., Rossier, B. C., Cotecchia, S., & Geering, K. (1994). Phosphorylation of the Na,K-ATPase α-subunit by protein kinase A and C *in vitro* and in intact cells. Identification of a novel motif for PKC-mediated phosphorylation. J. Biol. Chem. 269, 24437–24445.

Beguin, P., Beggah, A., Chibalin, A. V., Cotecchia, S., & Geering, K. (1996). Adrenergic, dopaminergic, and muscarinic receptor stimulation leads to PKA phosphorylation of Na,K-ATPase. Am. J. Physiol. 270, C131–C137.

Bennett, M. K., & Scheller, R. (1993). The molecular machinery for secretion is conserved from yeast to neurons. Proc. Natl. Acad. Sci. USA 90, 2559–2563.

Berberián, G., & Beaugé, L. (1992). Phosphatase activity and potassium transport in liposomes with Na^{+},K^{+}-ATPase incorporated. Biochim. Biophys. Acta 1103, 85–93.

Bertorello, A. M., Aperia, A., Walaas, S. I., Nairn, A. C., & Greengard, P. (1991). Phosphorylation of the catalytic subunit of Na^{+},K^{+}-ATPase inhibits the activity of the enzyme. Proc. Natl. Acad. Sci. USA 88, 11359–11362.

Blostein, R. (1983). Sodium pump-catalyzed sodium-sodium exchange associated with ATP hydrolysis. J. Biol. Chem. 258, 7948–7953.

Blostein, R. (1985). Proton-activated rubidium transport catalyzed by the sodium pump. J. Biol. Chem. 260, 829–833.

Blot-Chabaud, M., Wanstok, F., Bonvalet, J. P., & Farman, N. (1990). Cell sodium-induced recruitment of Na^{+},K^{+}-ATPase pumps in rabbit cortical collecting tubules is aldosterone-dependent. J. Biol. Chem. 265, 11676–11681.

Borlinghaus, R. T., Apell, H.-J., & Läuger, P. (1987). Fast charge translocations associated with partial reactions of the Na^{+},K^{+}-pump: I. Current and voltage transients after photochemical release of ATP. J. Memb. Biol. 97, 161–178.

Boulanger, B. R., Lilly, M. P., Hamlyn, J. L., Shurtleff, D., & Gann, D. S. (1993). Ouabain is secreted by the adrenal gland in awake dogs. Am. J. Physiol. 264, E413–E419.

Brown, T. A., Horowitz, B., Miller, R. P., McDonough, A. A., & Farley, R. A. (1987). Molecular cloning and sequence analysis of the $(Na^{+}+K^{+})$-ATPase β subunit from dog kidney. Biochim. Biophys. Acta 912, 244–253.

Campos, M., Berberián, G., & Beaugé, L. (1988). Some total and partial reactions of Na^{+}/K^{+}-ATPase using ATP and acetyl phosphate as a substrate. Biochim. Biophys. Acta 938, 7–16.

Canessa, C. M., Horisberger, J.-D., Louvard, D., & Rossier, B. C. (1992). Mutation of a cysteine in the first transmembrane segment of Na^{+},K^{+}-ATPase α subunit confers ouabain resistance. EMBO J. 11, 1681–1687.

Cantley, L. C., Cantley, L. G., & Josephson, L. (1978). A characterization of vanadate interactions with the (Na^{+},K^{+})-ATPase. Mechanistic and regulatory implications. J. Biol. Chem. 253, 7361–7368.

Capasso, J. M., Hoving, S., Tal, D. M., Goldshleger, R., & Karlish, S. J. D. (1992). Extensive digestion of Na^{+},K^{+}-ATPase by specific and nonspecific proteases with preservation of cation occlusion sites. J. Biol. Chem. 267, 1150–1158.

Caplan, M. J., Anderson, H. C., Palade, G., & Jamieson, J. D. (1986). Intracellular sorting and polarized cell surface delivery of $(Na^{+}+K^{+})$-ATPase, an endogenous component of MDCK cell basolateral plasma membranes. Cell 46, 623–631.

Caplan, M. J., Forbush, B., III, Palade, G. E., & Jamieson, J. D. (1990). Biosynthesis of the Na,K-ATPase in Madin-Darby kidney cells. J. Biol. Chem. 265, 3528–3534.

Cavieres, J. D., & Glynn, I. M. (1979). Sodium-sodium exchange through the sodium pump: The roles of ATP and ADP. J. Physiol. 297, 637–645.

Cayanis, E., Bayley, H., & Edelman, I. S. (1990). Cell-free transcription and translation of Na^+,K^+-ATPase α and β subunit cDNAs. J. Biol. Chem. 265, 10829–10835.

Charnock, J. S., & Post, R. L. (1963). Evidence of the mechanism of ouabain inhibition of cation activated adenosine triphosphatase. Nature 199, 910–911.

Chehab, F. F., Kan, Y. W., Law, M. L., Hartz, J., Kao, F.-T., & Blostein, R. (1987). Human placental Na^+,K^+-ATPase α subunit: cDNA cloning, tissue expression, DNA polymorphism, and chromosomal localization. Proc. Natl. Acad. Sci. USA 84, 7901–7905.

Chibalin, A. V., Lopina, O. D., Petukhov, S. P., & Vasilets, L. A. (1991). Phosphorylation of Na^+,K^+-ATPase by protein kinase C and cAMP-dependent protein kinase. Biol. Membr. 8, 1140–1141.

Chibalin, A. V., Vasilets, L. A., Hennekes, H., Pralong, D., & Geering, K. (1992). Phosphorylation of Na^+,K^+-ATPase α-subunit in microsomes and in homogenates of *Xenopus* oocytes resulting from the stimulation of protein kinase A and protein kinase C. J. Biol. Chem. 267, 22378–22384.

Clarke, D. M., Loo, T. W., Inesi, G., & MacLennan, D. H. (1989). Location of high affinity Ca^{2+}-binding sites within the predicted transmembrane domain of the sarcoplasmic reticulum Ca^{2+}-ATPase. Nature 339, 476–478.

Clausen, T. (1986). Regulation of active Na^+-K^+ transport in skeletal muscle. Physiol. Rev. 66, 542–580.

Clausen, T., & Everts, M. E. (1989). Regulation of the Na^+,K^+-pump in skeletal muscle. Kidney Int. 35, 1–13.

Clausen, T., Van Hardeveld, C., & Everts, M. E. (1991). Significance of cation transport in control of energy metabolism and thermogenesis. Physiol. Rev. 71, 733–774.

Conway, E. J. (1957). Nature and significance of concentration relations of potassium and sodium ion in skeletal muscle. Physiol. Rev. 37, 84–132.

Cornelius, F. (1989). Uncoupled Na^+-efflux on reconstituted shark Na^+,K^+-ATPase is electrogenic. Biochem. Biophys. Res. Commun. 160, 801–807.

Cornelius, F. (1990). Variable stoichiometry in reconstituted shark Na^+,K^+-ATPase engaged in uncoupled efflux. Biochim. Biophys. Acta 1026, 147–152.

Cornelius, F. (1991). Functional reconstitution of the sodium pump. Kinetics of exchange reactions performed by reconstituted Na^+/K^+-ATPase. Biochim. Biophys. Acta 1071, 19–66.

Cornelius, F. (1992). *Cis*-allosteric effects of cytoplasmic Na^+/K^+ discrimination at varying pH. Low-affinity multisite inhibition of cytoplasmic K^+ in reconstituted Na^+/K^+-ATPase engaged in uncoupled Na^+-efflux. Biochim. Biophys. Acta 1108, 190–200.

Cornelius, F., & Logvinenko, N. (1996). Functional regulation of reconstituted Na,K-ATPase by protein kinase A phosphorylation. FEBS Lett. 380, 277–280.

Cornelius, F., & Skou, J. C. (1985). Na^+–Na^+ exchange mediated by (Na^++K^+)-ATPase reconstituted into liposomes, evaluation of pump stoichiometry and response to ATP and ADP. Biochim. Biophys. Acta 818, 211–221.

Cornelius, F., & Skou, J. C. (1987). The sided action of Na^+ and K^+ on reconstituted shark (Na^++K^+)-ATPase engaged in Na^+–Na^+ exchange accompanied by ATP hydrolysis. I. The ATP activation curve. Biochim. Biophys. Acta 904, 353–364.

Cornelius, F., & Skou, J. C. (1988). The sided action of Na^+ on reconstituted shark Na^+/K^+-ATPase engaged in Na^+–Na^+ exchange accompanied by ATP hydrolysis. II. Transmembrane allosteric effects on Na^+ affinity. Biochim. Biophys. Acta 944, 223–232.

Cornelius, F., & Skou, J. C. (1991). The effect of cytoplasmic K^+ on the activity of the Na^+/K^+-ATPase. Biochim. Biophys. Acta 1067, 227–234.

Coutry, N., Blot-Chabaud, M., Mateo, P., Bonvalet, J. P., & Farman, N. (1992). Time course of sodium-induced Na^+,K^+-ATPase recruitment in rabbit cortical collecting tubule. Am. J. Physiol. 263, C61–C68.

Dean, R. (1941). Theories of electrolyte equilibrium in muscle. Biological Symposia 3, 331–348.

Denker, B. M., Smith, B. L., Kuhajda, F. P., & Agre, P. (1988). Identification, purification, and partial characterization of a novel Mr 28,000 integral membrane protein from erythrocytes and renal tubules. J. Biol. Chem. 263, 15634–15642.

De Weer, P. (1970). Effects of intracellular 5′ADP and orthophosphate on the sensitivity of sodium efflux from squid axon to external sodium and potassium. J. Gen. Physiol. 56, 583–620.

De Weer, P., & Rakowski, R. F. (1984). Current generated by backward-running electrogenic Na^+-pump in squid giant axons. Nature 309, 450–452.

De Weer, P., Gadsby, D. C., & Rakowski, R. F. (1988). Voltage dependence of the Na^+,K^+-pump. Annu. Rev. Physiol. 50, 225–241.

Dunham, E. T., & Glynn, I. M. (1961). Adenosine triphosphatase activity and the active movements of alkali metal ions. J. Physiol. 156, 274–293.

Dissing, S., & Hoffman, J. F. (1990). Anion-coupled Na^+ efflux mediated by the human red cell Na^+/K^+ pump. J. Gen. Physiol. 96, 167–193.

Dzhandzhungazyan, K., & Jørgensen, P. L. (1985). Asymmetric orientation of amino groups in the α-subunit and the β-subunit of (Na^++K^+)-ATPase in tight right-side-out vesicles of basolateral membranes from outer medulla. Biochim. Biophys. Acta 817, 165–173.

Dzhandzhugazyan, K., & Modyanov, N. (1985). Chemical and immunochemical approaches to structure of membrane-bound Na^+,K^+-ATPase. In: The Sodium Pump (Glynn, I. M., & Ellory, C., eds.) pp. 129–134. The Company of Biologists, Cambridge, UK.

Ellis-Davies, G. C. R., & Kaplan, J. H. (1990). Binding of Na^+ ions to the Na^+,K^+-ATPase increases the reactivity of an essential residue in the ATP binding domain. J. Biol. Chem. 265, 20570–20576.

Esmann, M. (1985). Occlusion of Rb^+ by detergent-solubilized (Na^+ + K^+)-ATPase from shark salt glands. Biochim. Biophys. Acta 815, 196–202.

Esmann, M., & Skou, J. C. (1985). Occlusion of Na^+ by the Na^+,K^+-ATPase in the presence of oligomycin. Biochem. Biophys. Res. Commun. 127, 857–863.

Esmann, M., & Sottrup-Jensen, L. (1992). Occlusion of Rb^+ after extensive tryptic digestion of shark rectal gland Na^+,K^+-ATPase. Biochim. Biophys. Acta 1108, 247–252.

Ewart, H. S., & Klip, A. (1995). Hormonal regulation of the Na^+,K^+-ATPase: Mechanisms underlying rapid and sustained changes in pump activity. Am. J. Physiol. 269, C295–C311.

Fahn, S., Koval, G. J., & Albers, R. W. (1966). Sodium–potassium-activated adenosine triphosphatase of *Electrophorus* electric organ. I. An associated sodium-activated *trans*-phosphorylation. J. Biol. Chem. 241, 1882–1889.

Fambrough, D. M. (1983). Studies on the (Na^++K^+)-ATPase of skeletal muscle and nerve. Cold Spring Harbor Symposium on Quantitative Biology 48, 297–304.

Fambrough, D. M. (1988). The sodium pump becomes a family. Trends Neurosci. 11, 325–328.

Fambrough, D. M., Wolitzky, B. A., Taormino, J. P., Tamkun, M. M., Takeyasu, K., Somerville, D., Renaud, K. J., Lemas, M. V., Lebovitz, R. M., Kone, B. C., Hamrick, M., Rome, J., Inman, E. M., & Barnstein, A. (1991). A cell biologist's perspective on sites of Na^+,K^+-ATPase regulation. In: The Sodium Pump: Structure, Mechanism, and Regulation (Kaplan, J. H., & De Weer, P., eds.), pp. 17–30. The Rockefeller University Press, New York.

Farley, R. A., Tran, M. C., Carilli, C. T., Hawke, D., & Shively, J. E. (1984). The amino acid sequence of a fluorescin-labeled peptide from the active site of (Na^+,K^+)-ATPase. J. Biol. Chem. 259, 9532–9535.

Fendler, K., Grell, E., Haubs, M., & Bamberg, E. (1985). Pump currents generated by the purified Na^+K^+-ATPase from kidney on black lipid membranes. EMBO J. 4, 3079–3085.

Féraille, E., Carranza, M.-L., Buffin-Meyer, B., Rousselot, M., Doucet, A., & Favre, H. (1995). Protein kinase C-dependent stimulation of Na^+,K^+-ATPase in rat proximal convoluted tubulus. Am. J. Physiol. 268, C1277–C1283.

Feschenko, M. S., & Sweadner, K. J. (1994). Conformation-dependent phosphorylation of Na,K-ATPase by protein kinase A and protein kinase C. J. Biol. Chem. 269, 30436–30444.

Feschenko, M. S., & Sweadner, K. J. (1995). Structural basis for species-specific differences in the phosphorylation of Na,K-ATPase by protein kinase C. J. Biol. Chem. 270, 14072–14077.

Fisone, G., Cheng, S., X-J., Nairn, A. C., Czernik, A. J., Hemmings, H. C., Jr., Höög, J.-O., Bertorello, A. M., Kaiser, R., Bergman, T., Jörnvall, H., Aperia, A., & Greengard, P. (1994). Identification of the phosphorylation site for cAMP-dependent protein kinase on Na^+,K^+-ATPase and effects of site-directed mutagenesis. J. Biol. Chem. 269, 9368–9373.

Forbush, B., III. (1987a). Rapid release of ^{42}K or ^{86}Rb from two distinct transport sites on the Na^+,K^+-pump in the presence of P_i or vanadate. J. Biol. Chem. 262, 11116–11127.

Forbush, B., III. (1987b). Rapid release of ^{42}K or ^{86}Rb from an occluded state of the Na^+,K^+-pump in the presence of ATP or ADP. J. Biol. Chem. 262, 11104–11115.

Forbush, B., III. (1988). Rapid ^{86}Rb release from an occluded state of the Na^+,K^+-pump reflects the rate of dephosphorylation and dearsenylation. J. Biol. Chem. 263, 7961–7969.

Forgac, M., & Chin, G. (1981). K^+-independent active transport of Na^+ by the (Na^+ and K^+)-stimulated adenosine triphosphatase. J. Biol. Chem. 256, 3645–3646.

Gadsby, D. C., Rakowski, R. F., & De Weer, P. (1993). Extracellular access to the Na^+,K^+ pump: Pathway similar to ion channel. Science 260, 100–103.

Garay, R. P., & Garrahan, P. J. (1973). The interaction of sodium and potassium with the sodium pump in red cells. J. Physiol. 231, 297–325.

Garrahan, P. J., & Glynn, I. M. (1967a). The stoichiometry of the sodium pump. J. Physiol. 192, 217–235.

Garrahan, P. J., & Glynn, I. M. (1967b). The behaviour of the sodium pump in red cells in the absence of external potassium. J. Physiol. 192, 159–174.

Garrahan, P. J., & Glynn, I. M. (1967c). The sensitivity of the sodium pump to external sodium. J. Physiol. 192, 175–188.

Garrahan, P. J., & Glynn, I. M. (1967d). The incorporation of inorganic phosphate into adenosine triphosphate by reversal of the sodium pump. J. Physiol. 192, 237–256.

Geering, K. (1990). Subunit assembly and functional maturation of Na^+,K^+-ATPase. J. Memb. Biol. 115, 109–121.

Geering, K. (1991). The functional role of the β-subunit in the maturation and intracellular transport of Na^+,K^+-ATPase. FEBS Lett. 285, 189–193.

Geering, K., Meyer, D. I., Paccolat, M. P., Kraehenbühl, J. P., & Rossier, B. C. (1985). Membrane insertion of α- and β-subunits of Na^+,K^+-ATPase. J. Biol. Chem. 260, 5154–5160.

Geering, K., Kraehenbühl, J. P., & Rossier, B. C. (1987). Maturation of the catalytic α-subunit of Na^+,K^+-ATPase during intracellular transport. J. Cell Biol. 105, 2613–2619.

Geering, K., Theulaz, I., Verrey, F., Häuptle, M. T., & Rossier, B. C. (1989). A role for the β-subunit in the expression of functional Na^+,K^+-ATPase in *Xenopus* oocytes. Am. J. Physiol. 257, C851–C858.

Gick, G. G., Ismail-Beigi, F., & Edelman, I. S. (1988). Hormonal regulation of Na^+,K^+-ATPase. Prog. Clin. Biol. Res. 268B, 277–295.

Gick, G. G., Melikian, J., & Ismail-Beigi, F. (1990). Thyroidal enhancement of rat myocardial Na^+,K^+-ATPase: Preferential expression of α2 activity and mRNA abundance. J. Memb. Biol. 115, 273–282.

Gick, G. G., Hatala, M. A., Chon, D., & Ismail-Beigi, F. (1993). Na^+,K^+-ATPase in several tissues of rat: Tissue-specific expression of subunit mRNAs and enzyme activity. J. Memb. Biol, 131, 229–236.

Gloor, S., Antonicek, H., Sweadner, K. J., Pagliusi, S., Frank, R., Moos, M., & Schacher, M. (1990). The adhesion molecule on glia (AMOG) is a homologue of the beta subunit of the Na^+,K^+-ATPase. J. Cell Biol. 110, 165–174.

Glynn, I. M. (1993). "All hands to the sodium pump." J. Physiol. 462, 1–30.

Glynn, I. M., & Hoffman, J. F. (1971). Nucleotide requirements for sodium–sodium exchange catalyzed by the sodium pump in human red cells. J. Physiol. 218, 239–256.

Glynn, I. M., & Karlish, S. J. D. (1976). ATP hydrolysis associated with an uncoupled sodium flux through the sodium pump: Evidence for allosteric effects of intracellular ATP and extracellular sodium. J. Physiol. 256, 465–496.

Glynn, I. M., & Richards, D. E. (1982). Occlusion of rubidium ions by the sodium–potassium pump: Its implication for the mechanism of potassium transport. J. Physiol. 330, 17–43.

Glynn, I. M., & Richards, D. E. (1989). Evidence for the ordered release of rubidium ion occluded within individual protomers of dog kidney Na^+,K^+-ATPase. J. Physiol. 408, 57–66.

Glynn, I. M., & Karlish, S. J. D. (1990). Occluded cations in active transport. Annu. Rev. Biochem. 59, 171–205.

Glynn, I. M., Lew, V. L., & Lüthi, V. (1970). Reversal of the potassium entry mechanism in red cells with and without reversal of the entire pump cycle. J. Physiol. 207, 371–391.

Glynn, I. M., Hoffman, J. F., & Lew, V. L. (1971). Some "partial reactions" of the sodium pump. Phil. Trans. R. Soc. Lond. B. 262, 91–102.

Glynn, I. M., Hara, Y., & Richards, D. E. (1984). The occlusion of sodium ions within the mammalian sodium–potassium pump: Its role in sodium transport. J. Physiol. 351, 531–547.

Glynn, I. M., Howland, J. L., & Richards, D. E. (1985). Evidence for the ordered release of rubidium ions occluded within the Na^+,K^+-ATPase of mammalian kidney. J. Physiol. 368, 453–469.

Glynn, I. M., Hara, Y., Richards, D. E., & Steinberg, M. (1987). Comparison of rates of cation release and of conformational change in dog kidney Na^+,K^+-ATPase. J. Physiol. 383, 477–485.

Goldshleger, R., Karlish, S. J. D., Rephaeli, A., & Stein, W. D. (1987). The effect of membrane potential on the mammalian sodium-potassium pump reconstituted into phospholipid vesicles. J. Physiol. 387, 331–355.

Goldshleger, R., Shahak, Y., & Karlish, S. J. D. (1990). Electrogenic and electroneutral transport modes of renal Na^+/K^+-ATPase reconstituted into proteoliposomes. J. Memb. Biol. 113, 139–154.

Goldshleger, R., Tal, D. M., Moorman, J., Stein, W. D., & Karlish, S. J. D. (1992). Chemical modification of Glu-953 of the α chain of Na^+,K^+-ATPase associated with inactivation of cation occlusion. Proc. Natl. Acad. Sci. USA 89, 6911–6915.

Good, P. J., Richter, K., & Dawid, I. B. (1990). A nervous system-specific isotype of the beta subunit of Na^+,K^+-ATPase expressed during early development of *Xenopus* laevis. Proc. Natl. Acad. Sci. USA 87, 9088–9092.

Gorga, F. R. (1985). Inhibition of (Na^++K^+)-ATPase by dicyclohexylcarbodiimide. Evidence for two carboxyl groups that are essential for enzymatic activity. Biochemistry 24, 6783–6788.

Green, M. N. (1989). Ions, gates and channels. Nature 339, 424–425.

Grell, E., Lewitski, E., Hoa, D. T. M., Gerhard, A., Ruf, A. et al. (1987). In: Ion Transport Through Membranes (Kunio, Y., & Pullman, B., eds.), pp. 41–60. Academic Press, New York.

Grenningloh, G., Reinitz, A., Schmitt, B., Methfessel, C., Zensen, M., Beyreuther, K., Gundelfinger, E. D., & Betz, H. (1987). The strychnine-binding subunit of the glycine receptor shows homology with nicotinic acetylcholine receptors. Nature 328, 215–220.

Halperin, J. A., & Cornelius, F. (1991). A voltage-activated cation transport pathway associated with the sodium pump. Biochim. Biophys. Acta 1070, 497–500.

Hamlyn, J. M., Ringel, R., Schaeffer, J., Levinson, P. D., Hamilton, B. P., Lowarski, A. A., & Blaustein, M. P. (1982). A circulating inhibitor of (Na^++K^+)-ATPase associated with essential hypertension. Nature 300, 650–652.

Hamlyn, J. M., Blaustein, M. P., Bova, S., DuCharme, D. W., Harris, D. W., Mandel, F., Mathews, W. R., & Ludens, J. H. (1991). Identification and characterization of a ouabain-like compound from human plasma. Proc. Natl. Acad. Sci. USA 88, 6259–6263.

Hara, Y., & Nakao, M. (1981). Sodium ion discharge from pig kidney Na^+,K^+-ATPase. Na^+-dependency of the $E_1P \leftrightharpoons E_2P$ equilibrium in the absence of KCl. J. Biochem. 90, 923–931.

Hara, Y., & Nakao, M. (1986). ATP-dependent proton uptake by proteoliposomes reconstituted with purified Na^+,K^+-ATPase. J. Biol. Chem. 261, 12655–12658.

Hara, Y., Urayama, O., Kawakami, K., Nojima, H., Nagamune, H., Kojima, T., Ohta, T., Nagano, K., & Nakao, M. (1987). Primary structure of two types of alpha-subunits of rat brain Na^+,K^+-ATPase deduced from cDNA sequences. J. Biochem. (Tokyo) 102, 43–48.

Hebert, H., Skriver, E., & Maunsbach, A. B. (1985). Three-dimensional structure of renal Na^+,K^+-ATPase determined by electron microscopy of membrane crystals. FEBS Lett. 187, 182–186.

Hebert, H., Skriver, E., Söderholm, M., & Maunsbach, A. B. (1988). Three-dimensional structure of renal Na^+,K^+-ATPase determined from two-dimensional membrane crystals of the p1 form. J. Ultrastruct. Mol. Struct. Res. 100, 86–93.

Herrera, V. L. M., Emanuel, J. R., Ruitz-Opazo, N., Levenson, R., & Nedal-Ginard, B. (1987). Three different expressed Na^+,K^+-ATPase alpha subunit isoforms: Structural and functional implications. J. Cell Biol. 105, 1855–1865.

Hinz, H. R., & Kirley, T. L. (1990). Lysine 480 is an essential residue in the putative ATP site of lamb kidney (Na^+,K^+)-ATPase. J. Biol. Chem. 265, 10260–10265.

Hoffmann, E. K., & Simonsen, L. O. (1989). Membrane mechanisms in volume and pH regulation in vertebrate cells. Physiol Rev. 69, 315–382.

Holzinger F., Frick, C., & Wink, M. (1992). Molecular basis for the insensitivity of the monarch (*Danaus plexippus*) to cardiac glycosides. FEBS Lett. 314, 477–480.

Homareda, H., Kawakami, K., Nagano, K., & Matsui, H. (1988). Membrane insertion of α and β subunits of human Na^+,K^+-ATPase. Prog. Clin. Biol. Res. 268B, 77–84.

Horisberger, J.-D., Lemas, V., Kraehenbühl, J.-P., & Rossier, B. C. (1991). Structure–function relationship of Na^+,K^+-ATPase. Annu. Rev. Physiol. 53, 565–584.

Horisberger, J.-D., Jaunin, P., Reuben, M. A., Lasater, L. S., Chow, D. C., Forte, J. B., Sachs, G., Rossier, B. C., & Geering, K. (1992). The H^+,K^+-ATPase β-subunit can act as a surrogate for the β-subunit of Na^+,K^+-pumps. J. Biol. Chem. 266, 19131–19134.

Horowitz, B., Eakle, K. A., Scheiner-Bobis, G., Randolph, G. R., Chen, C. Y., Hitzeman, R. A., & Farley, R. A. (1990a). Synthesis and assembly of functional mammalian Na^+,K^+-ATPase in yeast. J. Biol. Chem. 265, 4189–4192.

Horowitz, B., Hensley, C. B., Quintero, M., Axuma, K. K., Putnam, D., & McDonough, A. A. (1990b). Differential regulation of Na^+,K^+-ATPase $\alpha 1$, $\alpha 2$, and β subunit mRNA and protein levels by thyroid hormone. J. Biol. Chem. 265, 14308–14314.

Inesi, G., & Kitley, M. E. (1990). Coupling of catalytic and channel function in the Ca^{2+}-transport ATPase. J. Memb. Biol. 116, 1–8.

Ismail-Beigi, F., & Edelman, I. S. (1971). The mechanism of the calorigenic action of thyroid hormone. J. Gen. Physiol. 51, 710–722.

Jardetzky, O. (1966). Simple allosteric model for membrane pumps. Nature 211, 969–970.

Jaunin, P., Horisberger, J. D., Richter, K., Good, P. J., Rossier, B. C., & Geering, K. (1992). Processing, intracellular transport, and functional expression of endogenous and exogenous α-β_3 Na^+,K^+-ATPase complexes in *Xenopus* oocytes. J. Biol. Chem. 267, 577–585.

Jencks, W. P. (1989). How does a calcium pump calcium? J. Biol. Chem. 264, 18855–18858.

Jørgensen, P. L. (1974). Purification and characterization of (Na^++K^+)-ATPase. IV. Estimation of the purity and of the molecular weight and polypeptide content per enzyme unit in preparations from the outer medulla of rabbit kidney. Biochim. Biophys. Acta 356, 53–67.

Jørgensen, P. L. (1975). Purification and characterization of (Na^++K^+)-ATPase. V. Conformational changes in the enzyme. Transitions between the Na-form and the K-form studied with tryptic digestion as a tool. Biochim. Biophys. Acta 401, 399–415.

Jørgensen, P. L. (1977). Purification and characterization of (Na^++K^+)-ATPase. VI. Differential tryptic modification of catalytic functions of the purified enzyme in the presence of NaCl and KCl. Biochim. Biophys. Acta 466, 97–108.

Jørgensen, P. L. (1988). Purification of Na^+,K^+-ATPase: enzyme scorches, preparative problems, and preparation from mammalian kidney. Meth. Enzymol. 156, 29–43.

Jørgensen, P. L. (1992). Na^+,K^+-ATPase, structure and transport mechanism. In: Molecular Aspects of Transport Proteins (De Pont, J. J. H. H. M., ed.), pp. 1–26. Elsevier Science Publishers B. V.

Jørgensen, P. L., & Skou, J. C. (1971). Purification and characterization of (Na^++K^+)-ATPase. I. The influence of detergents on the activity of (Na^++K^+)-ATPase in preparation from the outer medulla of rabbit kidney. Biochim. Biophys. Acta 233, 366–380.

Jørgensen, P. L., & Karlish, S. J. D. (1980). Defective conformational response in a selectively trypsinized (Na^{+}+K^{+})-ATPase studied with tryptophane fluorescence. Biochim. Biophys. Acta 597, 305–317.

Jørgensen, P. L., & Collins, J. H. (1986). Tryptic and chymotryptic cleavage sites in sequence of α-subunit of (Na^{+}+K^{+})-ATPase from outer medulla of mammalian kidney. Biochim. Biophys. Acta 860, 570–576.

Jørgensen, P. L., & Andersen, J. P. (1988). Structural basis for E_1–E_2 conformational transitions in the Na^{+},K^{+}-pump and Ca^{2+}-pump proteins. J. Memb. Biol. 103, 95–120.

Kaplan, J. H. (1982). Sodium pump-mediated ATP:ADP exchange. The sided effects of sodium and potassium ions. J. Gen. Physiol. 80, 915–937.

Kaplan, J. H., & De Weer, P. (1991a). The Sodium Pump: Structure, Mechanism, and Regulation. The Rockefeller University Press, New York.

Kaplan, J. H., & De Weer, P. (1991b). The Sodium Pump: Recent Developments. The Rockefeller University Press, New York.

Karlish, S. J. D., & Yates, D. W. (1978). Tryptophan fluorescence of (Na^{+}+K^{+})-ATPase as a tool for study of the enzyme mechanism. Biochim. Biophys. Acta 527, 115–130.

Karlish, S. J. D., & Stein, W. D. (1982a). Effects of ATP or phosphate on passive rubidium fluxes mediated by Na^{+}–K^{+}-ATPase reconstituted into phospholipid vesicles. J. Physiol. 328, 317–331.

Karlish, S. J. D., & Stein, W. D. (1982b). Passive rubidium fluxes mediated by Na^{+}–K^{+}-ATPase reconstituted into phospholipid vesicles when ATP- and phosphate-free. J. Physiol. 328, 295–316.

Karlish, S. J. D., & Stein, W. D. (1985). Cation activation of the pig kidney sodium pump: Transmembrane allosteric effects of sodium. J. Physiol. 359, 119–149.

Karlish, S. J. D., Jørgensen, P. L., & Gitler, C. (1977). Identification of a membrane-embedded segment of the large polypeptide chain of (Na^{+}+K^{+})-ATPase. Nature 209, 715–717.

Karlish, S. J. D., Yates, D. W., & Glynn, I. M. (1978). Conformational transitions between Na^{+}-bound and K^{+}-bound forms of (Na^{+} + K^{+})-ATPase, studied with formycin nucleotides. Biochim. Biophys. Acta 525, 252–264.

Karlish, S. J. D., Lieb, W. R., & Stein, W. D. (1982). Combined effects of ATP and phosphate on rubidium exchange mediated by Na^{+},K^{+}-ATPase reconstituted into phospholipid vesicles. J. Physiol. 328, 333–350.

Karlish, S. J. D., Goldshleger, R., Shahak, Y., & Rephaeli, A. (1988). In: The Na^{+},K^{+}-Pump, Part A: Molecular Aspects (Skou, J. C., Nørby, J. G., Maunsbach, A. B., & Esmann, M., eds.), pp. 519–524. Alan R. Liss, Inc., New York.

Karlish, S. J. D., Goldshleger, R., & Stein, W. D. (1990). A 19-kDa C-terminal tryptic fragment of the α chain of Na^{+}/K^{+}-ATPase is essential for occlusion and transport of cations. Proc. Natl. Acad. Sci. USA 87, 4566–4570.

Karlish, S. J. D., Goldshleger, R., & Jørgensen, P. L. (1993). Location of Asn^{831} of the α chain of Na^{+}/K^{+}-ATPase at the cytoplasmic surface. Implication for topological models. J. Biol. Chem. 268, 3471–3478.

Kawakami, K., & Nagano, K. (1988). The transmembrane segment of the human Na^{+},K^{+}-ATPase β-subunit acts as the membrane incorporation signal. J. Biochem. 103, 54–60.

Kawakami, K., Noguchi, S., Noda, M., Takahashi, H., Ohta, T., Kawamura, M., Nojima, H., Nagano, K., Hirose, T., Inayama, S., Hayashida, H., Miyata, T., & Numa, S. (1985). Primary structure of the α-subunit of Torpedo californica (Na^{+}+K^{+})-ATPase deduced from cDNA sequence. Nature 316, 733–736.

Kawakami, K., Nojima, H., Ohta, T., & Nagano, K. (1986a). Molecular cloning and sequence analysis of human Na^{+},K^{+}-ATPase β-subunit. Nucleic Acids Res. 14, 2833–2844.

Kawakami, K., Ohta, T., Nojima, H., & Nagano, K. (1986b). Primary structure of the α-subunit of the human Na^{+},K^{+}-ATPase deduced from cDNA sequence. J. Biochem. (Tokyo) 100, 389–397.

Kelly, R. A., & Smith, T. W. (1989). The search for the endogenous digitalis: An alternative hypothesis. Am. J. Physiol. 256, C937–C950.

Kennedy, B. G., Lunn, G., & Hoffman, J. F. (1986). Effects of altering the ATP/ADP ratio on pump-mediated Na^+/K^+ and Na^+/Na^+ exchanges in resealed human red blood cell ghosts. J. Gen. Physiol. 87, 47–72.

Kirley, T. L. (1989). Determination of three disulfide bonds and one free sulfhydryl in the βsubunit of (Na^+,K^+)-ATPase. J. Biol. Chem. 264, 7185–7192.

Kirley, T. L., Wallic, E. T., & Lane, L. K. (1984). The amino acid sequence of the fluorescein isothiocyanate reactive site of lamb and rat kidney Na^+- and K^+-dependent ATPase. Biochem. Biophys. Res. Commun. 125, 767–773.

Klingenberg, M. (1981). Membrane protein oligomeric structure and transport function. Nature 290, 449–454.

Klodos, I.. & Nørby, J. G. (1987). $(Na^+ + K^+)$-ATPase: Confirmation of the three-pool model for the phosphointermediates of Na^+-ATPase activity. Estimation of the enzyme-ATP dissociation rate constant. Biochim. Biophys. Acta 897, 302–314.

Klodos, I., Nørby, J. G., & Plesner, I. W. (1981). The steady-state kinetic mechanism of ATP hydrolysis catalyzed by membrane-bound $(Na^+ + K^+)$-ATPase from ox brain. II. Kinetic characterization of phosphointermediates. Biochim. Biophys. Acta 643, 463–482.

Kone, B. C., Takeyasu, K., & Fambrough, D. M. (1991). Structure-function studies of Na^+/K^+-ATPase isozymes. In: The Sodium Pump: Recent Developments (Kaplan, J. H., & De Weer, P., eds.), pp. 265–269. The Rockefeller University Press, New York.

Kyte, J., & Doolitle, R. F. (1982). A simple method for displaying the hydropathic character of a protein. J. Mol. Biol. 157, 105–132.

Kyte, J., Xu, K., & Bayer, R. (1987). Demonstration that Lysine-501 of the α polypeptide of native sodium and potassium ion activated adenosinetriphosphatase is located on its cytoplasmic surface. Biochemistry 24, 8350–8360.

Last, T. A., Gantzer, M. L., & Tyler, C. D. (1983). Ion-gated channel induced in planar bilayers by incorporation of (Na^+,K^+)-ATPase. J. Biol. Chem. 258, 2399–2404.

Läuger, P. (1979). A channel mechanism for electrogenic ion pumps. Biochim. Biophys. Acta 552, 143–161.

Läuger, P. (1991a). Kinetic basis of voltage dependence of the Na^+,K^+-pump. In: The Sodium Pump: Structure, Mechanism, and Regulation (Kaplan, J. H., & De Weer, P., eds.) pp. 303–315. The Rockefeller University Press, New York.

Läuger, P. (1991b). Electrogenic Ion Pumps. Sinauer Associates, Inc., MA.

Läuger, P., & Apell, H.-J. (1986). A microscopic model for the current-voltage behaviour of the Na,K-pump. Eur. Biophys. J. 13, 309–321.

Läuger, P., & Apell, H.-J. (1988). Transient behaviour of the Na^+/K^+-pump: Microscopic analysis of nonstationary ion-translocation. Biochim. Biophys. Acta 944, 451–464.

Lebovitz, R. M., Takeyasu, K., & Fambrough, D. M. (1989). Molecular characterization and expression of the $(Na^+ + K^+)$-ATPase α-subunit in *Drosophila melanogaster*. EMBO J. 8, 193–202.

Lee, K. H., & Blostein, R. (1980). Red cell sodium fluxes catalyzed by the sodium pump in the absence of K^+ and ADP. Nature 285, 338–339.

Lemas, M. V., Takeyasu, K., & Fambrough, D. M. (1992). The carboxyl-terminal 161 amino acids of the Na^+,K^+-ATPase α-subunit are sufficient for assembly with the β-subunit. J. Biol. Chem. 267, 20987–20991.

Lescale-Matys, L., Putnam, D. S., & McDonough, A. A. (1993). Na^+,K^+-ATPase α_1- and β_1-subunit degradation: Evidence for multiple subunit specific rates. Am. J. Physiol. 264, C583–C590.

Ling, L., & Cantley, L. (1984). The (Na^+,K^+)-ATPase of Friend erythroleukemia cells is phosphorylated near the ATP hydrolysis by an endogeneous membrane-bound kinase. J. Biol. Chem. 259, 4089–4095.

Lingrel, J. B., Orlowski, J., Shull, M. M., & Price, E. (1990). Molecular genetics of Na^+,K^+-ATPase. Progress Nucleic Acid Res. Mol. Biol. 38, 37–89.

Lingrel, J. B., Orlowski, J., Price, E. M., & Pathak, B. G. (1991). Regulation of the α-subunit genes of the Na^+,K^+-ATPase and determinants of cardiac glycoside sensitivity. In: The Sodium Pump:

Structure, Mechanism, and Regulation (Kaplan, J. H., & De Weer, P., eds.), pp. 1–16. The Rockefeller University Press, New York.

Lowndes, J. M., Hokin-Neaverson, M., & Bertics, P. J. (1990). Kinetics of phosphorylation of Na^{+}/K^{+}-ATPase by protein kinase C. Biochim. Biophys. Acta 1052, 143–151.

Ludens, J. H., Clark, M. A., DuCharme, D. W., Harris, D. W., Lutzke, B. S., Mandel, F., Mathews, W. R., Sutter, D. M., & Hamlyn, J. M. (1991). Purification of an endogenous digitalis-like factor from human plasma for structural analysis. Hypertension 17, 923–929.

Luna, E. J., & Hitt, A. L. (1992). Cytoskeleton-plasma membrane interactions. Science 258, 955–964.

Macknight, A. D. C. (1987). Volume maintenance in isosmotic conditions. Curr. Top. Memb. Transp. 30, 3–43.

Maeda, M., Ishizaki, I., & Futai, M. (1988). cDNA cloning and sequence determination of pig gastric $(H^{+}+K^{+})$-ATPase. Biochem. Biophys. Res. Commun. 157, 203–209.

Martin, D. W., & Sachs, J. R. (1991). Simulation of dephosphorylation curves by an ensemble of enzyme molecules with different rate constants. J. Gen. Physiol. 98, 419–426.

Maruyama, K., & MacLennan, D. H. (1988). Mutation of aspartic acid-351, lysine-352, and lysine 515 alters the Ca^{2+}-transport activity of the Ca^{2+}-ATPase expressed in COS-1 cells. Proc. Natl. Acad. Sci. USA 85, 3314–3318.

Maruyama, K., Clarke, D. M., Fujii, J., Inesi, G., Loo, T. W., & MacLennan, D. H. (1989). Functional consequences of alterations to amino acids located in the catalytic center (isoleucine 348 to threonine 357) and nucleotide-binding domain of the Ca^{2+}-ATPase of sarcoplasmic reticulum. J. Biol. Chem. 264, 13038–13042.

Marver, D., Lear, S., Marver, L. T., Silva, P., & Epstein, F. H. (1986). Cyclic AMP-dependent stimulation of Na^{+},K^{+}-ATPase in shark rectal gland. J. Memb. Biol. 94, 205–215.

Matsui, H., & Homareda, H. (1982). Interaction of sodium and potassium ions with the Na^{+},K^{+}-ATPase. I. Ouabain-sensitive alternative binding of three Na^{+} or two K^{+} to the enzyme. J. Biochem. (Tokyo) 92, 193–217.

Matsui, H., & Schwartz, A. (1968). Mechanism of cardiac glycoside inhibition of the $(Na^{+}+K^{+})$-dependent ATPase from cardiac tissue. Biochim. Biophys. Acta 151, 655–663.

McDonough, A. A., Geering, K., & Farley, R. A. (1990). The sodium pump needs its β subunit. FASEB J. 4, 1598–1605.

McNeill, H., Ozawa, M., Kemler, R., & Nelson, W. J. (1990). Novel function of the cell adhesion molecule uvomorulin as an inducer of cell surface polarity. Cell 62, 309–316.

Mercer, R. W., Schneider, J. W., Savitz, A., Emmanuel, J., Benz, E. J., & Levenson, R. (1986). Rat brain Na^{+},K^{+}-ATPase β-chain gene: Primary structure, tissue-specific expression, and amplification in ouabain-resistant HeLa C^{+} cells. Mol. Cell. Biol. 6, 3884–3890.

Miller, R. P., & Farley, R. A. (1988). All three potential *N*-glycosylation sites of the kidney $(Na^{+}+K^{+})$-ATPase β-subunit contain oligosaccharide. Biochim. Biophys. Acta 954, 50–57.

Miller, R. P., & Farley, R. A. (1990). The beta subunit of $(Na^{+} + K^{+})$-ATPase contains three disulfide bonds. Biochemistry 29, 1524–1532.

Mironova, G. D., Bocharnikova, N. I., Mirsalikhova, N. M., & Mironov, G. P. (1986). Ion-transporting properties and ATPase activity of $(Na^{+}+K^{+})$-ATPase large subunit incorporated into bilayer lipid membranes. Biochim. Biophys. Acta 861, 224–236.

Mitchell, P., & Moyle, J. (1974). The mechanism of proton translocation in reversible proton-translocating adenosine triphosphatase. Biochem. Soc. Spec. Publ. 4, 91–111.

Moczydlowski, E. G., & Fortes, P. A. G. (1981). Inhibition of sodium and potassium triphosphatase by 2′3′-O-(2,4,6-trinitrocyclohexadienylidine) adenine nucleotides. Implications for the structure and mechanism of the $Na^{+}:K^{+}$ pump. J. Biol. Chem. 256, 2357–2366.

Modyanov, N., Lutsenko, S., Chertova, E., & Efremov, R. (1991). Architecture of the sodium pump molecule: probing the folding of the hydrophobic domain. In: The Sodium Pump: Structure, Mechanism, and Regulation (Kaplan, J. H., & De Weer, P., eds.) pp. 99–115. The Rockefeller University Press, New York.

Mohraz, M., Simpson, M. V., & Smith, P. R. (1987). The three-dimensional structure of the Na^+,K^+-ATPase from electron microscopy. J. Cell Biol. 105, 1–8.

Morrow, J. S., Cianci, C. D., Ardito, T., Mann, A. S., & Kashgarian, M. (1989). Ankyrin links fodrin to the alpha subunit of the Na^+,K^+-ATPase in Madin–Darby canine kidney cells and in intact renal tube cells. J. Cell Biol. 108, 455–465.

Mullins, L. J., & Noda, K. (1963). The influence of sodium-free solutions on the membrane potential of frog muscle fibers. J. Gen. Physiol. 47, 117–132.

Mårdh, S. (1975). Bovine brain Na^+,K^+-stimulated ATP phosphohydrolase, studied by a rapid-mixing technique. K^+-stimulated liberation of [^{32}P] orthophosphate from [^{32}P] phosphoenzyme and resolution of the dephosphorylation into two phases. Biochim. Biophys. Acta 391, 448–463.

Mårdh, S., & Zetterquist, Ö. (1974). Phosphorylation and dephosphorylation reactions of bovine brain (Na^++K^+)-stimulated ATP phosphohydrolase studied by rapid-mixing technique. Biochim. Biophys. Acta 350, 473–483.

Mårdh, S., & Post, R. L. (1977). Phosphorylation from adenosine triphosphate of sodium- and potassium-activated adenosine triphosphatase. J. Biol. Chem. 252, 633–638.

Nakao, M., & Gadsby, D. C. (1986). Voltage dependence of Na^+ translocation by the Na^+/K^+ pump. Nature 323, 628–630.

Navran, S. S., Allain, G., Garcia, H. F., Allen, J. C., & Seidel, C. L. (1991). Serotonin-induced Na^+/K^+ pump stimulation in vascular smooth muscle cells. Evidence for coupling to multiple receptor mechanisms. J. Pharmacol. Exp. Ther. 256, 297–303.

Nelson, W. J., & Veshnock, P. J. (1987). Ankyrin binding to (Na^+, K^+)-ATPase and implications for the organization of membrane domains in polarized epithelial cells. Nature 328, 533–535.

Nelson, W. J., & Hammerton, R. W. (1989). A membrane-cytoskeletal complex containing Na^+,K^+-ATPase, ankyrin and fodrin in Madin–Darby canine kidney (MDCK) cells. Implications for the biogenesis of epithelial cell polarity. J. Cell Biol. 108, 893–902.

Nielsen, S., Smith, B., Christensen, E. I., Knepper, M. A., & Agre, P. (1993). CHIP28 water channels are localized in constitutively water-permeable segments of the nephron. J. Cell Biol. 120, 371–383.

Nicholas, R. A. (1984). Purification of the membrane-spanning tryptic peptides of the α-polypeptide from sodium and potassium ion activated adenosinetriphosphatase labeled with 1-tritios-piro[adamentane-4,3′-diazirine]. Biochemistry 23, 890–897.

Noguchi, S., Noda, M., Takahashi, H., Kawakami, K., Ohta, T., Nagano, K., Hirose, T., Inayama, S., Kawamura, M., & Numa, S. (1986). Primary structure of the β-subunit of Torpedo californica (Na^++K^+)-ATPase deduced from the cDNA sequence. FEBS Lett. 196, 315–320.

Noguchi, S., Mishina, M., Kawamura, M., & Numa, S. (1987). Expression of functional (Na^++K^+)-ATPase from cloned cDNAs. FEBS Lett. 225, 27–32.

Noguchi, S., Higashi, K., & Kawamura, M. (1990). Assembly of the α-subunit of Torpedo californica Na^+/K^+-ATPase with pre-existing β-subunit in *Xenopus* oocytes. Biochim. Biophys. Acta 1023, 247–253.

Nørby, J. G., Klodos, I., & Christiansen, N. O. (1983). Kinetics of Na^+-ATPase activity by the Na^+,K^+ pump. Interactions of the phosphorylated intermediates with Na^+, $Tris^+$, and K^+. J. Gen. Physiol. 82, 725–759.

O'Connell, M. A. (1982). Exclusive labeling of the extracytoplasmic surface of sodium ion and potassium ion activated adenosinetriphosphatase and a determination of the distribution of surface area across the bilayer. Biochemistry 20, 5984–5991.

Ohta, T., Nagano, K., & Yshida, M. (1986a). The active site structure of Na^+/K^+-transporting ATPase: Location of the 5-(p-fluorosulfonyl)benzoyladenosine binding site and soluble peptides released by trypsin. Proc. Natl. Acad. Sci. USA 83, 2071–2075.

Ohta, T., Yshida, M., Nagano, K., Hirano, H., & Kawamura, M. (1986b). Structure of the extra-membranous domain of the β-subunit of (Na^+,K^+)-ATPase revealed by the sequence of its tryptic peptides. FEBS Lett. 204, 297–301.

Ohtsubo, M., Noguchi, S., Takeda, K., Morohashi, M., & Kawamura, M. (1990). Site-directed mutagenesis of Asp-376, the catalytic phosphorylation site, and Lys-570, the putative ATP binding site, of the α-subunit of Torpedo californica Na^+/K^+-ATPase. Biochim. Biophys. Acta 1021, 157–160.

Orlowski, J., & Lingrel, J. B. (1988). Tissue-specific and developmental regulation of rat Na^+,K^+-ATPase catalytic α isoform and β subunit mRNAs. J. Biol. Chem. 263, 19436–19442.

Ovchinnikov, Y. A., Modyanov, N. N., Broude, N. E., Petrukhin, K. E., Grishin, A. V., Arzamazova, N. M., Aldanova, N. A., Monastyrskaya, G. S., & Sverdlov, E. D. (1986). Pig kidney Na^+,K^+-ATPase. Primary structure and spatial organization. FEBS Lett. 201, 237–245.

Ovchinnikov, Y. A., Dzhandzugazyan, N. K., Lutsenko, S. V., Mustayev, A. A., & Modyanov, N. N. (1987). Affinity modification of E1-form of Na^++K^+-ATPase revealed Asp-710 in the catalytic site. FEBS Lett. 217, 111–116.

Ovchinikov, Y. A., Luneva, N. M., Arystarkhova, E. A., Gevondyan, N. M., Arzamazova, N. M., Kozhich, A. T., Mesmeyanov, V. A., & Modyanov, N. N. (1988). Topology of Na^+,K^+-ATPase. Identification of the extra- and intracellular hydrophilic loops of the catalytic subunit by specific antibodies. FEBS Lett. 227, 230–234.

Pagliusi, S. R., Antonicek, H., Gloor, S., Frank, R., Moos, M., & Schacher, M. (1989). Identification of a cDNA clone specific for the neural cell adhesion molecule AMOG. J. Neurosci. Res. 22, 113–119.

Pedemonte, C. H., & Kaplan, J. H. (1986a). Carbodiimide inactivation of Na^+,K^+-ATPase. A consequence of internal cross-linking and not carboxyl group modification. J. Biol. Chem. 261, 3632–3639.

Pedemonte, C. H., & Kaplan, J. H. (1986b). Carbodiimide inactivation of Na^+,K^+-ATPase, via intramolecular cross-link formation, is due to inhibition of phosphorylation. J. Biol. Chem. 261, 16660–16665.

Pedemonte, C. H., & Kaplan, J. H. (1990). Chemical modification as an approach to elucidation of sodium pump structure–function relations. Am. J. Physiol. 258, C1–C23.

Pedemonte, C. H., Kirley, T. L., Treuheit, M. J., & Kaplan, J. H. (1992). Inactivation of the Na^+,K^+-ATPase by modification of Lys-501 with 4-acetamido-4′-isothiocyanatostilbene-2,2′-disulfonic acid (SITS). FEBS Lett. 314, 97–100.

Pedersen, P. L., & Carafoli, E. (1987). Ion motive ATPases. I. Ubiquity, properties, and significance to cell function. Trends Biochem. Sci. 12, 146–150.

Plesner, L., & Plesner, I. W. (1981a). The steady-state kinetic mechanism of ATP hydrolysis catalyzed by membrane-bound ($Na^+ + K^+$)-ATPase from ox brain. I. Substrate identity. Biochim. Biophys. Acta 643, 449–462.

Plesner, I. W., & Plesner, L. (1981b). The steady-state kinetic mechanism of ATP hydrolysis catalyzed by membrane-bound ($Na^+ + K^+$)-ATPase from ox brain. IV. Rate-constant determination. Biochim. Biophys. Acta 648, 231–246.

Plesner, I. W., Plesner, L., Nørby, J. G., & Klodos, I. (1981). The steady-state kinetic mechanism of ATP hydrolysis catalyzed by membrane-bound ($Na^+ + K^+$)-ATPase from ox brain. III. A minimal model. Biochim. Biophys. Acta 643, 483–494.

Polvani, C., & Blostein, R. (1988). Protons as substitutes for sodium and potassium in the sodium pump reaction. J. Biol. Chem. 263, 16757–16763.

Post, R. L., & Kume, S. (1973). Evidence for an aspartyl phosphate residue at the active site of sodium and potassium ion transport adenosine triphosphatase. J. Biol. Chem. 248, 6993–7000.

Post, R. L., Kume, S.. Tobin, T., Orcutt, B., & Sen, A. K. (1969). Flexibility of an active center in sodium-plus-potassium adenosine triphosphatase. J. Gen. Physiol. 54, 306S–326S.

Post, R. L., Hegyvary, C., & Kume, S. (1972). Activation by adenosine triphosphate in the phosphorylation kinetics of sodium and potassium ion transport adenosine triphosphatase. J. Biol. Chem. 247, 6530–6540.

Preston, G. M., & Agre, P. (1991). Isolation of the cDNA for erythrocyte integral membrane protein of 28 kilodaltons: Member of an ancient channel family. Proc. Natl. Acad. Sci. USA 88, 11110–11114.

Price, E. M., & Lingrel, J. B. (1988). Structure–function relationship in the Na^+,K^+-ATPase α subunit: Site-directed mutagenisis of glutamine-111 to arginine and asparagine-122 to aspartic acid generates a ouabain-resistant enzyme. Biochemistry 27, 8400–8408.

Price, E. M., Rice, D. A., & Lingrel, J. B. (1989). Site-directed mutagenesis of a conserved, extracellular aspartic acid residue affects the ouabain sensitivity of sheep Na^+,K^+-ATPase. J. Biol. Chem. 264, 21902–21906.

Price, E. M., Rice, D. A., & Lingrel, J. B. (1990). Structure–function studies of Na^+,K^+-ATPase. Site-directed mutagenisis of the border residues from H1–H2 extracellular domain of the α subunit. J. Biol. Chem. 265, 6638–6641.

Rabon, E. C., Smillie, K., Seru, V., & Rabon, R. (1993). Rubidium occlusion within tryptic peptides of the H^+,K^+-ATPase. J. Biol. Chem. 268, 8012–8018.

Rakowski, R. F. (1991). Stoichiometry and voltage dependence of the Na^+/K^+ pump in squid giant axons and *Xenopus* oocytes. In: The Sodium Pump: Structure, Mechanism, and Regulation (Kaplan, J. H., & De Weer, P., eds.) pp. 339–353. Rockefeller University Press, New York.

Rakowski, R. F. (1993). Charge movement by the Na^+/K^+ pump in *Xenopus* oocytes. J. Gen. Physiol. 101, 117–144.

Rakowski, R. F., & Paxson, C. L. (1988). Voltage dependence of Na^+/K^+ pump current in *Xenopus* oocytes. J. Memb. Biol. 106, 173–182.

Rakowski, R. F., Vasilets, L. A., LaTona, J., & Schwarz, W. (1991). A negative slope in the current–voltage relationship of the Na^+/K^+ pump in *Xenopus* oocytes produced by reduction of external $[K^+]$. J. Memb. Biol. 121, 177–187.

Rephaeli, A., Richards, D. E., & Karlish, S. J. D. (1986). Electrical potential accelerates the $E_1P(Na) \rightarrow E_2P$ conformational transition of (Na^+,K^+)-ATPase in reconstituted vesicles. J. Biol. Chem. 261, 12437–12440.

Renaud, K. J., Inman, E. M., & Fambrough, D. M. (1991). Cytoplasmic and transmembrane domain deletions of Na^+,K^+-ATPase β-subunit. Effects on subunit assembly and intracellular transport. J. Biol. Chem. 266, 20491–20497.

Repke, K. R. H., & Schön, R. (1992). Role of protein conformation changes and transphosphorylations in the function of Na^+/K^+-transporting adenosine triphosphatase: An attempt at an integration into the Na^+/K^+ pump mechanism. Biol. Rev. 67, 31–78.

Robinson, J. D. (1974). Affinity of the (Na^++K^+)-dependent ATPase for Na^+ measured by Na^+-modified enzyme inactivation. Biochim. Biophys. Acta 648, 284–286.

Robinson, J. D., & Flashner, M. S. (1979). The (Na^++K^+)-activated ATPase. Enzymatic and transport properties. Biochim. Biophys. Acta 549, 145–176.

Robinson, J. D., & Pratap, P. R. (1993). Indicators of conformational changes in the Na^+/K^+-ATPase and their interpretation. Biochim. Biophys. Acta 1154, 83–104.

Rodriguez-Boulan, E., & Nelson, W. J. (1989). Morphogenesis of the polarized epithelial cell phenotype. Science 245, 718–725.

Sachs, J. R. (1980). The order of release of sodium and addition of potassium in the sodium–potassium pump reaction mechanism. J. Physiol. 302, 219–240.

Sachs, J. R. (1981). Mechanistic implications of the potassium–potassium exchange carried out by the sodium–potassium pump. J. Physiol. 316, 263–277.

Sachs, J. R. (1986). Potassium–potassium exchange as part of the over-all reaction mechanism of the sodium pump of the human red blood cell. J. Physiol. 374, 221–244.

Sachs, J. R. (1991). Successes and failures of the Albers–Post model in predicting ion flux kinetics. In: The Sodium Pump: Structure, Mechanism, and Regulation (Kaplan, J. H., & De Weer, P., eds.), pp. 249–266. The Rockefeller University Press, New York.

Satoh, T., & Endou, H. (1990). Inhibitory effect of phorbol ester on sodium transport in frog urinary bladder. Am. J. Physiol. 259, F425–F431.

Schatzmann, H-J. (1953). Cardiac glycosides as inhibitors of the active potassium and sodium transport across the red cell membrane. [*Hertzglycoside als Hemstoffe für den aktiven Kalium- und Natriumtransport durch die Erythrocytenmembran*]. Helv. Physiol. Acta 11, 346–354.

Scheiner-Bobis, G., Fahlbusch, K., & Schoner, W. (1987). Demonstration of cooperating α subunits in working (Na^++K^+)-ATPase by the use of the MgATP complex analogue cobalt tetramine ATP. Eur. J. Biochem. 168, 123–131.

Schofield, P. R., Darlison, M. G., Fujita, N., Burt, D. R., Stephenson, F. A., Rodriguez, H., Rhee, L. M., Ramachandran, J., Reale, V., Glencorse, T. A., Seeburg, P. H., & Barnard, E. A. (1987). Sequence and functional expression of the $GABA_A$ receptor shows a ligand-gated receptor super-family. Nature 328, 221–227.

Schultheis, P., & Lingrel, J. D. (1993). Substitution of transmembrane residues with hydrogen-bonding potential in the α subunit of Na^+,K^+-ATPase reveals alterations in ouabain sensitivity. Biochemistry 32, 544–550.

Schwartz, A., Matsui, H., & Laughter, A. H. (1968). Tritiated digoxin binding to (Na^++K^+)-activated adenosine triphosphatase: Possible allosteric site. Science 160, 323–325.

Shani-Sekler, M., Goldschleger, R., Tal, D. M., & Karlish, S. J. D. (1988). Inactivation of Rb^+ and Na^+ occlusion on (Na^+,K^+)-ATPase by modification of carboxyl groups. J. Biol. Chem. 263, 19331–19341.

Sharkey, R. G. (1983). Lactoperoxidase-catalyzed iodination of sodium and potassium-activated adenosine triphosphatase in the Madin–Darby canine kidney epithelial cell line and canine renal membranes. Biochim. Biophys. Acta 730, 327–341.

Shull, G. E., Schwartz, A., & Lingrel, J. B. (1985). Amino-acid sequence of the catalytic subunit of the (Na^++K^+)-ATPase deduced from a complementary DNA. Nature 316, 691–695.

Shull, G. E., Greeb, J., & Lingrel, J. B. (1986a). Molecular cloning of three distinct forms of the Na^+,K^+-ATPase α-subunit from rat brain. Biochemistry 25, 8125–8132.

Shull, G. E., Lane, L. K., & Lingrel, J. B. (1986b). Amino-acid sequence of the β-subunit of the (Na^++K^+)-ATPase deduced from cDNA. Nature 321, 429–431.

Simons, T. J. B. (1974). Potassium: Potassium exchange catalyzed by the sodium pump in human red cells. J. Physiol. 237, 123–155.

Simons, T. J. B. (1975). The interaction of ATP-analogues processing a blocked γ-phosphate group with the sodium pump in human red cells. J. Physiol. 244, 731–739.

Simons, K., & Fuller, D. (1985). Cell surface polarity in epithelia. Annu. Rev. Cell Biol. 1, 243–288.

Skou, J. C. (1957). The influence of some cations on an adenosine triphosphatase from peripheral nerves. Biochim. Biophys. Acta 23, 394–401.

Skou, J. C. (1960). Further investigations on a Mg^{2+} + Na^+-activated adenosine-triphosphatase, possibly related to the active, linked transport of Na^+ and K^+ across the nerve membrane. Biochim. Biophys. Acta 42, 6–23.

Skou, J. C. (1990). The energy coupled exchange of Na^+ and K^+ across the cell membrane. The Na^+,K^+-pump. FEBS Lett. 268, 314–324.

Skriver, E., Maunsbach, A. B., & Jørgensen, P. L. (1981). Formation of two-dimensional crystals in pure membrane-bound Na^+,K^+-ATPase. FEBS Lett. 131, 219–222.

Skriver, E., Maunsbach, A. B., Hebert, H., & Jørgensen, P. L. (1988). Crystallization of membrane-bound Na^+,K^+-ATPase in two dimensions. Meth. Enzymol. 156, 80–87.

Skriver, E., Maunsbach, A. B., Hebert, H., Scheiner-Bobis, G., & Schoner, W. (1989). Two-dimensional crystalline arrays of Na^+,K-ATPase with new subunit interaction induced by cobalt–tetramine-ATP. J. Ultrastruct. Molec. Struct. Res. 99, 234–243.

Skriver, E., Kavéus, U., Hebert, H., & Maunsbach, A. B. (1992). Three-dimensional structure of Na^+,K^+-ATPase determined from membrane crystals induced by cobalt–tetramine-ATP. J. Struct. Biol. 108, 176–185.

Smith, B. L., & Agre, P. (1991). Erythrocyte M_r 28,000 transmembrane protein exists as a multisubunit oligomer similar to channel proteins. J. Biol. Chem. 266, 6407–6415.

Stein, W. D. (1990). Energetics and the design principles of the Na^+/K^+-ATPase. J. Theor. Biol. 147, 145–159.

Stürmer, W., Apell, H.-J., Wuddel, I., & Läuger, P. (1989). Conformational transitions and charge translocation by the Na^+,K^+-pump: Comparison of optical and electrical transients elicited by ATP-concentration jumps. J. Memb. Biol. 110, 67–86.

Stürmer, W., Bühler, R., Apell, H.-J., & Läuger, P. (1991). Charge translocation by the Na^+,K^+-pump: II. Ion binding and release at the extracellular face. J. Memb. Biol. 121, 163–176.

Sweadner, K. J. (1989). Isozymes of the Na^+/K^+-ATPase. Biochim. Biophys. Acta 988, 185–200.

Sweadner, K. J. (1991). Subunit diversity in the Na^+,K^+-ATPase. In: The Sodium Pump: Structure, Mechanism, and Regulation (Kaplan, J. H., & De Weer, P., eds.), pp. 63–76. The Rockefeller University Press, New York.

Sweadner, K. J., & Gilkeson, R. C. (1985). Two isoforms of the Na,K-ATPase have distinct antigenic determinants. J. Biol. Chem. 260, 9016–9022.

Tahara, Y., Ohnishi, S., Fujiyoshi, Y., Kimura, Y., & Hayashi, Y. (1993). A pH induced two-dimensional crystal of membrane-bound Na^+,K^+-ATPase of dog kidney. FEBS Lett. 320, 17–22.

Takeyasu, K., Tamkun, M. M., Siegel, N. R., & Fambrough, D. M. (1987). Expression of hybrid (Na^++K^+)-ATPase molecules after transfection of mouse LtK^- cells with DNA encoding the β-subunit of an avian brain sodium pump. J. Biol. Chem. 262, 10733–10740.

Takeyasu, K., Tamkun, M. M., Renaud, K. J., & Fambrough, D. M. (1988). Ouabain-sensitive (Na^++K^+)-ATPase activity expressed in mouse L cells by transfection with DNA encoding the α-subunit of an avian sodium pump. J. Biol. Chem. 263, 4347–4354.

Takeyasu, K., Renaud, K. J., Taormino, J. P., Wolitzky, B. A., Barnstein, A., Tamkun, M. M., & Fambrough, D. M. (1989). Differential subunit and isoform expression are involved in regulation of the sodium pump in skeletal muscle. Curr. Top. Memb. Transp. 34, 143–165.

Takeyasu, K., Lemas, M. V., & Fambrough, D. M. (1990). Stability of the (Na^++K^+)-ATPase α-subunit isoforms in evolution. Am. J. Physiol. 259, C619–C630.

Tamkun, M. M., & Fambrough, D. M. (1986). The (Na^++K^+)-ATPase of chick sensory neurons: Studies on bisynthesis and intracellular transport. J. Biol. Chem. 261, 1009–1019.

Tanabe, T., Takeshima, H., Mikami, A., Klockerzi, V., Takahashi, H., Kangawa, K., Kojima, M., Matsuo, H., Hirose, T., & Numa, S. (1987). Primary structure of the receptor for calcium channel blockers from skeletal muscle. Nature 328, 313–318.

Tanford, C. (1982). Simple model for the chemical potential change of a transported ion in active transport. Proc. Natl. Acad. Sci. USA 79, 2882–2884.

Tanford, C. (1983). Translocation pathway in the catalysis of active transport. Proc. Natl. Acad. Sci. USA 80, 3701–3705.

Taormino, J. P., & Fambrough, D. M. (1990). Pre-translational regulation of the (Na^++K^+)-ATPase in response to demand for ion transport in cultured chicken skeletal muscle. J. Biol. Chem. 265, 4116–4123.

Tayler, W. R., & Green, N. M. (1989). The predicted secondary structure of the nucleotide binding sites of six cation-transporting ATPases lead to a probable tertiary fold. Eur. J. Biochem. 179, 241–248.

Teissie, J., & Tsong, T. Y. (1980). Evidence of voltage-induced channel opening in Na^+/K^+-ATPase of human erythrocyte membrane. J. Memb. Biol. 55, 133–140.

Thibault, D. (1993). The carboxy terminus of sodium and potassium ion transporting ATPase is located on the cytoplasmic surface of the membrane. Biochemistry 32, 2813–2821.

Tosteson, D. C., & Hoffman, J. F. (1960). Regulation of cell volume by active cation transport in high and low potassium sheep red cell. J. Gen. Physiol. 44, 169–194.

Van Huysse, J. W., Jewell, E. A., & Lingrel, J. B. (1993). Site-directed mutagenesis of a predicted cation binding site of Na^+,K^+-ATPase. Biochemistry 32, 819–826.

Vasilets, L. A., & Schwarz, W. (1992). Regulation of endogenous and expressed Na^+/K^+ pump in *Xenopus* oocytes by membrane potential and stimulation of protein kinases. J. Memb. Biol. 125, 119–132.

Vasilets, L. A., Omay, H. S., Ohta, T., Noguchi, S., Kawamura, M., & Schwarz, W. (1991). Stimulation of the Na^+/K^+ pump by external [K^+] is regulated by voltage-dependent gating. J. Biol. Chem. 266, 16285–16288.

Veech, R. L., Lowson, J. W. R., Cornell, N. W., & Krebs, H. A. (1979). Cytosolic phosphorylation potential. J. Biol. Chem. 254, 6538–6547.

Verrey, F., Kairouz, P., Schaerer, E., Fuentes, P., Kleyman, T. R., Geering, K., Rossier, B. C., & Kraehenbühl, J. P. (1989). Primary sequence of Xenopus laevis Na^+,K^+-ATPase and its localization in A6 kidney cells. Am. J. Physiol. 256, F1034–F1043.

Vilsen, B. (1992). Functional consequences of alterations to Pro^{328} and Leu^{332} located in the 4th transmembrane segment of the α-subunit of rat kidney Na^+,K^+-ATPase. FEBS Lett. 314, 301–307.

Vilsen, B., & Andersen, J. P. (1992). Mutational analysis of the role of Glu^{309} in the sarcoplasmic reticulum Ca^{2+}-ATPase of frog skeletal muscle. FEBS Lett. 306, 247–250.

Vilsen, B., & Andersen, J. P. (1993). CrATP-induced Ca^{2+} occlusion in mutants of the Ca^{2+}-ATPase of sarcoplasmic reticulum. J. Biol. Chem. 267, 25739–25743.

Vilsen, B., Andersen, J. P., Petersen, J., & Jørgensen, P. L. (1987). Occlusion of $^{22}Na^+$ and $^{86}Rb^+$ in membrane-bound and soluble protomeric αβ-units of Na^+,K^+-ATPase. J. Biol. Chem. 262, 10511–10517.

Von Heijne, G. (1991). Proline kinks in transmembrane α-helices. J. Mol. Biol. 218, 499–503.

Wang, K., & Farley, R. A. (1992). Lysine 480 is not an essential residue for ATP binding or hydrolysis by Na^+,K^+-ATPase. J. Biol. Chem. 267, 3577–3580.

Wierzbicki, W., & Blostein, R. (1993). The amino-terminal segment of the catalytic subunit of kidney Na^+,K^+-ATPase regulates the potassium deocclusion pathway of the reaction cycle. Proc. Natl. Acad. Sci. USA 90, 70–74.

Williams, K. A., & Deber, C. M. (1991). Proline residues in transmembrane helices: Structural dynamic role? Biochemistry 30, 8919–8923.

Williams, R. J. P. (1993). Are enzymes mechanical devices? Trends Biochem. Sci. 18, 115–117.

Wolitzky, B. A., & Fambrough, D. M. (1986). Regulation of the (Na^++K^+)-ATPase in cultured chick skeletal muscle: Modulation of expression by demand for ion transport. J. Biol. Chem. 261, 9990–9999.

Woolfson, D. N., Mortishire-Smith, R. J., & Williams, D. H. (1991). Conserved positioning of proline residues in membrane-spanning helices of ion-channel proteins. Biochem. Biophys. Res. Commun. 175, 733–737.

Xu, K. (1989). Any of several lysines can react with 5′-isothio-cyanato-fluorescin to inactivate sodium and potassium ion activated adenosinetriphosphatase. Biochemistry 28, 5764–5772.

Yoda, A., & Yoda, S. (1982a). Formation of ADP-sensitive phosphorylated intermediate in the electric eel Na^+,K^+-ATPase preparation. Mol. Pharm. 22, 693–699.

Yoda, A., & Yoda, S. (1982b). Interaction between ouabain and the phosphorylated intermediate of Na^+,K^+-ATPase. Mol. Pharmacol. 22, 700–705.

Yoda, S., & Yoda, A. (1986). ADP- and K^+-sensitive phosphorylated intermediate of Na^+,K^+-ATPase. J. Biol. Chem. 261, 1147–1152.

Yoda, A., & Yoda, S. (1987a). Two different phosphorylation-dephosphorylation cycles of Na^+,K^+-ATPase proteoliposomes accompanying Na^+ transport in the absence of K^+. J. Biol. Chem. 262, 110–115.

Yoda, S., & Yoda, A. (1987b). Phosphorylated intermediates of Na^+,K^+-ATPase proteoliposomes controlled by bilayer cholesterol. J. Biol. Chem. 262, 103–109.

Yool, A. J., & Schwarz, T. L. (1991). Alteration of ion selectivity of a K^+ channel by mutation of the H5 region. Nature 349, 700–704.

Young, R. M. & Lingrel, J. B. (1987). Tissue distribution of mRNAs encoding the α isoforms and β subunit of rat Na^+,K^+-ATPase. Biochem. Biophys. Res. Commun. 145, 52–58.

Young, R. M., Shull, G. E., & Lingrel, J. B. (1987). Multiple mRNAs from rat kidney and brain encode a single Na^+,K^+-ATPase β subunit protein. J. Biol. Chem. 262, 4905–4910.

Zamofing, D., Rossier, B. C., & Geering, K. (1988). Role of the Na^+,K^+-ATPase β-subunit in the cellular accumulation and maturation of the enzyme assessed by glycosylation inhibitors. J. Memb. Biol. 104, 69–79.

Zamofing, D., Rossier, B. C., & Geering, K. (1989). Inhibition of N-glycosylation affects transepithelial Na^+ but Na^+,K^+-ATPase transport. Am. J. Physiol. 256, C958–C966.

Zampighi, G., Kyte, J., & Freytag, W. (1984). Structural organization of (Na^++K^+)-ATPase in purified membranes. J. Cell Biol. 98, 1851–1856.

THE GASTRIC H^+/K^+-ATPASE

Jai Moo Shin, Dennis Bayle,
Krister Bamberg, and George Sachs

Biomembranes
Volume 5, pages 185–224.

ISBN: 1-55938-662-2.

I. INTRODUCTION

This review of the gastric H^+/K^+-ATPases focuses largely on the molecular and biochemical properties of the enzyme. This enzyme is responsible for the elaboration of HCl by the parietal cell of the gastric mucosa. It catalyses an electroneutral exchange of cytoplasmic protons for extracytoplasmic potassium. The extracytoplasmic potassium derives from cellular K^+ by K^+, Cl^- efflux. It is a member of an ever-growing family of P-type ion-motive ATPases that catalyze transport by means of conformational changes driven by cyclic phosphorylation and dephosphorylation of the catalytic subunit of the ATPase. The enzymes are represented in both prokaryotes and eukaryotes and are designed as primary ion transporters.

Structurally, these membrane embedded proteins can be divided into cytoplasmic, membrane, and extracytoplasmic domains. Since ion transport must occur across the membrane domain, much of this review is dedicated to an analysis of the nature of the membrane domain of the H^+/K^+-ATPase compared to the membrane domains of the Na^+/K^+- and Ca^{2+}-ATPases. The membrane domain is traversed by the ions transported by these ATPases, and this or the adjacent extracytoplasmic domain is the site of action of inhibitors such as ouabain in the case of the Na^+/K^+-ATPase and the substituted benzimidazoles and the imidazopyridines in the case of the H^+/K^+-ATPase.

II. CLASSES OF ION MOTIVE ATPASES

There are three main classes of ion transport ATPases, the F, V, and EP ATPases. Their function in the case of the F-type is ATP synthesis, in the case of the V-type, transport of protons, and in the case of the P-type, transport of various ions such as Na^+, Ca^{2+}, H^+, Mg^{2+}, AsO_3^{3-}, Cu^+, and K^+.

The F_1/F_0-ATPase is present in mitochondria and synthesizes ATP at the expense of the proton-motive force generated by electron transport (Futai et al., 1989). This enzyme shows a very similar organization and mechanism as compared to the V-ATPase. The F-ATPase has eight different subunits with a stoichiometry of 3α, 3β, 1τ, 1δ, 1ε, 1a, 2b, and 10–12 c. The α-subunit of the F_1/F_0-ATPases has homology with the B-subunit of the V-ATPase, and the β-subunit of F-ATPase is homologous to the A-subunit of V-ATPase. However, the τ subunit of the F_1/F_0-ATPase shows no homology with the C-subunit of the V-ATPase. The DCCD-binding c-subunit appears to be responsible for the proton transport (Hermolin and Fillingame, 1989). This protein complex is responsible for ATP synthesis in mitochondria.

The V-ATPase generates the proton-motive force for accumulating neurotransmitters in storage vesicles and energizing the vacuolar system of eukaryotes (Nelson, 1991). The V-ATPase is, therefore, responsible for acidification of the interior of organelles inside the cell. Such an enzyme is also involved in pH homeostasis by the kidney and in bone resorption by the osteoclast. The V-ATPase contains seven to ten subunits organized into two distinct domains. The cytoplasmic domain contains five different subunits with molecular weights between 72- and 26-kD, and the membrane domain contains two different subunits with molecular masses of about 20- and 16-kD. The A cytoplasmic subunit (molecular mass of 72 kD) in cooperation with the B cytoplasmic subunit (molecular mass of 57 kD) contains the catalytic site and the membrane subunits a and c conduct protons across the membrane.

The EP-type ATPase family in mammals contains the Ca^{2+}-ATPases, the Na^+/K^+-ATPases and the H^+/K^+-ATPases. These ATPases are membrane-bound enzymes with similar structural motifs (Jorgensen and Andersen, 1988; Glynn and Karlish, 1990; Rabon and Reuben, 1990). The Na^+/K^+-ATPase is an ubiquitous electrogenic pump mainly responsible for the maintenance of Na^+ and K^+ gradients between cells and their environment. The H^+/K^+-ATPase is the electroneutral pump responsible for producing gastric acid. The Ca^{2+}-ATPases maintain the Ca^{2+} gradient necessary for this ion's ability to act as a second messenger. The Na^+/K^+- and H^+/K^+-ATPases are composed of two subunits. The α-subunits, with molecular mass of about 100 kD, have the catalytic site and the β-subunits, with peptide mass of 35 kD, which are non-covalently bound to the α-subunit, are glycoproteins with most of their surface exposed to the extracytoplasmic surface. The Ca^{2+}-ATPases have a single catalytic protein of mass about 100 kD, which pumps Ca^{2+} in exchange for H^+ (Yamaguchi and Kanazawa, 1984). The SERCA gene family encodes intracellular Ca^{2+} pumps (Lytton et al., 1992). There are also several isoforms of the plasma membrane Ca^{2+} pump which contains a calmodulin binding domain absent from the SERCA family (Carafoli, 1992; Lytton et al., 1992). There is about 60% sequence homology between the Na^+/K^+-ATPase and the H^+/K^+-ATPase α-subunit, while the Ca^{2+}-ATPase of sarcoplasmic reticulum (SR) shows only about 15% overall homology with the H^+/K^+-ATPase. There is about 40% sequence homology between the β-subunit of the Na^+/K^+-ATPase and the similar subunit of the H^+/K^+-ATPase (Canfield et al., 1990).

The motif in both the F- and V-type ATPases appears to be a catalytic cytoplasmic domain and an ion transporting membrane domain. Although the membrane domain contains few subunits, there are multiple copies of the proton-transporting subunits suggesting that multiple membrane spanning helices are necessary for ion transport by these pumps. The ion selectivity of the membrane domain appears to determine the overall selectivity of transport. Whereas the F_1/F_0-ATPase of *E. coli* transports protons and the F_1/F_0-ATPase of *Propionigenium modestum* transports Na^+, chimeric pumps constructed of the cytoplasmic domain of one pump and the membrane domain of the other transport the ion determined by the membrane domain rather than the cytoplasmic domain (Laubinger et al., 1990). In the case of the P-type ATPases, they have several membrane spanning segments, many of them containing amino acids essential for ion transport. Chimeric constructs of these pumps appear to conform to what has been found for the F-type pump, namely that it is the membrane domain that determines general ion selectivity and the cytoplasmic domain energizes the transport of the ion (Blostein et al., 1993).

In this review, we focus on the properties of the gastric H^+/K^+-ATPase, which, along with the Na^+/K^+-ATPase has provided a therapeutic target in treatment of disease. Digitalis, which inhibits the Na^+/K^+-ATPase is probably the oldest drug still in use today, acting to selectively increase cell Ca^{2+} in cardiac muscle by allowing Na^+ loading and thereby decreasing Na^+/Ca^{2+} exchange. The substituted benzimidazoles which inhibit the H^+/K^+-ATPase are the newest class of anti-ulcer drugs used in the treatment of a variety of diseases of the upper GI tract. Imidazopyridines and arylquinolines are being investigated as an alternative means of inhibiting gastric acid secretion.

III. STRUCTURE OF THE H^+/K^+-ATPASE

A. General

The gastric H^+/K^+-ATPase consists of two subunits, a catalytic α-subunit and a heavily glycosylated β-subunit (Hall et al., 1990). While the Na^+K^+-ATPase β-subunit was identified soon after the discovery of the enzyme, the H^+/K^+-ATPase β-subunit was identified some years after the discovery of this enzyme because the β-subunit was not easy to detect on SDS gel electrophoresis using the Laemmli buffer system. The β-subunit was eventually identified by post-embedding staining techniques which showed that wheat germ agglutinin staining occurred on the extracellular face of the gastric vesicles and then the β-subunit was partially sequenced after deglycosylation followed by protease digestion (Hall et al., 1990). The partial amino acid sequence showed strong homology with the β-subunit of the Na^+/K^+-ATPase. Using lectin affinity chromatography, the H^+/K^+-ATPase α-subunit was co-purified with the β-subunit, showing that the α-subunit interacts with the β-subunit (Callaghan et al., 1990, 1992; Okamoto et al., 1990). By cross-linking with low concentrations of glutaraldehyde, the β-subunit was shown to be closely associated with the α-subunit (Rabon et al., 1990a; Hall et al., 1991).

Whereas two-dimensional crystals of the Ca^{2+}-ATPase have been diffracted to a resolution of 14 Å (Toyoshima et al., 1993) which allows visualization of the cytoplasmic, membrane, and extracytoplasmic domain, crystals of the H^+/K^+-ATPase have only been diffracted to a resolution of 25 Å (Rabon et al., 1986). The Ca^{2+}-ATPase data show the presence of a bird-like head, a stalk, and a transmembrane region which may contain eight or 10 membrane-spanning helices significantly tilted with respect to each other. The combination of electron diffraction, transmission electron microscopy using tannic acid staining, and freeze–fracture have allowed definition of the general dimensions of the H^+/K^+-ATPase showing a large cytoplasmic domain, a smaller membrane domain and an even smaller extracytoplasmic domain (Rabon et al., 1986; Mohraz et al., 1990). The dimensions calculated in this way are similar to those observed for the Ca^{2+}-ATPase. On the assumption that the H^+/K^+-ATPase has a shape quite similar to the Ca^{2+}-ATPase, a model of the shape of the H^+/K^+-ATPase is presented in Figure 1.

The only ion pump for which a detailed structure is available is bacteriorhodopsin which has been diffracted to a resolution of 2.8 Å. There are seven transmembrane helices, and the crucial all *trans*-retinal group that isomerizes to 13 *cis*-retinal with light is bound to Lys-216. Asp-85 and Glu-96 are on the cytoplasmic and extracytoplasmic side of the retinal group, respectively. Isomerization of the retinal alters the directionality of the lysine Schiff base allowing sided deprotonation. In addition, there appears to be tilting of the transmembrane helices which form a narrow channel on the cytoplasmic side and a broader channel on the extracytoplasmic side of the retinal moiety (Baldwin et al., 1988; Ceska and Henderson, 1990). All the helices participate on one side or the other of the ion pathway across bacteriorhodopsin. Presumably, protonation from the cytoplasmic side and deprotonation of the Schiff base to the extracytoplasmic face depends on the change in orientation of the Schiff base and a change in the peptide conformation on either side of the base.

The general concept which may be applicable to other membrane pumps is that pumping by bacteriorhodopsin involves a change of conformation in the central part of the membrane domain which acts like a switch sending protons from one

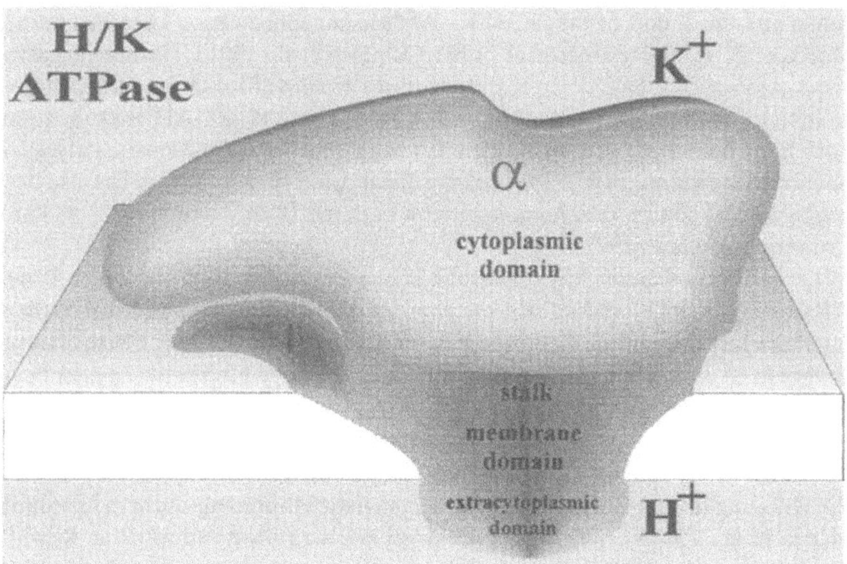

Figure 1. The postulated shape of the H^+/K^+-ATPase based on the crystal structure of the Ca^{2+}-ATPase (Toyoshima et al., 1993). Shown is the large cytoplasmic mass of the α-subunit, the stalk region, the transmembrane and extracytoplasmic domains. The small N terminal cytoplasmic domain of the β-subunit is also shown.

side of the membrane to the other. This change in conformation is driven by the isomerization of a bound photoabsorptive pigment. Changes in binding energy are presumably driven by the change in tilt of a significant number of transmembrane helices. EP-type pumps lack photoabsorptive pigments, but perhaps the cycle of phosphorylation and dephosphorylation determines both the switching and the conformational change in these pumps.

B. The α-subunit of the H^+/K^+-ATPase

General Aspects

The primary sequences of the α-subunits deduced from cDNA have been reported for pig (Maeda et al., 1988a), rat (Shull and Lingrel, 1986), and rabbit (Bamberg et al., 1992). The hog gastric H^+/K^+-ATPase α-subunit sequence deduced from its cDNA consists of 1,034 amino acids and has a M_r of 114,285 (Maeda et al., 1988a). The sequence based on the known N-terminal amino acid sequence is one less than the cDNA derived sequence (Lane et al., 1986). The rat gastric H^+/K^+-ATPase consists of 1,033 amino acids and has a M_r of 114,012 (Shull and Lingrel, 1986), and the rabbit gastric H^+/K^+-ATPase consists of 1,035 amino acids, showing a M_r of 114,201 (Bamberg et al., 1992). The degree of conservation among the α-subunits is extremely high (over 97% identity). In addition, the gene sequence for human and the 5′ part of the rat H^+/K^+-ATPase α-subunits have been determined (Maeda et al., 1990; Newman et al., 1990; Oshiman et al., 1991). The human gastric H^+/K^+-ATPase gene has 22 exons and encodes a protein of 1,035 residues including the initiator methionine residue (M_r = 114,047). These H^+/K^+-ATPase α-subunits show high homology (ca. 60% identity) with the Na^+/K^+-ATPase catalytic α-subunit (Maeda et al.. 1990). The putative distal colon H^+/K^+-ATPase has also been sequenced and shares 75% homology with both the H^+/K^+- and Na^+/K^+-ATPases (Crowson and Shull, 1992).

The gastric α-subunit has conserved sequences along with the other P-type ATPases for the ATP-binding site, the phosphorylation site, the pyridoxal 5′-phosphate binding site and the fluorescein isothiocyanate-binding site. These sites are thought to be within the ATP-binding domain in the large loop between membrane spanning segments 4 and 5.

In the case of the hog gastric H^+/K^+-ATPase, pyridoxal 5′-phosphate bound at Lys-497 of the α-subunit in the absence, but not the presence of ATP (Tamura et al., 1989), suggesting that Lys-497 is present in the ATP-binding site or in its vicinity (Maeda et al., 1988b). The phosphorylation site was observed to be at Asp-386 (Walderhaug et al., 1985). Fluorescein isothiocyanate (FITC) covalently labels the gastric H^+/K^+-ATPase in the absence of ATP (Jackson et al., 1983). The binding site of FITC was at Lys-518 (Farley and Faller, 1985). However, several additional lysines, such as those at positions 497 and 783, were shown to react with FITC during the inactivation of the Na^+,K^+-ATPase and to be protected from reaction with FITC when ATP was present in the incubation (Xu, 1989). Based on these data, similar lysines of the H^+/K^+-ATPase could be near or in the ATP binding site.

Secondary Structure of the α-subunit

There are various methods available for determining secondary structure. The most commonly used is to determine the location of membrane spanning α-helices using hydropathy plots (Kyte and Doolittle, 1982). These are based on determining a moving average of hydrophobic, neutral, and hydrophilic amino acids using a variety of scales. When the average is greater than a predetermined value for 11 or more amino acids, this sector is considered to be potentially membrane spanning.

The most exact structure can be obtained from the crystal structure derived from 2- or 3-dimensional crystals analyzed by electron diffraction or X-ray. As previously mentioned, this has been done for bacteriorhodopsin in terms of two-dimensional crystals and for the photosynthetic reaction center in terms of three-dimensional crystals (Baldwin et al., 1988; Ceska and Henderson, 1990). In the absence of crystal information, hydropathy plot information has to be buttressed by specific biochemical information, as discussed following.

Biochemical methods use sites of labeling with cytoplasmic-, extracytoplasmic-, or membrane-directed chemical probes, determination of the peptides remaining in the membrane following cleavage of the extramembranal domain (usually cytoplasmic), determination of the sidedness of epitopes for antibodies and site-directed mutagenesis. Most of these have been applied to the P-type ATPases.

Interpretation of the hydrophobicity plots of the α-subunit of the H^+/K^+-ATPase (Kyte and Doolittle, 1982) based on the primary amino acid sequence has suggested an eight- or 10-membrane spanning segment model for the secondary structure, as for the other mammalian P-type ATPases. There is agreement on the first four membrane-spanning segments in the N-terminal one-third of the α-subunit. In the C terminal one-third of the protein, a prediction from the hydropathy analysis is more difficult and, therefore, controversial.

The gastric enzyme is readily isolated as intact, ion-tight, uniformly inside-out vesicles. With such a sided preparation, it would seem straightforward to use a set of biochemical techniques that would establish unequivocally the location of each region of the enzyme. The techniques that have been used include protease cleavage of the cytoplasmic domain, and allocation of the sites of reaction of cytoplasmic ligands such as ATP, FITC, or pyridoxal phosphate. Extracytoplasmic reagents such as the substituted benzimidazoles or photoaffinity imidazopyridine analogs have also provided information as to the secondary structure of the enzyme. The topology of epitopes for either polyclonal antibodies directed against specific peptides or of monoclonal antibodies generated against the enzyme or even intact parietal cells has also provided topological data. Finally the fact that the C-terminal amino acids are Tyr–Tyr allows iodination in combination with carboxypeptidase cleavage to be used to determine the sidedness of the C-terminal domain.

An alternative is to determine the capability of putative membrane segments to act as signal anchor or stop transfer sequences in *in vitro* translation. This method has the additional advantage that the boundaries and determinant amino acids can be defined for each segment found either by direct experimentation or from hydropathy analysis.

Cytoplasmic ligand binding sites. The phosphorylation site, FITC-binding sites, and pyridoxal phosphate-binding site are cytoplasmic. Asp-386 is the phosphorylation site (Walderhaug et al., 1985), Lys-518 is the site of FITC binding (Farley and Faller, 1985), and Lys-497 the pyridoxal phosphate-binding site (Tamura et al., 1989). These are all predicted to be present in the loop between membrane segments M4 and M5, consistent with their cytoplasmic location and interaction with ATP.

Tryptic cleavage of the α-subunit. The key assumption in the use of protease digestion as a topological probe is that the protease cleaves only cytoplasmic residues while maintaining vesicle integrity. The digestion pattern is thus simplified in that fewer bonds are accessible and if the cytoplasmic fragments are removed by washing, the pattern is simplified even further. Thus, instead of 97 possible tryptic fragments of the H^+/K^+-ATPase α-subunit, only four or five membrane-spanning segment pairs should be left, along with the partially cleaved β-subunit.

In either an eight- or 10-transmembrane segment model, each transmembrane pair connecting the luminal loop has at least one cysteine, which allows fluorescent labeling of the cysteines left after complete cleavage of the cytoplasmic domain. The N-terminal position of each fluorescent peptide fragment is then defined by N-terminal sequencing. The size of the fragment is determined from molecular weight measurements in the tricine gradient gels used for the separation. This then allows the C-termination of the peptide to be identified using the Lys or Arg cleavage sites at this end. Although it is thought that there might be some chymotryptic activity in even the purest preparations of trypsin, in all the tryptic sequences we have analyzed, no evidence was obtained for chymotryptic cleavage sites.

Four transmembrane pairs connected by their luminal loop were detected in the hog gastric H^+/K^+-ATPase digest (Besancon et al., 1993; Shin et al., 1993). A tryptic peptide fragment beginning at Gln-104 represents the Ml/loop/M2 sector. The M3/loop/M4 sector was found at a single peptide beginning at Thr-291, and the M5/loop/M6 sector at a peptide beginning at Leu-776. The M7/loop/M8 region was found in a single peptide fragment of 11 kD, beginning at Leu-853. These represent the first four transmembrane segment pairs of the proposed 10 transmembrane segment model (Bamberg et al., 1992).

From these studies, four membrane segment/loop/membrane segment sectors were identified, corresponding to Ml/loop/M2 through M7/loop/M8. Although the hydropathy plot predicts two additional membrane spanning segments (H9 and

H10) at the C-terminal region of the enzyme, no evidence was obtained for this pair in spite of the fact that H9 is predicted to have four cysteine residues that are, in principle, able to bind F–M1. In the absence of direct evidence for the membrane spanning nature of the ninth and tenth hydrophobic domains, they may be membrane- or stalk-associated rather than membrane spanning, or another technique is necessary to demonstrate their presence in the membrane.

Iodination. The C-terminal amino acids of the α-subunit are Tyr–Tyr, which, therefore, can be iodinated with peroxidase-H_2O_2-^{125}I on the cytoplasmic side of intact hog gastric vesicles. Digestion with carboxypeptidase Y then released about 28% of the counts incorporated into the α-subunit, as would be predicted from a cytoplasmic location of the C-terminal tyrosines (Scott et al., 1992). These data show that there is an even number of transmembrane segments in the α-subunit.

Sided reagents. Ouabain is the best-known sided reagent for the Na^+/K^+-ATPase. High affinity binding depends on the boundary amino acids between M1 and M2 and their connecting extracytoplasmic loop (Price and Lingrel, 1988). In the case of the gastric H^+/K^+-ATPase, two sets of reagents are known that also react exclusively with the extracytoplasmic surface and hence, are useful in mapping the membrane positions of these sites.

The K^+-competitive photoaffinity reagent, [^3H]-MeDAZIP, was synthesized as an analog of SCH 28080, an imidazopyridine, and was shown to bind to the same general region of the H^+/K^+-ATPase as does ouabain on the Na^+/K^+-ATPase, namely the Ml/loop/M2 sector (Munson et al., 1991). Other reagents available as extracytoplasmic molecular probes are omeprazole, lansoprazole, and pantoprazole, which are all substituted benzimidazoles. These reagents are weak base, acid-activated compounds, which form cationic sulfenamides in acidic environments. The sulfenamides formed react with the SH-group of cysteines in proteins to form relatively stable disulfides. Since the pump generates acid on its extracytoplasmic surface, only those cysteines available from that surface would be accessible to these sulfenamides if labeling is carried out under acid-transporting conditions. The cysteines that are labeled, depending on the reagent used (see following) are Cys-321, Cys-813, Cys-822, and Cys-892. These data define the M3/M4, M5/M6, and M7/M8 segments of the enzyme. These reagents thus confirm the results of tryptic digestion providing evidence for eight membrane-spanning segments. Again, although a cysteine in the H9 sector is predicted to be close to the extracytoplasmic domain, these reagents do not demonstrate its presence in the mature pump.

K^+ competitive reagents. A series of K^+ competitive reagents has been developed, including imidazopyridines, that are known to react on the outside surface of the pump (Wallmark et al., 1987; Munson and Sachs, 1988; Mendlein and Sachs, 1990; Munson et al., 1991; Kaminski et al., 1991). [^3H]-MeDAZIP is an imidazopyridine that competitively inhibits before photolysis, and inhibits and binds in

a saturable manner after photolysis. This compound binds covalently in the same region of binding of the nonphotolyzed compound, proving that the sector binding this reagent was the M1/loop/M2 sector. Molecular modeling studies suggested that binding should be to Phe-124 and Asp-136 (Munson et al., 1991). This region of the H^+/K^+-ATPase is homologous to the region responsible for high affinity ouabain binding to the Na^+/K^+-ATPase (Price et al., 1990). These data verify the presence of the first two segments spanning the membrane in the α-subunit of the H^+/K^+-ATPase. A fluorescent K^+ competitive arylquinoline, MDPQ, shows enhanced hydrophobicity of its environment with formation of the E2–P.[I] conformer of the enzyme, as if the M1/loop/M2 segment moves further into the membrane in the E2 conformation (Rabon et al., 1991).

Substituted benzimidazoles. The substituted benzimidazole compounds available clinically with inhibitory activity on the H^+/K^+-ATPase are omeprazole, lansoprazole, and pantoprazole.

Omeprazole shows inhibitory activity only under the acidic conditions generated by the pump (Wallmark et al., 1984; Im et al., 1985; Lorentzon et al., 1985, 1987; Lindberg et al., 1986; Keeling et al., 1987). The action of omeprazole following acidification is to react with available cysteine-SH groups. The reaction with the cysteine-SH produces a disulfide, Cys–S–S–X, which covalently inhibits the enzyme in terms of ATPase activity and acid transport.

The enzyme was labeled using [^{3}H]-omeprazole under acid transporting conditions. The cysteines reacting in the ATPase were determined by trypsinolysis followed by SDS-PAGE electrophoretic separation of the tryptic fragments and sequencing. The two cysteines reacting were found to be in position 813 or 822 and 892. These should be on or close to the extracytoplasmic face of the enzyme and are predicted to be in the M5/loop/M6 and M7/loop/M8 sector of the enzyme, respectively (Besancon et al., 1993). This reagent, therefore, defines two pairs of membrane spanning segments. Since it does not react with the β-subunit, the six cysteines on this subunit predicted to be extracytoplasmic must be disulfide linked.

Lansoprazole, another substituted benzimidazole, labels at three positions on the enzyme, at Cys-321, Cys-813 or Cys-822, and Cys-892 (Sachs et al., 1993). Pantoprazole labels at both Cys-813 and Cys-822, with no labeling at either Cys-892 or Cys-321 (Shin et al., 1993). These cysteines are predicted to be in the M3/loop/M4 sector (Cys-321), in the M5/loop/M6 sector (Cys-813 and Cys-822), and in the M7/loop/M8 sector (Cys-892). The labeling by lansoprazole provides additional evidence for the existence of the M3/M4 membrane spanning pair.

Antibody epitope studies on the α-subunit. There are many monoclonal antibodies that have been generated reacting with the H^+/K^+-ATPase. With the sequence of the α-subunit deduced from cDNA, it is now possible to define the epitopes, and to determine the sidedness of these epitopes by staining intact or permeable cells with either fluorescent or immunogold techniques.

Antibody 95 inhibits ATP hydrolysis in the intact vesicles, and appears to be K^+ competitive (Bayle et al., 1992). Its epitope was identified by western blotting of an *E. coli* expression library of fragments of the cDNA encoding the α-subunit, as well as by western analysis of tryptic fragments (Bayle et al., 1992). The sequence recognized by this antibody was between amino acid positions 529 and 561. Since it inhibits intact vesicles this epitope must be cytoplasmic. Its epitope is close to the region known to bind the cytoplasmic reagent FITC, namely in the loop between M4 and M5. The ability of this antibody to immunoprecipitate intact gastric vesicles showed that it was on the cyroplasmic surface of the pump.

Antibody 1218 was shown by western analysis of tryptic fragments and by recombinant methodology to have its major epitope between amino acid positions 665 and 689 (Mercier et al., 1993). This epitope is on the cytoplasmic surface of the enzyme also in the loop between M4 and M5. This antibody does not inhibit ATPase activity. A second epitope for mAb 1218 was also identified. This epitope was between amino acid positions 853 and 946 according to western blot analysis of tryptic fragments. A synthetic peptide (888–907) apparently containing this second epitope displaced mAb 1218 from vesicles adsorbed to the surface of ELISA wells (Bayle et al., 1992).

Monoclonal antibody 146 was generated against intact parietal cells and subsequently purified and shown to react with rat H^+/K^+-ATPase. In cells, it reacts on the outside surface of the canaliculus as shown by immunogold electron microscopy (Mercier et al., 1989). Western analysis of rat, rabbit, and hog enzyme gave the surprising result that it was present on the β-subunit of rat and rabbit and absent from the β-subunit of hog. Disulfide reduction eliminated reactivity of this antibody. Comparing sequences of the different β-subunits, there is an Arg, Pro substitution in the hog for Leu, Val in the rat between the disulfide at position 161 and 178. This suggests that the β-subunit epitope of mAb 146 is contained within this region of the subunit. On the other hand, there was also an epitope on the α-subunit of all three species recognized by mAb 146 also using western analysis. This epitope was defined both by tryptic mapping and octamer walking to be between positions 873 and 877 of the hog α-subunit. This is on or close to the extracytoplasmic face of M7. The finding, using western blotting, that there was an epitope both on the α- and β-subunits was confirmed by expressing the rabbit subunits in SF9 cells using baculovirus transfection. The β-subunit reacted in the SF9 cells and on western blots with mAb 146. The α-subunit did not react in the cells, but did on western blots as if the epitope in the α-subunit was difficult to access in the absence of a denaturing detergent such as SDS (Mercier et al., 1993). These data may indicate tight binding between the α- and β-subunits in this region of the enzyme, as is discussed further below.

Molecular biological analysis. A molecular biological method was developed to analyze not only for the presence of the membrane segments previously defined, but also to explore the nature of membrane insertion (Bamberg and Sachs,

1993). A cDNA encoding a fusion protein of the 102 N-terminal amino acids of the rabbit H^+/K^+-ATPase α-subunit linked by a variable segment to the 177 most C-terminal amino acids of the rabbit H^+/K^+-ATPase β-subunit was transcribed and translated in a rabbit reticulocyte lysate system using labeled methionine in the absence or presence of microsomes. The cDNA for the variable region containing one or more putative membrane spanning segments is synthesized using selective primers in a PCR reaction and ligated into the cDNA construct. Since the β-subunit region has five consensus N-glycosylation sites, translocation of the C-terminal β-subunit part of the fusion protein into the interior of the microsomes can be determined by glycosylation. The presence of glycosylation is evidence for an odd number of transmembrane segments in the variable region preceding the β-subunit part. The absence of glycosylation shows either the presence of an even number of membrane spanning segments or absence of membrane insertion. Figure 2 illustrates the principle of the method.

Membrane segments can act either as an insertion signal, when the protein contains a signal anchor sequence, or can act as a stop transfer sequence when the protein contains a sequence which prevents further transport of the newly synthe-

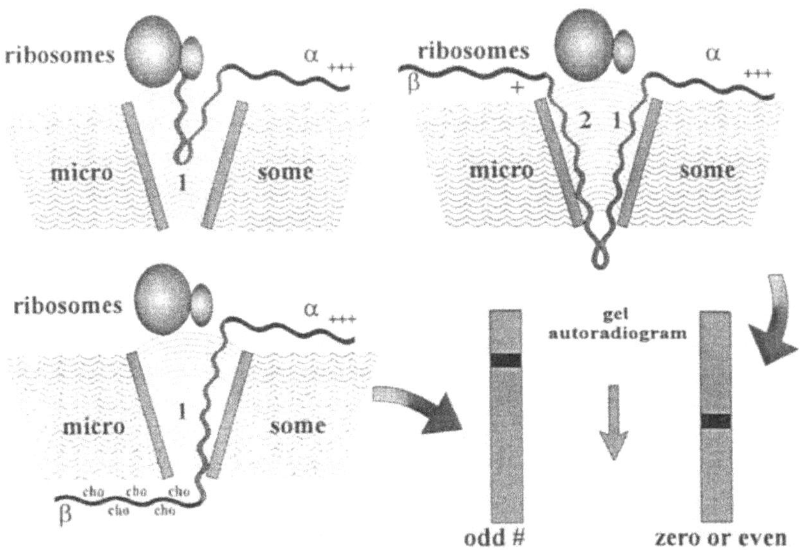

Figure 2. A simplified model illustrating the principle of *in vitro* translation used to determine membrane spanning segments of the α-subunit. Translation is carried out in the presence or absence of microsomes, and the molecular weight of the product determined by SDS-PAGE and autoradiography. With an odd number of membrane spanning segments, glycosylation alters the molecular weight of the product. In the absence of membrane spanning segments, or with an even number, no glycosylation is observed.

sized peptide across the microsomal membrane. To define a signal anchor sequence, that sequence is placed in the variable region. To define a stop transfer sequence, the sequence is placed after a known signal anchor sequence, either its natural companion sequence or after the sequence for M1 in the experiments performed on the H^+/K^+-ATPase.

The sequence inserted into the translation vector for the putative M1 segment (using the rabbit sequence beginning at methionine) began at Lys-102 and extended to Thr-136. When translation was carried out in the presence of microsomes, essentially all of the product was glycosylated as determined by an increase in molecular weight on SDS gel electrophoresis. Direct evidence for this was provided by digestion of the solubilized product with N-glycanase which reduced the molecular weight to that seen in the absence of microsomes.

The sequence used for the putative M2 membrane spanning segment began at Tyr-142 and extended to Ser-171. By itself, it not only acted as a signal anchor sequence, but in association with the M1 segment, also prevented glycosylation of M1, showing that it could act as a stop transfer sequence as well. These data demonstrate that translation of the M1/M2 sector of the enzyme agrees with the allocation of this membrane segment pair as determined from hydropathy and biochemical analysis. Thus, M1 acts as the signal anchor sequence and M2 as the stop transfer sequence in this region of the α-subunit. Figure 3 illustrates the SDS gel where translation of M0, M1, M2, and M1 + M2 was carried out. In the absence of a membrane signal sequence, there is no glycosylation. In the presence of either M1 or M2 alone, there is glycosylation. In the presence of both M1 and M2, glycosylation does not occur.

The sequence for the M3 membrane spanning segment used for insertion into the translation vector began at Val-303 and ended at Arg-330. When translated, the product was glycosylated, showing that this sequence could act as a signal anchor sequence. When inserted subsequent to the M1 segment, glycosylation was absent. This showed that this sequence could act as both a signal anchor and stop transfer sequence.

A sequence used for investigating the presence of the M4 segment began at Arg-330 and ended at Lys-360. This sequence was glycosylated on its own and acted as a stop transfer sequence when inserted in conjunction with M3 or even M1. These translation data again match the results of hydropathy analysis and biochemical cleavage and labeling data.

The data from translation thus allow the conclusion that it is likely that the first four membrane segment pairs are co-inserted during translation and form the membrane anchoring domain prior to translation of the large cytoplasmic loop between M4 and M5, which contains about 430 amino acids.

Various sequences were used to demonstrate the presence of M5 and M6 by this translation methodology. The putative M5 sequences tested began at Arg-777 or Ala-788 and ended at Leu-811, Ile-821, or Thr-825. There was some glycosylation evident with the sequence ending at Ile-821 and this sequence reduced, but did not

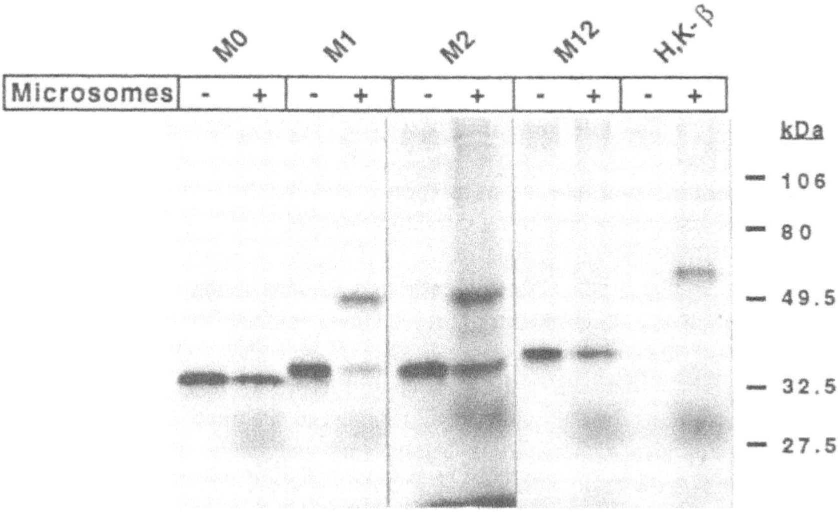

Figure 3. The results of *in vitro* transcription/translation of a DNA sequence containing the 100 first amino acids of the α-subunit, a linker sequence and the 200 C-terminal amino acids of the β-subunit. The translation is carried out in the absence or presence of microsomes. Inserted into the linker sequence is either no membrane segment (M0), the first (M1) or second membrane segment (M2), or both together (M1/2). It can be seen that with either membrane segment alone, there is glycosylation of the translation product, showing passage of either of the two sequences across the microsomal membrane. With a membrane segment pair (M1 + M2), no glycosylation is observed, showing that M2 can act as a stop transfer sequence. On the far right is translation of the β-subunit, showing absolute dependence on the presence of microsomes.

abolish glycosylation when inserted subsequent to M1. This can be considered as tentative evidence for an M5 membrane spanning segment, but the data are by no means as clear as those for the first four membrane sequences of the enzyme. None of the other presumed M5 sequences acted as either signal anchor or stop transfer sequences.

The M6 sequences tested began at Gly-814 and ended either at Glu-836 or Arg-852. These sequences did not act as either signal anchor or as stop transfer sequences. The translation system where co-insertion of M5 or M6 was investigated therefore, did not correlate with the trypsinolysis data nor with the labeling observed with the benzimidazoles.

Various sequences were also tested for translation of the supposed M7 spanning segments. The longest began at Arg-846 and ended at Arg-898. These acted neither as signal anchor sequences on their own nor as stop transfer sequences when associated with the M1 segment.

The M8 sequence selected for testing began at Tyr-924 and ended at Ile-947. This sequence did not act as a signal anchor sequence, but acted as a stop transfer sequence when translated in association with M1. Thus, of the membrane sequences M5 through M8, the only one conforming to its predicted role as a membrane insertion signal on its own is M8. The biochemical data demonstrating the existence of two pairs of membrane segments, M5 and M6, and M7 and M8, are strong. Thus M5, M6, M7, and M8 were all defined both by tryptic cleavage and by labeling by three substituted benzimidazoles which act as extracytoplasmic cysteine reagents. The negative *in vitro* translation data found in this region suggest that this region of the enzyme is post-translationally inserted and this may depend on preceding sequence or other factors.

The effect of preceding sequence was tested by generating longer insertion sequences beginning at M1 and ending with the different C-terminal sequences for M5 or M7. Although glycosylation would be anticipated from an odd number of membrane spanning segments, this was not observed. Hence the signal for membrane insertion of this region might be subsequent to M8.

Although no biochemical data have been obtained for the presence of a ninth or tenth membrane spanning sequence, hydropathy suggests the presence of such sequences. The translation vector containing the sequence indicated by hydropathy beginning at Phe-960 and ending at Phe-990 was glycosylated and prevented glycosylation when inserted subsequent to M1. Hence, this sequence can act both as a signal anchor sequence and as a stop transfer sequence.

The sequence representing the tenth hydrophobic domain begins at Pro-994 and ends at Arg-1017. When this was present alone in the translation vector, no glycosylation was detected. However, when present in association with M1 or the ninth hydrophobic segment, glycosylation was inhibited. This shows that whereas this sequence does not function as a signal anchor sequence, it does act as a stop transfer sequence.

The translation data, in contrast to previous data on the H^+/K^+-ATPase, show that there is likely to be an additional pair of membrane spanning sequences, M9 and M10, in the α-subunit of the H^+/K^+-ATPase.

From this approach, the first four membrane segments are present and are co-inserted with translation. The last two hydrophobic sequences also appear to be co-inserted. The information within the M5, M6, and M7 sequences is apparently insufficient for these to act on their own as either anchor or stop transfer sequences. The information within the M8 sequence is sufficient for this to act as a stop transfer sequence. That M5 through M7 are indeed membrane inserted in the mature enzyme is clear from the biochemical analyses. A reasonable hypothesis that may explain these translation data is that insertion of M5 through M7 is post-translational depending on insertion of M9/M10 and that M8 then acts as a stop transfer sequence. The presence of the additional pair of membrane segments predicted by hydropathy is strongly suggested by these translation data. Their absence using fluorescein maleimide as a cysteine labeling reagent may be due to covalent bonding of the 4 cysteines present in H9. These bonds are neither disulfide or thioester in nature since neither reduction nor hydroxylamine treatment elicit F–M1 labeling.

The combination of techniques described above, namely, sided proteolysis, epitope mapping, iodination, sided reactivity, and *in vitro* translation provide evidence for a 10 membrane segment model, with a large cytoplasmic loop between M4 and M5 and a large extracytoplasmic loop between M7 and M8.

Effect of disulfide reduction on the α-subunit. It has been shown that reduction of the disulfides of the β-subunit inhibits the activity of the H^+/K^+-ATPase (Chow et al., 1992). In the case of the Na^+/K^+-ATPase, the effect of reducing agents on the ability of the enzyme to hydrolyze ATP and bind ouabain was quantitatively correlated with the reduction of disulfide bonds in the β-subunit (Kirley, 1990). When these disulfides are disrupted in the H^+/K^+-ATPase, the pattern obtained following tryptic digestion changes. The normal tryptic cleavage site at the N-terminal end of M5 is at position 776. Following reduction, the cleavage site moves to position 792. This change of tryptic cleavage at the N-terminal end of M5 is similar to that seen with labeling of both cysteines in the M5/M6 region with pantoprazole (Shin et al., 1993).

C. The β-Subunit

The primary sequences of the β-subunits have been reported for rabbit (Reuben et al., 1990), hog (Toh et al., 1990), rat (Canfield et al., 1990; Shull, 1990; Newmann and Shull, 1991; Maeda et al., 1991), mouse (Canfield and Levenson, 1991; Morley et al., 1992), and human (Ma et al., 1991) enzyme. The hydropathy profile of the β-subunit appears less ambiguous than the α-subunit. There is one membrane spanning region predicted by the hydropathy analysis, located at the region between positions 38 and 63 near the N-terminus. Tryptic digestion of the intact gastric H^+/K^+-ATPase produces no visible cleavage of the β-subunit on SDS gels. Wheat germ agglutinin (WGA)-binding of the β-subunit is retained. These data indicate that most of the β-subunit is extracytoplasmic and glycosylated. When lyophilized hog vesicles are cleaved by trypsin followed by reduction, a small, non-glycosylated peptide fragment is seen on SDS gels with the N-terminal sequence AQPHYS which represents the C-terminal region beginning at position 236 (Mercier et al., 1993). This small fragment is not found either after trypsinolysis of intact vesicles nor in the absence of reducing agents. A disulfide bridge must, therefore, connect this cleaved fragment to the β-subunit containing the carbohydrates. The C-terminal end of the disulfide is at position 262. This leaves little room for an additional membrane spanning α-helix. Hence, it is likely that the β-subunit has only one membrane spanning segment.

D. Region of Association of the α and β Subunits

The β-subunit of both the Na^+/K^+-ATPase and the H^+/K^+-ATPases is necessary for targeting the complex from the endoplasmic reticulum to the plasma membrane (Renaud et al., 1991; Jaunin et al., 1992). It also stabilizes a functional form of both the gastric H^+/K^+-ATPase and Na^+/K^+-ATPase (Ackermann and Geering, 1990; Geering, 1991).

In the case of the Na^+/K^+-ATPase, the last 161 amino acids of the α-subunit are essential for effective association with the β-subunit (Lemas et al., 1992). Further, the last four or five C-terminal hydrophobic amino acids of the Na^+/K^+-ATPase β-subunit are essential for interaction with the α-subunit, whereas the last few hydrophilic amino acids are not (Geering, K., personal communication). Expression of the Na^+/K^+-ATPase α-subunit along with the β-subunit of either the Na^+/K^+-ATPase or H^+/K^+-ATPase in *Xenopus* oocytes has shown that the β-subunit of the gastric H^+/K^+-ATPase can act as a surrogate for the β-subunit of the Na^+/K^+-ATPase as far as membrane targeting and $^{86}Rb^+$ uptake is concerned, suggesting some homology in the associative domains of the β-subunits of the two pumps (Horisberger et al., 1991a). The H^+/K^+-ATPase α-subunit requires its β-subunit for efficient cell-surface expression and the C-terminal half of the α-subunit was shown to assemble with the β-subunit (Gottard and Caplan, 1993).

In order to specify the region of the α-subunit associated with the β-subunit, the tryptic digest was solubilized using non-ionic detergents such as NP40 or $C_{12}E_8$. These detergents allow the holoenzyme to retain ATPase activity. The soluble enzyme was then adsorbed to a WGA affinity column. Following elution of peptides not associated with the β-subunit binding to the WGA column, elution of the β-subunit with 0.1 M acetic acid also eluted almost quantitatively the M7/loop/M8 sector of the α-subunit. These data show that this region of the α-subunit is tightly associated with the β-subunit such that non-ionic detergents are unable to dissociate it from the β-subunit (Shin and Sachs, 1994).

If tryptic digestion is carried out in the presence of K^+, a fragment of 19–21 kD is produced which contains the M7 segment and continues to the C-terminal region of the enzyme. When this digest is solubilized and passed over the WGA column as outlined above, in addition to the 19–21 kD fragment, a fragment representing the M5/loop/M6 sector is now also retained by the β-subunit. Hence, provided there is no hydrolysis between M8 and H9, an additional interaction is present between the α and β subunits. The monoclonal antibody 146-14 also recognizes the region of the α-subunit at the extracytoplasmic face of the M7 segment as well as the β-subunit, a finding consistent with the association found by column chromatography.

E. Two-dimensional Structure

Reconstruction of two-dimensional crystals of the Ca^{2+}-ATPase showed that this P-type ATPase consisted of three distinct segments fitting into a box of 120 Å (height) by 50 Å (perpendicular to the dimer ribbons) by 85 Å (along the dimer ribbons; see also Chapter 1). The enzyme has a highly asymmetric mass distribution across the lipid bilayer. The cytoplasmic region comprises about 70% of the total mass, whereas the luminal region has only about 5%. The cytoplasmic domain was shown to have a complex structure similar to the shape of the head and neck of a bird. The "head" is responsible for forming dimer ribbons and contains the ATP-binding domain. The "neck" represents the stalk domain, is about 25 Å long, and consists of two segments. The transmembrane part consists of three segments defined as A, B, and C segments: the largest segment (A) consists of two parts, one vertical (A2) and the other inclined (A1); the B segment is also tilted and connected to A2 and to a third membrane segment (C; Toyoshima et al., 1993).

The structure of the two-dimensional crystals of the H^+/K^+-ATPase formed in an imidazole buffer containing VO_3^- and Mg^{2+} ions was resolved at about 25 Å (Rabon et al., 1986; Mohraz et al., 1990; Herbert et al., 1992). The average cell edge of the H^+/K^+-ATPase was 115 Å, containing four asymmetric protein units of 50 Å × 30 Å (Herbert et al., 1992), whereas the unit cell dimension of the $Co(NH_3)_4$ ATP-induced crystals of the Na^+/K^+-ATPase was 141 Å (Skriver et al., 1989, 1992). This suggests a more compact packing of the H^+/K^+-ATPase than of the Ca^{2+}-ATPase or the Na^+/K^+-ATPase. Recently, two-dimensional crystals of the Na+/K^+-ATPase were reported to be best formed at pH 4.8 in sodium citrate buffer and to represent an unique lattice (a = 108.7, b= 66.2, r = 104.2 Å) by electron cryomicroscopy. There are two high contrast parts in one unit cell (Tahara et al., 1993). The trypsinized Na^+/K^+-ATPase membranes were analyzed by electron microscopy (Ning et al., 1993). Both surface particles observed by negative staining and the protruding cytoplasmic portion of the α-subunit were removed, but general membrane structure was preserved. Intramembranal particles defined by freeze–fracture were preserved after trypsinolysis, showing that the remaining membrane protein fragments retained their native structure within the lipid bilayer after proteolysis.

F. A Model of the H^+/K^+-ATPase

The previously-mentioned data show the presence of probably 10 membrane spanning segments of the H^+/K^+-ATPase, consistent with those found for the Mg^{2+}-ATPase of *Salmonella typhimurium* using a fusion protein approach (Smith et al., 1993). Furthermore, the boundary amino acids are reasonably well defined by a combination of these techniques with alignment with other P-type ATPases and molecular modeling. A model is presented in Figure 4. Here, both subunits are represented. The α-subunit is shown with 10 membrane spanning segments, the β-subunit with a single membrane spanning segment. The largest extracytoplasmic loop of the α-subunit is between membrane segments M7 and M8.

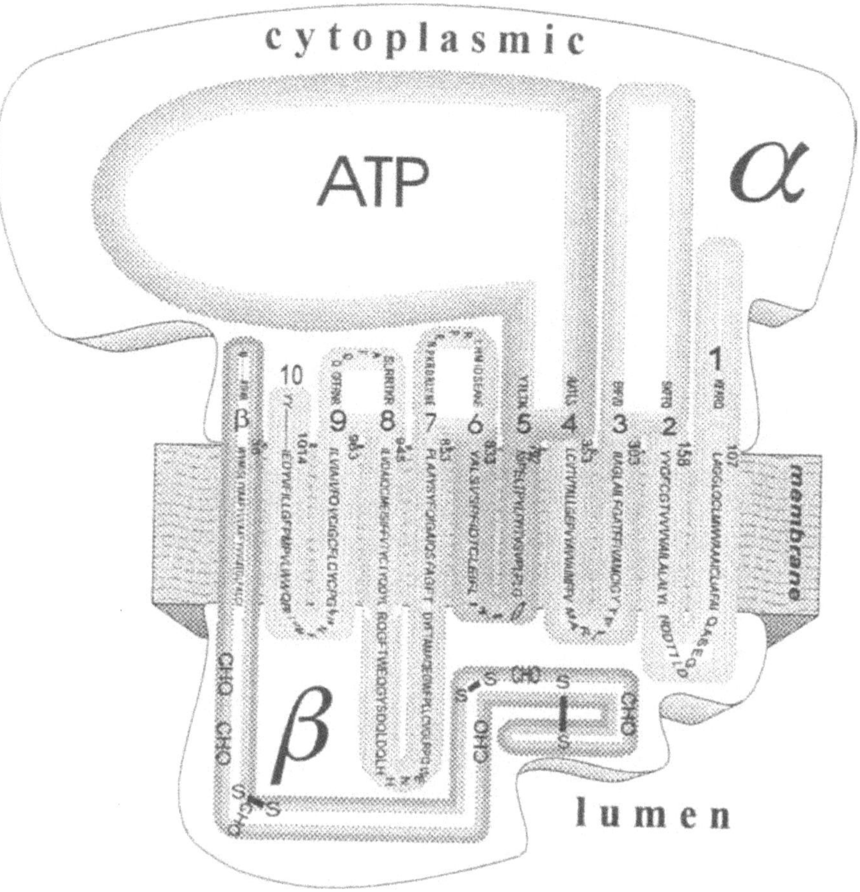

Figure 4. A two-dimensional representation of the H^+/K^+-ATPase heterodimer. The α-subunit is shown as a 10-membrane spanning segment model, based on biochemical and *in vitro* translation methods. The β-subunit is shown with a single membrane spanning segment.

IV. CONFORMATIONS OF THE H^+/K^+-ATPase

The H^+/K^+-ATPase generates HCl with an H^+ gradient of 4.10^6 and a K^+ gradient of greater than 10-fold. This transport is achieved by the electroneutral exchange of H^+ for K^+ which is dependent on conformation changes in the protein. The alteration of enzyme conformation changes the affinity and sidedness of the ion-binding sites during the cycle of phosphorylation and dephosphorylation.

The ions transported from the cytoplasmic side are H^+ or Na^+ at high pH. Since Na^+ is transported as a surrogate for H^+, it is likely that the hydronium ion, rather than the proton *per se* is the species transported. The ions transported inwards from the outside face of the pump are Tl^+, K^+, Rb^+, or NH_4^+ (Wallmark et al., 1980; Lorentzon et al., 1988). Presumably, the change in conformation changes a relatively small ion-binding domain in the outward direction into a larger ion-binding domain in the inward direction.

The E1 conformation of the H^+/K^+-ATPase binds the hydronium ion from the cytoplasmic side at high affinity. With phosphorylation, the conformation changes from the E1P–H_3O^+ to the E2P–H_3O^+ form, which has high affinity for K^+ and low affinity for H_3O^+ allowing release of H_3O^+ and binding of K^+ ion the extracytoplasmic surface of the enzyme. Breakdown of the E2P form requires K^+ or its congeners on the outside face of the enzyme. With dephosphorylation, the E1K^+ conformation is produced with a low affinity for K^+, releasing K^+ to the cytoplasmic side, allowing rebinding of H_3O^+ (Wallmark et al., 1980).

The mechanism of phosphorylation of the H^+/K^+-ATPase was studied by measuring the inorganic phosphate $P^{18}O_j^{16}O_{4-j}$ distribution as a function of time at different H^+, K^+, and $[^{18}O]P_i$ concentrations (Faller and Diaz, 1989). The formation of the E–P_i complex that exchanges ^{18}O with H_2O was slower at pH 5.5 than at pH 8 and is not diffusion controlled, suggesting unimolecular chemical transformation involving an additional intermediate in the phosphorylation mechanism such as, perhaps, a protein conformational change. From competitive binding between ATP and 2′,3′,-O-(2,4,6-trinitrophenylcyclohexadienylidine)adenosine 5′-phosphate (TNP–ATP), two classes of nucleotide-binding sites were suggested (Faller, 1989). TNP–ATP is not a substrate for the H^+/K^+-ATPase. However, TNP–ATP prevents phosphorylation by ATP and inhibits the K^+-stimulated pNPPase and ATPase activities. The number of TNP–ATP binding sites was twice the stoichiometry of phosphoenzyme formation.

Fluorescein isothiocyanate (FITC) binds to the H^+/K^+-ATPase at pH 9.0, inhibiting ATPase activity, but not pNPPase activity (Jackson et al., 1983). Fluorescence of the FITC-labeled enzyme, representing the E1 conformation, was quenched by K^+, Rb^+, and Tl^+ (Jackson et al., 1983; Markus et al., 1989). The quenching of the fluorescence by KCl reflects the formation of E2K^+. FITC binds at Lys-516 in the hog enzyme sequence (Farley and Faller, 1985). This FITC-binding site apparently becomes less hydrophobic when KCl binds to form the E2K^+ conformation.

The FITC labeled Na^+/K^+-ATPase has quite similar properties (Karlish, 1980; Farley et al., 1984; Smirnova and Faller, 1993). Two K^+ ions are required to cause the conformational change from E1 to E2 (Smirnova and Faller, 1993). The binding site of FITC was at Lys-501. However, several additional lysines at positions 480 and 766 were shown to react with FITC during inactivation of the Na^+/K^+-ATPase. These lysines were also protected from labeling in the presence of ATP (Xu, 1989).

A fluorescent compound, 1-(2-methylphenyl)-4-methylamino-6-methyl-2,3-dihydropyrrolo[3,2-c]quinoline (MDPQ) was shown to inhibit the H^+/K^+-ATPase and the K^+ phosphatase competitively with K^+ (Rabon et al., 1991). MDPQ fluorescence is quenched by the imidazopyridine, SCH 28080. The imidazopyridine Me-DAZIP binds to the M1/loop/M2 sector of the α-subunit (Munson et al., 1991). MDPQ binding to the extracytoplasmic surface of the pump enhances its fluorescence, suggesting that inhibitor binding occurs to a relatively hydrophobic region of the protein. The fluorescence was quenched by K^+, independently of Mg^{2+}. The binding of MgATP increased the fluorescence due to the formation of an E2P–[I] complex (Rabon et al., 1991).

The fluorescence changes with FITC and MDPQ may reflect relative motion of the cytoplasmic domain and the connecting loop between M1 and M2 with respect to the hydrophobic domain of the enzyme. In the E1 form, the FITC region is relatively closer to the membrane and the extracytoplasmic loop relatively hydrophilic. With the formation of the E2K form, the FITC region is more distant from the membrane, whereas with formation of the E2P form, the MDPQ-binding region between M1 and M2 moves toward the membrane. These postulated conformational changes are, therefore, reciprocal in the two major conformers of the enzyme.

Two irreversible inhibitors that form cysteine reactive sulfenamides in the acid space generated by the pump, omeprazole, and E3810, appear to inhibit the enzyme in different conformations (Morii and Takeguchi, 1993). Both omeprazole and EW3810, 1-{[4-(3-methoxypropoxy)-3-methylpyridin-2-yl]methylsulfinyl}-1H-benzimidazole, are acid activated at the luminal surface to form active sulfenamide derivatives, which can bind cysteines within the H^+/K^+-ATPase.

The omeprazole bound enzyme has a lower FITC fluorescence, perhaps due to a E2-like conformation. Both ATPase activity and steady-state phosphorylation was inhibited. The E3810 bound enzyme showed a high FITC fluorescence more like a E1 conformation. The fluorescence of the E3810 bound enzyme was quenched by K^+ in contrast to the omeprazole derivatized FITC labeled enzyme. It is not known whether these effects are due to differences in structure of the inhibitors or to differences in location of binding site or both.

Sodium ion is able to substitute for protons in the H^+/K^+-ATPase. Na^+ influx was observed in cytoplasmic side out vesicles at pH 8.5 (Polvani et al., 1989). This influx was inhibited by SCH 28080, required K^+, and was consequently due to the ability of Na^+ to act as a surrogate for H^+. When the fluorescence of FITC-labeled H^+/K^+-ATPase was quenched by K^+, Na^+ ions reversed the K^+-induced quench of the fluorescence (Rabon et al., 1990b). This demonstrates that Na^+ ions stabilize the E1 form by forming the E1–Na^+ conformation and K^+ ion stabilizes the E2 form to form the E2–K^+ conformation. With these ions, it was possible to show that the rate of the E2 to E1 conformation change in the H^+, K^+-ATPase was much faster than the equivalent step of the Na^+/K^+-ATPase, which accounted for the difficulty of demonstrating Rb^+ occlusion in the H^+/K^+-ATPase as compared to the Na^+/K^+-ATPase (Rabon et al., 1990b, 1993).

The effect of trypsin on the gastric H^+/K^+-ATPase provides evidence for conformational changes as a function of ligand binding (Saccomani et al., 1979; Helmich-de Jong et al., 1987). When essentially complete ATPase inhibition was observed, 34% of the α-subunit remained in the membrane. The tryptic digestion of the

H^+/K^+-ATPase in the presence of ATP revealed the appearance of a 78 kD and a 30 kD fragment, while the digestion in the absence of ATP produced a 87 kD and 47 kD peptide fragments. The difference of tryptic pattern can be interpreted in terms of two forms of phosphorylated protein (Saccomani et al., 1979). The conformation of the H^+/K^+-ATPase was carefully studied by limited proteolytic digestion (Helmich-de Jong, 1987). The α-subunit in the presence of K^+ (the E2–K^+ conformation) was cleaved into two fragments of 56- and 42-kD. In the presence of ATP (representing the E1–P conformation), tryptic cleavage produced two fragments of 67- and 35-kD. The 42- and 67-kD fragments were phosphorylated (Fellenius et al., 1981). Only K^+ of the ionic ligands provided significant protection against tryptic hydrolysis. Neither ATP nor ADP affected the tryptic pattern at high trypsin/protein ratios (Shun, J.M., unpublished observations).

However, only two large fragments of 67- and 33-kD were found in the presence of ATP, Mg^{2+}, and SCH 28080 as a K^+ surrogate; several fragments were produced in the absence of ligands. These data suggest that the E2–K^+ or more particularly the E2P–[SCH] conformation of the H^+/K^+-ATPase severely limits accessibility of trypsin to most of the lysines and arginines in the β-subunit (Munson et al., 1991; Besancon et al., 1993).

Extensive tryptic digestion of the gastric H^+/K^+-ATPase in the presence of KCl provided a C-terminal peptide fragment of 20 kD beginning at the M7 transmembrane segment, a peptide of 9.4 kD comprising the M1/loop/M2 sector beginning at Asp-84, and another peptide of 9.4 kD containing the M5/loop/M6 sector beginning at Asn-753 (Shin and Sachs, 1994). The C-terminal 20 kD peptide fragment was suggested to be capable of Rb^+ occlusion (Rabon et al., 1993).

In the case of the Na^+/K^+-ATPase, cation occlusion from the cytoplasmic surface was suggested to occur in two steps (Or et al., 1993). In an initial recognition step, transported cations interact with carboxyl groups. The second step is selective for transported cations and involves occlusion of cations which involves a conformational change forming a more compact structure.

Extensive trypsinolysis of the H^+/K^+-ATPase results in peptide fragments of 11 kD, 7.5 kD, and 6.5 kD, representing four pairs of membrane spanning segments corresponding to the M1/loop/M2, M3/loop/M4/, and M5/loop/M6 with M7/loop/M8 fragments. The band at 11 kD contains a single peptide beginning at Leu-853 which is derived from the M7/loop/M8 region. The band at 7.5 kD also contains a single peptide beginning at Thr-291 which is derived from the M3/loop/M4 region. The band at 6.5 kD consists of two peptides. One is a peptide beginning at Gln-104, derived from the M1/loop/M2 segments and the other is a peptide beginning at Leu-776, hence containing the M5/loop/M6 regions (Besancon et al., 1993; Shin et al., 1993).

When these digests are compared in the absence and presence of KCl, some regions near the membrane can be seen to be K^+ protected. The region between Gly-93 and Glu-104 near the M1 segment, the region between Asn-753 and Leu-776 near the cytoplasmic side of the M5 segment, and the region after the M8 segment, especially the region between Ile-945 and Ile-963 containing five arginines and one lysine, are protected from the trypsin digestion in the E2–K^+ conformation (Shin and Sachs, 1994). Further, there must be protection prior to M3 and for some distance subsequent to M4, since no fragment containing these segments was found at a molecular weight of less than 20 kD.

V. INHIBITORS OF THE H^+/K^+-ATPASE

The H^+/K^+-ATPase in the parietal cell secrets acid into the secretory canaliculus generating a pH of < 1.0 in the lumen of this structure. The acidity of this space is more than 1,000-fold greater than anywhere else in the body, and allows accumulation of weak bases. Weak bases of a pK_a less than 4.0 would be selectively accumulated in this acidic space.

Since the extracytoplasmic domain of the H^+/K^+-ATPase is composed of relatively few amino acids present in a highly acidic compartment, weak bases target to this domain. These bases could, for example, act as competitive or noncompetitive inhibitors of the enzyme or, following chemical conversion, as covalent inhibitors of enzyme function.

Since a substituted benzimidazole was first reported to inhibit the H^+/K^+-ATPase (Fellenius et al., 1981), many inhibitors of the H^+/K^+-ATPase have been synthesized. Most inhibitors can be classified into two groups: reversible and covalent. The covalent inhibitors all belong to the substituted benzimidazole family (Nagaya et al., 1989; Fujisaki et al., 1991; Sih et al., 1991; Beil et al., 1992; Kohl et al., 1992; Weidmann et al., 1992; Arakawa et al., 1993). Reversible inhibitors contain nitrogens that can be protonated and have a variety of structures. One type is represented by the imidazo-pyridine derivatives (Kaminski et al., 1991), others are piperidinopyridines (Hoiki et al., 1990), substituted 4-phenylaminoquinolines (Ife et al.. 1992), pyrrolo[3.2-c]quinolines (Leach et al.. 1992), guanidino-thiazoles (LaMattina et al., 1990), and scopadulcic acid (Asano et al., 1990). Some natural products, such as cassigarol A (Murakami et al., 1992) and naphthoquinone (Danzig et al., 1991), also showed inhibitory activity. Many of these reversible inhibitors show K^+-competitive characteristics, in contrast to the benzimidazole type.

A. Substituted Benzimidazoles

The first compound of this class with inhibitory activity on the enzyme and on acid secretion was the 2-(pyridylmethyl)sulfinylbenzimidazole, timoprazole (Fellenius et al., 1981), and the first pump inhibitor used clinically was 2-[[3,5-dimethyl-4-methoxypyridin-2-yl]methylsulfinyl]-5-methoxy-1H-benzimidazole, omeprazole (Wallmark et al., 1984).

Omeprazole is an acid-activated prodrug (Lindberg et al., 1986). Omeprazole can be accumulated in the acidic space and easily converted to a reactive cationic sulfenamide species, which binds to SH groups as shown in Figure 5 (Wallmark et

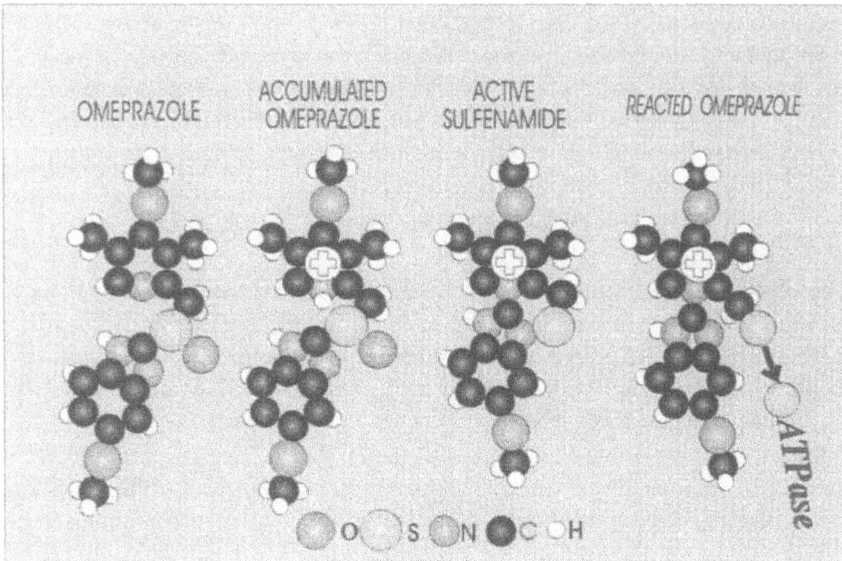

Figure 5. The mechanism of action of the benzimidazole, omeprazole, showing protonation and hence accumulation in acidic spaces with a pH < 4.0 (i.e., the pK_a of omeprazole). This is followed by an acid-catalyzed conversion to a tetracyclic cationic sulfenamide which then reacts with some of the extracytoplasmic cysteines of the α-subunit of the H^+/K^+-ATPase.

al., 1984; Im et al., 1985; Keeling et al., 1985, 1987; Lorrentzon et al., 1985, 1987; Lindberg et al., 1986). Omeprazole has a stoichiometry of 2 moles inhibitor per mole phosphoenzyme under acid-transporting conditions and bound only to the α-subunit (Keeling et al., 1987; Lorentzon et al., 1987).

Substituted benzimidazole inhibitors show slightly different effects depending on the inhibitor structure (Morii and Takeguchi, 1993). The omeprazole-bound enzyme is in the E2 form. Another inhibitor, E3810, 2-[[4-(3-methoxypropoxy)-3-methylpyridin-2-yl]methylsulfinyl]-1H-benzimidazole, produced the E1 form of the enzyme after binding. It is claimed that the K^+-dependent dephosphorylation from the phosphoenzyme was inhibited in the E3810-bound enzyme, but not in the omeprazole-bound enzyme, whereas phosphoenzyme formation in the absence of K^+ was inhibited in both the E3810- and omeprazole-bound enzymes (Morii and Takeguchi, 1993).

These inhibitors also have different binding sites. Omeprazole binds to cysteines in the extracytoplasmic regions of M5/M6 (Cys-813 or Cys-822) and M7/M8 (Cys-892) (Besancon et al., 1993). Pantoprazole binds only to the cysteines in M5/M6 (Shin et al., 1993) and lansoprazole binds to cysteines in M3/M4 (Cys-321), M5/M6, and M7/M8 (Sachs et al., 1993). These data suggest that, of the 28 cysteines in the α-subunit, only the cysteines present in the M5/M6 region are important for inhibition of acid secretion by the substituted benzimidazoles.

B. Substituted Imidazo[1,2a]pyridines

Imidazo[1,2a]pyridine derivatives were shown to inhibit gastric secretion (Keeling et al., 1989). SCH 28080, 3-cyanomethyl-2-methyl-8-(phenylmethoxy)imidazo[1,2a]pyridine, inhibited the H^+/K^+-ATPase competitively with K^+ (Wallmark et al., 1987). SCH 28080 binds to free enzyme extracytoplasmically in the absence of substrate to form E2(SCH 28080) complexes. SCH 28080 inhibits ATPase activity with high affinity in the absence of K^+. SCH 28080 has no effect on spontaneous dephosphorylation, but inhibits K^+-stimulated dephosphorylation, presumably by forming a E2–P[I] complex. Hence, SCH 28080 inhibits K^+-stimulated ATPase activity by competing with K^+ for binding E2P (Mendlein and Sachs, 1990). Steady-state phosphorylation is also reduced by SCH 28080, showing that this compound also binds to the free enzyme. Using a photoaffinity reagent, 8-[4-azidophenyl)methoxy]-1-tritiomethyl-2,3-dimethylimidazo[1,2a]pyridinium iodide, the binding site of this class of K^+-competitive inhibitor was identified to be in or close to the loop between the M1 and M2 segments (Munson et al., 1991).

VI. KINETICS OF THE H^+/K^+-ATPASE

The H^+/K^+-ATPase exchanges intracellular hydrogen ions for extracellular potassium ions. The H^+ for K^+ stoichiometry of the H^+/K^+-ATPase was reported to be one (Reenstra and Forte, 1981; Smith and Scholes, 1982; Mardh and Norberg, 1992) or two (Rabon et al., 1982; Skrabanja et al., 1987) per ATP hydrolyzed. The H^+/ATP ratio was independent of external KCl and ATP concentration (Rabon et al., 1982). Recently, a continuous flow method for determining the stoichiometry was developed, but delayed measurement of stoichiometry obtaining a 1:1:1 ratio of ATP and ions (Mardh and Norberg, 1992). If care is taken to measure initial rates in tight vesicles the ratio is 1 ATP:2 H^+:2 K^+. Since at full pH gradient the stoichiometry must fall to 1 ATP:1 H^+:1 K^+, this pump displays a variable stoichiometry.

Kinetic studies on the H^+/K^+-ATPase have defined some reaction steps (Wallmark et al., 1980). The rate of formation of the phosphoenzyme and the K^+-dependent rate of breakdown are sufficiently fast to allow the phosphoenzyme to be an intermediate in the overall ATPase reaction. The initial step is the reversible binding of ATP to the enzyme in the absence of added K^+ ion, followed by a Mg^{2+} (and proton)-dependent transfer of the terminal phosphate of ATP to the catalytic subunit (E1P–H). The Mg^{2+} remains occluded until dephosphorylation (Rabon et al., 1991). The addition of K^+ to the enzyme-bound acyl phosphate results in a two-step dephosphorylation. The faster initial step is dependent on the concentration of K^+, whereas the slower step is not affected by K^+

concentration. This showed the biphasic effect of K^+ on overall ATPase activity. The second phase of EP breakdown is accelerated in the presence of K^+ but, at K^+ concentrations exceeding 500 μM, the ratio becomes independent of K^+ concentration. This shows that two forms of EP exist. The first form, E1P, is K^+-insensitive and converts spontaneously in the rate-limiting step to E2P, the K^+-sensitive form. ATP binding to the H^+/K^+-ATPase occurs in both the E1 and E2 state, but with a lower affinity in the E2 state (2,000 times lower compared to E1; Brzezinski et al., 1988).

H^+ or K^+ interact competitively on the cytoplasmic surface of intact vesicles. The effects of H^+ and K^+ on formation and breakdown of phosphoenzyme were determined using transient kinetics (Stewart et al., 1981). Increasing hydrogen ion concentrations on the ATP-binding face of the vesicles accelerates phosphorylation, whereas increasing potassium ion concentrations inhibits phosphorylation. Increasing hydrogen ion concentration reduces this K^+ inhibition of the phosphorylation rate. Decreasing hydrogen ion concentration accelerates dephosphorylation in the absence of K^+, and K^+ on the luminal surface accelerates dephosphorylation. Increasing K^+ concentrations at constant ATP decreases the rate of phosphorylation and increasing ATP concentrations at constant K^+ concentration accelerates ATPase activity and increases the steady-state phosphoenzyme level (Lorentzon et al., 1988). Therefore, inhibition by cations is due to cation stabilization of a dephospho-form at a cytosolically accessible cation-binding site.

In order to determine the role of divalent cations in the reaction mechanism of the H^+/K^+-ATPase, calcium was substituted for magnesium, which is necessary for phosphorylation (Mendlein and Sachs, 1989). Calcium ion inhibits K^+ stimulation of the H^+/K^+-ATPase by binding at a cytoplasmic divalent cation site. The Ca–EP dephosphorylates 10–20 times more slowly than the Mg–EP in the presence of 10 mM KCl with either 8 mM CDTA or 1 mM ATP. The inability of the Ca–EP to dephosphorylate in the presence of K^+, compared to the Mg–EP, demonstrates that the type of divalent cation which occupies the catalytic divalent cation site required for phosphorylation is important for the conformational transition to a K^+-sensitive phosphoenzyme. Calcium is tightly bound to the divalent cation site of the phosphoenzyme and the occupation of this site by calcium causes slower phosphoenzyme kinetics. Since the presence of CDTA or EGTA does not change the dephosphorylation kinetics of the EP–Ca form of the enzyme, it is concluded that the divalent cation remains occluded in the enzyme until dephosphorylation occurs.

VII. RELATION WITH OTHER P-TYPE ENZYMES

One classification of P-type ATPases into five distinct types was based on ion specificity, biological occurrence, and sequence (Green, 1992). Alignments within and between classes show conserved regions of phosphorylation, nucleotide binding domains, and hinge regions as well as remarkable conservation of hydropathy profiles. However, only a few amino acids are conserved within the membrane domains themselves.

The topology of the Mg^{2+} transport ATPase encoded by the MgtB locus of *Salmonella typhimurium* was particularly suited for applying molecular biological methods (Smith et al., 1993). The membrane topology of MgtB was analyzed by measuring the activity of 35 sites of fusion between MgtB and the reporter enzymes BlaM and LacZ. The lactamase confers penicillin resistance when extracytoplasmic and the LacZ is functional only when cytoplasmic. The fusion protein reporter data showed that MgtB contains 10 transmembrane segments. An exception was when the BlaM protein was fused to Pro-766. This clone was ampicillin resistant although the 10 segment model predicts a cytoplasmic location for this site.

In the case of the sarcoplasmic reticulum (SR) Ca^{2+}-ATPase, an antipeptide polyclonal antibody against the region between 877 and 888 of SR Ca^{2+}-ATPase (Matthews et al., 1990) and a monoclonal antibody A20 against the region between 870–890 (Clarke et al., 1990b) bound to SR Ca^{2+}-ATPase after solubilization, but did not bind to intact vesicles, suggesting that this epitope region is located in the lumen between M7 and M8. This data is consistent with a 10-membrane segment model.

In the case of the Na^+/K^+-ATPase, from the trypsinolysis of the right side-out Na^+/K^+-ATPase, Asn-831 was identified as being located at the cytoplasmic surface, and this residue is present at the N-terminal end of the M7 segment (Karlish et al., 1993). This result also provides strong support for the 10 transmembrane segment model.

When the cDNA encoding the α-subunit of the Na^+/K^+-ATPase was linked to the β-subunit via a linker of 17 amino acids, it was expressed in a variety of mammalian cell lines as a single protein located primarily on the surface membrane. The β-subunit domain was exposed to the extracellular medium. The α–β fused protein functioned as a normal heterodimeric Na^+ pump. Hence, the C-terminal amino acid of the β-subunit and the N-terminal amino acid of the α-subunit are both cytoplasmic (Emerick and Fambrough, 1993).

As previously discussed, in the case of the H^+/K^+-ATPase, the first eight transmembrane segments were identified by biochemical methods. The likely existence of M9 and M10 was demonstrated by *in vitro* translation. Thus, to date, there is direct evidence for 10 membrane spanning segments in the Mg^{2+}-ATPase of *Salmonella* and for the α-subunit of the H^+/K^+-ATPase of gastric mucosa. A similar number of segments should also be present in the Na^+/K^+-ATPase and Ca^{2+}-ATPase.

The β-subunits of both the H^+/K^+-ATPase and Na^+/K^+-ATPase are glycoproteins (Reuben et al., 1990; Toh et al., 1990; Horisberger et al., 1991b). The β-subunits of both bind to WGA or tomato lectin (Callaghan et al., 1990; Okamoto et al., 1990; Treuheit et al., 1993). WGA-binding demonstrates the existence of N-acetylglucosamine and tomato lectin-binding shows the presence of polygalactosamine on the β-subunit. The carbohydrates of the β-subunit were identified in the case of the Na^+/K^+-ATPase. N-glycosylation sites are Asn at positions 157, 192, and 264. The glycans represent a sequence from the site of Asn as follows: two N-acetylglucosamine, mannose, N-acetylglucosamine, galactose, and additional glucosamine and galactose depending on the position (Treuheit et al., 1993). It appears that all potential glycosylation sites of the β-subunit of the H^+/K^+-ATPase are occupied (Reuben et al., 1990).

When expression of the cDNAs encoding β-subunits of the Na^+/K^+-ATPase that lack some or all of the cytoplasmic N-terminal domain, or that contain deletions in the transmembrane domain in cultured mouse L cells was analyzed, the β-subunit lacking the cytoplasmic domain can assemble with the α-subunit of the Na^+/K^+-ATPase, and the resulting hybrid sodium pumps are transported to the plasma membrane. All β-subunit mutants capable of membrane insertion were able to assemble with α-subunit (Renaud et al, 1991). This shows that the extracellular domain of the β-subunit is tightly bound to the α-subunit.

Chimeric cDNAs encoding regions of the Na^+/K^+-ATPase α-subunit and SR Ca^{2+}-ATPase were expressed with the avian Na^+/K^+-ATPase β-subunit cDNA in COS-1 cells to determine which region of the α-subunit is required for assembly with the β-subunit (Lemas et al., 1992). A chimera, in which 161 amino acids of the Na^+/K^+-ATPase C-terminus replaced the corresponding amino acids of SR Ca^{2+}-ATPase C-terminus, assembled with the β-subunit. These data suggest that the C-terminal 161 amino acid region of the Na^+/K^+-ATPase α-subunit is critical for subunit assembly with the β-subunit.

As discussed previously, trypsinolysis data and epitope mapping of the H^+/K^+-ATPase provided additional evidence for association of the β-subunit with the M7/M8 sector of the α-subunit (Shin and Sachs, 1994). Generally, similar results were obtained with tryptic digestion in the presence of Rb^+ of the Na^+/K^+-ATPase followed by WGA-affinity chromatography. The 10 kD-peptide fragment, containing the M5/loop/M6 sector, beginning at the N-terminal sequence QAADMI, was retained along with the 20 kD peptide fragment consisting of the C-terminus of the α-subunit beginning at M7 (Shin and Sachs, 1994; Shin, J.M., unpublished observations).

When trypsinolysis was carried out at pH 8.2 in the presence of RbCl, the Na^+/K^+-ATPase yielded a 10.8 kD-peptide fragment containing the M1/loop/M2 sector which begins at Asp-73, a 10 kD-peptide fragment containing the M5/loop/M6 sector which begins at Gln-742, and the C-terminal 19 kD-peptide fragment beginning at the M7 segment. A similar result was obtained from the trypsin digestion of the H^+/K^+-ATPase in the presence of KCl. The tryptic digest obtained from the trypsinolysis at pH 8.2 in the presence of KCl provided C-terminal 19–20 kD fragments and a 9.4 kD-peptide fragment which comprises two N-terminal sequences of peptides. One is a peptide beginning at Asp-84, which comprises the M1/loop/M2 sector, and the other is a peptide beginning at Asn-753, which comprises the M5/loop/M6 sector.

Tryptic digestion of the Na^+/K^+-ATPase in the presence of RbCl provided a stable 19 kD membrane fragment of the C-terminus beginning at the M7 sector, which can occlude cations (Karlish et al., 1990; Capasso et al., 1992). The C-terminal 20 kD peptide of the H^+/K^+-ATPase obtained from the trypsin digestion in the presence of KCl was also shown to be able occlude Rb^+ (Rabon et al., 1993). Site-directed mutagenesis of both glutamic acid residues 955 and 956 of the rat Na^+/K^+-ATPase α1-subunit gave a decrease of Na^+ affinity with a small effect on K^+ affinity, whereas single substitution at Glu-955 or Glu-956 had only slight effects on the cation stimulation of the Na^+/K^+-ATPase. DCCD binding to Glu-955 inhibited Rb^+ occlusion. These data are equivocal in determining the role of either Glu-955 and Glu-956 in cation binding (Goldshleger et al., 1992; van Huysse et al., 1993). These glutamic acid residues are predicted to be at the cytoplasmic end of M9.

When the tryptic fragments of both the hog gastric H^+/K^+-ATPase and the pig kidney Na^+/K^+-ATPase were compared, there is strong homology of the N-terminal sequences of the region located in the cytoplasmic domain before the transmembrane segments. For example, N-terminal sequences located in the cytoplasmic region near the M1 segment are DGPNALRPPRGTPEYVKFAR for the H^+/K^+-ATPase and DGPNALTPPPTTPEWVKFCR for the Na^+/K^+-ATPase. N-terminal sequences near the M5 segments show even higher homology between the Na^+/K^+-ATPase and the H^+/K^+-ATPase, namely NAADMILLDDNFASIVTGVEQGR-LIFD for the H^+/K^+-ATPase and QAADMILLDDNFASIVTGVEEGRLIFD for the Na^+/K^+-ATPase. These regions may be involved in cation transport into the membrane domain.

Trypsinolysis of intact SR vesicles from fast-twitch rabbit muscle also showed a similar pattern to that found for the H^+/K^+-ATPase. It was possible to identify regions corresponding to the first eight-membrane segment pairs. M9 was also clearly identified, trypsinolysis having occurred between M9 and M10. On the assumption that there was partial access of trypsin to this site, this is also direct evidence for eight and perhaps 10-membrane spanning pairs for the Ca^{2+}-ATPase of SR (Shin, J.M., and Sachs, G., unpublished data).

Part of the cation binding sites in the luminal surface of both the H^+/K^+-ATPase and the Na^+/K^+-ATPase are likely to be in the connecting loop between M1 and M2. In the case of the Na^+/K^+-ATPase, the double mutant that contained Asp-111 and Arg-122 showed the most resistance against ouabain inhibition (partially K^+-competitive, suggesting that these positions of the M1–M2 extracellular domain are important for high affinity ouabain binding (Price and Lingrel, 1988; Price et al., 1990). In addition to the identified H1/H2 ectodomain of the Na^+/K^+-ATPase, the H3/H4 ectodomain also participates in the ouabain binding site since a mutation in Tyr-317 also confers ouabain resistance (Canessa et al., 1993).

Mutant cDNAs of Ca^{2+}-ATPase were transfected into COS-1 cells, and ATP-dependent Ca^{2+} transport or partial reactions of the expressed Ca^{2+}-ATPase were measured (Clarke et al., 1990a). Possible Ca^{2+}-binding sites of the SR Ca^{2+}-ATPase were shown to be associated with Glu-309 in the M4 segment, Glu-771 in the M5 segment, Asn-796, Thr-799, and Asp-800 in the M6 segment, and Glu-908 in the M8 segment. Mutants that replaced Pro-312 with Ala or Gly were defective in the

E1P to E2P transition, suggesting that the M4 segment is important for conformational changes.

In HeLa cells expressing mRNA and protein of rat Na^+/K^+-ATPase α1, mutants that substituted both Asp-712 and Asp-716 to Asn cannot survive in 0.1 μM ouabain. The Na^+/K^+-ATPase mutants which represent Asn substitutions for Asp-712 or Asp-716 showed inhibition of ATP hydrolysis. However, only Asn substitution of Asp-716 inhibits phosphorylation (Lane et al., 1993). Probably, these Asp residues must be close to Asp-369 which is the phosphorylation site (Jorgensen and Andersen, 1988). The region of the SR Ca^{2+}-ATPase associated with ATP binding can be shown by the identification of the ATP-binding site, ADP-binding site, and phosphorylation site. Asp-351 is the phosphorylation site in this enzyme (Allen and Green, 1976). Lys-492 and Lys-684 reacted with the terminal part of an ATP analog, AP_3PL (Yamamoto et al., 1989). Lys-492 reacted with TNP-$8N_3$–ATP, a photoreactive ATP analog (McIntosh et al., 1992). Thr-532 and Thr-533 were labeled by 8-N^3-ADP (Lacapere et al., 1993). These data suggest that these amino acids of the Ca^{2+}-ATPase, namely Asp-351, Lys-492, Thr-532, Thr-533, and Lys-684, would be close to the phosphorylation site.

Illumination of sarcoplasmic reticulum vesicles by ultraviolet light in the presence of vanadate produced two photocleaved fragments. The cleavage patterns differed depending on the presence of calcium. The photocleavage in the absence of calcium ion occurs at the V-cleavage site near the phosphorylation site, Asp-351, producing fragments with molecular masses of 87 kD (V1) and 22 kD (V2). In the presence of calcium ions, the vanadate-catalyzed photocleavage occurs at the VC cleavage site near the FITC-binding site, Lys-515, producing fragments of 71 kD (VC1) and 38 kD (VC2). This suggests that, since vanadate is an analog of P_i, that perhaps there is a conformational change in the phosphorylation site in the presence of Ca^{2+} wherein it is brought closer to the FITC-binding region than in the absence of Ca^{2+}. Under similar conditions, the Na^+/K^+-ATPase was completely resistant to photocleavage, and the H^+/K^+-ATPase also did not provide any specific cleavage (Molnar et al., 1991).

VIII. ACID SECRETION AND THE ATPASE

The H^+/K^+-ATPase is present mainly in the gastric parietal cell. In the resting parietal cell, it is present in smooth surfaced cytoplasmic membrane tubules. Upon stimulation of acid secretion, the pump is found on the microvilli of the secretory canaliculus of the parietal cell. This morphological change results in a several-fold expansion of the canaliculus (Helander and Hirschowitz, 1972). In addition to this transition, there is activation of a K^+ and Cl^- conductance in the pump membrane which allows K^+ to access the extracytoplasmic face of the pump (Wolosin and Forte, 1983). This allows H^+ for K^+ exchange to be catalyzed by the ATPase (Sachs et al., 1976).

The covalent inhibitors of the H^+/K^+-ATPase that have been developed for the treatment of ulcer disease and esophagitis depend on the presence of acid secreted by the pump. They are also acid-activated pro-drugs that accumulate in the acid space of the parietal cell. Hence, their initial site of binding is only in the secretory canaliculus of the functioning parietal cell (Scott et al., 1993). These data show also that the pump present in the cytoplasmic tubules does not generate HCl.

IX. GENE EXPRESSION OF THE H^+/K^+-ATPASE

The upstream DNA sequence of the α-subunit contains both Ca^{2+} and cAMP-responsive elements in the case of the rat H^+/K^+-ATPase (Tamura et al., 1992). There are gastric nuclear proteins that bind selectively to a nucleotide sequence, GATACC, in this region of the gene (Maeda et al., 1991). These proteins have not been detected in other tissues.

Stimulation of acid secretion by histamine increases the level of mRNA for the α-subunit of the pump (Tari et al., 1991). Elevation of serum gastrin, which secondarily stimulates histamine release from the enterochromaffin-like cell in the vicinity of the parietal cell, also stimulates transiently the mRNA levels in the parietal cell (Tari et al., 1993). H2 receptor antagonists block the effect of serum gastrin elevation on mRNA levels (Tari et al., 1994). It seems, therefore, that activity of the H2 receptor on the parietal cell determines, in part, gene expression of the ATPase. It might be expected, therefore, that chronic stimulation of this receptor would up-regulate pump levels, whereas inhibition of the receptor would down-regulate levels of the ATPase.

However, chronic administration of these H2 receptor antagonists, such as famotidine, results in an increase in pump protein, whereas chronic administration of omeprazole (which must stimulate histamine release) reduces the level of pump protein in the rabbit (Crothers et al., 1993; Scott et al., 1994). Regulation of pump protein turnover downstream of gene expression must account for these observations.

X. THE H^+/K^+-ATPASE AND ACID-RELATED DISEASE

Secretion of acid by the gastric mucosa is required for the presence of duodenal and gastric ulcers and for the presence of esophageal reflux disease. It is, therefore, natural that medical treatment of these diseases has depended, in large measure, on reduction of the acidity of the stomach. The major change in medical treatment came with the introduction of H2 receptor antagonists (Black et al., 1972). These drugs have become the mainstay of treatment over the last twenty years. More recently, the substituted benzimidazoles, acting to inhibit the acid pump itself, have shown impressive clinical results in a series of comparative trials (Walan et al., 1989). These approaches to treatment have been a major medical success.

In the last decade, it has also been recognized that a second pathogenetic factor is essential in duodenal and gastric ulcer. This is infection of the gastric or duodenal mucosa by *H. pylori* (Marshall, 1983). Eradication of this organism prevents the recurrence of ulcers of these regions of the GI tract (Burget et al., 1990). The most likely contribution of this organism to duodenal and gastric ulcer is destruction of the tight junction. Thus, with an intact tight junction, even with acid present, there is clinically insignificant acid back diffusion. In the absence of acid, even with a disrupted tight junction, there is no acid to back diffuse and damage the epithelial cells. Both factors must be present for ulceration to occur. Future medical therapy of ulcer disease will be reduction of acid secretion and eradication of *H. pylori*. Prevention of ulcer disease will entail eradication of *H. pylori*.

ACKNOWLEDGMENTS

This work was supported by USVA-SMI and NIH grant R01 DK 40165 and R01 DK 41301.

REFERENCES

Ackermann, U., & Geering, K. (1990). Mutual dependence of Na^+,K^+-ATPase α- and β-subunits for correct posttranslational processing and intracellular transport. FEBS Lett. 269, 105–108.

Allen, G., & Green, N. M. (1976). A 31-residue tryptic peptide from the active site of the [Ca^{2+}]-transporting adenosine triphosphatase of rabbit sarcoplasmic reticulum. FEBS Lett. 63, 188–192.

Arakawa, T., Fukuda, T., Higuchi, K., Koike, K., Satoh, H., & Kobayashi, K. (1993). NC-1300, a proton pump inhibitor, requires gastric acid to exert cytoprotection in rat gastric mucosa. Jap. J. Pharmacol. 61, 299–302.

Asano, S., Mizutani, M., Hayashi, T., Morita, N., & Takeguchi, N. (1990). Reversible inhibitions of gastric H^+,K^+-ATPase by scopadulcic acid and diacetyl scoadol. New biochemical tools of H^+,K^+-ATPase. J. Biol. Chem. 265, 22167–22173.

Baldwin, J., Henderson, R., Beckman, E., & Zemlin, F. (1988). Images of purple membrane at 2.8 Å resolution obtained by cryo-electron microscopy. J. Mol. Biol. 202, 585–591.

Bamberg, K., & Sachs, G. (1993). Topology of the H^+,K^+-ATPase. J. Cell. Biochem. Suppl. 17C, H148.

Bamberg, K., Mercier, F., Reuben, M. A., Kobayashi, Y., Munson, K. B., & Sachs, G. (1992). cDNA cloning and membrane topology of the rabbit gastric H^+/K^+-ATPase α subunit. Biochim. Biophys. Acta 1131, 69–77.

Bayle, D., Robert, J. C., Bamberg, K., Benkouka, F., Cheret, A. M., Lewin, M. J. M., Sachs, G., & Soumarmon, A. (1992). Location of the cytoplasmic epitope for a K^+-competitive antibody of the H^+,K^+-ATPase. J. Biol. Chem. 267, 19060–19065.

Beil, W., Staar, U., & Sewing, K. F. (1992). Pantoprazole: A novel H^+/K^+-ATPase inhibitor with an improved pH stability. Eur. J. Pharmacol. 218, 265–271.

Besancon, M., Shin, J. M., Mercier, F., Munson, K., Miller, M., Hersey, S., & Sachs, G. (1993). Membrane topology and omeprazole labeling of the gastric H^+,K^+-adenosinetriphosphatase. Biochemistry 32, 2345–2355.

Black, J. W., Duncan, W. A. M., Durant, C. J., Ganellin, C. R., & Parsons, E. M. (1972). Definition and antagonism of histamine H2-receptors. Nature 236, 385.

Blostein, R., Zhang, R.-p., Gottardi, C. J., & Caplan, M. J. (1993). Functional properties of an H^+,K^+-ATPase/Na^+,K^+-ATPase chimera. J. Biol. Chem. 268, 10654–10658.

Brzezinski, P., Malmstrom, B. G., Lorentzon, P., & Wallmark, B. (1988). The catalytic mechanism of gastric H^+/K^+-ATPase: Simulations of pre-steady-state and steady-state kinetic results. Biochim. Biophys. Acta 942, 215–219.

Burget, D. W., Chiverton, S. G., & Hunt, R. H. (1990). Is there an optimal degree of acid suppression for healing of duodenal ulcers? Gastroenterology 99, 345–351.

Callaghan, J. M., Toh, B. H., Pettitt, J. M., Humphris, D. C., & Gleeson, P. A. (1990). Poly N-acetyllactosamine-specific tomato lectin interacts with gastric parietal cells. Identification of a tomato-lectin binding 60–90 KMr membrane glycoprotein of tubulovesicles. J. Cell Sci. 95, 563–575.

Callaghan, J. M., Toh, B. H., Simpson, R. J., Baldwin, G. S., & Gleeson, P. A. (1992). Rapid purification of the gastric H^+/K^+-ATPase complex by tomato-lectin affinity chromatography. Biochem. J. 283, 63–68.

Canessa, C. M., Horisberger, J.-D., & Rossier, B. C. (1993). Mutation of a tyrosine in the H3–H4 ectodomain of Na^+,K^+-ATPase α-subunit confers ouabain resistance. J. Biol. Chem. 268, 17722–17726.

Canfield, V. A., & Levenson, R. (1991). Structural organization and transcription of the mouse gastric H^+,K^+-ATPase β-subunit gene. Proc. Natl. Acad. Sci. USA 88, 8247–8251.

Canfield, V. A., Okamoto, C. T., Chow, D., Dorfman, J., Gros, P., Forte, J. G., & Levenson, R. (1990). Cloning of the H^+,K^+-ATPase β subunit. J. Biol. Chem. 265, 19878–19884.

Capasso, J. M., Hoving, S., Tal, D. M., Goldshleger, R., & Karlish, S. J. D. (1992). Extensive digestion of Na^+,K^+-ATPase by specific and nonspecific proteases with preservation of cation occlusion sites. J. Biol. Chem. 267, 1150–1158.

Carafoli, E. (1992). The Ca^{2+} pump of the plasma membrane. J. Biol. Chem. 267, 2115–2118.

Ceska, T. A., & Henderson, R. (1990). Analysis of high-resolution electron diffraction patterns from purple membrane labeled with heavy atoms. J. Mol. Biol. 213, 539–560.

Chow, D. C., Browning, C. M., & Forte, J. G. (1992). Gastric H^+,K^+-ATPase activity is inhibited by reduction of disulfide bonds in the β-subunit. Amer. J. Physiol. 263, C39–C46.

Clarke, D. M., Loo, T. W., & MacLennan, D. H. (1990a). Functional consequences of alterations to polar amino acids located in the transmembrane domain of the Ca^{2+}-ATPase of sarcoplasmic reticulum. J. Biol. Chem. 265, 6262–6267.

Clarke, D. M., Loo, T. W., & MacLennan, D. H. (1990b). The epitope for monoclonal antibody A20 (amino acids 870–890) is located on the luminal surface of the Ca^{2+}-ATPase of sarcoplasmic reticulum. J. Biol. Chem. 265, 17405–17408.

Crothers, J. M., Chow, D. C., & Forte, J. G. (1993). Omeprazole decreases H^+,K^+-ATPase protein and increases permeability of oxyntic secretory membrane in rabbits. Amer. J. Physiol. 265, G231–G241.

Crowson, M. S., & Shull, G. E. (1992). Isolation and characterization of a cDNA encoding the putative distal colon H^+,K^+-ATPase. Similarity of deduced amino acid sequence to gastric H^+,K^+-ATPase and Na^+,K^+-ATPase and mRNA expression in distal colon, kidney, and uterus. J. Biol. Chem. 267, 13740–13748.

Danzig, A. H., Minor, P. L., Garrigus, J. L., Fukuda, D. S., & Mynderse, J. S. (1991). Studies on the mechanism of action of A80915, a semi-naphthoquinone natural product, as an inhibitor of gastric H^+,K^+-ATPase. Biochem. Pharmacol. 42, 2019–2026.

Emerick, M., & Fambrough, D. M. (1993). Intramolecular fusion of Na pump subunits assures exclusive assembly of the fused α and β subunit domains into a functional enzyme in cells also expressing endogenous Na pump subunits. J. Biol. Chem. 268, 23455–23459.

Faller, L. D. (1989). Competitive binding of ATP and the fluorescent substrate analogue 2′,3′-O-(2,4,6-trinitrophenylcyclohexadienylidine)adenosine 5′-phosphate to the gastric H,K ATPase: Evidence for two classes of nucleotide sites. Biochemistry 28, 6771–6778.

Faller, L. D., & Diaz, R. A. (1989). Evidence from ^{18}O exchange measurements for steps involving a weak acid and a slow chemical transformation in the mechanism of phosphorylation of the gastric H^+,K^+-ATPase by inorganic phosphate. Biochemistry 28, 6908–6914.

Farley, R. A., & Faller, L. D. (1985). The amino acid sequence of an active peptide from the H^+,K^+-ATPase of gastric mucosa. J. Biol. Chem. 260, 3899–3901.

Farley, R. A., Tran, C. M., Carilli, C. T., Hawke, D., & Shively, J. E. (1984). The amino acid sequence of a fluorescein-labeled peptide from the active site of Na^+-K^+-ATPase. J. Biol. Chem. 259, 9532–9535.

Fellenius, E., Berglindh, T., Sachs. G., Olbe, L., Elander, B., Sjostrand. S-E., & Wallmark, B. (1981). Substituted benzimidazoles inhibit gastric acid secretion by blocking (H^+,K^+)-ATPase. Nature 290, 159–161.

Fujisaki, H., Shibata, H., Oketani, K., Murakami, M., Fujimoto, M., Wakabayashi, T., Yamatsu, I., Yamaguchi, M., Sakai, H., & Takeguchi, N. (1991). Inhibition of acid secretion by E3810 and omeprazole, and their reversal by glutathione. Biochem. Pharmacol. 42, 321–328.

Futai, M., Noumi., T., & Maeda, M. (1989). ATP synthase (H^+-ATPase): Results by combined biochemical and molecular biological approaches. Ann. Rev. Biochem. 58, 111–136.

Geering, K. (1991). The functional role of the β-subunit in the maturation and intracellular transport of Na^+,K^+-ATPase. FEBS Lett. 285, 189–193.

Glynn, I. M., & Karlish, S. J. D. (1990). Occluded cations in active transport. Ann. Rev. Biochem. 59, 171–205.

Goldshleger, R., Tal, D. M., Moorman, J., Stein, W. D., & Karlish, S. J. D. (1992). Chemical modification of Glu-953 of the α chain of Na^+,K^+-ATPase associated with inactivation of cation occlusion. Proc. Natl. Acad. Sci. USA 89, 6911–6915.

Gottard, C. J., & Caplan, M. J. (1993). Molecular requirements for the cell-surface expression of multisubunit ion-transporting ATPases. J. Biol. Chem. 268, 14342–14347.

Green, N. M. (1992). Evolutionary relationships within the family of P-type cation pumps. Ann. New York Acad. Sci. 671, 104–112.

Hall, K., Perez, G., Anderson, D., Gutierrez, C., Munson, K., Hersey, S. J., Kaplan, J. H., & Sachs, G. (1990). Location of the carbohydrates present in the H^+,K^+-ATPase vesicles isolated from hog gastric mucosa. Biochemistry 29, 701–706.

Hall, K., Perez, G., Sachs, G., & Rabon, E. (1991). Identification of H^+,K^+-ATPase α,β-heterodimers. Biochim. Biophys. Acta 1077, 173–179.

Helander, H. F., & Hirschowitz, B. I. (1972). Quantitative ultrastructural studies on gastric parietal cells. Gastroenterology 63, 951.

Helmich-de Jong, M. L., Vries, van E-D., & de Pont, J. J. H. H. M. (1987). Conformational states of (H^++K^+-ATPase studied using tryptic digestion as a tool. Biochim. Biophys. Acta 905. 358–370.

Herbert, H., Xian, Y., Hacksell, I., & Mardh, S. (1992). Two-dimensional crystals of membrane-bound gastric H^+,K^+-ATPase. FEBS Lett. 299, 159–162.

Hermoulin, J., & Fillingame, R. H. (1989). H^+-ATPase activity of *Escherichia coli* F_1F_0 is blocked after reaction with dicyclohexylcarbodiimide with a single proteolipid (subunit c) of the F_0 complex. J. Biol. Chem. 264, 3896–3903.

Hoiki, Y., Takada, J., Hidaka, Y., Takeshita, H., Hosoi, M., & Yano. M. (1990). A newly synthesized pyridine derivative, (Z)-5-methyl-2-[2-(1-naphtyl)ethenyl]-4-piperidinopyridine hydrochloride (AU-1421), as a reversible gastric pump inhibitor. Arch. Intern. Pharmacol. Therap. 305, 32–44.

Horisberger, J-D., Jaunin, P., Reuben, M. A., Lasater, L. S., Chow, D. C., Forte, J. G., Sachs, G., Rossier, B. C., & Geering, K. (1991a). The H^+,K^+-ATPase β-subunit can act as a surrogate for the β-subunit of Na^+,K^+-pumps. J. Biol. Chem. 266, 19131–19134.

Horisberger, J-D., Lemas, V., Krachenbuhl, J-P., & Rossier, B. C. (1991b). Structure-function relationship of Na^+,K^+-ATPase. Ann. Rev. Physiol. 53, 565–584.

Ife, R. J., Brown, T. H., Keeeling, D. J., Leach, C. A., Meeson, M. L., Parsons, M. E., Reavill, D. R., Theobald, C. J., & Wiggall, K. J. (1992). Reversible inhibitors of the gastric H^+,K^+-ATPase.3-substituted-4-(phenylamino)quinolines. J. Med. Chem. 35, 3413–3422.

Im, W. B., Blakeman, D. P., & Sachs, G. (1985). Reversal of antisecretory activity of omeprazole by sulfhydryl compounds in isolated rabbit gastric glands. Biochim. Biophys. Acta 845, 54–59.

Jackson, R. J., Mendlein, J., & Sachs, G. (1983). Interaction of fluorescein isothiocyanate with the H^+,K^+-ATPase. Biochim. Biophys. Acta 731, 9–15.

Jaunin, P., Horisberger, J. D., Richter, K., Good, P. J., Rossier, B. C., & Geering, K. (1992). Processing, intracellular transport, and functional expression of endogenous and exogenous α-β3 Na^+,K^+-ATPase complexes in *Xenopus Oocytes*. J. Biol. Chem. 267, 577–585.

Jorgensen, P. L., & Andersen, J. P. (1988). Structural basis for E1–E2 conformational transitions in Na^+,K^+-pump and Ca^{2+}-pump proteins. J. Membrane Biol. 103, 95–120.

Kaminski, J. J., Wallmark, B., Briving, C., & Andersson, B. M. (1991). Antiulcer agents. 5. Inhibition of gastric H^+/K^+-ATPase by substituted imidazo[1,2a]pyridines and related analogues and its implication in modeling the high affinity potassium ion binding site of the gastric proton pump enzyme. J. Med. Chem. 34, 533–541.

Karlish, S. J. D. (1980). Characterization of conformational changes in (Na^+,K^+)-ATPase labeled with fluorescein at the active site. J. Bioenerg. Biomembr. 12, 111–136.

Karlish, S. J. D., Goldshleger, R., & Stein, W. D. (1990). A 19-kD C-terminal fragment of the α chain of Na^+/K^+-ATPase is essential for occlusion and transport of cations. Proc. Natl. Acad. Sci. USA 87, 4566–4570.

Karlish, S. J. D., Goldshleger, R., & Jorgensen, P. L. (1993). Location of Asn-831 of the α chain of Na^+/K^+-ATPase at the cytoplasmic surface. Implication for topological models. J. Biol. Chem. 268, 3471–3478.

Keeling, D. J., Fallowfield, C., Milliner, K. J., Tingley, S. K., Ife, R. J., & Underwood, A. H. (1985). Studies on the mechanism of action of omeprazole. Biochem. Pharmacol. 34, 2967–2973.

Keeling, D. J., Fallowfield, C., & Underwood, A. H. (1987). The selectivity of omeprazole as an H^+,K^+-ATPase inhibitor depends upon the means of its activation. Biochem. Pharmacol. 36, 339–344.

Keeling, D. J., Taylor, A. G., & Schudt, C. (1989). The binding of a K^+ competitive ligand, 2-methyl, 8-(phenylmethoxy)imidazo[1,2-a]pyridine 3-acetonitrile, to the gastric H^+,K^+-ATPase. J. Biol. Chem. 264, 5545–5551.

Kirley, T. L. (1990). Inactivation of Na^+,K^+-ATPase by β-mercaptoethanol. Different sensitivity to reduction of the three β subunit disulfide bonds. J. Biol. Chem. 265, 4227–4232.

Kohl, B., Sturm, E., Senn-Bilfinger, J., Simon, W. A., Kruger, U., Schaefer, Rainer, G., Figala, V., & Klemm, K. (1992). H^+,K^+-ATPase inhibiting 2-[(2-pyridylmethyl)sulfinyl]benzimidazoles. A novel series of dimethoxypyridyl-substituted inhibitors with enhanced selectivity. The selection of pantoprazole as a clinical candidate. J. Med. Chem. 35, 1049–1057.

Kyte, J., & Doolittle, R. F. (1982). A simple method for displaying the hydropathic character of a protein. J. Mol. Biol. 157, 105–132.

Lacapere, J-J., Garin, J., Trinnaman, B., & Green, N. M. (1993). Identification of amino acid residues photolabeled with 8-azidoadenosine 5′-diphosphate in the catalytic site of sarcoplasmic reticulum Ca^{2+}-ATPase. Biochemistry 32, 3414–3421.

LaMattina, J. L., McCarthy, P. A., Reiter, L. A., Holt, W. F., & Yeh, L. A. (1990). Antiulcer agents. 4-Substituted 2-guanidinothiazoles: Reversible, competitive, and selective inhibitors of gastric H^+,K^+-ATPase. J. Med. Chem. 33, 543–552.

Lane, L. K., Kirley, T. L., & Ball, W. J., Jr. (1986). Structural studies on H^+,K^+-ATPase: Determination of the NH_2-terminal amino acid sequence and immunological cross-reactivity with Na^+,K^+-ATPase. Biochem Biophys. Res. Commun. 138, 185–192.

Lane, L. K., Feldmann, J. M., Flarsheim, C. E., & Rybczynski, C. L. (1993). Expression of rat α1 Na^+,K^+-ATPase containing substitutions of "essential" amino acids in the catalytic center. J. Biol. Chem. 268, 17930–17934.

Laubinger, W., Deckers-Hebestreit, G., Altendorf, K., & Dimroth, P. (1990). A hybrid adenosinetriphosphatase composed of F_1 of *Escherichia coli* and F_0 of *Propionigenium modestum* is a functional sodium ion pump. Biochemistry 29, 5458–5463.

Leach, C. A., Brown, T. H., Ife, R. J., Keeling, D. J., Laing, S. M., Parsons, M. E., Price, C. A., & Wiggall, K. J. (1992). Reversible inhibitors of the gastric H^+,K^+-ATPase. 2. 1-Arylpyrrolo[3,2-c]quinolines: Effect of the 4-substituent. J. Med. Chem. 35, 1845–1852.

Lemas, M. V., Takeyasu, K., & Fambrough, D. M. (1992). The carboxyl-terminal 161 amino acids of the Na^+,K^+-ATPase α-subunit are sufficient for assembly with the β-subunit. J Biol. Chem. 267, 20987–20991.

Lindberg, P., Nordberg, P., Alminger, T., Brandstrom, A., & Wallmark, B. (1986). The mechanism of action of the gastric acid secretion inhibitor, omeprazole. J. Med. Chem. 29, 1327–1329.

Lorentzon, P., Eklundh, B., Brandstrom, A., & Wallmark, B. (1985). The mechanism for inhibition of gastric H^+,K^+-ATPase by omeprazole. Biochim. Biophys. Acta 817, 25–32.

Lorentzon, P., Jackson, R., Wallmark, B., & Sachs, G. (1987). Inhibition of H^+,K^+-ATPase by omeprazole in isolated gastric vesicles requires proton transport. Biochim. Biophys. Acta 897, 41–51.

Lorentzon, P., Sachs, G., & Wallmark, B. (1988). Inhibitory effects of cations on the gastric H^+,K^+-ATPase. A potential sensitive step in the K^+ limb of the pump cycle. J. Biol. Chem. 263, 10705–10710.

Lytton, J., Westlin, M., Burk, S. E., Shull, G. E., & MacLennan, D. H. (1992). Functional comparisons between isoforms of the sarcoplasmic or endoplasmic reticulum family of calcium pumps. J. Biol. Chem. 267, 14483–14489.

Ma, J. Y., Song, Y. H., Sjostrand, S. E., Rask, L., & Mardh, S. (1991). cDNA cloning of the β subunit of the human gastric H^+,K^+-ATPase. Biochem. Biophys. Res. Commun. 180, 39–45.

Maeda, M., Ishizaki, J., & Futai, M. (1988a). cDNA cloning and sequence determination of pig gastric H^+,K^+-ATPase. Biochem. Biophys. Res. Commun. 157, 203–209.

Maeda, M., Tagaya, M., & Futai, M. (1988b). Modification of gastric H^+,K^+-ATPase with pyridoxal 5′-phosphate. J. Biol. Chem. 263, 3652–3656.

Maeda, M., Oshiman, K-I., Tamura, S., & Futai, M. (1990). Human gastric H^+,K^+-ATPase gene. Similarity to Na^+,K^+-ATPase genes in exon/intron organization but difference in control region. J. Biol. Chem. 265, 9027–9032.

Maeda, M., Oshiman, K-I., Tamura, S., Kaya, S., Mahmood, S., Reuben, M.A., Lasater, I.S., Sachs, G., & Futai, M. (1991). The rat H^+,K^+-ATPase β subunit gene and recognition of its control region by gastric DNA binding protein. J. Biol. Chem. 266, 21584–21588.

Mardh, S., & Norberg, L. (1992). A continuous flow technique for analysis of the stoichiometry of the gastric H^+,K^+-ATPase. Acta Physiol. Scand. 146, 259–263.

Markus, S., Priel, Z., & Chipman, D. M. (1989). Interaction of calcium and vanadate with fluorescein isothiocyanate labeled Ca^{2+}-ATPase from sarcoplasmic reticulum: Kinetics and equilibria. Biochemistry 28, 793–799.

Marshall, B. (1983). Unidentified curved bacilli on gastric epithelium in active chronic gastritis. Lancet i. 1273–1275.

Matthews, I., Sharma, R. P., Lee, A. G., & East, J. M. (1990). Transmembrane organization of (Ca^{2+}-Mg)-ATPase from sarcoplasmic reticulum. Evidence for luminal location of residues 877–888. J. Biol. Chem. 265, 18737–18740.

McIntosh, D. B., Woolley, D. G., & Berman, M. C. (1992). 2′,3′,-O-(2,4,6-Trinitrophenyl)-8-azido-AMP and -ATP photolabel Lys-492 at the active site of sarcoplasmic reticulum Ca^{2+}-ATPase. J. Biol. Chem. 267, 5301–5307.

Mendlein, J., & Sachs, G. (1989). The substitution of calcium for magnesium in H^+,K^+-ATPase catalytic cycle. J. Biol. Chem. 264, 18512–18519.

Mendlein, J., & Sachs, G. (1990). Interaction of a K^+-competitive inhibitor, a substituted imidazo[1,2a]pyridine, with the phospho- and dephosphoenzyme forms of H^+,K^+-ATPase. J. Biol. Chem. 265, 5030–5036.

Mercier, F., Reggio, H., Devilliers, G., Bataille, D., & Mangeat, P. (1989). Membrane-cytoskeleton dynamics in rat parietal cells: Mobilization of actin and spectrin upon stimulation of gastric acid secretion. J. Cell Biol. 108, 441–453.

Mercier, F., Bayle, D., Besancon, M., Joys, T., Shin, J. M., Lewin, M. J. M., Prinz, C., Reuben, A. M., Soumarmon, A., Wong, H., Walsh, J. H., & Sachs, G. (1993). Antibody epitope mapping of the gastric H^+/K^+-ATPase. Biochim. Biophys. Acta 1149, 151–165.

Mohraz, M., Sathe, S., & Smith, P. R. (1990). Proc. XIIth Internat. Congr. Electr. Microscopy, (Peachley, L.D., & Williams, D.B., eds.), Vol. 1, pp. 94–95. San Francisco Press Inc.

Molnar, E., Varga, S., & Martonosi, A. (1991). Difference in the susceptibilty of various cation transport ATPases to vanadate-catalyzed photocleavage. Biochim. Biophys. Acta 1068, 17–26.

Morii, M., & Takeguchi, N. (1993). Different biochemical modes of action of two irreversible H^+,K^+-ATPase inhibitors, omeprazole and E3810. J. Biol. Chem. 268, 21553–21559.

Morley, G. P., Callaghan, J. M., Rose, J. B., Toh, B. H., Gleeson, P. A., & van Driel, I. R. (1992). The mouse gastric H^+,K^+-ATPase β subunit. Gene structure and co-ordinate expression with the α subunit during ontogeny. J. Biol. Chem. 267, 1165–1174.

Munson, K. B., & Sachs, G. (1988). Inactivation of H^+,K^+-ATPase by a K^+-competitive photoaffinity inhibitor. Biochemistry 27, 3932–3938.

Munson, K. B., Gutierrez, C., Balaji, V. N., Ramnarayan, K., & Sachs, G. (1991). Identification of an extracytoplasmic region of H^+,K^+-ATPase labeled by a K^+-competitive photoaffinity inhibitor. J. Biol. Chem. 266, 18976–18988.

Murakami, S., Araim, I., Muramatsu, M., Otomo, S., Baba, K., Kido, T., & Kozawa, M. (1992). Effect of stilbene derivatives on gastric H^+,K^+-ATPase. Biochem. Pharmacol. 44, 33–37.

Nagaya, H., Satoh, H., Kubo, K., & Maki, Y. (1989). Possible mechanism for the inhibition of gastric H^+,K^+-adenosinetriphosphatase by the proton pump inhibitor AG-1749. J. Pharmacol. Exper. Therapeutics 248, 799–805.

Nelson, N. (1991). Structure and pharmacology of the proton-ATPases. TIPS 12, 71–75.

Newman, P. R., & Shull, G. E. (1991). Rat gastric H^+,K^+-ATPase β-subunit gene: Intron/exon organization, identification of multiple transcription initiation sites, and analysis of the 5′-flanking region. Genomics 11, 252–262.

Newman, P. R., Greeb, J., Keeton, T. P., Reyes, A. A., & Shull, G. E. (1990). Structure of the human gastric H^+,K^+-ATPase gene and comparison of the 5′-flanking sequences of the human and rat genes. DNA and Cell Biol. 9, 749–762.

Ning, G., Maunsbach, A. B., & Esmann. M. (1993). Ultrastructure of membrane-bound Na^+,K^+-ATPase after extensive tryptic digestion. FEBS Lett. 330, 19–22.

Okamoto, C. T., Karpilow, J. M., Smolka, A., & Forte, J. G. (1990). Isolation and characterization of gastric microsomal glycoproteins. Evidence for a glycosylated subunit of the H^+/K^+-ATPase. Biochim. Biophys. Acta 1037, 360–372.

Or, E., David, P., Shainskaya, A., Tal, D. M., & Karlish, S. J. D. (1993). Effects of competitive sodium-like antagonists on Na^+,K^+-ATPase suggest that cation occlusion from the cytoplasmic surface occurs in two steps. J. Biol. Chem. 268, 16929–16937.

Oshiman, K., Motojima, K., Mahmood, S., Shimada, A., Tamura, S., Maeda, M., & Futai, M. (1991). Control region and gastric specific transcription of the rate H^+,K^+-ATPase β gene. FEBS Lett. 281, 250–254.

Polvani, C., Sachs, G., & Blostein, R. (1989). Sodium ions as substitutes for protons in the gastric H^+,K^+-ATPase. J. Biol. Chem. 264, 17854–17859.

Price, E. M., & Lingrel, J. B. (1988). Structure–function relationship in the Na^+,K^+-ATPase α subunit: Site-directed mutagenesis of glutamine-111 to arginine and asparagine-122 to aspartic acid generates a ouabain-resistant enzyme. Biochemistry 27, 8400–8408.

Price, E. M., Rice, D. A., & Lingrel, J. B. (1990). Structure–function studies of Na^+,K^+-ATPase. Site-directed mutagenesis of the border residues from the H1–H2 extracellular domain of the α subunit. J. Biol. Chem. 265, 6638–6641.

Rabon, E., & Reuben, M. A. (1990). The mechanism and structures of the gastric H^+,K^+-ATPase. Ann. Rev. Physiol. 52, 321–344.

Rabon, E. C., McFall, T. L., & Sachs, G. (1982). The gastric H^+,K^+-ATPase: H^+/ATP stoichiometry. J. Biol. Chem. 257, 6296–6299.

Rabon, E., Wilke, M., Sachs, G., & Zamphigi, G. (1986). Crystallization of the gastric H^+,K^+-ATPase. J. Biol. Chem. 261, 1434–1439.

Rabon, E. C., Bassilian, S., & Jakobsen, L. J. (1990a). Glutaraldehyde crosslinking analysis of the $C_{12}E_8$ solubilized H^+,K^+-ATPase. Biochim. Biophys. Acta 1039, 277–289.

Rabon, E. C., Bassilian, S., Sachs, G., & Karlish, S. J. D. (1990b). Conformational transitions of the H^+,K^+-ATPase studies with sodium ions as surrogates for protons. J. Biol. Chem. 265, 19594–19599.

Rabon, E., Sachs, G., Bassilian, S., Leach, C., & Keeling, D. (1991). K^+-competitive fluorescent inhibitor of the H^+,K^+-ATPase. J. Biol. Chem. 266, 12395–12401.

Rabon, E. C., Smillie, K., Seru, V., & Rabon, R. (1993). Rubidium occlusion within tryptic peptides of the H^+,K^+-ATPase. J. Biol. Chem. 268, 8012–8018.

Reenstra, W., & Forte, J. (1981). H^+/K^+ ATP stoichiometry for gastric H^+,K^+-ATPase. J. Memb. Biol. 61, 55–60.

Renaud, K. J., Inman, E. M., & Fambrough, D. M. (1991). Cytoplasmic and transmembrane domain deletions of Na^+,K^+-ATPase β-subunit. Effects on subunit assembly and intracellular transport. J. Biol. Chem. 266, 20491–20497.

Reuben, M. A., Lasater, L. S., & Sachs, G. (1990). Characterization of a β subunit of the gastric H^+/K^+-ATPase. Proc. Natl. Acad. Sci. USA 87, 6767–6771.

Saccomani, G., Dailey, D. W., & Sachs, G. (1979). The action of trypsin on the (H^++K^+)-ATPase. J. Biol. Chem. 254, 2821–2827.

Sachs, G., Chang, H. H., Rabon, E., Schackman, R., Lewin, M., & Saccomani, G. (1976). A nonelectrogenic H^+ pump in plasma membranes of hog stomach. J. Biol. Chem. 251, 7690–7698.

Sachs, G., Shin, J. M., Besancon, M., & Prinz, C. (1993). The continuing development of gastric acid pump inhibitors. Alimentary Pharmacology and Therapeutics 7, 4–12.

Scott, D., Munson, K., Modyanov, N., & Sachs, G. (1992). Determination of the sidedness of the C-terminal region of the gastric H^+,K^+-ATPase β subunit. Biochim. Biophys. Acta 1112, 246–250.

Scott, D. R., Helander, H. F., Hersey, S. J., & Sachs, G. (1993). The site of acid secretion in the mammalian parietal cell. Biochim. Biophys. Acta 1146, 73–80.

Scott, D., Besancon, M., Sachs, G., & Helander, H. F. (1994). Effects of anti-secretory agents on parietal cell structure and H^+/K^+-ATPase levels in rabbit gastric mucosa *in vivo*. Amer. J. Dig. Dis. 39, 2118–2126.

Shin, J. M., & Sachs, G. (1994). Identification of a region of the H^+,K^+-ATPase α subunit associated with the β subunit. J. Biol. Chem., 269, 8642–8646.

Shin, J. M., Besancon, M., Simon, A., & Sachs, G. (1993). The site of action of pantoprazole in the gastric H^+/K^+-ATPase. Biochim. Biophys. Acta 1148, 223–233.

Shull, G. E. (1990). cDNA cloning of the β-subunit of the rat gastric H^+,K^+-ATPase. J. Biol. Chem. 265, 12123–12126.

Shull, G. E., & Lingrel, J. B. (1986). Molecular cloning of the rat stomach H^+,K^+-ATPase. J. Biol. Chem. 261, 16788–16791.

Sih, J. C., Im, W. B., Robert, A., Graber, D. R., & Blakeman, D. P. (1991). Studies on H^+,K^+-ATPase inhibitors of gastric acid secretion. Prodrugs of 2-[(2-pyridylmethyl)sufinyl]benzimidazole proton-pump inhibitors. J. Med. Chem. 34, 1049–1062.

Skrabanja, A. T. P., van der Hijden, H. T. W. M., & de Pont, J. J. H. H. M. (1987). Transport ratios of reconstituted H^+,K^+-ATPase. Biochim. Biopys. Acta 903, 434–440.

Skriver, E., Maunsbach, A. B., Herbert, H., Scheiner-Bobis, G., & Schoner, W. (1989). Two-dimensional crystalline arrays of Na^+,K^+-ATPase with new subunit interactions induced by cobalt–tetrammine-ATP. J. Ultrastructure and Molecular Structure Research 102, 189–195.

Skriver, E., Kaveus, U., Herbert, H., & Maunsbach, A. B. (1992). Three-dimensional structure of Na^+,K^+-ATPase determined from membrane crystals induced by cobalt–tetrammine-ATP. J. Struct. Biol. 108, 176–185.

Smirnova, I., & Faller, L. D. (1993). Mechanism of K^+ interaction with fluorescein 5′-isothiocyanate-modified Na^+,K^+-ATPase. J. Biol. Chem. 268, 16120–16123.

Smith, D. L., Tao, T., & Maguire, M. E. (1993). Membrane topology of a P-type ATPase. The MgtB magnesium transport protein of *Salmonella typhimurium*. J. Biol. Chem. 268, 22469–22479.

Smith, G., & Scholes, P. (1982). The H^+/K^+ ATP stoichiometry of the H^+,K^+-ATPase of dog gastric microsomes. Biochim. Biophys. Acta 688, 803–807.

Stewart, B., Wallmark, B., & Sachs, G. (1981). The interaction of H^+ and K^+ with the partial reactions of gastric H^+,K^+-ATPase. J. Biol. Chem. 256, 2682–2690.

Tahara, Y., Ohnishi, S-I., Fujiyoshi, Y., Kimura, Y., & Hayashi, Y. (1993). A pH induced two-dimensional crystal of membrane-bound Na^+,K^+-ATPase of dog kidney. FEBS Lett. 320, 17–22.

Tamura, S., Tagaya, M., Maeda, M., & Futai, M. (1989). Pig gastric H^+,K^+-ATPase. Lys-497 conserved in cation transporting ATPases is modified with pyridoxal 5′-phosphate. J. Biol. Chem. 264, 8580–8584.

Tamura, S., Oshiman, K-I., Nishi, T., Mori, M., Maeda, M., & Futai, M. (1992). Sequence motif in control regions of the H^+,K^+-ATPase alpha and beta subunit genes recognised by gastric specific nuclear proteins. FEBS Lett. 298, 137–141.

Tari, A., Wu, V., Sumii, M., Sachs, G., & Walsh, J. H. (1991). Regulation of rat gastric H^+,K^+-ATPase α subunit mRNA by omeprazole. Biochim. Biophys. Acta 1129, 49–56.

Tari, A., Yamamoto, G., Sumii, K., Sumii, M., Takehara, Y., Haruma, K., Kajiyama, G., Wu, V., Sachs, G., & Walsh, J. H. (1993). The role of the histamine-2 receptor in the expression of rat gastric H^+,K^+-ATPase α subunit. Amer. J. Physiol. 265, G752–G758.

Tari, A., Yamamoto, G., Yonei, Y., Sumii, M., Sumii, K., Haruma, K., Kajiyama, G., Wu, V., Sachs, G., & Walsh, J. H. (1994). Stimulation of H^+,K^+-ATPase alpha subunit expression by histamine. Amer. J. Physiol. 266, G444–G450.

Toh, B-H., Gleeson, P. A., Simpson, R. J., Moritz, R. L., Callaghan, J. M., Goldkorn, I., Jones, C. M., Martinelli, T. M., Mu, F-T., Humphris, D. C., Pettitt, J. M., Mori, Y., Masuda, T., Sobieszczuk, P., Weinstock, J., Mantamadiotis, T., & Baldwin, G. S. (1990). The 60- to 90-kD parietal cell autoantigen associated with autoimmune gastritis is a β subunit of the gastric H^+,K^+-ATPase (proton pump). Proc. Natl. Acad. Sci. USA 87, 6418–6422.

Toyoshima, C., Sasabe, H., & Stokes, D. (1993). Three-dimensional cryo-electron microscopy of the calcium ion pump in the sarcoplasmic reticulum. Nature 362, 469–471.

Treuheit, M. J., Costello, C. E., & Kirley, T. L. (1993). Structure of the complex glycans found on the β subunit of Na^+,K^+-ATPase. J. Biol. Chem. 268, 13914–13919.

van Huysse, J. W., Jewell, E. A., & Lingrel, J. B. (1993). Site-directed mutagenesis of a predicted cation binding site of Na^+,K^+-ATPase. Biochemistry 32, 819–826.

Walan, A., Bader, J. P., Classen, M., Lamers, B. H. W., Piper, D. W., Rutgersson, K., & Erikson, S. (1989). Effect of omeprazole and ranitidine on ulcer healing and relapse rates in patients with benign gastric ulcer. New England J. Med. 320, 69.

Walderhaug, M. O., Post, R. L., Saccomani, G., Leonard, R. T., & Briskin, D. P. (1985). Structural relatedness of three ion-transport adenosine triphosphatases around their active sites of phosphorylation. J. Biol. Chem. 260, 3852–3859.

Wallmark, B., Stewart, H. B., Rabon, E., Saccomani, G., & Sachs, G. (1980). The catalytic cycle of gastric $(H^+ + K^+)$-ATPase. J. Biol. Chem. 255, 5313–5319.

Wallmark, B., Brandstrom, A., & Lassen, H. (1984). Evidence for acid-induced transformation of omeprazole into an active inhibitor of H^+,K^+-ATPase within the parietal cell. Biochim. Biophys. Acta 778, 549–558.

Wallmark, B., Briving, C., Fryklund, J., Munson, K., Jackson, R., Mendlein, J., Rabon, E., & Sachs, G. (1987). Inhibition of gastric H^+,K^+-ATPase and acid secretion by SCH 28080, a substituted pyridyl[1,2a]imidazole. J. Biol. Chem. 262, 2077–2084.

Weidmann, K., Herling, A. W., Lang, H. J., Scheunemann, K. H., Rippel, R., Nimmesgern, H., Scholl, T., Bickel, M., & Metzger, H. (1992). 2-[(2-Pyridylmethyl)sulfinyl]-1H-thieno[3,4-d]imidazoles. J. Med. Chem. 35, 438–450.

Wolosin, J. M., & Forte, J. G. (1983). Kinetic properties of the KCl transport at the secreting apical membrane of the oxyntic cell. J. Memb. Biol. 71, 195–207.

Xu, K-Y. (1989). Any of several lysines can react with 5′-isothiocyanatofluorescein to inactivate sodium and potassium ion activated adenosine triphosphatase. Biochemistry 28, 5764–5772.

Yamaguchi, M., & Kanazawa, T. (1984). Protonation of the sarcoplasmic reticulum Ca^{2+}-ATPase during ATP hydrolysis. J. Biol. Chem. 259, 9526–9531.

Yamamoto, H., Imamura, Y., Tagaya, M., Fukui, T., & Kawakita, M. (1989). Ca^{2+}-dependent conformational change of the ATP-binding site of Ca^{2+}-transporting ATPase of sarcoplasmic reticulum as revealed by an alteration of the target-site specificity of adenosine triphosphopyridoxal. J. Biochem. 106, 1121–1125.

THE PLASMA MEMBRANE H^+-ATPASE OF FUNGI AND PLANTS

Francisco Portillo, Pilar Eraso, and Ramón Serrano

I. INTRODUCTION

The plasma membrane H^+-ATPase of fungi and plants (EC 3.6.1.35) is a proton pump which plays a central role in the physiology of these organisms.

Biochemical and physiological studies have provided a first characterization of the mechanism, regulation, and physiological role of the enzyme (Goffeau and Slayman, 1981; Leonard, 1983; Serrano, 1984, 1985). However, the limitations of these approaches with fungal and plant systems have not allowed advances comparable to those obtained with animal ATPases (Stein, 1986). On the other hand, during the last years the powerful tools of molecular biology have been successfully

Biomembranes
Volume 5, pages 225–240.

ISBN: 1-55938-662-2.

used in fungi and plants, allowing rapid progress on the molecular basis of the fungal and plant proton pump.

This review will provide some basic information on the proton pump, but it will mostly focus on recent results on the active sites and regulation of plasma membrane H^+-ATPase obtained by molecular approaches.

II. PHYSIOLOGICAL AND BIOCHEMICAL PROPERTIES OF THE H^+-ATPASE

Fungal and plant cells actively extrude protons across the plasma membrane. This activity is mediated by an ATP-driven proton pump. The H^+-ATPase is a major component of both fungal and plant plasma membranes and it can generate a pH gradient of at least four units (Serrano, 1984). This proton gradient has two major physiological roles: it provides the energy for secondary active transport and it regulates both the intracellular and extracellular pH (Serrano, 1989).

The proton gradient drives the transport of three types of nutrients (Figure 1; Serrano, 1985). Uncharged molecules (sugars, neutral amino acids) may cross the membrane by proton symport. Transport of both anionic (chloride, phosphate, sulfate, lactate, acetate, anionic amino acids) and cationic substrates (K^+, NH_4^+, Na^+, Ca^{2+}, Mg^{2+}, cationic amino acids) may also be mediated by proton symports, but ionic channels driven by the membrane potential (uniports) are also operative. In addition, the extrusion of undesirable compounds is driven by the proton gradient and mediated by proton antiports.

The role of the ATPase in intracellular pH regulation has been suggested by measurements of internal pH in yeast expressing either low-activity mutant ATPases or reduced amounts of wild-type ATPase (Portillo and Serrano, 1989; Vallejo and Serrano, 1989). In such studies, a linear correlation could be established between ATPase activity and intracellular pH when yeast was growing in acid media (pH 3–4). Under these conditions, the growth rate also correlated with ATPase

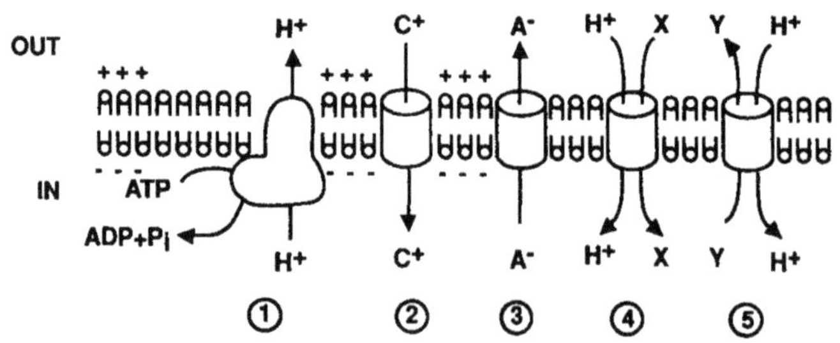

Figure 1. Diagram of the active transport processes driven by the proton gradient in fungal and plant plasma membranes. (**1**) Primary proton pump. (**2**) Cation (C^+) channel. (**3**) Anion (A^-) channel. (**4**) Proton-symport. (**5**) Proton-antiport.

activity levels. Therefore, at least at low environmental pH, the ATPase is rate-limiting for growth and essential for pH homeostasis.

The same decreases in ATPase activity caused by point mutations (Portillo and Serrano, 1989) and by reduced expression (Vallejo and Serrano, 1989) have different effects on intracellular pH and on growth rate. The fact that point mutations are more deleterious than reduced expression could be explained if the ATPase itself formed the proton leakage pathway at the plasma membrane. In this case, reduced ATPase would mean not only reduced pumping, but also reduced leakage. Point mutations would reduce pumping without affecting leakage and, therefore, would be more deleterious.

In plant cells, the regulation of extracellular (apoplastic) pH by the ATPase provides the basis for the so called "acid growth theory" (Hager et al., 1971, 1991). According to this theory, the acidification of the cell wall space resulting from proton pumping would activate polysaccharidases involved in loosening the cell wall during turgor-driven growth. The plant growth hormone, auxin, and the phytotoxin, fusicoccin, seem to promote growth because of activation of the ATPase. Of course, sustained growth also requires biosynthetic events only indirectly dependent on the ATPase.

The fungal and plant H^+-ATPase belong to the E–P (Serrano, 1991) or P-type (Perdesen and Carafoli, 1987) family of ion pumps. This family includes enzymes such as the bacterial K^+-ATPases, Ca^{2+}-ATPases from animal and plant cells and animal Na^+,K^+- and H^+,K^+-ATPases. These ATPases contain a single catalytic subunit of 70–130 kD, although accessory subunits have been described for the Na^+,K^+-ATPase of animal cells (Jorgensen and Andersen, 1988), for the K^+-ATPase of *Escherichia coli* (Hesse et al., 1984), and more recently for the yeast H^+-ATPase (Navarre et al., 1992). The enzymes of this family form an aspartyl–phosphate intermediate during the catalytic cycle and are sensitive to vanadate. The biochemical properties of the fungal and plant H^+-ATPases have already been reviewed (Goffeau and Slayman, 1981; Leonard, 1983; Serrano, 1985, 1991).

III. STRUCTURE OF THE H^+-ATPASE

Most of the structural information about fungal and plant H^+-ATPases is based on their inferred amino acid sequence and on comparative studies with other members of the ATPase family. The alignment of the amino acid sequence of all these enzymes (Serrano, 1989; Goffeau and Green, 1990; Wach et al., 1992) support the notion that all (E–P) ATPases have a common evolutionary origin (Serrano, 1989; Jorgensen and Andersen, 1988). Homology between distant cation pumps of the family is only 15–20%, while H^+-ATPases from plants and fungi have 32–36% amino acid identity (Serrano, 1989). Homology between different ATPases is maximal in five regions covering about 200 amino acids (Serrano and Portillo, 1990). These regions include six especially conserved motifs (I–VI) that could correspond to the basic catalytic machinery preserved by evolution (Figure 2).

MOTIFS:		I		II		III		IV		V		VI
		• • ****		++******** +		*+++		*+++		* ***		+**********++*
Isc	225	IDQSAITGESL	374	ILCSDKTGTLTKNK	472	CVKGAP	532	CMDPPRDDT	555	KMLTGDA	630	AMTGDGVNDAPSLKKA
Isp	223	VDQSAITGESL	372	VLCSDKTGTLTKNK	470	CVKGAP	529	CSDPPRHDT	553	KMLTGDA	628	AMTGDGVNDAPSLKKA
Inc	225	VDQSAITGESL	374	ILCSDKTGTLTKNK	472	CVKGAP	532	CMDPPRHDT	555	KMLTGDA	630	AMTGDGVNADPSLKKA
Izr	228	VDQSSITGESL	376	ILCSDKTGTLTKNK	474	CVKGAP	534	CMDPPRDDT	558	KMLTGDA	632	AMTGDGVNDAPSLKKA
Ica	203	VDQSAITGESL	351	ILCSDKTGTLTKNK	449	CVKGAP	509	CMDPPRDDT	532	KMLTGDA	607	AMTGDGVNDAPSLKKA
Iat	177	KDQSALTGESL	326	VLCSDKTGTLTLNK	422	VSKGAP	486	LFDPPRHDS	509	KMITGDQ	585	GMTGDGVNDAPALKKA
Inp	183	IDQSALTGESL	330	VLCSDKTGTLTLNK	426	VSKGAP	491	LFDPPRHDS	513	KMVTGDQ	589	GMTGDGVNDAPALKKA
Ile	182	IDQSALTGESL	328	VLCSDKTGTLTLNK	425	VSKGAP	489	LFDPPRHDS	512	KMITGDQ	588	GMTGDGVNDAPALKKA
IaK	211	VDNSSLTGESE	370	VICSDKTGTLTTNQ	504	VMKGAP	589	MIDPPRAAP	612	IMVTGDH	711	AVTGDGVNDSPALKKA
IK	222	VDNSSLTGESE	381	VICSDKTGTLTTNQ	515	VMKGAP	600	MIDPPRATV	623	IMVTGDH	722	AVTGDGVNDSPALKKA
:aSr	175	VDQSILTGESV	347	VICSDKTGTLTTNQ	513	FVKGAP	588	MLDPPRIEV	621	IMITGDN	698	AMTGDGVNDAPALKKA
:apm	238	IDESSLTGESD	471	AICSDKTGTLTMNR	599	FSKGAS	682	IEDPVRPEV	705	RMVTGDN	793	AVTGDGTNDGPALKKA
Ksf	127	VDESAVTGESK	274	VIMLDKTGTLTQGK	330	EKKITP	403	LGDVIKPEA	426	VMLTGDN	472	IMVGDGINDAPSLARA
Kec	153	VDESAITGESA	303	VLLLDKTGTITLGN	393	IRKGSV	445	LKDIVKGGI	468	VMITGDN	514	AMTGDGTNDAPALAQA

Most conserved motifs (I to VI) of (E–P) ATPases. Conserved amino acids in either all the sequences (*) o
nces (+) are indicated. The position of the first amino acid of each motif is also indicated. Hsc: H^+-ATPase fro
(PMA1 gene). Hsp: H^+-ATPase from *Schizosaccharomyces pombe* (PMA1 gene). Hnc: H^+-ATPase from N
TPase from *Zygosaccharomyces rouxii*. Hca: H^+-ATPase from *Candida albicans*. Hat: H^+-ATPase from *Ar*
ne). Hnp: H^+-ATPase from *Nicotiana plumbaginifolia* (PMA1 gene). Hle: H^+-ATPase from *Lycopersicom*
Pase from sheep kidney (α-subunit). HK: H^+,K^+-ATPase from rat stomach. CaSr: Ca^{2+}-ATPase from rabbit mu
Capm: Ca^{2+}-ATPase from human plasma membrane. Ksf: K^+-ATPase from *Streptococcus faecalis*. Kec
a coli (subunit B).

Although limitations of methods for transmembrane segment prediction made the identification of membrane spanning helices somewhat uncertain, a comparative hydropathy analysis of (E–P) ATPases (Goffeau and Green, 1990; Serrano and Portillo, 1990) suggest a consensus structure with 10 membrane-spanning helices in all eukaryotic ATPases (Figure 3).

This model is in agreement with the cytoplasmic location of the N- and C-termini of the enzyme as determined using specific antibodies (Mandala and Slayman, 1989; Monk et al., 1991) and by analyzing tryptic peptides released from proteoliposomes (Scarborough and Hennessey, 1990).

All the conserved motifs are located within cytoplasmic domains (Figure 3). Conserved motif I lies within the small hydrophilic region and motifs II–VI are placed within the large hydrophilic central domain. Alignment of ATPase sequences based on secondary structure prediction (Serrano, 1989) suggest that all the conserved motifs are coils or loops that connect elements of the secondary structure and predict a surface location of all of them because of their high polarity.

Several charged amino acids conserved among all the H^+-ATPases and predicted to be at the membrane surface or buried in the membrane helices could form the polar channel across the membrane (Figure 3). Site-directed mutagenesis of some of these polar groups has demonstrated their essential role (see following).

Epitope mapping studies of yeast ATPase (Serrano et al., 1993) indicate that two cytoplasmic regions within the amino-terminal part of the ATPase (at amino acid positions 5–105 and 168–255) contain most of the antigenic determinants. Only

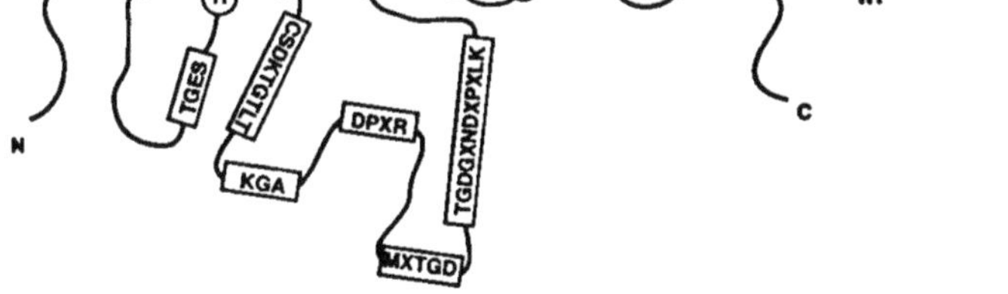

Figure 3. A model for the membrane-spanning domain of the H^+-ATPase. M1–M10 denote the ten predicted transmembrane helices. Charged amino acids predicted to be at the membrane surface or buried in the transmembrane helices and conserved among all the H^+-ATPases are encircled. The position of the most conserved motifs are indicated by boxes. The length of the bar representing the polypeptide chain is not scaled.

the epitopes at amino acids 24–56 were accessible in ATPase preparations not treated with detergents or organic solvents. This region contains an acidic stretch of amino acids conserved in fungal ATPases. By deletion analysis, it has been shown to be essential for the appearance of the enzyme at the plasma membrane (Portillo et al., 1989). The role of this region is unknown, but it is ideally suited for recognition by other proteins, for example by receptors needed for plasma membrane targeting.

IV. MODEL FOR THE ACTIVE SITE AND MECHANISM OF (E–P) ATPASES

In vitro site-directed mutagenesis has been exploited to generate a large collection of mutant ATPases in yeast H^+-ATPase (see Gaber, 1992 for a compilation) and Ca^{2+}-ATPase from sarcoplasmic reticulum (MacLennan et al., 1992). Analysis *in vivo* and *in vitro* of the yeast mutants had lead to the identification of several conserved residues essential for activity. Table 1 summarizes the proposed function of the conserved motifs and Figure 4 depicts a model for the active site and mechanism of (E–P) ATPases deduced from the yeast studies.

In the E_I conformation, the conserved motifs III to VI could form the ATP binding site. Aspartic acids 534, 560, and 638 (in motifs IV, V, and VI, respectively) seem to participate in adenine binding because mutations at these amino acids decrease the nucleotide specificity of the enzyme (Portillo and Serrano, 1988). Lysine 474 and aspartic acid 634 are also components of the ATP binding site because in the case of animal Na^+,K^+-ATPase the equivalent residues are labeled with ATP derivatives and protected from this labeling by ATP (Ovchinnikov et al., 1987). Similar results are obtained for lysine 474 with *Neurospora* ATPase (Pardo and Slayman, 1988). Mutagenesis of these residues result in ATPases with reduced phosphorylating activity, but no change in nucleotide specificity (Portillo and Serrano, 1989), suggesting that these amino acids may bind to the phosphate group of ATP.

Table 1. Consensus Sequence of the Conserved Motifs of Eukaryotic Cation-ATPases and Their Predicted Function

Motifs	*Proposed Function*
I. DXSX(I,L)TGES	Hydrolysis of intermediate, coupling of ATP hydrolysis.
II. (I,L)CSDKTGTLTXN	Phosphorylated intermediate, energy transduction.
III. KGA	ATP binding (phosphate part)
IV. DPXR	ATP binding (adenine part)
V. MXTGD	ATP binding (adenine part)
VI. TGDGXNDXPXLKKAXXGXA	ATP binding (adenine and phosphate part), hydrolysis of intermediate.

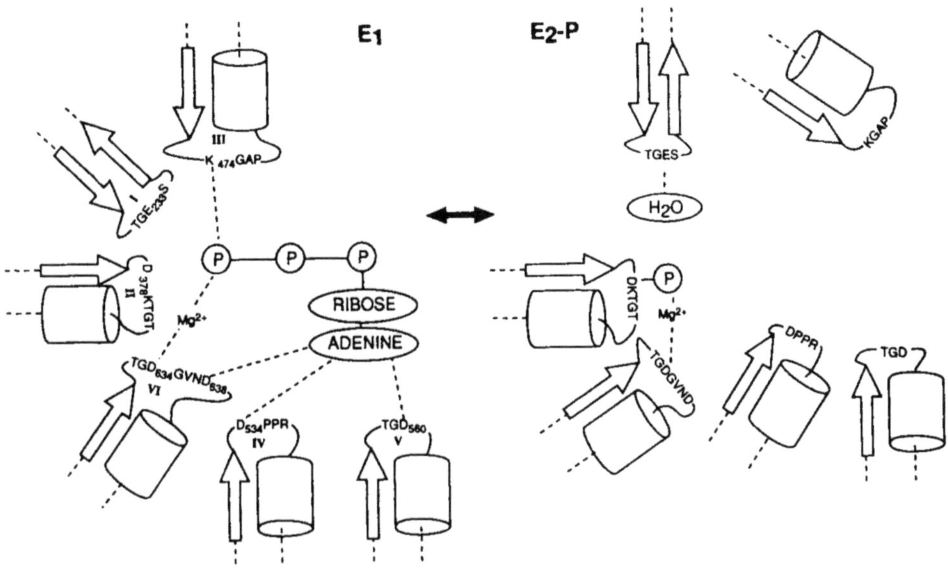

Figure 4. Model for the active site and mechanism of (E–P)ATPases. The motifs despicted correspond to the conserved motifs of Figure 2 and Table 1. Cylinders represent α-helices and arrows represent β-strands. E_1 is the conformation catalyzing the formation of the phosphorylated intermediate and E_2 is the conformation catalyzing the hydrolysis of the intermediate.

After transfer of the γ-phosphate group of ATP to aspartic acid 378 in conserved motif II, the enzyme changes to the E_2 conformation. This amino acid is fully essential for activity. Our previous report that the substitution of this aspartic acid by asparagine resulted in a functional ATPase (Portillo and Serrano, 1988) was in error because the mutation was lost in the plasmid used for expression (R. Serrano, unpublished). The conserved proline 335 on transmembrane helix 4 is essential for activity (Portillo and Serrano, 1988). It may participate in the E_1 to E_2 conformational change because of the possibility of *cis-trans* isomerization (Brandl and Deber, 1986).

In the E_2 conformation, the phosphorylated intermediate is hydrolyzed and the proton is pumped to the exterior. Site-directed mutagenesis has shown that glutamic acid 233 at conserved motif I is important for the hydrolysis of the phosphorylated intermediate (Portillo and Serrano, 1988). Other site-directed mutants in the same motifs (aspartic acid 226 and serine 234) result in an enzyme with uncoupled ATP hydrolysis (Portillo and Serrano, 1989). This result suggests that conserved motif I is also essential for coupling ATP hydrolysis to proton transport. In addition, aspartic acid 634 could also be involved in the hydrolysis of the phosphorylated intermediate because the phospho-enzyme formed by an ATPase with the mutation aspartic acid 634 to asparagine also exhibited slow turnover (Portillo and Serrano, 1989).

Finally, although not indicated in Figure 4, arginine 695 within transmembrane helix 5 (F. Portillo, unpublished) and aspartic acid 730 within transmembrane helix 6 (F. Portillo and R. Serrano, 1989) (Figure 3) could form part of the proton transport pathway across the membrane because they are fully essential for activity. However, an effect on the correct folding of the enzyme is also possible. Other polar residues predicted to be membrane-buried (Figure 3) proved not to be essential (Portillo, unpublished data).

In the case of mutants of sarcoplasmic reticulum Ca^{2+}-ATPase, a more sophisticated analysis of the catalytic cycle and E_1–E_2 conformations is available (MacLennan et al., 1992). A comparison of the H^+-ATPase and Ca^{2+}-ATPase studies shows both agreements and discrepancies. The glutamic acid of motif I (TGES) and the first aspartic acid of motif VI (TGDGXND) are required for hydrolysis of the phosphorylated intermediate in both ATPases. We interpreted these results to indicate that the mutated residues participate in the chemical hydrolysis of the intermediate (Portillo and Serrano, 1988; Serrano and Portillo, 1990), but in the case of the Ca^{2+}-ATPase, these amino acids are important for the $E_1 \rightarrow E_2$ conformational change, prior to the hydrolysis step (MacLennan et al., 1992). The lysine of conserved motif III (KGA), the aspartic acid of conserved motif IV (DPXR), conserved motif V (MXTGD), and the last aspartic acid of motif VI (TGDGXND) are important for the formation of the phosphorylated intermediate in both ATPases. In this case, there is agreement about the participation of all these residues in ATP binding and phosphorylation. A discrepancy, however, is the relative importance of the lysine of motif III, which is much more essential for H^+-ATPase (Portillo and Serrano, 1989) than for Ca^{2+}-ATPase (MacLennan et al., 1992). The conserved aspartic acid in the putative transmembrane stretch 6 is essential in both enzymes; however, in the H^+-ATPase, it is involved in the hydrolysis of the phosphorylated intermediate (Portillo and Serrano, 1989), whereas in the Ca^{2+}-ATPase, it is involved in the phosphorylation reaction (MacLennan et al., 1992). Also, the glutamic acid present in putative transmembrane stretch 8 is essential for the Ca^{2+}-ATPase (MacLennan et al., 1992), but not for the H^+-ATPase (Portillo, unpublished data). Another discrepancy is that the conserved proline in the middle of putative transmembrane stretch 4 is essential for the H^+-ATPase (Portillo and Serrano, 1988), but not for the Ca^{2+}-ATPase, where a second, nonconserved proline turns out to be essential (MacLennan et al., 1992).

These discrepancies indicate that not all conserved residues between H^+-ATPase and Ca^{2+}-ATPase may be strictly equivalent in terms of catalytic roles. This can be explained by the low level of overall homology between the two enzymes. It would be interesting to extend this mutational analysis to other (E–P) ATPases. The recent expression of plant ATPases in the yeast system (Villalba et al., 1992) will soon make possible the identification of essential residues for the plant enzyme.

Vanadate-resistant mutations of yeast ATPase occur at three different positions: (a) threonine 231 (Portillo and Serrano, 1989), lysine 250 (Goffeau and De Meis, 1990), and glycine 268 (Ghislain et al., 1987) within the postulated phosphatase domain; (b) alanine 608 (Van Dyck et al., 1990), aspartate 634 (Portillo and Serrano, 1989), and proline 640 (Perlin et al., 1989) within the ATP binding domain; and (c) serine 368 (Harris et al., 1991) and cysteine 376 (Portillo and Serrano, 1989) close to the phosphorylation site. This is in agreement with the existence of a phosphatase-like site involving motifs I, II, and part of motif VI (Figure 4). The ATPase defect in the serine 368 to phenylalanine mutation can be partially corrected by mutations at putative transmembrane stretches 1, 2, 3, and 8, suggesting a coupling between the phosphorylation/phosphatase site and the transmembrane domain of the ATPase (Harris et al., 1991).

V. REGULATION OF THE PLASMA MEMBRANE ATPASE

In fungal and plant cells, the activity of the proton pump is regulated by a large number of environmental factors. In plant cells, hormones, such as auxin, or phytotoxins, such as fusicoccin, increase the activity of the enzyme (Altabella et al., 1990; Johansson et al., 1993). In yeast, the activity of the ATPase is increased by glucose metabolism (Serrano, 1983) and acidic growth media (Eraso and Gancedo, 1987).

Some recent insights on the mechanism of ATPase regulation have provided a preliminary model. Removal of the carboxyl-terminus of either fungal ATPase (by deletions at the gene level; Portillo et al., 1989) or plant ATPase (by trypsin treatment; Palmgren et al., 1990) render an enzyme with the properties of the activated state. These results suggest that fungal and plant ATPases may be regulated by an inhibitory domain at the carboxyl-terminus of the enzyme.

In yeast, the inhibitory domain has been localized to the last 11 amino acids of the carboxyl-terminus (Portillo et al., 1989). The homology of the C-termini of fungal and plant ATPases is very low (Figure 5) and only six residues are fully conserved within the region proposed to be the yeast inhibitory domain. Nevertheless, it seems that the same region of plant ATPases constitutes an inhibitory domain because a synthetic peptide covering part of the region of homology is able to inhibit a plant ATPase where the C-terminus has been removed by tryptic cleavage (Palmgren et al., 1991). Therefore, the differences between fungal and plant C-termini might reflect the regulation of the ATPases by different effectors instead of by differences between molecular mechanism.

Several amino acids within the yeast carboxyl-terminus that are important for ATPase regulation have been identified using site-directed mutagenesis. The fully conserved arginine 909 and threonine 912 seem to be essential because mutant enzymes at these amino acids exhibit a defective activation by glucose (Table 2). Although nothing is known about the mechanism of activation, recent results suggest that it could be mediated by phosphorylation (Chang and Slayman, 1991). Interestingly, arginine 909 and threonine 912 could define a potential phosphorylation site for calmodulin-dependent protein kinase II (Kemp and Pearson, 1990).

```
             +        *  *        **  * *
STSEAFDRLMNGK.PMKEKKST.RSVEDFMAAMQRVSTQHEKET.

SESAGFDRMMNGK.P.KESRNQ.RSIEDLVVALQRTSTRHEKGDA.

QDSVGFDNLMHGKSP.KGNQKQ.RSLEDFVVSLQRVSTQHEKSQ.

SESETFDRLMNGK.PLKENKST.RSVEDFLASMRRVSTQHEKGN.

STSEAFDNFCNGRKP.QQHTDK.RSLEDFLVSMQRVSTQHEKST.

KAWLNLFENKTAFTMKKDYGKEER..EAQWALAQR..TLHGLQPKEAVNIFPEKGSYRELSEIAEQAKRRAEIARLRELHTLKGHVESVVKLK

RAWDLVLEQRIAFTRKKDFGKEQR..ELQWAHAQR..TLHGLQVPDT.KLFSEATNFNELNQLAEEAKRRAEIARQRELHTLKGHVESVVKLK

KAWDLVLEQRIAFTRKKDFGKELR..ELQWAHAQR..TLHGLQVPDP.KIFSETTNFNELNQLAEEAKRRAEIARLRELHTLKGHVESVVKLK
```

quence alignment of the carboxyl-termini of fungal and plant plasma membrane H⁺-ATPases. Amino ac
ces are indicated by an asterisk. Essential amino acids for the activation of the *Saccharomyces cerevisiae* A
d. The synthetic peptide used as inhibitor of plant ATPase is underlined in the *Arabidopsis* sequence. Th
d of the carboxyl-terminus is also indicated. Hsc: H⁺-ATPase from *Saccharomyces cerevisiae* (PMA1 gene).
ccharomyces pombe* (PMA1 gene). Hnc: H⁺-ATPase from *Neurospora crassa*. Hzr: H⁺-ATPase from *Zyg*
H⁺-ATPase from *Candida albicans*. Hat: H⁺-ATPase from *Arabidopsis thaliana* (AHA2 gene). Hnp: H⁺-ATPas
lia (PMA1 gene). Hle: H⁺-ATPase from *Lycopersicom esculetum*.

Table 2. ATP Hydrolysis and Kinetic Properties of Mutant ATPases from Glucose Starved (GS) and Glucose Fermenting (GF) Yeast Cells

Allele	*ATPase Activity*[(a)]		K_m[(b)]		V_{max}[(c)]	
	GS	*GF*	*GS*	*GF*	*GS*	*GF*
Wild-type	0.10	0.80	4.0	1.0	0.4	1.1
Thr912→Ala	0.15	0.40	4.0	1.0	0.2	0.3
Thr912→Asp	0.10	0.40	4.0	1.0	0.2	0.5
Ser899→Ala	0.20	1.70	4.0	4.0	0.2	2.9
Ser899→Asp	0.12	1.20	1.4	1.4	0.2	2.0

Notes: [(a)]$\mu mol\ min^{-1}\ mg^{-1}$.
[(b)]mM^{-1}.
[(c)]$\mu mol\ min^{-1}\ mg^{-1}$.

Glucose metabolism modifies the kinetic properties of the enzyme (Serrano, 1983). It increases the maximal rate about three times, it reduces the K_m from 4- to about 1-mM, it shifts the optimum pH to more alkaline values, and it increases the sensitivity of the enzyme to vanadate.

The molecular mechanism of activation of yeast ATPase is clearly more complex than just a phosphorylation of the potential site defined by arginine 909 and threonine 912. Mutant enzyme with threonine 912 replaced by alanine (Table 2) is defective in the V_{max} increase induced by glucose, but the K_m had a normal change upon glucose metabolism (Portillo et al., 1991). This suggests that other unidentified amino acids may be involved in the activation of the enzyme by glucose.

There is another potential phosphorylation site at the carboxyl-terminus of the yeast ATPase. Conserved serine 899 and the nearby glutamic acid 901 and aspartic acid 902 define a site that could be phosphorylated by casein kinase II (Kemp and Pearson, 1990). Site-directed mutagenesis of serine 899 (Eraso and Portillo, unpublished data) greatly reduces the degree of activation of the ATPase, and in this case, only the K_m change induced by glucose is affected (Table 2).

These results suggest that serine 899 is important for the K_m decrease and threonine 912 for the V_{max} increase induced by glucose metabolism. In addition, genetic evidence supports the fact that serine 899 could be phosphorylated during the activation of the enzyme: when serine 899 was mutated to aspartic acid (Eraso and Portillo, unpublished data) the ATPase from glucose-starved and glucose-fermenting cells exhibits a K_m similar to that of the wild-type enzyme after activation by glucose (Table 2). Apparently, the negative charge introduced by mutagenesis has replaced the requirement for glucose, pointing to a glucose-triggered phosphorylation of serine 899. In contrast, when threonine 912 is replaced by aspartic acid (Eraso and Portillo, unpublished data), the V_{max} of the ATPase from glucose-starved cells was not increased. This suggests that phosphorylation of threonine 912, if occurring, may increase the V_{max} of the enzyme by a mechanism more complex that simply providing negative charge.

The effects of the double mutation—serine 911 to alanine and threonine 912 to alanine (which render the ATPase nonactivable by glucose)—are suppressed by the alanine 547 to valine mutation, located within the predicted ATP binding domain (Cid and Serrano, 1988; Portillo et al., 1991). In addition, the carboxyl-terminus of the ATPase is less accessible to specific antibodies in glucose-starved cells than in glucose-fermenting cells (Monk et al., 1991). This suggests a model where the carboxyl-terminus interacts with the active site of the enzyme to inhibit H^+-ATPase activity. Glucose would trigger a modification of the ATPase which would release this interaction.

Additional amino acids involved in the interaction between the inhibitory domain and the active site have been identified by an intragenic suppression analysis of the double mutation mentioned above (serine 911 to alanine and threonine 912 to alanine; Eraso and Portillo, unpublished data). Figure 6 shows the location of the second-site mutations able to suppress the original double mutation. Alanine 165, valine 169, aspartic acid 170, alanine 350, and alanine 351 are located at the end of transmembrane stretches 2 and 4, within the so-called stalk region (MacLennan et al., 1985). Proline 536, alanine 565, glycine 587, glycine 648, proline 669, and glycine 670 lie within the predicted ATP binding domain.

Based on these results, it is tempting to speculate that, in glucose-starved cells, the carboxyl-terminus is interacting with both the binding site for the transported proton (at the stalk region) and with the ATP binding site. After glucose metabolism, and probably because of phosphorylation of the ATPase, these interactions are weakened, allowing the ATP and proton binding sites to change to a more active conformation.

A similar model has been proposed (Carafoli, 1992) for the Ca^{2+}-ATPase of red blood cells. In this case, there is direct evidence for: (a) an interaction between the carboxyl-terminus of the ATPase and its active site (from cross-linking studies); and (b) release of inhibition by phosphorylation (James et al., 1989) or calmodulin binding to the inhibitory carboxyl-terminus (Falchetto et al., 1991).

Future research involving site-directed mutagenesis of the plant ATPase expressed in yeast will clarify the molecular mechanism of activation of the plant enzyme. In addition, the interactions proposed on the basis of mutational studies should be confirmed at the protein level by cross-linking studies.

VI. ISOFORMS AND TISSUE DISTRIBUTION OF PLANT ATPASE

Several expressed isoforms of plant plasma membrane H^+-ATPase have been identified in *Arabidopsis thaliana* (Pardo and Serrano, 1989; Harper et al., 1989, 1990), tobacco (Perez et al., 1992) and tomato (Ewing et al., 1990). The physiological role of this diversity is not completely understood. The isoforms may have

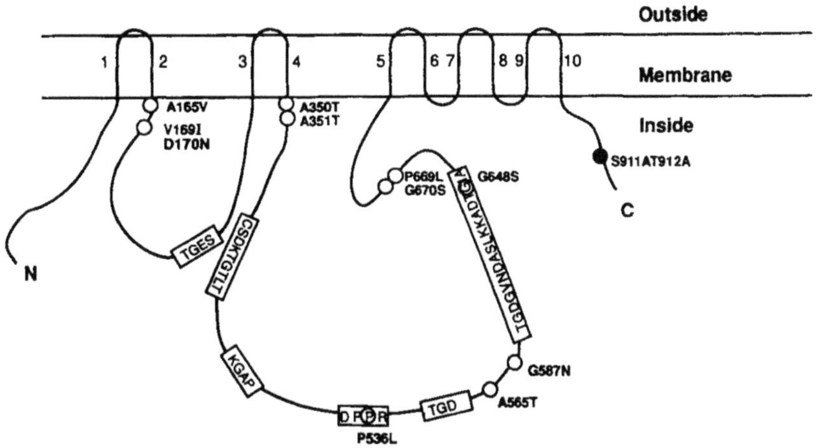

Figure 6. Localization of the second-site mutations (white circle) able to suppress the nonactivable phenotype of yeast ATPase with the double mutation serine 911 to alanine and threonine 912 to alanine (black circle). Boxes denote the position of the conserved motifs. The length of the bar representing the polypeptide chain is not scaled. Data from Eraso and Portillo, unpublished observations.

different catalytic or regulatory properties suited for special purposes. Alternatively, or in addition, the most significant difference between isoforms could be the promoters of the corresponding genes, which could determine a differential pattern of developmental expression and tissue distribution. The plant ATPase, although probably essential for all cells, is concentrated in tissues specialized for active transport such as the phloem, the stomatal guard cells and the root epidermis (Parets-Soler et al., 1990; Villalba et al., 1991; Samuels et al., 1992). One of the *Arabidopsis* isoforms has been shown to be phloem-specific (DeWitt et al., 1991) and there is evidence for kinetic and regulatory differences between the different isoforms expressed in yeast (M.G. Palmgren, unpublished). Therefore, both explanations for the existence of isoforms are open and a more detailed study is needed to understand the physiological role of this phenomenon.

REFERENCES

Altabella, T., Palazón, J., Ibarz, E., Piñol, M. T., & Serrano, R. (1990). Effect of auxin concentration and growth phase on the plasma membrane H^+-ATPase of *tobacco calli*. Plant. Sci. 70, 209–219.

Brandl, C. J., & Deber, C. M. (1986). Hypothesis about the function of membrane-buried proline residues in transport proteins. Proc. Natl. Acad. Sci. USA 83, 917–921.

Carafoli, E. (1992). The Ca^{2+}-pump of the plasma membrane. J. Biol. Chem. 267, 2115–2118.

Chang, A., & Slayman, C. W. (1991). Maturation of the yeast plasma membrane H^+-ATPase involves phosphorylation during intracellular transport. J. Cell. Biol. 115, 289–295.

Cid, A., & Serrano, R. (1988). Mutations of the yeast plasma membrane H^+-ATPase which cause thermosensitivity and altered regulation of the enzyme. J. Biol. Chem. 263, 14134–14139.

DeWitt, N. D., Harper, J., & Sussman, M. R. (1991). Evidence for a plasma membrane proton pump in phloem cells of higher plants. Plant J. 1, 121–128.

Eraso, P., & Gancedo, C. (1987). Activation of yeast plasma membrane ATPase by acid pH during growth. FEBS Lett. 224, 187–192.

Ewing, N., Wimmers, L. E., Meyer, D. J., Chetelat, R. T., & Bennett, A. B. (1990). Molecular cloning of tomato plasma membrane H^+-ATPase. Plant Physiol. 94, 1874–1881.

Falchetto, R., Vorherr, T., Brunner, J., & Carafoli, E. (1991). The plasma membrane Ca^{2+}-pump contains a site that interacts with its calmodulin-binding domain. J. Biol. Chem. 266, 2930–2936.

Gaber, R. F. (1992). Molecular genetics of yeast ion transport. International Review of Cytology 137A, 299–353.

Ghislain, M., Schlesser, A., & Goffeau, A. (1987). Mutation of a conserved glycine residue modifies the vanadate sensitivity of the plasma membrane H^+-ATPase from *Schizosaccharomyces pombe*. J. Biol. Chem. 262, 17549–17555.

Goffeau, A., & Slayman, C. W. (1981). The proton-translocating ATPase of the fungal plasma membrane. Biochim. Biophys. Acta 639, 197–223.

Goffeau, A., & De Meis, L. (1990). Effects of phosphate and hydrophobic molecules on two mutations in the β-strand sector of the H^+-ATPase from the yeast plasma membrane. J. Biol. Chem. 265, 15503–15505.

Goffeau, A., & Green, N. M. (1990). The H^+-ATPase from fungal plasma membrane. In: Monovalent Cations in Biological Systems (Pasternak, C. A., ed.). CRC Press, Boca Raton, FL.

Hager, A., Menzel, H., & Krauss, A. (1971). Experiments and hypothesis on the primary action of auxin in ellongation. [*Versuche und Hypothese zur Primarwirkung des Auxins beim Streckungswachstum*]. Planta 100, 47–75.

Hager, A., Debus, G., Edel, H.-G., Stransky, H., & Serrano, R. (1991). Auxin induces exocytosis and the rapid synthesis of a high-turnover pool of plasma membrane H^+-ATPase. Planta 185, 527–537.

Harper, J. F., Surowy, T. K., & Sussman, M. R. (1989). Molecular cloning and sequence of cDNA encoding the plasma membrane proton pump (H^+-ATPase) of *Arabidopsis thaliana*. Proc. Natl. Acad. Sci. USA 86, 1234–1238.

Harper, J. F., Manney, L., DeWitt, N. D., Yoo, M. H., & Sussman, M. R. (1990). The *Arabidopsis thaliana* plasma membrane H^+-ATPase multigene family. Genomic sequence and expression of a third isoform. J. Biol. Chem. 265, 13601–13608.

Harris, S. L., Perlin, D. S., Seto-Young, D., & Haber, J. E. (1991). Evidence for coupling between membrane and cytoplasmic domains of the yeast plasma membrane H^+-ATPase. J. Biol. Chem. 266, 24439–24445.

Hesse, J. E., Wieczorek, L., Altendorf, K., Reicin, A. S., Dorus, E., & Epstein, W. (1984). Sequence homology between two membrane transport ATPases, the Kdp-ATPase of *Escherichia coli* and the Ca^{2+}-ATPase of sarcoplasmic reticulum. Proc. Natl. Acad. Sci. USA 81, 4746–4750.

James, P., Inui, M., Tada, M., Chiesi, M., & Carafoli, E. (1989). Nature and site of phospholamban regulation of the Ca^{2+}-pump of sarcoplasmic reticulum. Nature 342, 90–92.

Johanson, F., Sommarin, M., & Larsson, C. (1993). Fusicoccin activates the plasma membrane H^+-ATPase by a mechanism involving the C-terminal inhibitory domain. Plant Cell 5, 321–327.

Jorgensen, P. L., & Andersen, J. P. (1988). Structural basis for E1–E2 conformational transitions in Na,K-pump and Ca-pump proteins. J. Membrane. Biol. 103, 95–120.

Kemp, B. E., & Pearson, R. B. (1990). Protein kinase recognition sequence motifs. Trends in Biochem. Sci. 15, 342–346.

Leonard, R. T. (1983). Potassium transport and the plasma membrane ATPase in plants. In: Metals and Micronutrients: Uptake and Utilization by Plants (Robb, D. A., & Pierpoint, W. S., eds.), pp. 71–86. Academic Press, London.

MacLennan, D. H., Brandl, C. J., Korczak, B., & Green, N. M. (1985). Amino-acid sequence of a Ca^{2+}–Mg^{2+}-dependent ATPase from rabbit muscle sarcoplasmic reticulum, deduced from its complementary DNA sequence. Nature 316, 696–700.

MacLennan, D. H., Clarke, D. M., Loo, T. W., & Skerganc, I. S. (1992). Site-directed mutagenesis of the Ca^{2+}-ATPase of sarcoplasmic reticulum. Acta Physiol. Scand. 146, 141–150.

Mandala, S. M., & Slayman, C. W. (1989). The amino and carboxyl-termini of the *Neurospora* plasma membrane H^+-ATPase are cytoplasmically located. J. Biol. Chem. 264, 16276–16281.

Monk, B. C., Montesinos, C., Ferguson, C., Leonard, K., & Serrano, R. (1991). Immunological approaches to the transmembrane topology and conformational changes of the carboxyl-terminal regulatory domain of the yeast plasma membrane H^+-ATPase. J. Biol. Chem. 266, 18097–18103.

Navarre, C., Ghislain, M., Leterme, S., Ferroud, C., Dufour, J. P., & Goffeau, A. (1992). Purification and complete sequence of a small proteolipid associated with the plasma membrane H^+-ATPase of *Saccharomyces cerevisiae*. J. Biol. Chem. 267, 6425–6428.

Ovchinnikov, Y. A., Dzhandzugaryan, K. N., Lutsenko, S. V., Mustayev, A. A., & Modyanov, N. N. (1987). Affinity modification of E_1 form for Na^+, K^+-ATPase revealed Asp-710 in the catalytic site. FEBS Lett. 217, 111–116.

Palmgren, M. G., Larsson, C., & Sommarin, M. (1990). Proteolytic activation of the plant plasma membrane H^+-ATPase by removal of a terminal fragment. J. Biol. Chem. 265, 13423–13426.

Palmgren, M. G., Sommarin, M., Serrano, R., & Larsson, C. (1991). Identification of an autoinhibitory domain in the C-terminal region of the plant plasma membrane H^+-ATPase. J. Biol. Chem. 266, 20470–20475.

Pardo, J. M., & Serrano, R. (1989). Structure of a plasma membrane H^+-ATPase gene from the plant *Arabidopsis thaliana*. J. Biol. Chem. 264, 8557–8562.

Pardo, J. P., & Slayman, C. W. (1988). The fluorescein isothiocyanate-binding site of the plasma membrane H^+-ATPase of *Neurospora crassa*. J. Biol. Chem. 264, 18664–18668.

Parets-Soler, A., Pardo, J. M., & Serrano, R. (1990). Immunocytolocalization of plasma membrane H^+-ATPase. Plant Physiol. 93, 1654–1658.

Pedersen, P. L., & Carafoli, E. (1987). Ion motive ATPases. I. Ubiquity, properties and significance to cell function. Trends in Biochem. Sci. 12, 146–150.

Perez, C., Michelet, B., Ferrant, V., Bogaerts, P., & Boutry, M. (1992). Differential expression within a three-gene subfamily encoding a plasma membrane H^+-ATPase in *Nicotiana plumbaginifolia*. J. Biol. Chem. 267, 1204–1211.

Perlin, D. S., Harris, S. L., Seto-Young, D., & Haber, J. E. (1989). Defective H^+-ATPase of hygromycin B-resistant *pma*1 mutants from *Saccharomyceas cerevisiae*. J. Biol. Chem. 264, 21657–21864.

Portillo, F., & Serrano, R. (1988). Dissection of functional domains of the yeast proton-pumping ATPase by directed mutagenesis. EMBO J. 7, 1793–1798.

Portillo, F., & Serrano, R. (1989). Growth control strength and active site of yeast plasma membrane ATPase studied by site-directed mutagenesis. Eur. J. Biochem. 186, 501–507.

Portillo, F., de Larrinoa, I. F., & Serrano, R. (1989). Deletion analysis of yeast plasma membrane H^+-ATPase and identification of a regulatory domain at the carboxyl-terminus. FEBS Lett. 247, 381–385.

Portillo, F., Eraso, P., & Serrano, R. (1991). Analysis of the regulatory domain of yeast plasma membrane H^+-ATPase by directed mutagenesis and intragenic suppression. FEBS Lett. 287, 71–74.

Samuels, A. L., Fernando, M., & Glass, A. D. M. (1992). Immunofluorescent localization of plasma membrane H^+-ATPase in barley roots and effects of K nutrition. Plant Physiol. 99, 1509–1514.

Scarborough, G. A., & Hennessey, J. P., Jr. (1990). Identification of the major cytoplasmic regions of the *Neurospora crassa* membrane H^+-ATPase using protein chemical techniques. J. Biol. Chem. 265, 16145–16149

Serrano, R. (1983). *In vivo* glucose activation of the yeast plasma membrane ATPase. FEBS Lett. 156, 11–14.

Serrano, R. (1984). Plasma membrane ATPase of fungi and plants as a novel type of proton pump. Curr. Top. Cell. Reg. 23, 87–126.

Serrano, R. (1985). Plasma Membrane ATPase of Plants and Fungi. CRC Press, Boca Raton, FL.

Serrano, R. (1989). Structure and function of plasma membrane ATPase. Annu. Rev. Plant. Physiol. Plant. Mol. Biol. 40, 61–94.

Serrano, R. (1991). Transport across yeast vacuolar and plasma membrane. In: The Molecular and Cellular Biology of Yeast *Saccharomyces*. Genome Dynamics, Protein Synthesis and Energetics (Broach, J. R., Pringle, J. R., & Jones, E. W., eds.), pp. 523–585. Cold Spring Harbor Laboratory Press, Cold Spring Harbor, NY.

Serrano, R., & Portillo, F. (1990). Catalytic and regulatory sites of yeast plasma membrane H^+-ATPase studied by directed mutagenesis. Biochim. Biophys. Acta 1018, 195–199.

Serrano, R., Monk, B. C., Villalba, J. M., Montesinos, C., & Weiler, E. W. (1993). Epitope mapping and accessibility of immunodominant regions of yeast plasma membrane H^+-ATPase. Eur. J. Biochem. 212, 737–744.

Stein, W. D. (1986). Transport and Diffusion across Cell Membranes. Academic Press, Orlando, FL.

Vallejo, C. G., & Serrano, R. (1989). Physiology of mutants with reduced expression of plasma membrane H^+-ATPase. Yeast 5, 307–319.

Van Dyck, L., Petretski, J. H., Wolosker, H., Rodrigues, G., Schlesser, A., Guislain, M., & Goffeau, A. (1990). Molecular and biochemical characterization of the Dio-9 resistant *pma1-1* mutation of the H^+-ATPase from *Saccharomyces cerevisiae*. Eur. J. Biochem. 194, 785–790.

Villalba, J. M., Lützelschwab, M., & Serrano, R. (1991). Immunocytolocalization of plasma membrane H^+-ATPase in maize coleoptiles and enclosed leaves. Planta 185, 458–461.

Villalba, J. M., Palmgren, M. G., Berberián, G. E., Ferguson, C., & Serrano, R. (1992). Functional expression of plant plasma membrane H^+-ATPase in yeast endoplasmic reticulum. J. Biol. Chem. 267, 12341–12349.

Wach, A., Schlesser, A., & Goffeau, A. (1992). An alignment of 17 deduced protein sequences from plant, fungi and ciliate H^+-ATPase genes. J. Bioenerg. Biomembr. 24, 309–317.

ANION-TRANSLOCATING ATPASES

Barry P. Rosen, Saibal Dey, and Dexian Dou

Biomembranes
Volume 5, pages 241–269.

ISBN: 1-55938-662-2.

I. INTRODUCTION

Arsenicals can be categorized into three groups: (1) inorganic salts of the arsenical oxyanions arsenate (pentavalent) and arsenite (trivalent), (2) organic arsenicals, both trivalent and pentavalent, and (3) arsine gas. Because the environment is mostly oxidizing, pentavalent arsenic compounds predominate in nature. Most of the compounds are water soluble and are rapidly absorbed through mucosal membranes and disseminated in most body tissues. Although arsenate is less toxic than arsenite, most tissues are capable of reducing it to the more toxic arsenite (Schoolmaster and White, 1980).

Arsenic is an active ingredient in a variety of commonly used insecticides, rodenticides, and herbicides (Abernathy, 1983; Alden, 1983; Knowles and Benson, 1983; Smith and Oehme, 1991), for example calcium methylarsenate is the sole active ingredient of Ortho Crabgrass Killer, Formula II. Organic arsenicals such as methylarsenate and dimethylarsenate are also used in forestry as herbicides (Peterson and Rumack, 1977). Lead, calcium, and magnesium arsenate are used primarily in pesticidal sprays and in ripening sprays for citrus crops. Because of the continuing use of herbicidal and insecticidal arsenicals, arsenic contamination of fruits and vegetables is not an uncommon event. Arsenic is also present in wallpaper, paint, ceramics, glass, and certain metal alloys. Thus, there exist innumerable opportunities for exposure to arsenicals.

The solution chemistry of arsenicals of the +5 oxidation state is fairly well understood. In solution, arsenic acid (H_3AsO_4) ionizes to $H_2AsO_4^-$ ($pK_1 = 2.25$), to $HAsO_4^{2-}$ ($pK_2 = 6.77$) and AsO_4^{3-} ($pK_3 = 11.60$). These forms are quite similar chemically to various ionization states of phosphate, which is why arsenate acts as a phosphate analog in many enzymatic reactions.

Arsenite solution chemistry is less well understood (Smith, 1973). The oxyanionic form of arsenite as (+3) is frequently written as AsO_2^-, but from the Raman spectra it is clear that this form of arsenite does not exist in solution (Loehr and Plane, 1968). There are a number of hydrated species in solution, but the predominant form is the pyramidal $As(OH)_3$, which is in equilibrium with the ionized form $As(OH)_2O^-$, with a pK of 9.23. Since the predominant form in neutral and acidic solutions is the un-ionized species, the concentration of oxyanion at physiological pH (7.5) is only about 2% of the total arsenite concentration. However, the pK of arsenite *in vivo* could be quite different, and thus it is difficult to predict the biologically relevant form of arsenite.

As will be discussed following, the transport Ars ATPase that produces bacterial arsenite resistance uses antimonite as the preferred substrate. The solution chemistry of oxyanions of antimony (+3) is obscure (Smith, 1973). The common form of the compound is potassium antimonyl tartrate ($K(SbO)C_4H_4O_6$). One would expect that an enzyme that has both arsenite and antimonite as substrates would recognize a common structural and ionization species. By analogy with arsenite, it is possible that one form of antimonite in solution is the pyramidal $Sb(OH)_3$, which would be in equilibrium with $Sb(OH)_2O^-$. The 10-fold preference of the Ars ATPase for antimonite might reflect a lower pK for the antimonial. If the true substrate of the pump were the oxyanion, then an antimonite pK one unit lower than the arsenite pK would produce a 10-fold increase in the concentration of the antimonite oxyanion over arsenite oxyanion at the same total concentration. This is a critical question, since understanding the biochemical mechanism of enzymes that react with arsenite or antimonite requires understanding the chemical nature of the substrates.

There are several mechanism by which oxyanions of arsenic can exert their toxic effects. Arsenite reacts in a reversible manner with vicinal sulfhydryl groups of enzyme complexes such as the lipoyl dehydrogenase component of the pyruvate and succinate dehydrogenases, eventually leading to cell death (Massy and Veeger, 1960, 1961; for a review of the action of arsenicals, see Knowles and Benson, 1983). As a phosphate analog, arsenate can substitute for phosphate in many enzymatic reactions. When the stable phosphoryl group is replaced with the less stable arsenyl group, the concentrations of glycolytic intermediates are reduced, and oxidative phosphorylation is uncoupled. Arsenic also inactivates other enzymes, including monoamine oxidase, lipases, acid phosphatase, liver arginase, cholinesterase, and adenyl cyclase (Vaziri et al., 1980).

In medicine, there is a long history of the use of arsenicals. The ancient Greek Hippocrates pioneered the use of arsenic as a medicinal agent (Bryson, 1989). In 1649, Schroeder developed a method for preparing elemental arsenic from "mispickel" (arsenopyrite FeSAs). The first antimicrobial agent specifically developed for chemotherapy was the arsenical Salvarsan, the *Magic Bullet* against syphilis and trypanosomal diseases such as sleeping sickness. For his development of this arsenical drug, Paul Ehrlich was awarded the Nobel Prize in Medicine in 1908. With the advent of antimicrobial chemotherapy came the rise of microbial drug resistance. In his Nobel Lecture Paul Ehrlich stated,

> If a certain substance is able to kill . . . , this can happen only because it accumulates in [cells]. We must, therefore, assume that the arsenoceptor of the cell is only able to take up an arsenic radicle . . . The test-tube experiments seemed to indicate that, although the arsenoceptor had been retained in the [arsenic resistant cells], it had, nevertheless, suffered a reduction in its avidity, which was revealed by the fact that it was only by the use of much stronger solutions than the concentration needed for a lethal action was reached; whereas the normal arsenoceptor of the original strain, in consequence of its higher original avidity, attracts to itself the same amount even from weaker solutions (Ehrlich, 1960).

As is discussed following, resistance to arsenicals and antimonials in both prokaryotes and eukaryotes results from acquisition or amplification of genes that extrude the toxic oxyanions. Although Ehrlich had no knowledge of modern molecular biology, it is evident that, almost a century ago, Ehrlich understood that resistance in living organisms could result from inability to accumulate the drug. One of the major mechanisms of microbial resistance to drugs and heavy metals is now known to be through expression of genes for transport systems that lower the intracellular concentration of the toxic compound (Tisa and Rosen, 1990a; Kaur and Rosen, 1992c).

II. BACTERIAL RESISTANCE TO ARSENICALS AND ANTIMONIALS

A. The *ars* Operon

The topic of bacterial resistance to arsenic and antimony compounds has been recently reviewed (Kaur and Rosen, 1992c). Plasmid-mediated arsenical resistance in bacteria was first reported in the Gram-positive bacterium *Staphylococcus aureus* (Novick and Roth, 1968). The resistance genes were shown to be carried by plasmid pI258. Similarly, the conjugative resistance factor R773 was shown to mediate arsenical resistance in the Gram-negative bacterium *Escherichia coli* (Hedges and Baumberg, 1973). Plasmid-bearing cells of both *S. aureus* and *E. coli* exhibited reduced accumulation of $^{74}AsO_4^{-3}$, indicative of active extrusion from cells (Silver et al., 1981). Extrusion from both Gram-positive and Gram-negative cells was energy dependent (Mobley and Rosen, 1982; Silver and Keach, 1982). R773-mediated extrusion of arsenate and arsenite from cells of *E. coli* was shown to be coupled to chemical, but not electrochemical energy, suggestive of a plasmid-encoded primary pump (Mobley and Rosen, 1982; Rosen and Borbolla, 1984). The data indicated a temporal relationship between efflux and intracellular ATP concentrations, but the demonstration of an ATP-coupled process required genetic and biochemical analysis.

Genetic and molecular biological analysis of the resistance genes on R-factor R773 revealed that a single operon, termed the *ars* operon, was responsible for the resistant phenotype (Mobley et al., 1983). The operon consists of five genes, two regulatory (*arsR* and *arsD*) and three structural genes (*arsA*, *arsB*, and *arsC*) (Figure 1) (Chen et al., 1986b). The ArsR protein forms a substrate-inducible dimeric repressor that controls the basal level of expression of the operon (Wu and Rosen, 1991, 1993a). The ArsD protein apparently serves as a low affinity substrate-independent repressor that controls the upper level of operon expression (Wu and Rosen, 1993b). The ArsD protein is postulated to prevent overproduction of the ArsB protein, which is toxic when produced in high amounts (Wu et al., 1992; Wu and Rosen, 1993b).

Resistance results from the activity of the ArsA, ArsB, and ArsC proteins. The ArsA and ArsB proteins form a membrane-bound oxyanion-translocating ATPase that confers resistance to arsenite and antimonite (the +3 oxidation state of oxyanions of arsenic and antimony) (Figure 2). In addition, the operon produces resistance to arsenate, the oxyanion of the +5 oxidation state of arsenic. This requires the ArsC protein in addition to the ArsA and ArsB proteins. The related ArsC protein from the Gram-positive plasmid pI258 exhibits arsenate reductase activity *in vitro* (Ji and Silver, 1992b), and cells expressing either the pI258 or R773 ArsC proteins

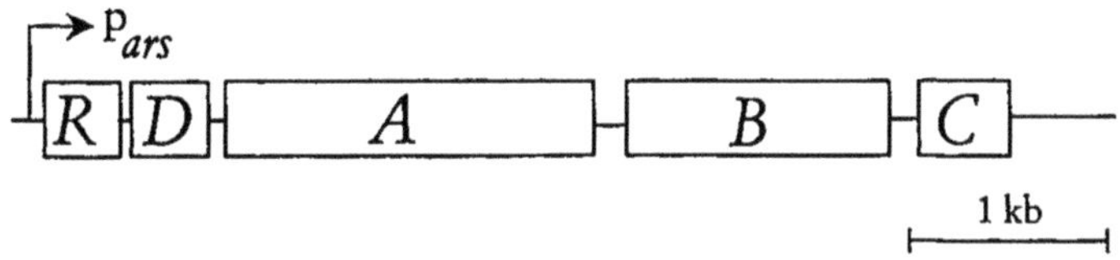

Protein	ArsR	ArsD	ArsA	ArsB	ArsC
Total residues	117	120	583	429	141
Molecular weight	13,198	13,218	63,188	45,598	15,830

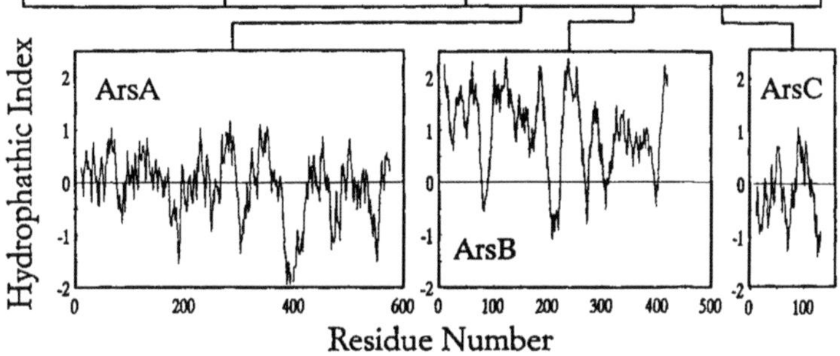

Figure 1. Physical map of the *ars* operon. In the top line, the five genes of the operon are shown with the direction of transcription indicated by the arrow, starting with the promoter, p_{ars}. The genes are indicated by boxes, with the length of the DNA in kilobase pairs (kb). Restriction endonuclease sites are (**B**) *Bam*HI, (**E**) *Eco*RI, (**H**) *Hind*III, (**P**) *Pst*I. In the middle portion, the five gene products are listed with the number of amino acid residues and molecular masses in daltons (Da). In the bottom panel are shown the hydropathy plots of the ArsA, ArsB, and ArsC proteins.

reduce arsenate to arsenite (Ji and Silver, 1992b; Oden et al., 1994). As discussed following, it is postulated that the ArsC protein interacts with the ArsA–ArsB complex to reduce arsenate on the membrane, so that the more toxic arsenite can be immediately extruded by the Ars pump (Figure 2). In the next sections, each of the three Ars proteins will be discussed in more detail.

B. The ArsA Protein

From the nucleotide sequence, the ArsA protein was predicted to have evolved by duplication and fusion of a gene half the size of the existing *arsA* gene (Chen et al., 1986b). The first (A1) half of the protein exhibits sequence similarity to the second (A2) half (Figure 3). The family of proteins that includes the ArsA protein are soluble nucleotide-binding proteins with a variety of intracellular functions (Figure 4). Each is approximately half the size of the ArsA protein, and, except for the ArsA protein, none has a transport function, as far as is known. They include

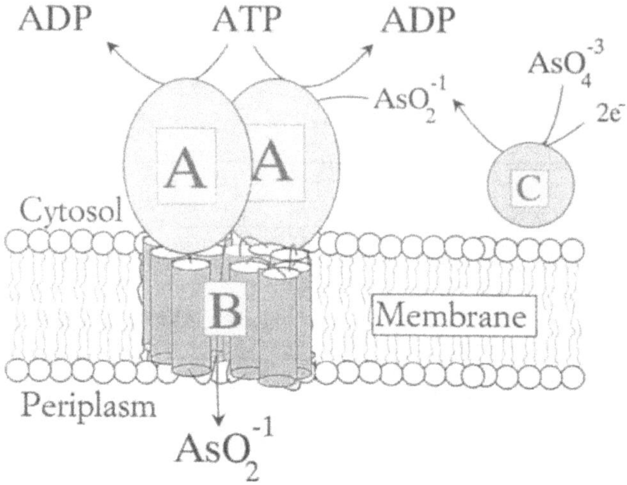

Figure 2. The plasmid-encoded oxyanion pump. The complex of the ArsA and ArsB proteins is an oxyanion-translocating ATPase for extrusion of anions of arsenic and antimony. The model shows two subunits of the ArsA protein, its active structure in solution. The stoichiometry of the ArsA and ArsB proteins in the complex has not been determined. The ArsA protein is the catalytic subunit, with oxyanion-stimulated ATPase activity. The ArsB protein is an integral membrane protein located in the inner membrane of *E. coli*. It is both the membrane anchor for the ArsA protein and the subunit of the complex with anion conductivity. The ArsC protein is shown as an arsenate reductase, converting arsenate to arsenite in the vicinity of the pump to allow for its immediate extrusion.

Figure 3. Internal homology of the A1 and A2 halves of the ArsA protein. Portions of the N-terminal (**A1**) (top) and C-terminal (**A2**) (bottom) halves of the ArsA protein are aligned with identical residues (:) and conservative replacements (.) identified.

ArsA N-term	6	NIPPYLFFTGKGGVGKTSISCATAIRLAEQGKRVLLV	STDPA
ArsA C-term	325	NEHGLIMLMGKGGVGKTTMAAAIAVRLADMGFDVHLT	TSDPA
C. elegans orf		KTLKWIFVGGKGGVGKTTCSCSLAAQLSKVRERVLLI	STDPA
MinD	1	MARIIVVTSGKGGVGKTTSSAAIATGLAQKGKKTVID	FDFDI
SopA	105	VFPPVIGAAHKGGVYKTSVSVHLAQDLALKGLRVLLV	EGNDPQ
FrxC	1	MKIAVYGKGGIGKSTTSCNISIALARRGKKVLQI	GCDPK
NifH	2	AALRQIAFYGKGGIGKSTTSQNTLAALVDLGQKILIV	GCDPK
ParA	107	EAYVIFISNLKGGVSKTVSTVSLAHAMRAHPHLLMED	LRILV
RepA	118	QLQVISVMNFKGGSGKTTTAAHLAQYLAMRGYRVLAI	DLDPG
IncC	105	TMKTLVTANQKGGVGKTSTLVHLAFDFFERGLRVAVI	DLDPQ
DnaB	159	GKGKGLYLYGKFGVGKTFMLAALAQDLAEKEYSSMIV	YVPEF

Figure 4. Relatedness of the ArsA protein to other nucleotide binding proteins. Proteins related to the ArsA protein include an open reading frame from *C. elegans* (Sulston et al., 1992); the *E. coli* MinD protein (de Boer et al., 1989), the FrxC protein of chloroplasts (Lidholm and Gustafsson, 1991); NifH, the dinitrogen reductase of nitrogenase (Mevarech et al., 1980), the ParA protein of plasmid P1 (Davis et al., 1992); the RepA protein of plasmid pRiA4b (Nishiguchi et al., 1987); the IncC protein of plasmid RK2 (Bechhofer and Figurski, 1983); the *E. coli* DnaB protein (Nakayama et al., 1984); and the F plasmid protein SopA (Nakayama et al., 1984). Residues in ArsA which have extensive similarity with homologs are shaded.

plasmid- and chromosomally-encoded bacterial proteins involved in plasmid replication (Nishiguchi et al., 1987), formation of the division septum (Davis et al., 1992), and nitrogen fixation (Mevarech et al., 1980). A reading frame of unknown function, but with striking similarity to the ArsA protein was recently identified in the eukaryote *Caenorhabditis elegans* (Sulston et al., 1992). Each of the members of this family have a consensus nucleotide-binding sequence (Walker et al., 1982). The *nifH* gene product, dinitrogen reductase, is a subunit of the nitrogen-fixing enzyme nitrogenase and, like the ArsA protein, is an energy-transducing ATPase (Mevarech et al., 1980). The *E. coli* MinD protein (de Boer et al., 1989) and the related plasmid-encoded ParA protein have also been shown to exhibit ATPases activity (Davis et al., 1992).

In the ArsA protein, the sequence GKGGVGKT is repeated in the A1 and A2 halves, and the purified protein exhibits ATPase activity that requires the presence of the oxyanionic substrate antimonite or arsenite (Hsu and Rosen, 1989). Although the consensus nucleotide-binding sequences are common with many other proteins, including many transport ATPases, there is otherwise no similarity between the ArsA protein and any known transport ATPase. This suggests that the Ars oxyanion-translocating ATPase is unrelated evolutionarily to other known classes of transport ATPases. Members of other families of transport ATPases have catalytic subunits that may have evolved similarly, including the α and β subunits of the F_1-ATPase (Futai and Kanazawa, 1983) and the N- and C-terminal halves of the P-glycoprotein (Chen et al., 1986a). A hypothesis for the evolution of these pumps from parallel evolution of the genes for soluble ATPases and membrane proteins (Rosen et al., 1992) will be discussed in more detail later.

Functionally, the ArsA protein forms a complex with the ArsB protein in the inner membrane of *E. coli* (Tisa and Rosen, 1990b). However, when expressed in high amounts, most of the ArsA protein is found in the cytosol, from which it can be easily purified (Rosen et al., 1988; Hsu and Rosen, 1989). Purified ArsA protein is an antimonite- and arsenite-stimulated ATPase. ATP is the only nucleotide substrate, with a K_m of 10^{-4}M. Inhibitors of other families of transport ATPases had no effect on ArsA ATPase activity, including N,N'-dicyclohexylcarbodiimide, azide, vanadate, and nitrate. The optimal pH range for ATP hydrolysis was 7.5 to 7.8, and Mg^{2+} was optimal at a molar ratio of 2 ATP:1 Mg^{2+}.

Oxyanions that are not pump substrates did not stimulate ATPase activity (Hsu and Rosen, 1989). Antimonite is the preferred oxyanion; the concentration at which half maximal ATP hydrolysis occurred was 10^{-5}M for antimonite and 10^{-4}M for arsenite, and the V_{max} was 5- to 10-fold greater with antimonite over arsenite. This assumes that the chemical forms recognized by the ArsA protein are present in the same amounts for both when both are dissolved at the same nominal molarity. However, as discussed previously, if the true substrates are the oxyanions, the concentration of oxyanion in solution depends on the pK_as, which are not known with certainty. If the two pK_as differ by a factor 10 (8 for antimonite and 9 for arsenite), then at pH 7.5, the concentration of oxyanions would also differ by a factor of about 10, which could explain the apparent preference for antimonite.

The existence of multiple ATP-binding sites is found in other transport ATPases, frequently as a result of duplication and fusion of genes for proteins with single nucleotide binding sites. Other proteins which have two ATP-binding motifs tandemly arranged in a linear sequence include members of the *mdr* family of transport ATPases, including the *mdr* gene product itself (Chen et al., 1986a), the CFTR protein (Riordan et al., 1989), the *rbsA* gene product (Bell et al., 1986), and the yeast STE6 gene product (McGrath and Varshavsky, 1989). The F_0F_1 H^+-translocating ATPase has two similar, but not identical, types of ATP-binding sites in the α and β subunits (Futai and Kanazawa, 1983). Tandem binding sites may permit cooperative interactions, imparting a regulatory role to nucleotide binding. This multiplicity of nucleotide binding sites in the F_1 ATPase produces an acceleration of catalysis (Futai and Kanazawa, 1983; Senior, 1985). It is reasonable to ask whether there is a function for multiple nucleotide binding sites in the ArsA protein besides simply increasing the number of catalytic sites, such as cooperative interactions of catalytic sites and regulatory domains. The role of the A1 (G_{15}KGGVGKTS) and A2 (G_{334}KGGVGKTT) sequences of the ArsA protein has been studied by a variety of approaches. Both consensus sequences appear to be involved in adenylate binding; by use of the fluorescent ATP analog, 2′,3′-O-(2,4,6-trinitrophenylcyclohexadienylidene)adenosine-5′-triphosphate (TNP–ATP), it was shown that there are two nucleotide-binding sites per molecule of wild-type ArsA protein and only one TNP–ATP binding site per molecule of ArsA proteins with mutations in the A1 site (Karkaria and Rosen, 1991). Following covalent reaction with α-[^{32}P]ATP in a UV-catalyzed reaction, the protein forms a photoadduct (Rosen et al., 1988), which has been localized to the A1 half of the protein by mutagenesis of the two consensus sequences (Karkaria et al., 1990; Kaur and Rosen, 1992a). Why only one of the two adenylate-binding sites should form a photoadduct probably reflects differences in the local environment of the two ATP-binding sites rather than a fundamental difference in the binding sites themselves. The glycine-rich sequence has been suggested to form a flexible loop that interacts with ATP at the phosphoryl groups. Although the exact chemistry of the UV catalyzed photoadduct formation is not known, reaction through the adenine ring seems reasonable. The site of the ^{32}P-label has been shown to be contained in a single cyanogen bromide fragment that contains residues 283–304 of the ArsA protein (Kaur and Rosen, 1994). This sequence is more than 260 residues away from the phosphate-binding loop. When the two halves of the ArsA protein are aligned with each other, this sequence corresponds to a region of non-alignment at the end of the ArsA1 half of the protein. One possibility is that it may be a linker polypeptide that bridges the two large domains. Assuming that the nucleotide-binding site includes the glycine-rich phosphate binding loop at the N-terminus, the adenine ring of ATP must be sticking out into the region between the two halves of the protein and is only 30 residues in the primary sequence from the second glycine-rich phosphate binding loop.

From the results of mutagenesis of the two glycine rich loops, it is clear that both are required for resistance and ATPase activity (Karkaria et al., 1990; Kaur and Rosen, 1992b). However, a genetic complementation analysis has shown that the two nucleotide-binding domains need not be contained in one polypeptide (Kaur and Rosen, 1992b). A combination of various mutant genes or partial clones were expressed separately from compatible plasmids in one cell. Co-expression of an *arsA1* mutant gene with an *arsA2* mutant gene resulted in resistance, even though neither alone did so. Thus, the two defective ArsA proteins together formed a functional ArsA complex, with the wild-type A1 site on one interacting with the wild-type A2 site on the other. Co-expression of a gene encoding a peptide containing the sequence of the N-terminal A1 half of the ArsA protein complemented a clone expressing only the C-terminal half of the protein. In this respect, the ArsA protein is similar to a heterodimer in which two homologous, but nonidentical subunits are held together by a short linker polypeptide.

Those genetic results could be interpreted as interaction of subunits. The ArsA protein was shown to form a homodimer by two different methods, chemical cross-linking and by light scattering (Hsu et al., 1991). In both cases, the dimer was observed only when the protein was incubated with arsenite or antimonite. No other oxyanion tested (arsenate, phosphate, sulfate, sulfite, nitrate, and nitrite) induced dimer formation, nor did nucleotides. These results suggest that soluble ArsA protein is in an equilibrium of inactive monomer and active dimer, where binding of the oxyanionic substrate shifts the equilibrium in favor of the catalytically

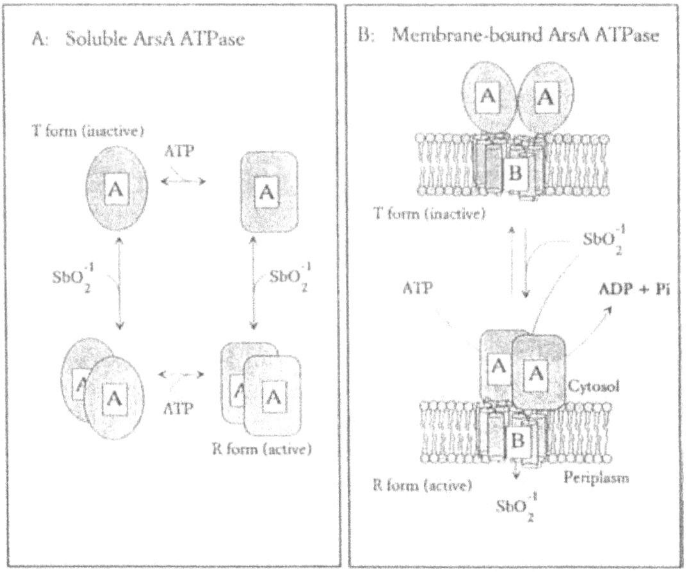

Figure 5. An allosteric model for the Conformation of the ArsA protein. (**A**) Soluble ArsA protein: The 63 kD ArsA has independent binding sites for antimonite and ATP. There is an equilibrium between monomer and dimer. In the absence of substrates, the equilibrium favors the inactive monomer (T form). The oxyanion acts as an allosteric activator, binding preferentially to the dimer and increasing by mass action dimer concentration. Binding of ATP converts the dimer into the catalytically active R conformation. (**B**) Membrane-bound ArsA protein: As a subunit of the pump, the ArsA protein exists at all times as a dimer. In the absence of oxyanion the ArsA subunits are primarily in the inactive T form; binding of oxyanion stabilizes the R conformation, promoting catalysis.

competent dimer (Figure 5). These results suggest that the oxyanion is an allosteric activator of the ATPase activity as well as the pump substrate.

C. Structure and Function of the ArsB Protein

Only recently have homologs of the ArsB protein been identified (Figure 6). The ArsB proteins of the Gram-positive plasmids pI258 (Ji and Silver, 1992a) and pSX267 (Rosenstein et al., 1992) are 58% identical and serve the same physiological function. Two other potential homologs include a reading frame for a 45 kD hydrophobic protein of unknown function identified from DNA sequence analysis in *Mycobacterium leprae* (Oskam, Hermans, Jarings, Klatser, and Hartskeerl, unpublished data) and a gene product of unknown function involved in the human disease, type II oculocutaneous albinism (Rinchik et al., 1993). Since the roles of these latter two homologs are undefined, the similarities with the ArsB proteins

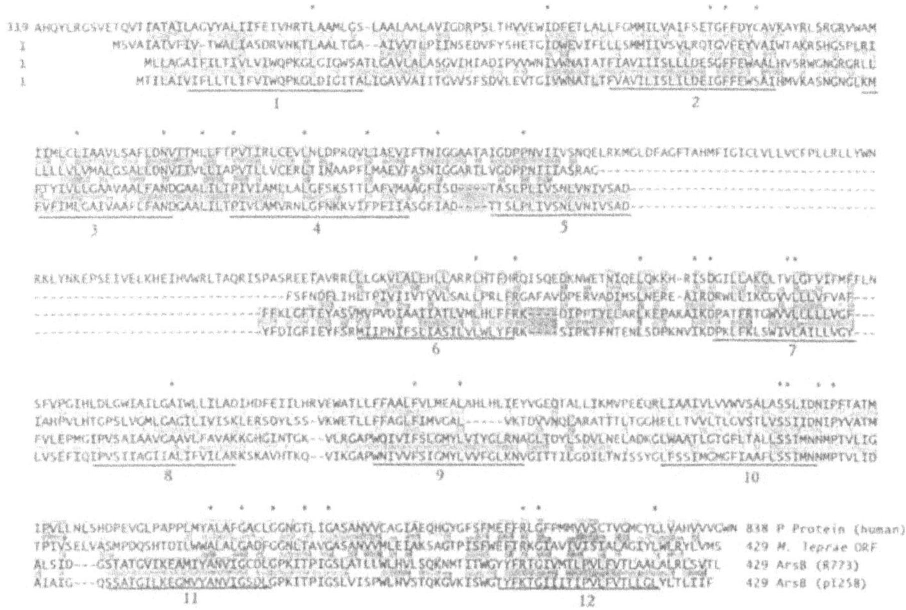

Figure 6. The ArsB family of membrane proteins. The ArsB proteins from Gram-negative plasmid R773 (Chen et al., 1986) and the Gram-positive plasmids pI258 (Ji and Silver, 1992a) and pSX267 (not shown; Rosenstein et al., 1992) are closely related in function and exhibit 58% sequence identity. The other two homologs are the P protein involved in human type II oculocutaneous albinism (Rinchik et al., 1993) and the translation product of a *Mycobacterium leprae* open reading frame (Oskam, L., Hermans, C., Jarings, G., Klatser, P. R., and Hartskeerl, R. A., unpublished). Similarities are shaded. Residues identical in all four proteins are indicated (*).

shed no light on the mechanism of oxyanion transport. However, there are 34 invariant residues in these homologs (Figure 6), and some of these may be involved in common functions. Intriguingly, there are only two conserved charged residues, a positively charged arginyl residue and a negatively charged aspartyl residue, and both are located in the C3 cytoplasmic loop connecting the two sets of six membrane spanning α-helices.

Production of sufficient ArsB protein for biochemical studies has been difficult. Although all of the *ars* genes are transcribed as a polycistronic message, they are differentially expressed (Owolabi and Rosen, 1990). Of the three structural genes, the *arsB* gene is poorly expressed, and its amount apparently limits the level of resistance. Northern analysis of the *ars* transcript demonstrated a 4,400-nucleotide mRNA that was rapidly converted into two smaller species, one of 2,700 nucleotides containing the *arsR*, *arsD*, and *arsA* genes, and the other of 500 nucleotides containing the *arsC* sequence. The half-life of the initial transcript was only 4 min, and no transcript corresponding to the *arsB* region could be detected at longer times.

The 2700- and 500-nucleotide derivatives had half-lives of 10 min or longer, so that the ArsR, ArsD, ArsA, and ArsC proteins were synthesized in much greater amounts than the ArsB protein. The reason for selective instability of the *arsB* message has been attributed to a potential stem–loop structure beginning with the third codon of the *arsB* coding sequence, and the fact that the second codon is a rarely used leucine codon. Furthermore, the ribosomal binding site of the *arsB* gene was found to be the weakest of all five *ars* genes. All these could collectively lead to ribosomal pausing, leaving this region susceptible to endonuclease digestion, resulting in low production of the ArsB protein. Because of its poor level of expression or due to poor dye binding to hydrophobic proteins, the ArsB protein was never detected as a Coomassie-stained band in sodium dodecyl sulfate-polyacrylamide gel electrophoresis. It could only be visualized as a 36 kD protein in gels when labeled with [^{35}S]methionine under control of the T7 promoter (San Francisco et al., 1989). As a fusion protein to β-galactosidase, the ArsB protein was localized in the inner membrane of cells of *E. coli*.

From its hydropathic profile, the ArsB protein was predicted to have 10 to 12 membrane spanning α-helices (Chen et al., 1986b). A topological analysis of the ArsB protein was conducted by construction of gene fusions to three different types of reporter genes (Wu et al., 1992). Complementary information was obtained from in-frame fusions with the *phoA* gene for alkaline phosphatase, the *lacZ* gene for β-galactosidase and the *blaM* gene for the mature form of β-lactamase. Interestingly, the analysis indicated 12 membrane spanning α-helices with six periplasmic and five cytoplasmic loop regions and the N-, and C-terminus oriented to the cytoplasmic side of the inner membrane (Figure 7), a structure much more similar to that of secondary porters than primary pumps (Marger and Saier, 1993). The relevance of this structure to the possible evolution of the Ars pump is discussed in more detail following.

D. ArsA–ArsB Interaction

In addition to its putative role as the anion-conducting pathway, the ArsB protein also functions as the membrane anchor for the ArsA protein (Tisa and Rosen, 1990b). From its apparent function in resistance, the ArsA protein was predicted to be a peripheral membrane protein; however, expression of the *arsA* gene from multicopy plasmids resulted in accumulation as a soluble protein in the cytosol (Rosen et al., 1988). When an *E. coli* lysate was fractionated into cytosol, inner, and outer membrane, a significant amount of ArsA protein was found in the inner membrane fraction (Tisa and Rosen, 1990b). Association of the ArsA protein to the inner membrane required expression of the *arsB* gene, presumably because the two proteins form a membrane-bound complex. Everted membrane vesicles containing the ArsB protein were reconstituted with purified ArsA protein into a tightly bound complex which can only be dissociated by treatment with high concentrations of urea or KCl. The membrane-bound complex also exhibited oxyanion-stimulated

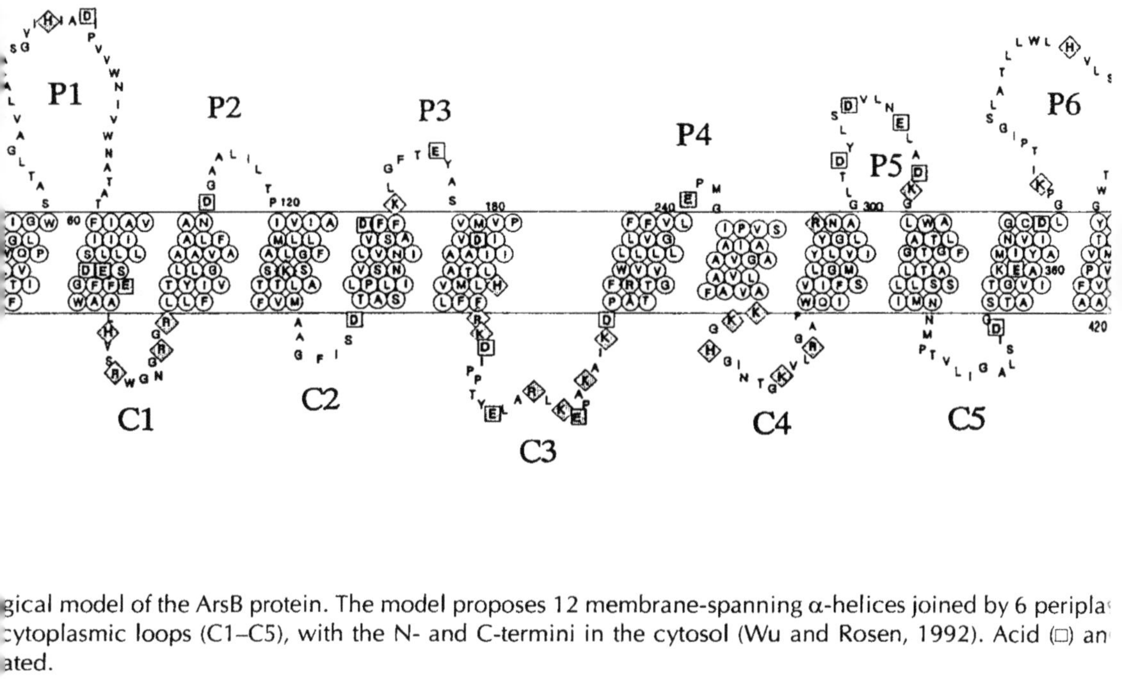

gical model of the ArsB protein. The model proposes 12 membrane-spanning α-helices joined by 6 peripla
cytoplasmic loops (C1–C5), with the N- and C-termini in the cytosol (Wu and Rosen, 1992). Acid (□) an
ated.

ATPase activity. More recently, an immunodot blot assay has been developed for quantitative determination of the ArsA protein bound to ArsB protein-containing vesicles (Dey et al., 1994b). Binding of ArsA protein is a saturable process and exhibits sigmoidal binding with respect to the amount of added ArsA protein. The binding isotherm is converted to a hyperbolic relationship in presence of the anionic substrates, arsenite or antimonite, with a 10-fold reduction in the concentration of ArsA protein required for half-maximal saturation of the membranes. As discussed above, the ArsA protein is a functional homodimer, and the concentration of dimer would be expected to increase with total ArsA concentration. These results suggest that the homodimer binds to the ArsB protein with higher affinity than the monomer. The membrane-bound ArsA–ArsB complex would, therefore, be expected to include a homodimer of the ArsA protein at all times, and it would also be expected that the catalytic cycle might include conformational movement of the subunits without actual dissociation of the complex (Figure 5).

The ArsB protein has not been amenable to over-expression or antibody production. To circumvent these problems, a gene fusion was created that produced a functional chimeric protein containing an N-terminal portion of the ArsA protein and, on the C-terminal side, the entire ArsB sequence except for the initial methionyl residue (Dou et al., 1992). This 94 kD chimeric polypeptide, termed the ArsAB2 protein, was expressed in substantial amounts. Since it contains part of the ArsA protein, antiserum against the ArsA protein could be used to detect the chimera. When a wild-type *arsA* gene was co-expressed with the *arsAB2* gene, the ArsA protein was bound to the membrane in amounts proportional to the amount of ArsAB2 protein in the membrane. In addition, soluble ArsA protein could be reconstituted with membranes containing the ArsAB2 protein (S. Dey and B.P. Rosen, unpublished). When saturating amounts of ArsA protein were added, the amount of membrane-bound ArsA was proportional to the amount of the ArsAB2 protein present in the membrane, consistent with a specific interaction of the two proteins.

As discussed previously, topological analysis of the ArsB protein indicated five cytoplasmic regions (Wu and Rosen, 1992) that are potential sites for ArsA protein binding. Three of the five cytoplasmic sequences (C1–C5) contain more than one charged amino-acyl residue, with ten of them in the C3 sequence. When a portion of the C1 sequence was deleted, there was no effect on binding of the ArsA protein, but the construct did not confer resistance to arsenite (Dey et al., 1994a). Thus, while the C1 sequence is not required for ArsA binding, it is required for function. A series of C-terminal deletions of the ArsB protein were constructed (Chen et al., 1986b). A truncated ArsB protein lacking the C5 sequence was incorporated into the membrane, but did not bind ArsA protein (Tisa and Rosen, unpublished results). Construction of additional deletions and insertions into the ArsB protein will be required to determine regions of ArsA interaction.

E. *In Vitro* Transport of $^{73}AsO_2^-$

Recently, an *in vitro* assay for arsenite transport has been developed (Dey, Dou, and Rosen, manuscript in preparation). Everted membrane vesicles prepared from *E. coli* cells co-expressing the *arsA* and *arsAB2* genes exhibited ATP-dependent uptake of $^{73}AsO_2^-$. Uptake in these inside-out vesicles is the equivalent of extrusion from intact cells. The activity of the pump required both subunits of the ATPase complex; everted vesicles prepared from cells expressing either of the two genes had no ATP-dependent uptake of $^{73}AsO_2^-$. Removal of the ArsA protein from the membrane-bound complex with 0.8 M KCl eliminated transport. AsO_4^{3-}, the oxyanion of +5 oxidation state of arsenic, was not transported, consistent with the specificity of the pump. The apparent K_i for inhibition of $^{73}AsO_2^-$ transport by antimonite was 10-fold less than the K_m for arsenite, indicating that the pump has higher affinity for antimonite than arsenite.

The energetics of $^{73}AsO_2^-$ accumulation indicates obligatory coupling to ATP hydrolysis. In the absence of ATP or in the presence of ATPγS, a nonhydrolyzable analog of ATP, no detectable transport was observed. Hydrolysis of ATP by purified ArsA protein requires Mg^{2+} (Hsu and Rosen, 1989). Uptake in everted membrane vesicles similarly exhibited a strict Mg^{2+} requirement, and neither Ca^{2+} nor Mn^{2+} could replace Mg^{2+}. No accumulation was observed at 4° C, consistent with a requirement ATP hydrolysis. When vesicles were prepared from an *unc*⁻ strain, which lacks the H^+-translocating F_0F_1, respiratory substrates, such as lactate or NADH, were unable to support $^{73}AsO_2^-$ uptake. In addition, ATP-driven transport was insensitive to the protonophore, CCCP. These results indicate that an electrochemical gradient is neither necessary nor sufficient for AsO_2^- transport, consistent with energetics observed in intact cells (Mobley and Rosen, 1983; Rosen and Borbolla, 1984).

ATP-dependent anion transport was unaffected by NaN_3 or orthovanadate, inhibitors of F-Type and P-Type ATPases, respectively (Pedersen and Carafoli, 1987). N-ethylmaleimide was a potent inhibitor of $^{73}AsO_2^-$ uptake in vesicles. N-ethylmaleimide similarly inhibits oxyanion-stimulated ATPase activity by the purified ArsA protein (M. Ksenzenko and B.P. Rosen, manuscript in preparation).

F. The ArsC Protein

The *ars* operon of the broad host-range plasmid R773 carries a third structural gene, the *arsC* gene, that is required for resistance to arsenate in addition to the *arsA* and *arsB* genes (Rosen and Borbolla, 1984; Chen et al., 1985). While arsenate and arsenite are both oxyanion salts of arsenic, they contain the +5 and +3 oxidation states, respectively, of arsenic and are thus chemically dissimilar. The way in which expression of the *arsC* gene expands the range of resistance of the determinant has been studied. The ArsC protein, composed of 141 residues with a molecular mass of 15,811 daltons, has been purified and crystallized (Rosen et al., 1991). However,

```
          *  *                 *
  MSNITIYHNPACGTSRNTLEMIRNSGTEPTIILYLENPPSRDELVKLIADMGISV
     MEIWINPACSKCRSAVQLLDAEGADYTVRRYLEDVPSEDEIRQVLDRLGLEP
MTSTSIVFYEKPGCATNRLQKQLLRSAGLHLEVRNLLSEPWTPERLRPFFGDR--PV
  MADVLFFEKPGCRNNTRQKALLIASGHRVEAHDIRQQPWTAETLRPYFGDK--PV
                    MDKKTIYFICTGNSCRSQMAEGWGKEILGEGWNVYS

                                        ** **
RALLRKNVEPYEQLGL---AEDKFTDDQLIDFMLQHPILINRPIVVTPLGTRLC-RP
WDITRTQEAEAKELGVKEWARDASARDQWIKALAEHPKLIQRPIITADDGTAVVGRT
VEWFNRSAPAIKYGEL---DPTQLSASQALELMLAHPILIRRPLIHYGSQYMAGFDL
AQWINPAAPRVKAGEV---RPEALDESEALALMVKDALLIRRPLMAVGQTKTCGFDR
AGIETHGVNPKAIEAM---KEVDIDISNHTSDLIDNDILKQSDLVVTLCSDADNNCP

SEVVLDILQDAQKGAFTKEDGEKVVDEAGKRLK       R773 ArsC protein
DEAVRDALSR                              Orf,Streptomyces gln locus
VKLCTHLPAKNDLPDAGEDLEKRSVGQNPIGCLAAGNQ  Orf, A. vinlandii nif gene cluster
AAVNAWIGLIAQDPGDIETCPSQATNHRCAEEPGAA    Orf, R. rubrum  nif gene cluster
ILPPNVKKEHWGFDDPAGKEWSEFQRVRDEIKLAIEKFKLR pI258 ArsC protein
```

Figure 8. The ArsC family. The ArsC proteins from Gram-negative plasmid R773 (Chen et al., 1986b) and the Gram-positive plasmids pl258 (Ji and Silver, 1992a) and pSX267 (not shown; Rosenstein et al., 1992) are related in function, but the level of sequence similarity is not statistically significant. Other sequences more closely related to the R773 ArsC protein include reading frames of unknown function from the *Streptomyces gln* locus (Behrmann et al., 1990) and from *nif* loci of *A. vinelandii* (Mevarech et al., 1980) and *R. rubrum* (Fitzmaurice et al., 1989). Similarities are shaded.

the *in vivo* mechanism of functioning of this hydrophilic polypeptide is not yet clear. There are no data to indicate that the ArsC protein is part of the membrane-bound ATPase complex in *E. coli.*

Recently, the Gram-positive counterparts of the *ars* operon from the staphylococcal plasmids pI258 and pSX267 have been identified (Ji and Silver, 1992a; Rosenstein et al., 1992). The plasmids carry *arsC* genes that are nearly identical to each other, but whose slightly larger product (151 residues) is only distantly related to the R773 *arsC* gene product (Figure 8). Although the sequences of the Gram-positive and Gram-negative ArsC proteins are quite different, both confer resistance to arsenate. Resistant cells of both *S. aureus* and *E. coli* reduced arsenate to arsenite (Ji and Silver, 1992b), which was presumed to be extruded from the cells by the ArsA–ArsB complex in *E. coli* and by the ArsB protein alone in *S. aureus*. The pI258 ArsC protein exhibited reductase activity *in vitro*, indicating that it is an intracellular reductase, converting arsenate to arsenite (Ji and Silver, 1992b). However, in *E. coli*, arsenate was reduced to arsenite at almost the same rate in cells lacking the *ars* operon (Oden et al., 1994), so that the relationship between reduction and resistance appears to be more complex than the *in vitro* data suggest. The source of reductant *in vivo* is not yet clear. *In vitro* purified *S. aureus* ArsC protein could use *E. coli* thioredoxin (Holmgren, 1989) as a reductant in a heterologous system (Ji and Silver, 1992b). Glutaredoxin did not replace thioredoxin *in vitro*. Recently, Bröer et al. (1993) have shown that chemical energy is required for the overall process of arsenate reduction and extrusion of the derived arsenite in *S. aureus*. Whether chemical energy is required solely for reductive energy or for the transport process was not determined in that study.

In contrast, the R773 ArsC protein did not exhibit thioredoxin-coupled arsenate reduction, and *E. coli* mutants defective in thioredoxin reductase or in the structural genes for thioredoxin or glutaredoxin exhibited normal *ars* operon-mediated arsenate resistance (Oden et al., 1994). On the other hand, mutants defective in the genes for either glutathione synthetase or glutathione reductase were sensitive to arsenate, but not to arsenite in the presence of all three *ars* structural genes. Thus, glutathione is an obligatory intermediate in the pathway of reduction of arsenate to arsenite in *E. coli*. These studies do not eliminate glutaredoxin as a potential reductant, only as an obligatory component; alternate pathways are certainly possible. However, since in *E. coli* reduction of thioredoxin does not require glutathione, these results would tend to eliminate thioredoxin as a reductant.

If the two ArsC proteins are reductases *in vivo*, the mechanism of electron transfer is not obvious from the primary sequence of either protein. The R773 ArsC protein does not contain prosthetic groups, molybdenum cofactors or iron–sulfur cages (Oden et al., 1994). It is possible that both ArsC proteins contain redox-active cysteine residues. The pI258 ArsC protein contains four cysteinyl residues, two near the amino-terminus, and two near the carboxyl-terminus, separated by 4 and 6 amino acid residues, respectively. In contrast, the R773 ArsC protein has only two cysteinyl residues, one near either terminus of the protein. Neither of these cysteines appear to be in regions of the protein conserved in the pI258 protein. Mutagenic replacement of the cysteinyl residues in both proteins is in progress and should shed light on the involvement of these residues in function.

Arsenite poses a greater threat to the viability of the cell than arsenate since it covalently inactivates enzymes, while arsenate simply competes with phosphate in enzymatic reactions (Knowles and Benson, 1983). Thus, reduction within the cytosol is undesirable. It is possible that the ArsC protein is a part of the membrane bound Ars complex and is involved in reduction of arsenate on the membrane, where it would be immediately extruded from the cells. This could explain why the rates of reduction *in vivo* in *E. coli* are nearly the same whether or not the *arsC* gene is expressed, but only those cells expressing the gene are resistant (Oden et al., 1994).

Turner et al. (1992) found that in *E. coli*, the ArsC protein is also required for resistance to tellurite, the +4 oxidation state oxyanion of tellurium. Cells harboring the *ars* operon, deleted of the *arsC* gene, show reduced growth in the presence of inhibitory concentrations of potassium tellurite. The exact role of the ArsC protein in tellurite resistance is unknown. However, here too, reduction of tellurite to metallic tellurium occurs at almost the same rate, whether or not the *arsC* gene is expressed. Just as with arsenate, it could be postulated that a membrane bound complex of the ArsA, ArsB, and ArsC proteins is required for resistance, with the ArsC protein reducing tellurite to a more reduced oxyanion that is then extruded by the pump.

Other homologs of the R773 ArsC protein have been identified (Figure 8). The most closely related (31% identity) is a potential open reading frame of unknown function near the *Streptomyces biaphalos* glutamine biosynthesis operon (Behrmann et al., 1990). The ArsC protein also exhibited limited similarity (19% identity) to an open reading frame for a protein of 147 residues of unknown function in an *nif* gene cluster from *Azotobacter vinelandii* (Joerger and Bishop, 1988). A very low degree of similarity has also been suggested between the *arsC* and *amiB* genes (Alloing et al., 1990).

G. Evolution of an Ion Pump

The operons from the *E. coli* plasmid R773 (Chen et al., 1986b) and from the two staphylococcal plasmids, pI258 (Ji and Silver, 1992a) and pSX267, (Rosenstein et al., 1992) have similar functions and show similarities. Strikingly, however, is the absence of the *arsA* gene in the Gram-positive operons. The question arises as to how a transport ATPase can function without the ATPase subunit.

One possibility is that both are transport ATPases, but that the gene for the ATPase is chromosomal in Gram-positive organisms. However, the pI258 ArsB protein has recently been shown to be sufficient to produce arsenite resistance in *E. coli* (Bröer et al., 1992; Dou et al., 1994; Dey and Rosen, 1995). Unless *E. coli* also carries the same chromosomal gene as *Staphylococcus aureus*, this possibility seems unlikely.

Another possibility is that the ArsB protein by itself acts as a secondary anion transporter. Again, it is striking that the 12-membrane spanning α-helical structure of the ArsB protein is overall very similar to many secondary porters (Marger and Saier, 1993). The positive exterior membrane potential could drive anion extrusion. Similarly the H^+-translocating F_0F_1 ATPase is an ATP-dependent H^+ pump, but when the complex is dissociated, the two portions function independently. Dissociated F_1 is a soluble ATPase. By itself, F_0 is a proton conducting pathway which transports protons into the cell, responding to the $\Delta\psi$ (Figure 9).

If the Gram-positive ArsB protein is a secondary porter, could the R773 ArsB protein also catalyze arsenite transport by itself? Proteins as structurally related as the two ArsB proteins would be expected to have similar mechanisms. The R773 ArsB protein has recently shown to be sufficient for arsenite resistance (Dou et al., 1994; Dey and Rosen, 1995). Moreover, co-expression of an enzymatically inactive ArsA mutant protein inhibits the function of the wild-type ArsB protein (Kaur and Rosen, 1992b), just as mutant F_1 prevents proton conduction by the F_0 (Futai and Kanazawa, 1983). These results suggest that the ArsB protein is a secondary anion extrusion system coupled to the $\Delta\psi$, while, like the F_0F_1, the ArsA–ArsB complex functions only as a transport ATPase (Figure 9). Since ATP levels drop more slowly

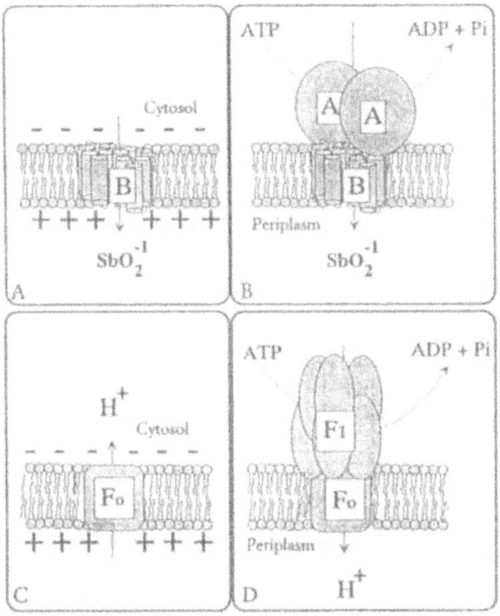

Figure 9. Primary and secondary transport properties of ion translocating systems. The components of ion transport systems such as the oxyanion pump (**A** and **B**) or the H^+-translocating F_0F_1 (**C** and **D**) function either as secondary porters (**A** and **C**) or primary ATP-driven pumps (**B** and **D**), depending on association of the catalytic subunits with the intrinsic membrane subunits.

than the membrane potential under conditions of stress, an ATPase would be a more effective mechanism of providing resistance to environmental toxins.

This leads to a hypothesis for the evolution of transport ATPases. Multicomponent transport ATPases could not have emerged *de novo*. Rather, they arose through the parallel and independent evolution of the genes for soluble ATPases and membrane proteins, with primordial physiological roles quite different from their present transport function. The soluble ATPases and membrane proteins gradually evolved affinity for each other, with the interactions gradually becoming functional to produce transport ATPases such as the Ars oxyanion pump and the F_0F_1 (Figure 10). Other solute transport ATPases, such as the histidine (Ames, 1990) and maltose (Davidson and Nikaido, 1991) permeases of *E. coli*, may have evolved similarly. The genes for the eukaryotic members of that family, including the P-glycoprotein (Chen et al., 1986a; Juranka et al., 1989) and cystic fibrosis transmembrane regulator (CFTR) protein (Riordan et al., 1989) subsequently fused to produce single large genes encoding multifunctional proteins. It is of interest that the overall similarity between the members of that family seems to be restricted to the portions of the proteins involved in energy utilization. Confining the analysis to the regions

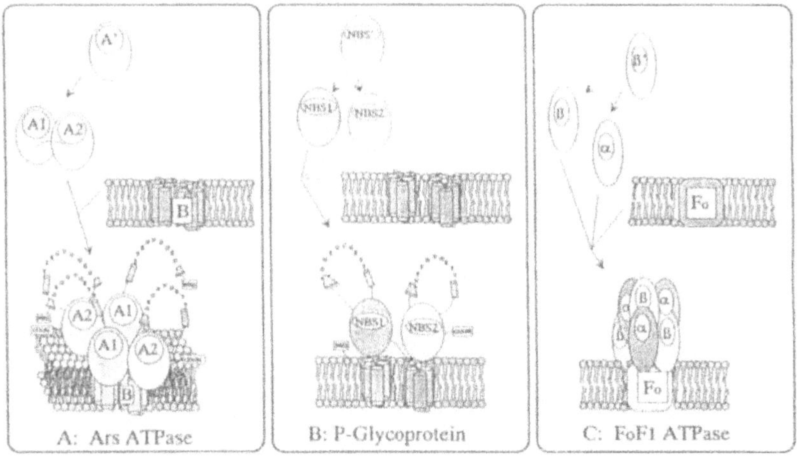

Figure 10. Parallel evolution of genes for the components of transport ATPases. The genes for membrane proteins such as the ArsB protein, the F_0, or the ancestor of the membrane sector of the P-glycoprotein may have evolved as secondary carriers or ion channels. In parallel, the genes for the smaller nucleotide binding proteins ancestral to the N- and C-terminal halves of the ArsA protein or the P-glycoprotein or to the α and β subunits of the F_1 would have evolved through duplications, with divergence of the descendants. Genes for the present day proteins could have evolved with multiple nucleotide binding sites by gene fusions, as in the case of ArsA, or could remain as linked genes within a single operon, as in the case of the genes for the α and β subunits of F_1. Gradually the membrane proteins would have developed binding sites for soluble ATP-binding proteins. In additional evolutionary stages, this association would progress to become functional, resulting in the development of transport ATPases such as the oxyanion pump (**A**) and the H^+-translocating F_0F_1. Subsequent fusions in eukaryotes would have created single genes for large multifunctional proteins such as the P-glycoprotein (**C**).

around the ATP-binding sites results in a family tree that includes the entire superfamily (Reizer et al., 1992). However, it does not appear that the membrane proteins or the membrane portions of the larger eukaryotic proteins are similarly related in a single superfamily (Reizer et al., 1992). For example, the integral membrane proteins of the Nod or Cys systems are related to each other, but unrelated to the analogous membrane constituents of the Cys or Nos systems, even though the subunits with the ATP-binding sites are closely related. Similarly, analysis of the membrane proteins of the histidine permease demonstrates no relationship to the membrane sectors of the P-glycoprotein. In turn, neither of them show a relationship to the *drrB* gene product, the putative membrane sector of the *Streptomyces* daunorubicin efflux system (Guilfoile and Hutchinson, 1991). Different membrane proteins may have formed association with a family of soluble

ATPase to give rise to a heterogeneous group of solute ATPases such as the ABC superfamily. Evolution of pumps through association with different types of membrane proteins could also account for the observation that some members of the family are ion channels while most are probably not.

III. OTHER ANION-TRANSLOCATING ATPASES

Most known ion-translocating ATPase are cation specific (Pedersen and Carafoli, 1987). Only a few anion-translocating ATPases have been reported, all in eukaryotes. Although molecular biological information is not currently available for these systems, biochemical characterizations have been reported.

A. A Cl^--translocating ATPase from *Aplysia* Gut

An ATP-dependent Cl^- transport activity has been reported by Gerencser (1988) in the enterocyte basolateral membrane of *Aplysia*. Plasma membrane vesicles from *Aplysia* enterocytes exhibited ATP-dependent Cl^- transport that was sensitive to vanadate. Transport was electrogenic, as measured by the distribution of the lipophilic cation triphenylmethylphosphonium across the vesicular membrane. Cl^--stimulated ATPase activity was inhibited by N-ethylmaleimide and p-chloromercuribenzenesulfonate. The fact that dithiothreitol reversed p-chloromercuribenzenesulfonate inhibition suggested that surface sulfhydryl ligands of the ATPase participated in the catalytic activity (Gerencser, 1990a). The ATPase has been solubilized and reconstituted in liposomes (Gerencser, 1990b). The reconstituted Cl^- transport activity in proteoliposomes retained sensitivity to vanadate.

B. An ATP-driven Cl^- Pump from Rat Brain

An inwardly directed Cl^- gradient was shown to be maintained in neuronal cells (Lux, 1971). The concentration of intracellular Cl^- was lower than expected from passive distribution, and the resulting hyperpolarization produces postsynaptic potentials that regulate a variety of neuronal activities (Eccles et al., 1964; Barker and Owen, 1986; Haefely and Polc, 1986). Inagaki and Shiroya (1988) first reported an ATP-driven Cl^- pump from the plasma membrane of rat brain cells. In addition, EDTA-treated microsomes, rich in Cl^--ATPase activity, were prepared from rat brain and shown to accumulate Cl^- in an ATP-dependent manner (Inagaki and Shiroya, 1988). Uptake was dependent on the concentrations of ATP and Cl^-, with K_m values of 1.5 mM and 7.4 mM, respectively. Among the other nucleotides, GTP, ITP, and UTP partially stimulated Cl^- uptake, whereas CTP, ADP, and AMP did not. The nonhydrolyzable ATP analog, β,γ-methylene ATP, also was unable to stimulate Cl^- uptake, consistent with a requirement for ATP hydrolysis. ATP-dependent Cl^- uptake in microsomes was insensitive to ouabain, an inhibitor of the Na^+,K^+-ATPase, and to DCCD, which inhibits the H^+-translocating F_0F_1 ATPase. These results tend to rule out Na^+ or H^+ as the coupling ion for Cl^- transport. On the other hand, Cl^- uptake was sensitive to orthovanadate, similar to the Cl^- transport system in *Aplysia* intestine. These results suggest that both chloride pumps may have phosphorylated intermediates in their reactions. Ethacrynic acid, an inhibitor of glutathione S-transfereases, and N-ethylmaleimide also completely inhibited ATP-dependent Cl^- transport.

C. ATP-dependent Efflux of Methotrexate in Leukemia Cells

Methotrexate is an antifolate widely used as an antimicrobial and anti-neoplastic agent. It is a potent inhibitor of the enzyme, dihydrofolate reductase. At physiological pH, methotrexate is a divalent anion. Energy-dependent efflux of methotrexate in tumor cells was first proposed by Hakala (1965) using the sarcoma 180 cell line. A similar methotrexate efflux system producing resistance to methotrexate was also observed in L1210 leukemia cells (Goldman, 1969; Dembo and Sirotnak, 1976; Dembo et al., 1984; Sirotnak and O'Leary, 1991). In everted plasma membrane vesicles of leukemia cells, extrusion of this folate analog was ATP-dependent (Schlemmer and Sirotnak, 1992). No efflux was observed with ATPγS, the nonhydrolyzable analog of ATP. Methotrexate efflux in everted membrane vesicles was sensitive to the sulfhydryl inhibitor, N-ethylmaleimide, and to orthovanadate. Although these are properties of P-type ATPases, no P-type ATPase has ever been demonstrated to transport anything but a cation. Other compounds, including bromosulfopthalein, verapamil, reseprine, quinidine, and cholate were all potent inhibitors of ATP-dependent methotrexate transport (Schlemmer and Sirotnak, 1992). This property is similar to that of the multidrug resistance-related P-glycoprotein (Juranka et al., 1989), but these are apparently different systems. Tumor cells that express P-glycoprotein did not show cross-resistance to methotrexate (Biedler et al., 1983). However, the methotrexate-translocating ATPase may be part of the same family of ATP-coupled solute pumps (Juranka et al., 1989; Hyde et al., 1990).

IV. ARSENITE RESISTANCE IN EUKARYOTES

A. Arsenite Resistance in *Leishmania*

Because of their toxicity, the use of arsenical and antimonial drugs is now limited to the treatment of a few topical protozoan diseases. Pentavalent antimony in the form of pentostam is still a widely used chemotherapeutic agent in the treatment of leishmaniasis. Treatment failure results from emergence of pentostam-resistant *Leishmania*. Pentostam-resistant clinical isolates were cross-resistant to trivalent antimony when cultured in macrophages (Berman, 1988). Several workers have proposed that pentavalent antimony derivatives were metabolized *in vivo* into trivalent antimony compounds, which may be the active form (Croft et al., 1981). Thus, resistance to trivalent antimonials *in vitro* may reflect resistance to pentavalent antimonials during infection *in vivo* (Callahan and Beverley, 1991). As discussed above, arsenite and antimonite are chemically similar and are both substrates of the Ars ATPase in *E. coli*. As might be expected, cross-resistance to arsenite occurs in antimony-resistant *Leishmania*. Amplification of the genes *lmpgpA* and *ltpgpA* from *L. major* and *L. tarantolae*, respectively, has been correlated with resistance to arsenite and trivalent antimonials (Callahan and Beverley, 1991; Ouellette and Borst, 1991). Grogl et al. (1991) showed a fivefold increase in accumulation of ^{125}Sb-pentostam in drug-sensitive over drug-resistant clones of *Leishmania*, and some degree of specificity in the binding of Pentostam to the *Leishmania* P-glycoprotein-like components. This differential accumulation of drugs in resistant versus sensitive *Leishmania* clones could be due to an efflux mechanism similar to that found in chloroquine-resistant *Plasmodium* and multidrug resistant neoplastic cells (Grogl et al., 1991).

However, the relationship between amplification of the *pgp*-like genes and pentostam resistance is not entirely clear. Recently, four classes of arsenite-resistant mutants were obtained (Dey et al., 1994c; Papadopoulou et al., 1994). The first contained an H locus-derived circular amplicon alone. A second contained a linear amplicon alone. A third had both amplicons. Finally, a fourth class had no detectable DNA amplification. Using a $^{73}AsO_2^-$ transport assay similar to that developed for arsenite-resistant *E. coli*, the steady-state accumulation of radiolabeled arsenite was measured in a representative of each of the four classes of arsenite-resistant mutants and in trivalent and pentavalent antimony resistant cell lines. In all these mutants, whether or not they had gene amplification, a marked decreased in the accumulation of arsenite was observed. Following treatment of the cells with metabolic inhibitors, an increase in arsenite accumulation was observed in the resistant parasite, consistent with efflux of arsenite from the cells. Thus, high level resistance is apparently related to the emergence of an efflux system that is seemingly unrelated to the LtpgpA protein.

The role of the amplicons in oxyanion resistance was investigated using gene transfection experiments (Dey et al., 1994c; Papadopoulou et al., 1994). The results of transfection experiments in *Leishmania* suggested that *ltpgpA* is involved in low level resistance to arsenite and antimonite, as well as to pentavalent antimony. Transfection of either complete large amplicons or alleles of *ltpgpA* isolated from mutants selected for arsenite resistance did not give higher resistance. No evidence for reduced accumulation or active efflux of arsenite in cells transfected with the *ltpgpA* gene was found. Either the low resistance conferred by *ltpgpA* is the result of a low level of arsenite efflux or resistance is by a different mechanism.

B. Arsenite Resistance in Mammalian Cells

Resistance to arsenite and antimonite has been reported in Chinese hamster cell lines (Wang and Rossman, 1993; Lo et al., 1992). Active efflux has been suggested as a mechanism (Wang and Rossman, 1993), although a role for the enzyme, glutathione S-transferase π, has been suggested (Lo et al., 1992). The two possibilities are not mutually inconsistent. An efflux system for glutathione S-conjugants has been characterized in vesicles from rat heart and liver (Ishikawa, 1992). This is a orthovanadate-sensitive transport ATPase involved in extrusion of xenobiotics, such as aflatoxin B1, and release of biologically important substances, such as leukotrienes. This system has not been reported to be involved in arsenite detoxification (Ishikawa, 1992), nor has the glutathione S-transferase been shown to form a glutathione-arsenite conjugate (Lo et al., 1992). However, biliary excretion of arsenic after administration of either arsenite or arsenate has been shown to be dependent on the hepatobiliary transport of glutathione, suggesting that transport of arsenic as a glutathione complex may account for the glutathione dependence of biliary arsenic excretion (Gyurasics et al., 1991).

REFERENCES

Abernathy, J. R. (1983). Role of arsenical chemicals in agriculture. In: Arsenic (Lederer, W. H., & Fensterheim, R. J., eds.), pp. 57–62. Van Nostrand Reinhold Co., New York.

Alden, J. C. (1983). The continuing need for inorganic arsenical pesticides. In: Arsenic (Lederer, W. H., & Fensterheim, R. J., eds.), pp. 63–71. Van Nostrand Reinhold Co., New York.

Alloing, G., Trombe, M. C., & Claverys, J. P. (1990). The *amiB* locus of the Gram-positive bacterium *Streptococcus pneumoniae* is similar to binding protein-dependent transport operons of Gram-negative bacteria. Mol. Microbiol. 4, 633–644.

Ames, G. F. L. (1990). Energetics of periplasmic transport systems. The Bacteria 12, 225–246.

Barker, J. L., & Owen, D. B. (1986). Electrophysiological pharmacology of GABA and diazepam in cultered CNS neurons. In: Receptor Biochemistry and Methodology, Volume 5: Benzodiazprine/GABA Receptors and Chloride Channels, Structural and Functional Properties (Venter, J. C., & Harrison, L. C., series eds.; Olsen, R. W., & Venter, J. C., volume eds.), pp. 135–165. Alan R. Liss, New York.

Bechhofer, D. H., & Figurski, D. H. (1983). Map location and nucleotide sequence of *korA*, a key regulatory gene of promiscuous plasmid RK2. Nucleic Acids Res. 11, 7453–7469.

Behrmann, I., Hillemann, D., Puhler, A., Strauch, E., & Wohlleben, W. (1990). Overexpression of a *Streptomyces viridochromogenes* gene (*glnII*) encoding a glutamine synthase similar to those of eukaryotes confers resistance against the antibiotic phosphinothricylalanyl-alanine. J. Bacteriol. 172, 5326–5334.

Bell, A. W., Buckel, S. D., Groarke, J. M., Hope, J. N., Kingsley, D. H., & Hermodson, M. A. (1986). The nucleotide sequence of the *rbsD*, *rbsA*, and *rbsC* genes of *Escherichia coli* K12. J. Biol. Chem. 261, 7652–7658.

Berman, J. D. (1988). Chemotherapy for leishmaniasis: Biochemical mechanisms, clinical efficacy and future strategies. Rev. Infect. Dis. 10, 560–586.

Biedler, J. L., Chang, I. D., Meyers, M. B., Peterson, R. H., & Spengler, B. A. (1983). Drug resistance in Chinese hamster lung and mouse tumor cells. Canc. Treat. Rep. 67, 859–867.

Bröer, S., Ji, G., Bröer, A., & Silver, S. (1993). Arsenic efflux governed by the arsenic resistance determinant of *Staphylococcus aureus* plasmid pI258. J. Bacteriol. 175, 3480–3485.

Bryson, P. D. (1989). Arsenic. Comprehensive Review in Toxicology, 2nd ed., pp. 501–508. Aspen Publishers, Rockville, MD.

Callahan, H. L., & Beverley, S. M. (1991). Heavy metal resistance: A new role for P-glycoprotein in *Leishmania*. J. Biol. Chem. 266, 18427–18430.

Chen, C. J., Chin, J. E., Ueda, K., Clark, D. P., Pastan, I., Gottesman, M. M., & Roninson, I. B. (1986a). Internal duplication and homology with bacterial transport proteins in the *mdr1* (P-glycoprotein) gene from multidrug-resistant human cells. Cell 47, 381–389.

Chen, C. M., Mobley, H. L. T., & Rosen, B. P. (1985). Separate resistances to arsenate and arsenite (antimonite) encoded by the arsenical resistance operon of R-factor R773. J. Bacteriol. 161, 758–763.

Chen, C. M., Misra, T., Silver, S., & Rosen, B. P. (1986b). Nucleotide sequence of the structural genes for an anion pump: The plasmid-encoded arsenical resistance operon. J. Biol. Chem. 261, 15030–15038.

Croft, S. L., Neame, K. D., & Homewood, C. A. (1981). Accumulation of [^{125}Sb]sodium stibogluconate by *Leishmania mexicana amazonensis* and *Leishmania donovani in vitro*. Comp. Biochem. Physiol. 68C, 95–98.

Davidson, A. L., & Nikaido, H. (1991). Purification and characterization of the membrane-associated components of the maltose transport system from *Escherichia coli*. J. Biol. Chem. 266, 8946–8951.

Davis, M. A., Martin, K. A., & Austin, S. J. (1992). Biochemical activities of the ParA partition protein of P1 plasmid. Molec. Microbiol. 6, 1141–1147.

de Boer, P. A. J., Crossley, R. E., & Rothfield, L. I. (1989). A division inhibitor and a topological specificity factor encoded for by the minicell locus determine proper placement of the division septum in *E. coli*. Cell 56, 641–649.

Dembo, M., & Sirotnak, F. M. (1976). Antifolate transport in L1210 leukemia cells. Kinetic evidence for the non-identity of carriers for influx and efflux. Biochim. Biophys. Acta 448, 505–516.

Dembo, M., Sirotnak, F. M., & Moccio, D. M. (1984). Effects of metabolic deprivation on methotrexate transport in L1210 leukemia cells: Further evidence for separate influx and efflux systems with different energetic requirements. J. Mem. Biol. 78, 9–17.

Dey, S., & Rosen, B. P. (1995). Dual mode of energy coupling by the oxyanion-translocating ArsB protein. J. Bacteriol. 177, 385–389.

Dey, S., Dou, D., & Rosen, B. P. (1994a). ATP-dependent transport in everted membrane vesicles of *Escherichia coli*. J. Biol. Chem. 269, 25442–25446.

Dey, S., Dou, D., Tisa, L. S., & Rosen, B. P. (1994b). Interaction of the catalytic and the membrane subunits of an oxyanion-translocation ATPase. Arch. Biochem. Biophys. 311, 418–424.

Dey, S., Papadopoulou, B., Roy, G., Grondin, K., Dou, D., Rosen, B. P., & Ouellette, M. (1994c). High level arsenite resistance in *Leishmania tarentolae* is mediated by an active extrusion system. Molec. Biol. Parasitol. 67, 49–57.

Dou, D., Owalabi, J. B., Dey, S., & Rosen, B. P. (1992). Construction of a chimeric ArsA–ArsB protein for overexpression of the oxyanion-translocating ATPase. J. Biol. Chem. 267, 25768–25775.

Dou, D., Dey, S., & Rosen, B. P. (1994). A functional chimeric membrane subunit of an ion-translocating ATPase. Antonie van Leeuwenhoek 65, 359–368.

Eccles, J., Eccles, R. M., & Ito, M. (1964). Effects produced on inhibitory postsynaptic potentials by the coupled injections of cations and anions into motor neurons. Proc. Roy. Soc. London 160, 197–210.

Ehrlich, P. (1960). On the partial function of the cell. In: Collected Papers of Paul Ehrlich (Himmelweit, F., ed.), pp. 183–194. Pergammon Press, London.

Fitzmaurice, W. P., Saari, L. L., Lowery, R. G., Ludden, P. W., & Roberts, G. P. (1989). Genes coding for the reversible ADP-ribosylation system of dinitrogenase reductase from *Rhodospirillum rubrum*. Mol. Gen. Genet. 218, 340–347.

Futai, M., & Kanazawa, H. (1983). Structure and function of proton-translocating ATPase (F_0F_1): Biochemical and molecular biological approaches. Microbiol. Rev. 47, 285–313.

Gerenscer, G. A. (1988). Electrogenic ATP-dependent Cl^- transport by plasma membrane vesicles from *Aplysia* intestine. Am. J. Physiol. 254, 127–133.

Gerenscer, G. A. (1990a). Inhibition of Cl^--stimulated ATPase activity in isolated basolateral membranes from *Aplysia* gut. Biochim. Biophys. Acta 1023, 475–477.

Gerenscer, G. A. (1990b). Reconstitution of a chloride-translocating ATPase from *Aplysia californica gut*. Biochim. Biophys. Acta 1030, 301–303.

Goldman, I. D. (1969). Transport energetics of folic acid analogue, methotrexate, in L1210 leukemia cells. Enhanced accumulation by metabolic inhibitors. J. Biol. Chem. 243, 5007–17.

Grogl, M., Martin, R. K., Oduola, A. M., & Kyle, D. E. (1991). Characteristics of multidrug resistance in *Plasmodium* and *Leishmania*: Detection of P-glycoprotein like components. Am. J. Trop. Med. Hyg. 45, 98–111.

Guilfoile, P. G., & Hutchinson, C. R. (1991). A bacterial analog of the *mdr* gene of mammalian tumor cells is present in *Streptomyces peucetius*, the producer of daunorubicin and doxorubicin. Proc. Natl. Acad. Sci. USA 88, 8553–8557.

Gyurasics, A., Varga, F., & Gregus, Z. (1991). Glutathione-dependent biliary excretion of arsenic. Biochem. Pharmacol. 42, 465–468.

Haefely, W., & Polc, P. (1986). Physiology of GABA enhancement by benzodiazepines and barbiturates. In: Receptor Biochemistry and Methodology, Volume 5: Benzodiazprine/GABA Receptors and Chloride Channels, Structural and Functional Properties (Venter, J. C., & Harrison, L. C., series eds.; Olsen, R. W., & Venter, J. C., volume eds.), pp. 97–133. Alan R. Liss, New York.

Hakala, M. T. (1965). On the nature of permeability of sarcoma-180 cells to amethopterin *in vivo*. Biochim. Biophys. Acta 102, 210–225.

Hedges, R. W., & Baumberg, S. (1973). Resistance to arsenic compounds conferred by a plasmid transmissible between strains of *Escherichia coli*. J. Bacteriol. 115, 459–460.

Holmgren, A. (1989). Thioredoxin and glutoredoxin systems. J. Biol. Chem. 264, 13963–13966.

Hsu, C. M., & Rosen, B. P. (1989). Characterization of the catalytic subunit of an anion pump. J. Biol. Chem. 264, 17349–17354.

Hsu, C. M., Kaur, P., Karkaria, C. E., Steiner, R. F., & Rosen, B. P. (1991). Substrate-induced dimerization of the ArsA protein, the catalytic component of an anion-translocating ATPase. J. Biol. Chem. 266, 2327–2332.

Hyde, S. C., Emsley, P., Hartshorn, M. J., Mimmack, M. M., Gileadi, U., Pearce, S. R., Gallagher, M. P., Gill, D. R., Hubbord, R. E., & Higgins, C. F. (1990). Structural model of ATP-binding protein associated with cystic fibrosis, multidrug resistance and bacterial transport. Nature 346, 362–365.

Inagaki, C., & Shiroya, T. (1988). ATP-dependent Cl^- uptake by plasma membrane vesicles from rat brain. Biochem. Biophys. Res. Commun. 154, 108–112.

Ishikawa, T. (1992). ATP-dependent glutathione S-conjugate export pump. Trend. Biochem. Sci. 17, 463–468.

Ji, G., & Silver, S. (1992a). Regulation and expression of the arsenic resistance operon from *Staphylococcus aureus* plasmid pI258. J. Bacteriol. 174, 3684–3694.

Ji, G., & Silver, S. (1992b). Reduction of arsenate to arsenite by the ArsC protein of the arsenic resistance operon of the *Staphylococcus aureus* plasmid pI258. Proc. Natl. Acad. Sci. USA 89, 9474–9478.

Joerger, R. D., & Bishop, P. E. (1988). Nucleotide sequence and genetic analysis of the *nifB–nifQ* region from *Azotobacter vinelandii*. J. Bacteriol. 170, 1475–1487.

Juranka, P. F., Zastawny, R. L., & Ling, V. (1989). P-glycoprotein: Multidrug-resistance and a superfamily of membrane associated transport proteins. FASEB J. 3, 2583–2592.

Karkaria, C. E., Chen, C. M., & Rosen, B. P. (1990). Mutagenesis of a nucleotide binding site of an anion-translocating ATPase. J. Biol. Chem. 265, 7832–7836.

Karkaria, C. E., & Rosen, B. P. (1991). Trinitrophenyl-ATP binding to the wild type and mutant ArsA proteins. Arch. Biochem. Biophys. 288, 107–111.

Kaur, P., & Rosen, B. P. (1992a). Mutagenesis of the C-terminal nucleotide binding site of an anion-translocating ATPase. J. Biol. Chem. 267, 19272–19277.

Kaur, P., & Rosen, B. P. (1992b). Complementation between nucleotide binding domains in an anion-translocating ATPase. J. Bacteriol. 175, 351–357.

Kaur, P., & Rosen, B. P. (1992c). Plasmid-encoded resistance to arsenic and antimony. Plasmid 27, 29–40.

Kaur, P., & Rosen, B. P. (1994). Identification of the site of α-[^{32}P]ATP adduct formation in th ArsA protein. Biochemistry 33, 6456–6461.

Knowles, F. C., & Benson, A. A. (1983). The biochemistry of arsenic. Trends in Biochem. Sci. 8, 178–180.

Lidholm, J., & Gustafsson, P. (1991). Homologues of the green algal *gidA* gene and the liverwort *frxC* gene are present on the chloroplast genomes of conifers. Plant Mol. Biol. 17, 787–798.

Lo, J. F., Wang, H. F., Tam, M. F., & Lee, T. C. (1992). Glutathione S-transferase pi in an arsenic-resistant Chinese hamster ovary cell line. Biochem. J. 288, 977–982.

Loehr, T. M., & Plane, R. A. (1968). Raman spectra and structures of arsenious acid and arsenites in aqueous solution. Inorganic Chemistry 7, 1708–1714.

Lux, H. D. (1971). Ammonium and chloride extrusion: Hyperpolarizing synaptic inhibition in spinal motor neurons. Science 173, 555–557.

Marger, M. D., & Saier, M. H. (1993). A major superfamily of transmembrane facilitators that catalyze uniport, symport and antiport. Trends in Biochem. Sci. 18, 13–20.

Massy, V., & Veeger, C. (1960). On the reaction mechanism of lipoyl dehydrogenase. Biochim. Biophys. Acta 40, 184–185.

Massy, V., & Veeger, C. (1961). Studies on the reaction mechanism of lipoyl dehydrogenase. Biochim. Biophys. Acta 48, 33–47.

McGrath, J. P., & Varshavsky, A. (1989). The yeast *STE6* gene encodes a homologue of the mammalian multidrug resistance P-glycoprotein. Nature 340, 400–404.

Mevarech, M., Rice, D., & Haselkorn, R. (1980). Nucleotide sequence of a cyanobacterial *nifH* gene coding for nitrogenase reductase. Proc. Natl. Acad. Sci. USA 77, 6476–6480.

Mobley, H. L. T., & Rosen, B. P. (1982). Energetics of plasmid-mediated arsenate resistance in *Escherichia coli*. Proc. Natl. Acad. Sci. USA 79, 6119–6122.

Mobley, H. L. T., Chen, C.-M., Silver, S., & Rosen, B. P. (1983). Cloning and expression of R-factor mediated arsenate resistance in *Escherichia coli*. Mol. Gen. Genet. 191, 421–426.

Nakayama, N., Arai, N., Bond, M. W., Kaziro, Y., & Arai, K. (1984). Nucleotide sequence of *dnaB* and the primary structure of the *dnaB* protein from *Escherichia coli*. J. Biol. Chem. 259, 97–101.

Nishiguchi, R., Takanami, M., & Oka, A. (1987). Characterization and sequence determination of the replicator region in the hairy-root-inducing plasmid pRiA4b. Mol. Gen. Genet. 206, 1–8.

Novick, R. P., & Roth, C. (1968). Plasmid-linked resistance to inorganic salts in *Staphylococcus aureus*. J. Bacteriol. 95, 1335–1342.

Oden, K. L., Gladysheva, T. B., & Rosen, B. P. (1994). Arsenate reduction mediated by the plasmid-encoded ArsC protein is coupled to glutathione. Molec. Microbiol. 12, 301–306.

Ouellette, M., & Borst, P. (1991). Drug resistance and P-glycoprotein gene amplification in the protozoan parasite *Leishmania*. Res. Microbiol. 142, 737–746.

Owolabi, J. B., & Rosen, B. P. (1990). Differential mRNA stability controls relative gene expression within the plasmid-encoded arsenical resistance operon. J. Bacteriol. 172, 2367–2371.

Papadopoulou, B., Roy, G., Dey, S., Rosen, B. P., & Ouellette, M. (1994). Contribution of *Leishmania* P-glycoprotein related *ltpgpA* to oxyanion resistance. J. Biol. Chem. 269, 11980–11986.

Pedersen, P. L., & Carafoli, E. (1987). Ion motive ATPases. I. Ubiquity, properties, and significance to cell function. Trends in Biochem. Sci. 12, 146–150.

Peterson, R., & Rumack, B. (1977). D-Penicillamine therapy of acute arsenic poisoning. J. Pediatr. 91, 661–666.

Reizer, J., Reizer, A., & Saier, M. H. (1992). A new subfamily of bacterial ABC-type transport systems catalyzing export of drugs and carbohydrate. Prot. Sci. 1, 1326–1332.

Rinchik, E. M., Bultman, S. J., Horsthemke, B., Lee, S. T., Strunk, K. M., Spritz, R. A., Avidano, K. M., Jong, M. T., & Nicholls, R. D. (1993). A gene for the mouse pink-eyed dilution locus and for human type II oculocutaneous albinism. Nature 361, 72–76.

Riordan, J. R., Rommens, J. M., Kerem, B.-S., Alon, N., Rozmahel, R., Grzelczak, Z., Zielenski, J., Lok, S., Plavsic, N., Chou, J.-L., Drumm, M. L., Iannuzzi, M. C., Collins, F. S., & Tsui, L.-C.

(1989). Identification of the cystic fibrosis gene: Cloning and characterization of complementary DNA. Science 245, 1066–1073.

Rosen, B. P., & Borbolla, M. G. (1984). A plasmid-encoded arsenite pump produces arsenite resistance in *Escherichia coli*. Biochem. Biophys. Res. Commun. 124, 760–765.

Rosen, B. P., Weigel, W., Karkaria, C., & Gangola, P. (1988). Molecular characterization of an anion pump. The *arsA* gene product is an arsenite(antimonate)-stimulated ATPase. J. Biol. Chem. 263, 3067–3070.

Rosen, B. P., Weigel, U., Monticello, R. A., & Edwards, B. P. F. (1991). Molecular characterization of an anion pump: Purification of the ArsC protein. Arch. Biochem. Biophys. 284, 381–385.

Rosen, B. P., Dey, S., Dou, D., Ji, G., Kaur, P., Ksenzenko, M., Silver, S., & Wu, J. (1992). Evolution of an ion-translocating ATPase. Ann. New York Acad. Sci. 671, 257–272.

Rosenstein, R., Peschel, P., Wieland, B., & Götz, F. (1992). Expression and regulation of the *Staphylococcus xylosus* antimonite, arsenite and arsenate resistance operon. J. Bacteriol. 174, 3676–3683.

San Francisco, M. J. D., Tisa, L. S., & Rosen, B. P. (1989). Identification of the membrane component of the anion pump encoded by the arsenical resistance operon of R-factor R773. Mol. Microbiol. 3, 15–21.

Schlemmer, S. R., & Sirontak, F. M. (1992). Energy dependent efflux of methotrexate in L1210 leukemia cells. Evidence for the role of an ATPase obtained with inside-out plasma membrane vesicles. J. Biol. Chem. 267, 14746–14752.

Senior, A. E. (1985). The proton-ATPase of *Escherichia coli*. Curr. Topics Memb. Trans. 23, 135–151.

Silver, S., & Keach, D. (1982). Energy-dependent arsenate efflux: The mechanism of plasmid mediated resistance. Proc. Natl. Acad. Sci. USA 79, 6114–6118.

Silver, S., Budd, K., Leahy, K. M., Shaw, W. V., Hammond, D., Novick, R. P., Willsky, G. R., Malamy, M. H., & Rosenberg, H. (1981). Inducible plasmid-determined resistance to arsenate, arsenite and antimony (III) in *Escherichia coli* and *Staphylococcus aureus*. J. Bacteriol. 146, 983–996.

Silver, S., Ji, G., Broer, S., Dey, S., Dou, D., & Rosen, B. P. (1993). Orphan enzyme or patriarch of a new tribe: The arsenic resistance ATPase of bacterial plasmids. Mol. Microbiol. 8, 637–642.

Schoolmaster, W. L., & White, D. R. (1980). Arsenic poisoning. South. Med. J. 73, 198–208.

Sirotnak, F. M., & O'Leary, D. F. (1991). The issues of transport multiplicity and energetics pertaining to methotrexate efflux in L1210 cells addressed by an analysis of cis and trans effects of inhibitors. Cancer. Res. 51, 1412–1417.

Smith, E. A., & Oehme, F. W. (1991). A review of selected herbicides and their toxicities. Vet. Hum. Toxicol. 33, 596–608.

Smith, J. D. (1973). Arsenic, antimony and bismuth. In: Comprehensive Inorganic Chemistry, Volume 2 (Bailar, J. C., Emeléus, H. J., Nyholm, R., & Trotman-Dickerson, A. F., eds.) pp. 547–683. Pergammon Press, Oxford.

Sulston, J., Du, Z., Thomas, K., Wilson, R., Hiller, L., Staden, R., Halloran, N., Green, P., Thierry-Mieg, J., Qiu, L., Dear, S., Coulson, A., Craxton, M., Durbin, R., Berks, M., Metzstein, M., Hawkins, T., Ainscough, R., & Waterson, R. (1992). The *C. elegans* genome sequencing project: A beginning. Nature 356, 37–41.

Szymanski, H. A., Marabella, L., Hoke, J., & Harter, J. (1968). Infrared and raman studies of arsenic compounds. Applied Spectroscopy 22, 297–304.

Tisa, L. S., & Rosen, B. P. (1990a). Plasmid-encoded transport mechanisms. J. Bioenerg. Biomemb. 22, 493–507.

Tisa, L. S., & Rosen, B. P. (1990b). Molecular characterization of an anion pump: The ArsB protein is the membrane anchor for the ArsA protein. J. Biol. Chem. 265, 190–194.

Turner, R. J., Hou, Y., Weiner, J. H., & Taylor, D. E. (1992). The arsenical ATPase efflux pump mediates tellurite resistance. J. Bacteriol. 174, 3092–3094.

Vaziri, N., Upham, T., & Barton, C. (1980). Hemodialysis clearance of arsenic. Clinic. Toxicol. 17, 451–456.

Walker, J. E., Saraste, M., Runswick, M. J., & Gay, N. J. (1982). Distantly related sequences in the α- and β-subunits of the ATP synthase, myosin kinases and other ATP-requiring enzymes and a common nucleotide binding fold. EMBO J. 1, 945–951.

Wu, J. H., & Rosen, B. P. (1991). Regulation of the *ars* operon: The *arsR* gene product is a negative regulatory protein. Molec. Microbiol. 5, 1331–1336.

Wu, J. H., & Rosen, B. P. (1993a). Metalloregulated expression of the *ars* operon. J. Biol. Chem. 268, 52–58.

Wu, J. H., & Rosen, B. P. (1993b). The *arsD* gene encodes a second *trans*-acting regulatory protein of the plasmid-encoded arsenical resistance operon. Molec. Microbiol. 8, 615–623.

Wu, J. H., Tisa, L. S., & Rosen, B. P. (1992). Membrane topology of the ArsB protein, the membrane component of an anion-translocating ATPase. J. Biol. Chem. 267, 12570–12576.

Wang, Z., & Rossman, T. G. (1993). Stable and inducible arsenite resistance in Chinese hamster cells. Toxicol. & Appl. Pharmacol. 118, 80–86.

THE MAGNESIUM TRANSPORT ATPASES OF *SALMONELLA TYPHIMURIUM*

Tao Tao and Michael E. Maguire

Biomembranes
Volume 5, pages 271–285.

ISBN: 1-55938-662-2.

I. INTRODUCTION

For both prokaryotic and eukaryotic cells, Mg^{2+} is one of the most abundant cations, second to potassium, and the most abundant divalent cation (Grubbs and Maguire, 1987; Reinhart, 1988; Rotevatn et al., 1991). Mg^{2+} is important for many aspects of cellular function. Structurally, it is a major stabilizing factor for all biomembranes and the ribosomal translation apparatus, and it binds to nucleic acids and proteins to help maintain specific conformations (White and Hartzell, 1989; Matsuda, 1991). Enzymatically, it is required by all enzymes using ATP as the energy source. Mg^{2+} has modulating activity on important cellular activities such as photosynthesis, neuronal excitation, and cell cycle control (Walker, 1986; Grubbs and Maguire, 1987). Besides these, the physico-chemical properties of magnesium make it unique among biological cations. It is the most charge-dense of all the biological cations (Diebler et al., 1969). The volume ratio between the hydrated cation and the atomic form is about 360, over an order of magnitude larger than the volume ratio for Ca^{2+}, Na^{+}, and K^{+}. Since they must see and interact with the huge hydrated form and transport only the atomic form (Hille, 1992), this obviously imposes great challenges to any Mg^{2+} transport system. However, in contrast to its importance, the understanding of magnesium homeostasis is very limited. Partly, this is because of the previous misconceptions that Mg^{2+} plays only a passive role in cellular function. The technical difficulties involved in measurement of magnesium biochemistry have also contributed greatly to this lack of understanding.

As a essential first step toward understanding the biological importance of magnesium, its transmembrane flux must be understood. Knowledge of this process would answer not only specific questions regarding Mg^{2+} alone, but would provide interesting and potentially revealing comparisons to systems mediating the transport of other biological cations, especially Ca^{2+}.

Studies on Mg^{2+} transport in eukaryotes have been performed in a wide variety of cells, including the red blood cells of several species (Gunther and Vormann, 1985; Feray and Garay, 1986; Flatman and Smith, 1991), squid axon, and barnacle muscle (Brinley et al., 1977). Observations from these experiments indicate that Na^{+}-dependent Mg^{2+} efflux is probably the major extrusion mechanism in many eukaryotic cells. In contrast, for influx, no such consensus has been found. The proteins involved in eukaryotic Mg^{2+} flux are completely unknown since the conventional route of purification and reconstitution is not feasible in these systems due to the lack of a functional assay to follow purification. Even modern molecular genetics can not overcome the lack of an assay on function. This difficulty arises because no reconstitution assays for Mg^{2+} transport have been developed, due both to a lack of dyes adequately selective for Mg^{2+} and to the extreme expense and short half-life of $^{28}Mg^{2+}$.

In our laboratory, the Gram negative bacterium *Salmonella typhimurium* is used as the model system for the study of transmembrane Mg^{2+} transport because i) in general, bacterial transport systems are homologous in structure and mechanism to eukaryotic systems (Maloney, 1990; Henderson and Maiden, 1990), ii) *Salmonella typhimurium* has been studied in great detail and the genetic techniques for prokaryotic systems are unmatched by any eukaryotic system, and iii) previous studies of Mg^{2+} transport in the related bacterium, *E. coli*, provided background knowledge (Silver, 1969; Lusk and Kennedy, 1969; Nelson and Kennedy, 1971, 1972; Silver and Clark, 1971).

II. PREVIOUS STUDIES ON Mg^{2+} TRANSPORT IN ENTERIC BACTERIA

A. Previous Studies in *Escherichia coli*

Early studies of Mg^{2+} transport in *E. coli* were performed about two decades ago by two groups—those of Silver (Silver, 1969; Silver and Clark, 1971) and Kennedy (Lusk and Kennedy, 1969; Nelson and Kennedy, 1971, 1972), with the following findings: (1) *E. coli* has active Mg^{2+} transport since this process is blocked by the addition of cyanide to the growth medium, which blocks energy metabolism; (2) under normal laboratory culture conditions, Mg^{2+} transport is competitively inhibited by Co^{2+}, and Co^{2+} transport could be competitively inhibited by Mg^{2+}, suggesting that these two cations are being transported by the same system; (3) for cells grown in a low versus a high extracellular Mg^{2+} concentration, the influx of Mg^{2+} has different kinetic parameters, and the influx measured after growth at low Mg^{2+} concentration is not Co^{2+}-sensitive, suggesting the presence of at least one additional system; and (4) genetic analysis of the mutant strains indicated that distinct loci are responsible for the two types of Mg^{2+} transport.

B. Studies in *Salmonella typhimurium*

In our laboratory, transmembrane Mg^{2+} flux was studied in *Salmonella typhimurium*. Initial kinetic studies using $^{28}Mg^{2+}$ demonstrated that, similar to *E. coli*, there were two classes of Mg^{2+} transport systems with different kinetic characteristics (Hmiel et al., 1986, 1989; Snavely et al., 1989a, b). The high capacity system was also inhibited by Co^{2+}. By selecting for Co^{2+} resistance, mutant strains were isolated which demonstrated Co^{2+}-insensitive transport characteristics and had decreased Mg^{2+} transport. This transport system was thus designated as *corA*, for cobalt resistance (Hmiel et al., 1986). The other class of transport system(s) was designated as *mgt*, for Mg^{2+} transport (Hmiel et al., 1989) (Table 1).

Starting with a *corA* mutant strain, strains containing mutations in all the Mg^{2+} transport systems were isolated (Hmiel et al., 1989). This was accomplished by ethylmethane sulfonate chemical mutagenesis, growth on plates containing very high Mg^{2+} concentrations, and subsequent screening for lack of growth in low Mg^{2+} plates. One such strain, MM77, demonstrated a Mg^{2+} dependent phenotype requiring 100mM added Mg^{2+} for growth; transport assays with MM77 showed no

Table 1. Properties of the Mg^{2+} Transport Systems of *Salmonella typhimurium*

Transport System	*mgtA*	*mgtB*	*corA*
Chromosomal location (min)	98	80.5	83.5
Function	Influx only	Influx only	Influx and efflux
Cations Transported	Mg^{2+} and Ni^{2+}	Mg^{2+} and Ni^{2+}	Mg^{2+}, Co^{2+}, and Ni^{2+}
Mg^{2+} K_m (20°/37°, in μM)	12/29	ND[a]/6	15/?
V_{max} (20°/37°) (pmol min^{-1} 10^8 $cells^{-1}$)	20/>100	<2/>100	300/>1500
Cation Inhibition	$Zn^{2+}Mg^{2+}>Ni^{2+} = Co^{2+}>Ca^{2+}$	$Mg^{2+} = Co^{2+} = Ni^{2+} > Mn^{2+} >> Ca^{2+}$	$Mg^{2+}>Mn^{2+}>Co^{2+} > Ni^{2+}>Ca^{2+}$
Mg^{2+} regulation	$[Mg^{2+}]$, Ca^{2+}, O_2, carbon source	$[Mg^{2+}]$, Ca^{2+}, carbon source	None?

Note: [a]ND = Not Detectable. Transport via MgtB appears very temperature-sensitive and cannot be detected at 20°C. Conversely, transport via CorA at 37°C is so rapid that accurate kinetics cannot be performed.

detectable transmembrane Mg^{2+} flux. Using MM77 as recipient strain, a *S. typhimurium* genomic library was screened for complementation of the Mg^{2+} dependent phenotype of this strain. Three classes of clones with the ability to restore Mg^{2+} influx were isolated. One group abolished the Mg^{2+} dependent phenotype, restored both Mg^{2+} and Co^{2+} uptake, and restored Co^{2+} sensitivity. Its transport characteristics resembled those obtained under the normal growth condition. These findings indicated that it was likely a clone of the CorA transporter system. The remaining clones contained very large inserts with different restriction patterns that fell into two classes (Hmiel et al., 1989). Chromosomal mapping demonstrated that they were linked to different markers from each other and from *cor*A, and thus resided at different positions on the *S. typhimurium* chromosome. Kinetic studies clearly indicated that these two clones represented two distinct Mg^{2+} transporters different from CorA. They were designated as *mgtA* and *mgtB*, respectively.

This was confirmed by reconstruction of a Mg^{2+}-dependent strain via genetic transduction of these three cloned loci. The three cloned genes were inactivated by an Mu *dJ* insertion (Smith et al., 1993a) and then used to replace the wild type alleles in the chromosome via *pol*A-mediated forced recombination. A strain with all three Mg^{2+} transport genes replaced by their insertionally inactivated alleles showed the same characteristics as MM77. The construction of this strain, MM281, referred to as the "triple mutant", confirmed that *S. typhimurium* contained only three Mg^{2+} transport systems (Hmiel et al., 1989; Snavely et al., 1989b). This strain was used exclusively in our later studies.

III. THE Mg^{2+} TRANSPORTING ATPASES OF *SALMONELLA TYPHIMURIUM*

Of the three Mg^{2+} transport systems from *Salmonella typhimurium,* CorA is a novel transport system as indicated by the deduced protein sequence which shares no similarity to any currently known transport system. The *corA* gene encodes a small and highly charged membrane protein with distinct membrane topology (Smith et al., 1993a) and is likely the dominant Mg^{2+} influx system of gram negative bacteria. It will not be covered further. Of the remaining two clones carrying Mg^{2+} transporters restriction mapping, complementation assay, and protein labeling using *E. coli* maxicells, all indicated that *mgtA* and *mgtB* each independently encoded large, integral membrane proteins with apparent molecular masses of 101- and 91-kD, respectively (Hmiel et al., 1989; Snavely et al., 1989a) and that they were integral membrane proteins. Sequence comparison using the deduced amino acid sequences of MgtA and MgtB (Snavely et al., 1991a; Tao et al., 1995) revealed a high degree of homology to the superfamily of cation transporting P-type ATPases.

A. Cation Transport

The MgtA and MgtB systems mediate the influx of Mg^{2+}, having no role in the efflux of Mg^{2+}. In strains carrying wild type alleles of *mgtA* and/or *mgtB*, but lacking *corA*, no Mg^{2+} efflux could be observed even at very high extracellular Mg^{2+} concentrations. Efflux is observed only in strains carrying the CorA system, indicating that efflux is mediated only by the CorA system. This efflux process is not simple; several other loci have since been found to be relevant to Mg^{2+} efflux (Gibson et al., 1992), but none appear to have influence on the MgtA or MgtB systems.

The MgtA and MgtB systems also mediate the influx of Ni^{2+} (Snavely et al., 1991b), but the efficiency, as reflected by the relative V_{max}, is lower as compared with Mg^{2+} transport. The difference in K_m for Ni^{2+} versus Mg^{2+} transport by these two systems is much less. The K_m is virtually the same for both cations when transport via MgtB was assayed; when influx via MgtA was measured, the K_m for Ni^{2+} transport is 5-fold less than that for Mg^{2+}. Mg^{2+} and Ni^{2+} are competitive inhibitors of the other's transport (Snavely et al., 1989b, 1991b). The transport of both cations is sensitive to inhibition by other divalent cations, even though no other cations tested can be transported by either system. For these inhibitory cations, the order of potency are different for each system. For Mg^{2+} transport via MgtA, the inhibitory order is $Zn^{2+}>Ni^{2+}>Co^{2+}>Ca^{2+}>Mn^{2+}$, while for Mg^{2+} transport via MgtB the order is $Co^{2+}>Ni^{2+}>Mn^{2+}$ with Ca^{2+} and Zn^{2+} being noninhibitory. Similar orders are seen for inhibition of Ni^{2+} uptake.

Ni^{2+} transport via these two systems is unlikely to have physiological relevance. Aerobically grown bacteria should have little use for Ni^{2+}, and at the concentration required for detectable Ni^{2+} uptake to occur, this cation is very toxic. The transport of Ni^{2+} by these two systems does, however, provide an alternative in the functional assay of these two systems. Instead of using the expensive and short lived $^{28}Mg^{2+}$, the cheaper alternative $^{63}Ni^{2+}$ can be used in their kinetic and functional characterization.

B. Genetics

Of these two loci, *mgtB* has been studied more comprehensively. The locus was mapped to 80.5 min of the *S. typhimurium* chromosome. Restriction mapping of the initial clone demonstrated that this locus contained two genes transcribed in the same direction (Figure 1). Protein expression profiles revealed that two proteins were produced with apparent molecular masses of 22.5- and 101-kD (Snavely et al., 1989a). The DNA sequence predicted two open reading frames with calculated molecular weights of 18.5–24 (depending on the start site used) and 102 kD, respectively (Snavely et al., 1991a). These genes were designated *mgtC* and *mgtB*, respectively.

DNA sequencing revealed that there were only 250 bp between the *mgtC* and *mgtB* structural genes. This left little space to accommodate a promoter plus other regulatory sites suggesting the possible presence of a bicistronic operon. To confirm this, promoter location and activity were mapped using luciferase reporter genes. Results from these experiments indicated that the sequence between *mgtC* and *mgtB* indeed had no promoter activity, while the sequence upstream of *mgtC* had strong promoter activity that was Mg^{2+} modulated (Tao et al., 1995).

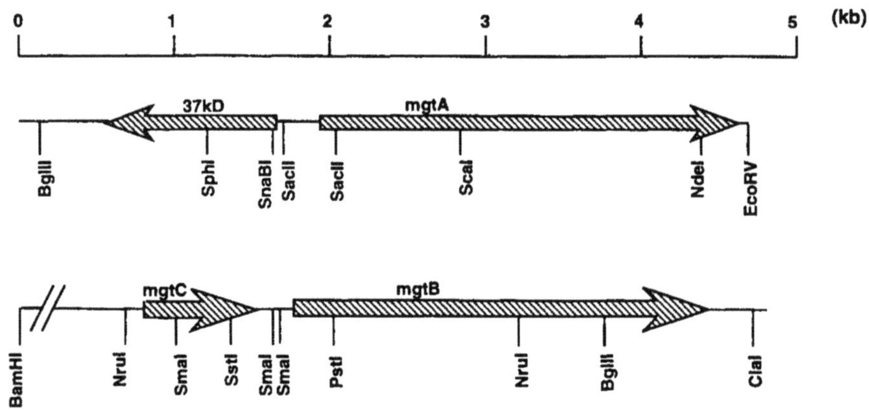

Figure 1. The genetic structure of *mgtA* and *mgtB* loci. The cloned *mgtB* locus consists of two genes, *mgtC* and *mgtB,* which are orientied in the same direction. These two genes form an operon. The cloned *mgtA* locus has two genes oriented in opposite orientations, *mgtA* and the gene coding for a 37 kD protein. The length of these genes and some of the restriction sites are given.

The MgtA system is relatively understudied in comparison to MgtB. It was mapped to 98 min on the *S. typhimurium* chromosome. Two genes transcribed in different orientations were found in the initial clone, producing two products with a size of 37- and 91-kD, respectively (Hmiel et al., 1989). The larger gene complements the Mg^{2+} growth dependence of the "triple mutant" and was designated *mgtA*. DNA sequence data for this clone predicted an open reading frame with 901 residues and a calculated molecular mass of 95- to 98-kD depending on the start codon used (Hmiel et al., 1989). Promoter mapping demonstrated that the promoter for *mgtA* was located in the 400 base pair intergenic region between *mgtA* and 37 kD gene and that it was highly Mg^{2+} responsive. For the 37 kD gene, no significant Mg^{2+}-dependent promoter activity has been detected under various conditions although the sequence does exhibit modest intrinsic promoter activity (Tao et al., 1995). The chromosomal relationships of *mgtA* and *mgtB* are illustrated in Figure 1 (top).

C. Protein Structure

Protein sequence analysis and comparison against the database for all four proteins was very informative. Analysis on the deduced protein sequence for MgtC showed that it was a very hydrophobic protein with 4 or 5 possible transmembrane domains (Snavely et al., 1991a). The only meaningful similarity found in the global sequence comparison was a weak homology to SEC7 from yeast. For the 37 kD protein, preliminary sequence indicated that it shares a high degree of sequence similarity with many known bacterial operon regulators such as LacI and MalR (Tao et al., 1995). The possible functions of these two proteins will be addressed below.

The two larger genes, *mgtB* and *mgtA*, share extensive sequence similarity to each other; both encode proteins homologous to the many known proteins from the superfamily of P-type ATPases (Snavely et al., 1991a and Tao and Maguire, manuscript in preparation). P-type ATPases (Pederson and Carafoli, 1987) are found in essentially all eukaryotic and probably all prokaryotic species, and are the major cation-transporting systems for the biologically important cations such as Na^+, H^+, K^+, Mg^{2+}, and Ca^{2+}. The transport mechanism involves direct phosphorylation of a conserved aspartyl residue by ATP and its subsequent hydrolysis coupled with the final translocation of transported cations. Most of P-type ATPases appear to be single subunit membrane bound proteins though some do have a second, smaller subunit with unknown function (Pederson and Carafoli, 1987; Snavely et al., 1991a).

Careful sequence alignment of multiple P-type ATPases gave very interesting results with regard to MgtA and MgtB. These two proteins were found to be more similar to those isolated from eukaryotes than to the limited number isolated from prokaryotes (Hesse et al., 1984). In fact, MgtA and MgtB demonstrated about 50% overall similarity to the sarcoplasmic reticulum Ca^{2+} ATPase from rabbit skeletal muscle. In contrast, only 15–30% general similarity was found between MgtA or MgtB and other prokaryotic P-type ATPases (Snavely et al., 1991a). Computer aided phylogenetic analysis on MgtA and MgtB versus other known P-type ATPases gave a similar answer, suggesting that MgtA and MgtB, two prokaryotic P-type ATPases, were most closely related to the H^+-transporting P-type ATPases of yeasts and fungi and to the Ca^{2+}-transporting P-type ATPases of mammals (Figure 2). The obviously fallacious implication of the computer analysis, that *S.*

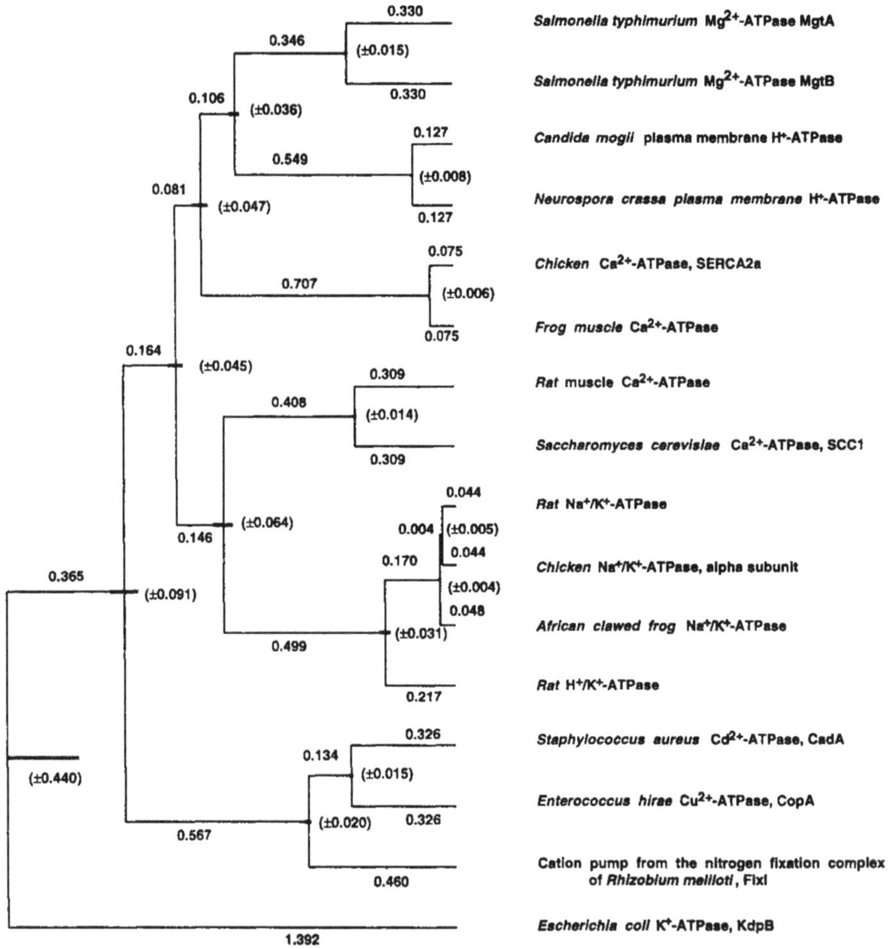

Figure 2. Phylogenetic relationship among selected P-type ATPases from several species. The sequences of P-type ATPases, including MgtA and MgtB from *Salmonella typhimurium,* were analyzed using GeneWorks (Intelligenetics, Mountain View, CA) for their phylogenetic relationship. The result is presented as the arbitrarily rooted tree with relative evolutionary distances.

typhimurium is a closer relative to mammals than to *E. coli* and other prokaryotes, is due primarily to the fact that only a few prokaryotic P-type ATPases sequences have been reported. Clearly, the P-type ATPase superfamily should be separated into subfamilies based on size, structure, function, and other criteria to allow a more precise analysis. This latter argument is exemplified by a similar finding from eukaryotic systems in which the human Cu^{2+}-transporting P-type ATPase associated with Menke's disease was found to have great similarity to that of the Cu^{2+} efflux P-type ATPase from *Enterococcus hirae* (Odermatt et al., 1993; Vulpe et al., 1993), but significantly lesser similarity to other eukaryotic P-type ATPases or to MgtA/MgtB. Nonetheless, it is clear that MgtB and MgtA belong to a new class of prokaryotic P-type ATPase with high sequence homology to mammalian Ca^{2+} ATPases. This similarity makes the studies on these two Mg^{2+} transporting P-type ATPases interesting not only as the model systems to understand transmembrane Mg^{2+} flux, but also as models for mammalian divalent cation transporting ATPases. This is important since genetic manipulation in prokaryotic systems is technically much easier and experimentally far more rapid.

D. Membrane Topology of MgtB

In order to understand transmembrane flux of Mg^{2+}, it is necessary to know the transport protein's arrangement within the membrane. Unlike eukaryotic systems, the determination of this arrangement, membrane topology as it is usually called, is relatively straightforward in bacteria. Because of the complexity and difficulty of the topological determinations in eukaryotic systems, numerous conflicting models have been proposed for the topology of P-type ATPases (Figure 3, top). To date, the only consensus on the membrane topology for eukaryotic P-type ATPases is that both N- and C-termini reside in the cytosol (Mandala and Slayman, 1989; Hennessey and Scarborough, 1990).

The membrane topology of MgtB has recently been determined in our laboratory using random gene fusion techniques (Smith et al., 1993b). In this process, a gene for a reporter protein, such as PhoA, BlaM, or LacZ is fused to N-terminal portions of MgtB to produce a hybrid protein. The enzymatic activity of the reporter protein in such fusions depends on its position relative to the membrane. Fusion junctions and their position can be determined by the combination of sequencing and reporter protein activity. Results from these experiments clearly indicate that MgtB adopts a topology containing 10-membrane spanning segments rather similar to the model proposed for sarcoplasmic reticular Ca^{2+} ATPase by MacLennan et al. (1985). Our data argue strongly against the various eight membrane spanning segment models (Serrano et al., 1986) as well as alternative ten and twelve segment models (see discussion in Smith et al., 1993b). These data provide a solid ground for analysis of previous site-directed mutagenesis experiments and provide a more defined set of targets for future structural analysis.

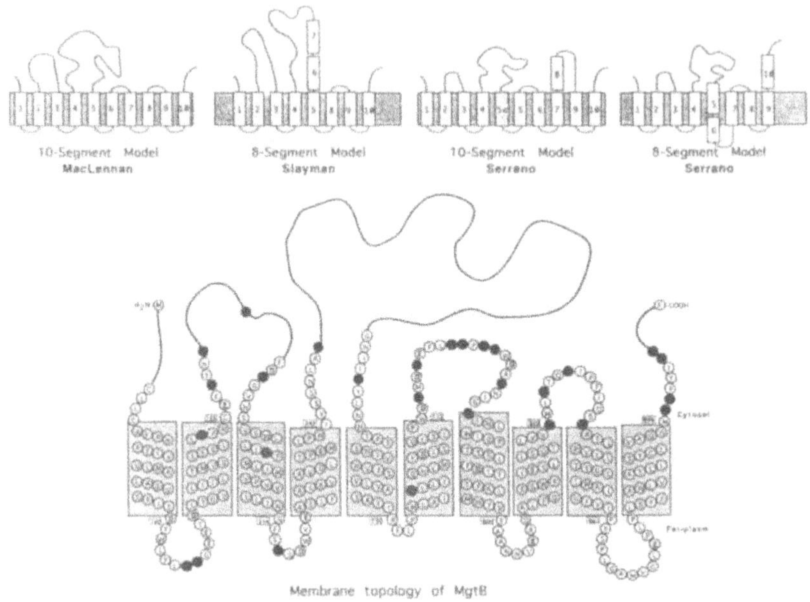

Figure 3. Selected literature membrane topology models forcation-transporting P-type ATPases compared with the experimental model derived from fusion protein data. The drawings in the upper panel are membrane topologies proposed by various groups primarily from hydropathy analysis. Each membrane segment is shown as a numbered bar within the membrane relative to the 10 membrane segment model proposed by MacLennan et al. (1985). The other three models to the right show the respective topologies proposed by other laboratories but using the same membrane segments for sake of comparison. These are the models of Mandala and Slayman (1989) and both 8- and 10-segment models from Serrano et al. (1986). The lower figure is the membrane topology of the P-type ATPases as derived from our extensive genetic fusion data (Smith et al., 1993b) for the Mg^{2+} transporting P-type ATPase MgtB from *Salmonella typhimurium*.

E. Regulation of MgtA and MgtB

A characteristic that further sets these two *mgt* loci apart from the CorA Mg^{2+} transport system is that they are repressed under the normal (laboratory) growth conditions where CorA accounts for the vast majority of the Mg^{2+} transport. This is primarily because the Mg^{2+} in the growth medium strongly represses transcription of *mgtA* and *mgtB*. However, when extracellular Mg^{2+} concentration is lowered, CorA becomes relatively less effective for Mg^{2+} influx, and the transcription of *mgtA* and *mgtB* increases sufficiently that the MgtB system, in particular, plays the major role in uptake of Mg^{2+} from the medium. We have performed initial studies on this transcriptional regulation (Snavely et al., 1991b). For this purpose, transcriptional fusions to *lac*Z were constructed for *corA*, *mgtB*, and *mgtA*. Measurement of β-galactosidase activity from cells grown in medium containing millimolar concentrations of Mg^{2+} indicated that *corA* is constitutively expressed, while the expression of *mgtA* and *mgtB* is repressed. This repression of *mgtA* and *mgtB* transcription can be relieved by several factors, but Mg^{2+} is, by far, the most effective and dramatic. With 10 mM Mg^{2+} in the growth medium, transcription of *mgtA* and *mgtB* is virtually undetectable. Transcription increases 10-fold for both systems when the medium Mg^{2+} concentration is lowered to 10 μM. The increase in transcription for *mgtB* is blocked by extracellular Ca^{2+} although Ca^{2+} is not transported by, nor does it inhibit, Mg^{2+} uptake via *mgtB*. When the extracellular Mg^{2+} concentration is lowered further to around 1 μM, transcription of *mgtA* increases another 3- to 5-fold. However, transcription of *mgtB* increases about 100-fold; this latter increase was not Ca^{2+} sensitive, suggesting that an additional regulatory mechanism might be involved. Overall, *mgtA*'s transcription increased approximately 40-times above baseline, while for *mgtB* the overall increase was almost 1,000-fold.

The results show large increases in transcription, but this experimental approach cannot determine translation and membrane incorporation. For MgtB, this issue was addressed using Ni^{2+} transport in the "triple mutant" strain carrying the *mgtB* locus on a plasmid (Smith et al., 1993b). When this strain was grown in 10 mM Mg^{2+}, Ni^{2+} transport was virtually undetectable. However, Ni^{2+} transport increased dramatically when such strains were grown in medium with less than 50 μM added Mg^{2+} for several hours. The peak of transport capacity, occurring after 8–20 hours incubation in the low Mg^{2+} medium, was more than 1,000-fold greater than the baseline. During this incubation period, cells are in a vegetative state because of the lack of Mg^{2+} and cell density doubled only once. This observation is consistent with the gene fusion results, indicating that a large increase in transcription results in a comparable functional incorporation of MgtB protein into the membrane.

Since there appears to be much similarity between the MgtB and Mgt*A* systems, distinct physiological functions of each system are unclear. One possible clue to differentiate them may come from preliminary studies with a pathogenic strain of *S. typhimurium* containing an *mgt*B::*LacZ* fusion (Garcia-del-Portillo et al., 1992). When the bacterium invades cultured mouse epithelial cells by phagocytosis, the *LacZ* activity and thus *mgtB* transcription is induced more than 100-fold, suggesting a possible role of *mgtB* in virulence.

F. Role of the Accessory Proteins: MgtC and 37 kD

The relevance of the *mgtC* gene 5′ to *mgtB* and the gene encoding a 37 kD protein 5′ to *mgtA* are unclear. mgt*C* is an small integral membrane protein, with only limited sequence similarity to the SEC7 protein of yeast. It shows no homology to the known subunits of eukaryotic P-type ATPases. However, the close association with *mgtB* in an operon and its membrane location argue for some functional relationship of the MgtC and MgtB proteins.

Theoretically, MgtC may have any of the following activities: (1) α-subunit of MgtB, since eukaryotic P-type ATPases mediating cation influx have a large α- and a small β-subunit, both of which are required for activity and/or functional insertion of the ATPase subunit into the membrane (McDonough et al., 1990; Geering, 1992); (2) part for a regulatory system acting in sensing extracellular Mg^{2+} concentration and then transducing the signal into cytosol; and (3) no relationship to MgtB function at all.

Preliminary experiments with a mutant strain carrying a chromosomal *mgtBC* and *corA* deletions demonstrate that this strain can respond to changes in extracellular Mg^{2+} concentration (Tao and Maguire, manuscript in preparation), as reflected by the response of an introduced plasmid carrying only the *mgtC* promoter region to changes in extracellular Mg^{2+}. Such results effectively rule out a primary regulatory function for *mgtC*.

If *mgtC* encoded a subunit of the MgtB ATPase, it could function in one or all of three aspects: (1) to act as an MgtB specific chaperone to guide MgtB into the membrane; (2) to bind Mg^{2+} and present it to MgtB for transport; and/or (3) to act collectively with MgtB during the transport process. The first possibility was tested using a mutant strain with a chromosomal *mgtBC* deletion and a plasmid with an *Mgt*B::*Bla*M in frame fusion. Preliminary data suggest that *mgtC* is not required for membrane insertion process since this strain showed ampicillin resistance mediated by the *mgtB*::*blaM* fusion product in the absence of a functional *mgtC*. Similarly, plasmids carrying an intact *mgtBC* operon or a deletion of most of the *mgtC* gene were tested for their ability to allow Mg^{2+} uptake. Since the plasmid carrying the *mgtC* deletion and thus expressing only *Mgt*B was able to support Mg^{2+} uptake, expression of *mgtC* is not absolutely required for the function of the MgtB transport system and thus it is unlikely to be a subunit. Experiments are in progress to distinguish the remaining possibilities.

The sequence of the 37 kD protein indicates that this gene has a high similarity to many known operon repressors. One obvious possibility is that this gene may function in the Mg^{2+}-mediated regulation of the two *mgt* loci, however, the lack of significant Mg^{2+} regulation of its promoter makes this possibility unlikely. Chromosomal deletion constructs of the 37 kD gene are currently being made to test its potential role(s) in *mgtA* and/or *mgtB* function.

IV. CONCLUSIONS

Our studies in *Salmonella typhimurium* have contributed significantly to the understanding of transmembrane Mg^{2+} flux. Two classes of transporters have been identified, the *corA* system, belonging to a novel class of transporter, and the *mgtA* and *mgtB* belonging to the known P-type ATPase superfamily of transporters. The identification and cloning of the two Mg^{2+} transporting P-type ATPases have

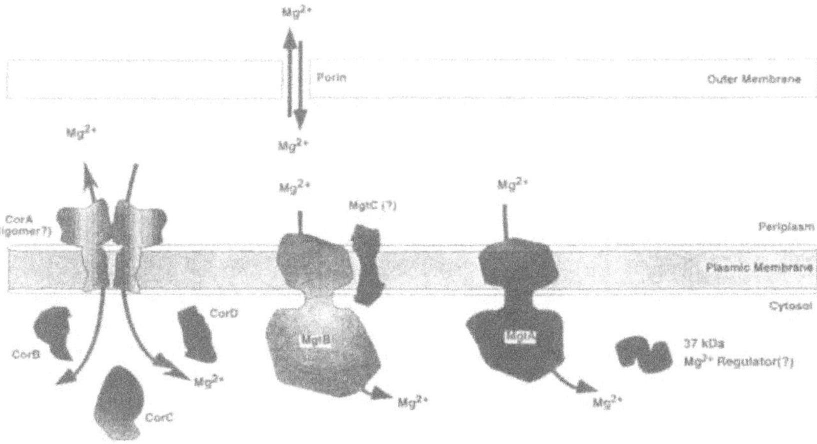

Figure 4. Cartoon summary of the Mg^{2+} transport systems in *Salmonella typhimurium.* CorA is the major constituitively expressed Mg^{2+} transport system and mediates the influx as well as efflux of Mg^{2+}. The functional form is probably an homo-oligomer. MgtB is one of the inducible Mg^{2+} transport systems, with MgtC as its possible α-subunit. It mediates only the influx of Mg^{2+}. MgtA is also an inducible Mg^{2+} transport system mediating only Mg^{2+} influx. The 37 kD protein is a potential DNA binding protein, based on its sequence homology to know DNA binding proteins. Its relationship to MgtA and/or MgtB function is unknown.

provided manipulable systems to study the detailed process of Mg^{2+} influx. It has also opened the possibility for cloning Mg^{2+} transport ATPases from eukaryotic systems which would provide a route for study the physiological role(s) Mg^{2+} may play in the higher organisms. Finally, study of the regulation of Mg^{2+} transport genes by extracellular Mg^{2+} should provide more information on the Mg^{2+} homeostasis and the controls Mg^{2+} may exert on the cellular functions (Walker, 1986; Grubbs and Maguire, 1987).

REFERENCES

Brinley, F. J., Scarpa, A., & Tiffert, T. (1977). The concentration of ionized magnesium in barnade muscle fibres. J. Physiol. 266, 545–565.

Diebler, H., Eigen, M., Ilgenfritz, G., & Winkler, R. (1969). Kinetics and mechanism of reactions of main group metal ions with biological carriers. Pure Appl. Chem. 20, 93–115.

Feray, J. C., & Garay, R. (1986). An Na^+-stimulated Mg^{2+}-transport system in human red blood cells. Biochim. Biophys. Acta 856, 76–84.

Flatman, P. W., & Smith, W. L. (1991). Sodium-dependent magnesium uptake by ferret red cells. J. Physiol. 443, 217–230.

Garcia-del-Portillo, F., Foster, J. W., Maguire, M. E., & Finlay, B. B. (1992). Characterization of the micro-environment of *Salmonella typhimurium* containing vacuoles within MDCK epithelial cells. Mol. Microbiol. 6, 3289–3297.

Geering, K. (1992). The assembly and posttranslational processing of Na^+-pumps. Acta Physiol. Scand. 146 Suppl. 607, 177–181.

Gibson, M. M., Bagga, D. A., Miller, C. G., & Maguire, M. E. (1992). Magnesium transport in *Salmonella typhimurium*: The influence of new mutations conferring Co^{2+} resistance on the CorA Mg^{2+} transport system. Mol. Microbiol. 5, 2753–2762.

Grubbs, R. D., & Maguire, M. E. (1987). Magnesium as a regulatory cation: Criteria and evolution. Magnesium 6, 113–127.

Gunther, T., & Vormann, J. (1985). Mg^{2+} efflux is accomplished by an amiloride-sensitive Na^+–Mg^{2+} antiport. Biochem. Biophys. Res. Commun. 130, 540–545.

Henderson, P. J. F., & Maiden, M. C. J. (1990). Homologous sugar transport proteins in *Escherichia coli* and their relatives in both prokaryotes and eukaryotes. Philos. Trans. Royal Soc. Lond. [Biol.] 326, 391–410.

Hennessey, J. P., Jr., & Scarborough, G. A. (1990). Direct evidence for the cytoplasmic location of the NH_2- and -COOH terminal ends of the *Neurospora crassa* plasma membrane H^+-ATPase. J. Biol. Chem. 265, 532–537.

Hesse, J. E., Weiczorek, L., Altendorf, K., Reicin, A. S., Dorus, E., & Epstein, W. (1984). Sequence homology between two membrane transport ATPases, the Kdp-ATPase of *Escherichia coli* and the Ca^{2+}-ATPase of sarcoplasmic reticulum. Proc. Natl. Acad. Sci. USA 81, 4746–4750.

Hille, B. (1992). Ionic Channels of Excitable Membranes, 2nd ed. Sinauer, Sunderland, MA.

Hmiel, S. P., Snavely, M. D., Miller, C. G., & Maguire, M. E. (1986). Magnesium transport in *Salmonella typhimurium*: Characterization of magnesium influx and cloning of a transport gene. J. Bacteriol. 168, 1444–1450.

Hmiel, S. P., Snavely, M. D., Florer, J. B., Maguire, M. E., & Miller, C. G. (1989). Magnesium transport in *Salmonella typhimurium*: Genetic characterization and cloning of three magnesium transport loci. J. Bacteriol. 171, 4742–4751.

Lusk, J. E., & Kennedy, E. P. (1969). Magnesium transport in *Escherichia coli*. J. Bacteriol. 144, 1653–1655.

MacLennan, D. H., Brandl, C. J., Korczak, B., & Green, N. M. (1985). Amino-acid sequence of a Ca^{2+} + Mg^{2+} dependent ATPase from rabbit muscle sarcoplasmic reticulum, deduced from its complementary DNA sequence. Nature 316, 696–700.

Maloney, P. C. (1990). Microbes and membrane biology. FEMS Microbiol. Rev. 87, 91–102.

Mandala, S. M., & Slayman, C. W. (1989). The amino and carboxyl termini of the *Neurospora* plasma membrane H^+-ATPase are cytoplasmically located. J. Biol. Chem. 264, 16276–16281.

Matsuda, H. (1991). Magnesium gating of the inwardly rectifying K^+ channel. Ann. Rev. Physiol. 53, 289–298.

McDonough, A. A., Geering, K., & Farley, R. A. (1990). The sodium pump needs its β-subunit. FASEB J. 4, 1598–1605.

Nelson, D. L., & Kennedy, E. P. (1971). Magnesium transport in *Escherichia coli*. Inhibition by cobaltous ion. J. Biol. Chem. 246, 3042–3049.

Nelson, D. L., & Kennedy, E. P. (1972). Transport of magnesium by a repressible and a nonrepressible system in *Escherichia coli*. Proc. Natl. Acad. Sci. USA 69, 1091–1093.

Odermatt, A., Suter, H., Krapf, R., & Solioz, M. (1993). Primary structure of two P-type ATPases involved in copper homeostasis in *Enterococcus hirae*. J. Biol. Chem. 268, 12775–12779.

Pederson, P. L., & Carafoli, E. (1987). Ion motive ATPases. I. Ubiquity, properties, and significance to cell function. TIBS 12, 146–150.

Reinhart, R. A. (1988). Magnesium metabolism: A review with special reference to the relationship between intracellular content and serum level. Arch. Intern. Med. 148, 2415–2420.

Rotevatn, S., Sarheim, H., & Murphy, E. (1991). Intracellular free magnesium concentration: Relevance to cardiovascular medicine. Acta Physiol. Scand. 142 Suppl. 599, 125–133.

Serrano, R., Kielland-Brandt, C., & Fink, G. R. (1986). Yeast plasma membrane ATPase is essential for growth and has homology with Na^+/K^+-, K^+-, and Ca^{2+}-ATPases. Nature 319, 689–693.

Silver, S. (1969). Magnesium transport in *Escherichia coli*. Proc. Natl. Acad. Sci. USA 62, 764–771.

Silver, S., & Clark, D. (1971). Magnesium transport in *Escherichia coli*. Interference by manganese with magnesium metabolism. J. Biol. Chem. 246, 569–571.

Smith, R. L., Banks, J., Snavely, M., & Maguire, M. E. (1993a). Sequence and topology of the CorA magnesium transport systems of *Salmonella typhimurium* and *Escherichia coli*: Identification of a new class of transport protein. J. Biol. Chem. 268, 14071–14080.

Smith, D. L., Tao, T., & Maguire, M. E. (1993b). Membrane topology of a P-type ATPase: The MgtB magnesium transport protein of *Salmonella typhimurium*. J. Biol. Chem. 268, 22469–22479.

Snavely, M. D., Florer, J. B., Miller, C. G., & Maguire, M. E. (1989a). Magnesium transport in *Salmonella typhimurium*: Expression of cloned genes for three distinct Mg^{2+} transport systems. J. Bacteriol. 171, 4752–4760.

Snavely, M. D., Florer, J. B., Miller, C. G., & Maguire, M. E. (1989b). Magnesium transport in *Salmonella typhimurium*: $^{28}Mg^{2+}$ transport by the CorA, MgtA, and MgtB systems. J. Bacteriol. 171, 4761–4766.

Snavely, M. D., Miller, C. G., & Maguire, M. E. (1991a). The MgtB Mg^{2+} transport locus of *Salmonella typhimurium* encodes a P-type ATPase. J. Biol. Chem. 266, 815–823.

Snavely, M. D., Gravina, S. A., Cheung, T. T., Miller, C. G., & Maguire, M. E. (1991b). Magnesium transport in *Salmonella typhimurium*: Regulation of *mgt*A and *mgt*B expression. J. Biol. Chem. 266, 824–829.

Tao, T., Snavely, M. D., Farr, S. G., & Maguire, M. E. (1995). Magnesium transport in *Salmonella typhimurium*: *mgtA* encodes a P-Type ATPase and is regulated by Mg^{2+} in a manner similar to that of the *mgtB* P-type ATPase. J. Bacteriology 177, 2663–2672.

Vulpe, C., Levinson, B., Whitney, S., Packman, S., & Gitschier, J. (1993). Isolation of a candidate gene for Menke's disease and evidence that it encodes a copper-transporting ATPase. Nature Genetics 3, 7–13.

Walker, G. M. (1986). Magnesium and cell cycle control: An update. Magnesium 5, 9–23.

White, R. E., & Hartzell, H. C. (1989). Magnesium in cardiac function. Regulator of ion channels and second messengers. Biochem. Pharmacol. 38, 859–867.

THE *ACHOLEPLASMA LAIDLAWII* $(Na^{+}+Mg^{2+})$-ATPASE

Ronald N. McElhaney

Biomembranes
Volume 5, pages 287–316.

ISBN: 1-55938-662-2.

I. INTRODUCTION

The mycoplasmas are a diverse group of procaryotic microorganisms that lack a cell wall. Since the mycoplasmas are genetically and morphologically the simplest organisms capable of autonomous replication, they provide useful models for the study of a number of problems in molecular and cellular biology. Mycoplasmas are particularly valuable for studies of the structure and function of cell membranes. Being non-photosynthetic procaryotes, as well as lacking a cell wall or outer membrane, mycoplasma cells possess only a single membrane—the limiting or plasma membrane. This membrane contains essentially all the cellular lipid and, because these cells are small, a substantial fraction of the total cellular protein as well. Because of the absence of a cell wall, substantial quantities of highly pure membranes can usually be easily prepared by gentle osmotic lysis followed by differential centrifugation, a practical advantage not offered by other procaryotic microorganisms.

Another useful property of mycoplasmas is the ability to induce dramatic, yet controlled, variations in the fatty acid composition of their membrane lipids. Thus, relatively large quantities of a number of exogenous saturated, unsaturated, branched-chain, or alicyclic fatty acids can be biosynthetically incorporated into the membrane phospho- and glyco-lipids of these organisms. In cases in which *de novo* fatty acid biosynthesis is either exhibited or absent, fatty acid-homogeneous membranes (membranes whose glycerolipids contain only a single species of fatty acyl-chain) can sometimes be produced. Moreover, by growing mycoplasmas in the presence or absence of various quantities of cholesterol or other sterols, the amount of these compounds present in the membrane can be dramatically altered. The ability to manipulate membrane lipid fatty acid composition and cholesterol content, and thus to alter the phase state and fluidity of the membrane lipid bilayer, makes these organisms ideal for studying the roles of lipids in biological membranes.

The unique advantages of mycoplasmas for membrane studies, especially for studies of membrane lipid organization and dynamics, have induced a large number of investigators to study these microorganisms by using a wide variety of physical techniques. For this reason, we probably know more about the structural roles of lipids in mycoplasma membranes in general, and in the *Acholeplasma laidlawii* membrane in particular, than in any other biological membrane (see McElhaney, 1992a). In general, however, our understanding of membrane function in this group of organisms is less well advanced (see McElhaney, 1992b). Nevertheless, the structure and function of the (Na^++Mg^{2+})-ATPase has been the focus of considerable study and, in this chapter, I will summarize what is currently known about this enzyme.

II. GENERAL PROPERTIES OF THE *A. LAIDLAWII* (Na^++Mg^{2+})-ATPASE

A. Temperature Dependence

The temperature dependence of the membrane-bound (Na^++Mg^{2+})-ATPase activity of *A. laidlawii* has been studied both in isolated native membranes and after purification of the enzyme and reconstitution into lipid vesicles. Rottem and Razin (1966) originally reported that the activity of this enzyme in isolated membranes is maximal at 37°C. However, Jinks et al. (1978) found that the ATP hydrolytic activity of isolated membranes increases continuously over the entire maximum growth temperature range of this organism (8–44°C) and Silvius and McElhaney (1980, 1982) reported maximal enzyme activity between 45–50°C in isolated membranes enriched in various exogenous fatty acids. Moreover, George et al. (1987b, 1989) and George and McElhaney (1992) also found that the activity of the purified enzyme reaches a maximum between 45–50°C when reconstituted into a variety of phospholipid vesicles. Since Rottem and Razin (1966) also reported that heating *A. laidlawii* membranes at 47°C for 10 minutes does not reduce ATPase activity, and Hwang et al. (1992) have found that the thermal denaturation of this enzyme in native membranes commences at about 48°C, the preponderance of evidence favors a temperature optimum near these values.

The thermal stability of the purified and phospholipid-reconstituted *A. laidlawii* (Na^++Mg^{2+})-ATPase depends to some extent on the physical properties of the host phospholipid bilayer (unpublished observations from this laboratory). Specifically, the activity of this enzyme decreases more rapidly and to a greater extent upon exposure to 55°C temperatures for a given time period when reconstituted with unsaturated as compared to saturated phospholipids, or when reconstituted with short-chain saturated as opposed to moderate- or long-chain saturated phospholipids. Thus, a high degree of membrane lipid fluidity or a relatively thin host phospholipid bilayer are apparently associated with an increased tendency for the thermal denaturation of this ATPase. This and related phenomena may explain why the optimal and maximal growth temperatures of this organism are reduced when its membrane lipids are enriched in exogenous unsaturated fatty acids (McElhaney, 1974a, 1974b), and why this organism is unable to grow well on exogenous short-chain fatty acids when endogenous fatty acid biosynthesis and the chain elongation of exogenous fatty acids are inhibited (Silvius and McElhaney, 1978).

B. pH Dependence

The pH dependence of the (Na^{+}+Mg^{2+})-ATPase activity of isolated *A. laidlawii* membranes has been studied by Rottem and Razin (1966) and by Jinks et al. (1978). Both groups report the absence of a sharp pH optimum in their preparations. ATP hydrolytic is essentially constant between pH 6.7–7.0 and pH 8.0–8.4, but declines steeply at pH values either above or below this range.

C. Cation Requirements

Rottem and Razin (1966) and Jinks et al. (1978) have studied the monovalent and divalent cation requirements of the (Na^{+}+Mg^{2+})-stimulated ATPase in isolated membranes of this organism. Both groups agree that Mg^{2+} is essential for ATP hydrolytic activity and that Mn^{2+}, and to a lesser extent Co^{2+}, can partially fulfill this divalent cation requirement, but that Ca^{2+} is either relatively ineffective (Rottem and Razin, 1966) or completely ineffective (Jinks et al., 1978) in this regard. The divalent cations Ba^{2+}, Zn^{2+}, Sn^{2+}, Cu^{2+}, and Pb^{2+} are also ineffective and actually inhibit ATPase activity in the presence of normally optimal levels of Mg^{2+} (Rottem and Razin, 1966). Both groups also report that the concentration of Mg^{2+} required for maximum ATPase activity depends on the ATP concentration with highest activity being obtained at Mg^{2+}-ATP ratios between 1:1 and 1:2. This observation, plus the fact that the apparent affinity constant for Mg^{2+} is roughly equal to the K_m for ATP in the presence of excess Mg^{2+} (Jinks et al., 1978), suggests that Mg^{2+} interacts with this enzyme as a complex with ATP (the true substrate) and not directly as the free ion.

Rottem and Razin (1966) found that the monovalent cations Na^{+}, K^{+}, or NH_4^{+} do not stimulate the ATPase activity of isolated *A. laidlawii* membranes at concentrations of 5–100 mM and that higher concentrations of Na^{+} actually decrease enzyme activity. In contrast, Jinks et al. (1978), using membranes first dialyzed against EDTA and washed in inorganic monovalent and divalent cation-free buffer, report a 3–4 fold stimulation of ATP hydrolytic activity by 10 mM Na^{+}, with Li^{+} being only slightly less effective. A rapid, hyperbolic rise in enzyme activity is observed between Na^{+} concentrations of 0- to 0.5-mM, followed by a more gradual increase in activity up to Na^{+} concentrations of 50 mM. Equivalent concentrations of Rb^{+} or Cs^{+} are completely ineffective in this regard and K^{+} and NH_4^{+} actually inhibit basal ATPase activity. Moreover, the addition of 10 mM K^{+} to a reaction medium containing 50 mM Na^{+} also inhibits ATP hydrolytic activity slightly. Jinks et al. (1978) ascribe the lack of Na^{+} stimulation reported by Rottem and Razin (1966) to the presence of monovalent and divalent cations in their membrane preparation and ATPase assay system. Since subsequent work provided both indirect (Jinks et al., 1978) and direct (Mahajan et al., 1988, 1990) evidence that this enzyme functions as an ATP-dependent Na^+ pump in both intact cells and reconstituted model membranes, the designation of this enzyme as a (Na^++Mg^{2+})-ATPase rather than simply a (Mg^{2+})-ATPase seems amply justified.

D. Nucleotide Specificity

Rottem and Razin (1966) studied the hydrolysis of a variety of nucleotides and related compounds by isolated *A. laidlawii* B membranes. Although maximum hydrolytic rates are observed with ATP, other nucleotide triphosphates such as GTP, CTP, and UTP are also degraded at rates between 40–55% of that observed for ATP. Some hydrolysis of ADP and, to a lesser extent AMP (but not of GMP, CTP, or UTP), also occurs. Pyrophosphate, ribose-5-phosphate, and glucose-6-phosphate are not hydrolyzed by isolated membranes. The addition of hydrolyzable nucleotides to membranes already containing an excess of ATP does not increase the amount of inorganic phosphate released, suggesting that a single enzyme hydrolyzes all the different nucleotides tested. The apparent K_m for ATP was determined to be about 20 mM at a temperature of 37°C.

Jinks et al. (1978) and Silvius et al. (1978) have reported that the apparent K_m for ATP of the (Na^++Mg^{2+})-ATPase in isolated membranes is markedly, but non-linearly related to the assay temperature (see Figure 1). These workers report that K_m values for ATP increase from about 1 mM at lower temperatures (5–10°C) to about 3–5 mM at higher temperatures (30–40°C), depending somewhat on the fatty acid composition of the membrane lipids. Moreover, at all temperatures, ATP concentrations several-fold higher than the K_m values strongly inhibit this enzyme. Thus at room temperature, 10 mM ATP inhibits (Na^++Mg^{2+})-ATPase activity by 98%. The reasons for the large discrepancy in K_m values for ATP reported by these two groups probably results from the much greater concentration of *A. laidlawii* membranes and the absence of Na^+ in the ATPase assay utilized by Rottem and Razin (1966). As might be expected, the purified and phospholipid-reconstituted (Na^++Mg^{2+})-ATPase exhibits a markedly lower K_m value for ATP (0.1–0.2 mM at a temperature of 37°C). As with the native membrane preparations, however, ATP concentrations above about 2 mM progressively inhibit this enzyme, which becomes essentially completely inactive at ATP concentrations of 10 mM. In contrast to the results of Rottem and Razin (1966), Jinks et al. (1978) find that ADP is hydrolyzed by isolated membranes very poorly and behaves as a simple competitive inhibitor with a K_i of about 2 mM.

Lewis and McElhaney (1983) also investigated the nucleotide triphosphate substrate specificity of the purified (Na^++Mg^{2+})-ATPase. As observed for native membranes of this organism, hydrolytic activity is highest for ATP and declines progressively for ITP, GTP, and UTP; the purified enzyme, however, does not

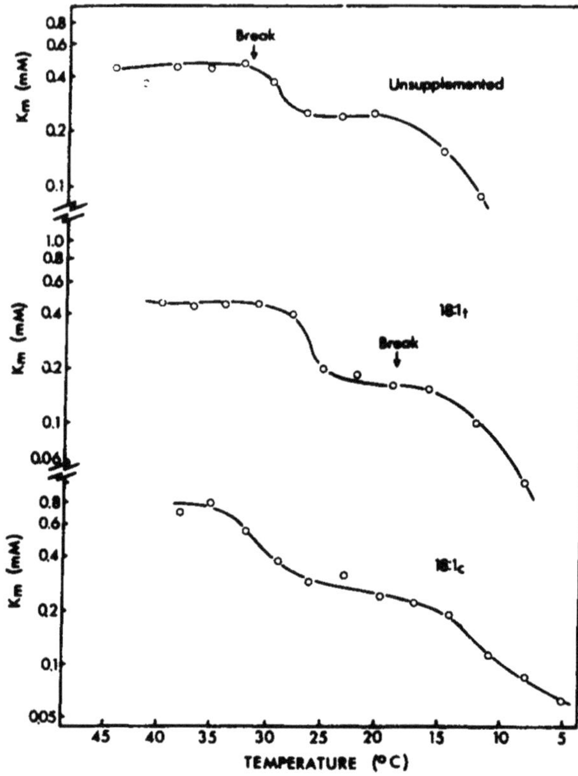

Figure 1. Variation in the K_m values for ATP as a function of the temperature of assay for the (Na^++Mg^{2+})-ATPase of *A. laidlawii* B in membranes enriched in different fatty acids. The position of the break in the Arrhenius plot of ATPase activity for unsupplemented and elaidic acid-enriched membranes is indicated by the arrows. Fatty acids are designated by the number of carbon atoms followed by the number of double bonds, if any, present in the hydrocarbon chain. The letters c and t denote the *cis*- and *trans*-configurations, respectively, of the double bond(s) (from Jinks et al., 1978, with permission).

hydrolyze CTP at all. The ATP analog, 2-deoxy-ATP, is also a good substrate, being hydrolyzed by this enzyme at a rate about 80% that observed for ATP itself. Interestingly, p-nitrophenyl phosphate is quite a poor substrate for the purified enzyme in contrast to isolated membrane preparations, suggesting that the (Na^++Mg^{2+})-ATPase and p-nitrophenyl phosphatase activities present in isolated native membranes arise from distinct enzymes, as also indicated by the work of Chen et al. (1984). ADP and the non-hydrolyzable ATP analog, AdoPP[NH]P, are not hydrolyzed at detectable rates and are both competitive inhibitors of the purified ATPase with K_i values of 2 mM and 70 μM, respectively.

E. Inhibitor Spectrum

Rottem and Razin (1966) report that several sulfhydryl reagents, including p-chloromercuribenzoate sulfate (PCMBS), N-ethylmaleimide (NEM), and mercurous chloride, markedly inhibit the ATPase activity of isolated *A. laidlawii* membranes. The compound 2,4-dinitrophenol (DNFB), a stimulator of mitochondrial and bacterial F_1F_0 ATPases, is a weak inhibitor of the *A. laidlawii* ATPase. Potassium cyanide, sodium azide, sodium nitrite, and diisopropyl phosphate do not inhibit the ATP hydrolytic activity of isolated membranes except at very high concentrations. Ouabain, an inhibitor of eucaryotic (Na^++K^+)-ATPases, has no effect on the *A. laidlawii* (Na^++Mg^{2+})-ATPase. The inhibitory effects of NEM and DNFB, but not PCMBS, are markedly diminished when excess ATP is added to the isolated membranes just prior to these inhibitors. Similarly, the prior addition of β-mercaptoethanol to the ATPase assay system almost completely abolishes the inhibitory effect of PCMB, but only partly counteracts the inhibition caused by NEM and DNFB.

Jinks et al. (1978) also studied the effects of various inhibitors on the ATPase activity of isolated *A. laidlawii* membranes. These workers also report that the sulfhydryl reagents, PCMBS and NEM, strongly inhibit ATP hydrolysis and that sodium azide is without effect. The potent bacterial and mitochondrial F_1F_0-ATPase inhibitor, dicyclohexylcarbodiimide (DCCD), inhibits the ATPase activity of this organism only at very high concentrations. The calculated affinity constants for PCMBS, NEM, and DCCD are about 0.2-, 2.2-, and 0.2-mM, respectively. The inhibition of ATP hydrolytic activity by increasing concentrations of PCMBS and DCCD is adequately described by a classical single-site inactivation model, implying that only one major species of ATPase is present in the isolated membranes of this organism. The inactivation by NEM is not completely described by single-hit kinetics. However, the NEM inhibition pattern observed is more compatible with the presence of multiple sites of NEM action on a single enzyme species than with the presence of several different enzyme species differing in NEM sensitivity. The ATP hydrolytic activities assayed in the presence and absence of 50 mM Na^+ exhibit essentially identical dose-response curves for these three inhibitors, suggesting that the (Mg^{2+})-ATPase and (Na^++Mg^{2+})-ATPase activities arise from the same enzyme. In contrast, variations in ATP and DCCD levels have quite different effects on ATP and p-nitrophenyl phosphate hydrolysis, again indicating that the ATPase and p-nitrophenyl phosphate hydrolysis activities arise from two distinct enzymes. This latter conclusion is consistent with the very different dependencies of these two enzyme activities on the fluidity and phase-state of the membrane lipid (de Kruijff et al., 1973), which will be discussed in more detail later in this chapter. Chen et al. (1984) also found that the ATPase activity of isolated membranes is only slightly inhibited by DCCD, oligomycin, and potassium thiocyanate, even at high concentrations, and is completely insensitive to ouabain.

Lewis and McElhaney (1983) have investigated the sensitivity of the (Na^++Mg^{2+})-ATPase purified from *A. laidlawii* membranes to a variety of inhibitors. The classical inhibitors of bacterial and mitochondrial F_1F_0 ATPases either have no effect on ATP hydrolytic activity (sodium azide, aurovertin, quercetin) or are poor inhibitors even at relatively high concentrations (DCCD, efrapeptin, leucinostatin, oligomycin). The purified enzyme is also unaffected by ouabain, as is the ATPase activity of isolated membranes. Inhibition by sodium orthovanadate, a potent inhibitor of ATPases of eucaryotic cell membranes, is not complete even at very high concentrations (10 mM). Generally speaking, the inhibitor spectra of the isolated and purified (Na^++Mg^{2+})-ATPase and of the (Na^++Mg^{2+})-ATPase activity in isolated membranes are similar.

III. STRUCTURE AND DISPOSITION OF THE *A. LAIDLAWII* (Na^++Mg^{2+})-ATPASE

A. Overall Arrangement in the Native Membrane

A magnesium-dependent adenosine 5′-triphosphatase (Mg^{2+}-ATPase) activity has been detected in the membrane of every mycoplasma species studied so far (Razin, 1982), but the *A. laidlawii* enzyme is by far the best characterized. The *A. laidlawii* Mg^{2+}-ATPase cannot be released by the techniques usually employed to detach peripheral membrane proteins (Ne'eman et al., 1971) and is firmly bound to the membrane (Pollack et al., 1965; Ne'eman et al., 1972), indicating that it is an integral membrane protein. Since no Mg^{2+}-ATPase activity can be detected in intact cells, the ATP hydrolytic site is presumed to be located on the cytoplasmic side of the membrane (Pollack et al., 1965; Rottem and Razin, 1966). Because the Mg^{2+}-ATPase of this organism is inactivated by detergents and organic solvents, early attempts at its isolation and characterization failed (Ne'eman et al., 1972). In its native, membrane-associated form, however, it was reported that the *A. laidlawii* ATPase is not activated by Na^+ or K^+ and that it is insensitive to ouabain, a specific inhibitor of eucaryotic (Na^++K^+)-ATPases (Rottem and Razin, 1966). However, the activity of this enzyme in isolated membranes is sensitive to the fluidity and phase-state of the membrane lipids and this enzyme pumps Na^+ in native and in reconstituted model membranes (see following), indicating that it is an integral, transmembrane, lipid-requiring enzyme.

B. Subunit Composition and Stoichiometry

Lewis and McElhaney (1983) have succeeded in isolating the (Na^++Mg^{2+})-ATPase from the *A. laidlawii* B membrane in a relatively pure and enzymatically active form. This enzyme has been shown to consist of five subunits (α, β, γ, δ, ε) of apparent molecular weight from 68,000, 55,000, 35,000, 27,000, and 16,000 daltons; the probable subunit stoichiometry is $\alpha_3\beta_3\gamma\delta_2\varepsilon_3$. Similar results have been obtained by Chen et al. (1984) employing a different procedure to isolate and purify this enzyme.

C. Subunit Disposition in Lipid Bilayers

All of the subunits of the purified, phospholipid-reconstituted ATPase can be labeled by water-soluble, group-specific reagents, suggesting that at least a portion of each of these subunits is exposed to the aqueous phase (George and McElhaney, 1985; Lewis et al., 1986). On the other hand, the α- (and possibly the ε-) subunit can be photolabeled by phospholipids containing a photo-sensitive fatty acyl group, indicating that portions of at least this subunit(s) penetrate into or transverse the hydrophobic core of the phospholipid bilayer (George et al., 1985).

D. Amino Acids Involved in the ATP-Binding Site

Studies with a variety of reagents that specifically modify various amino acid residues indicate that the modification of one reactive lysine, one reactive arginine, or two reactive tyrosine residues completely inactivates this enzyme. A partial inactivation of enzyme activity by certain sulfhydryl- and carboxyl-group reagents also occurs, suggesting that these groups may also be involved in catalysis, although nonspecific effects such as subunit cross-linking may be responsible for these effects (Lewis et al., 1986). The (Na^++Mg^{2+})-ATPase can also be labeled by the 2′, 3′-dialdehyde derivative of ATP, however, since all subunits except the ε are labeled, localization of the ATP-binding subunit is not achieved (George and McElhaney, 1985). However, a nucleotide analog that reacts with phenolic- or sulfhydryl-groups, or both, specifically labels the α-subunit, suggesting that it contains the active site (Lewis et al., 1986). However, only the γ-subunit is phosphorylated by radiolabeled ATP (unpublished observation from this laboratory).

Some recent unpublished work from this laboratory has confirmed previously published results suggesting that an arginine residue plays a key role in the functioning of the *A. laidlawii* (Na^++Mg^{2+})-ATPase. In the absence of Na^+ and ATP, the chemical modification of a single reactive arginine residue by the arginine-specific reagent phenyl glyoxal results in the complete inactivation of this enzyme whether or not Mg^{2+} is present. Moreover, the presence of ATP alone provides nearly complete protection from phenyl glyoxal inhibition, but the presence of ATP *and* Mg^{2+} provides only partial protection. Interestingly, the presence of Na^+ also provides partial protection from phenyl glyoxal inactivation in the absence of ATP and Mg, but when ATP is present, Na^+ actually potentiates inactivation by this reagent. These results suggest that the *A. laidlawii* (Na^++Mg^{2+})-ATPase must exist in at least four different conformational states, depending on the presence or absence of ATP and activating cations. The active arginine residue appears to be relatively exposed to phenyl glyoxal in those conformation states existing in the absence of substrate and activating cations or in the presence of Mg^{2+} alone, whereas this reactive arginine is partially or fully protected in the conformation states existing in the presence of Na^+ alone or ATP alone, respectively.

Additional unpublished results from this laboratory provide further support for the existence of multiple conformational states for this enzyme. In the absence of ATP and activating cations, the limited trypsin digestion of the *A. laidlawii* (Na^++Mg^{2+})-ATPase results in essentially complete inhibition of this enzyme in 15 minutes. However, the presence of Na^+ alone, Mg^{2+} alone, or ATP alone provides increasingly more effective protection from trypsin inactivation such that anywhere from 40% (Na^+ alone) to 75% (ATP alone) of the initial ATP hydrolytic activity remains after 30 minutes. Moreover, this limited trypsin treatment yields a different and distinct pattern of proteolytic cleavage products under each of these four experimental conditions. These results strongly suggest that the binding of substrate and each of the individual activating cations changes the conformation of this enzyme such that different trypsin-sensitive sites are exposed on its surface.

IV. FUNCTION OF THE *A. LAIDLAWII* (Na^++Mg^{2+})-ATPASE

A. Cell Volume Regulation

Jinks et al. (1978) reported that whole cells of *A. laidlawii* incubated in NaCl- or KCl-containing buffers swell and lyse if deprived of glucose or if treated with several (Na^++Mg^{2+})-ATPase inhibitors in the presence of glucose. In contrast, cells incubated in $MgSO_4$ buffer or sucrose do not swell or lyse under these same conditions. Moreover, there is a good correlation between the concentrations of the ATPase sulfhydryl group inhibitors, NEM and PCMBS, required to inhibit the ATP hydrolytic activity of this enzyme and the amount required to produce cell lysis in NaCl- or KCl-containing buffers in the presence of glucose. These and other results led these workers to postulate that this enzyme is intimately involved in the regulation of the intracellular osmolarity and cellular volume of this organism, probably by functioning as an ion pump which actively extrudes monovalent cations (probably Na^+) from cells. This hypothesis has recently been confirmed by Mahajan et al. (1988, 1990), who showed that the purified (Na^++Mg^{2+})-ATPase catalyzes the ATP-dependent translocation of Na^+ when reconstituted into liposomes. Moreover, the amount of Na^+ transported increases with the concentration of ATP added and both Na^+ transport and ATPase activity are inhibited to comparable extents by inhibitors of the *A. laidlawii* (Na^++Mg^{2+})-ATPase, such as vanadate and leucinostatin. Although some Na^+/Na^+ exchange is observed in the reconstituted proteoliposomes studied, a net accumulation of Na^+ is observed with a stoichiometry (corrected for leakage and exchange) of about 1 mol of Na^+ per mole of ATP hydrolyzed.

B. Sodium Ion Extrusion

Mahajan et al. (1988) have also studied Na^+ transport in live *A. laidlawii* cells using $^{23}Na^+$-nuclear magnetic resonance (NMR) spectroscopy and a nonpermeable shift reagent to permit the differentiation of intra- and extra-cellular Na^+ pools. When energy-depleted, sodium-loaded cells are provided with glucose, a time-dependent extrusion of Na^+ from cells is observed, with about half of the intracellular Na^+ removed in five minutes and a steady-state level reached in about 20 minutes (see Figure 2). In the presence of excess glucose, intracellular Na^+ concentrations of roughly 15 mM are obtained in isotonic buffer containing 156 mM Na^+. Moreover, non-metabolizable glucose analogs cannot substitute for glucose, and the presence of glycolytic metabolic poisons, such as sodium fluoride and merthiolate, inhibit Na^+ extrusion. These results clearly show that Na^+ efflux in this organism is an active transport process ultimately driven by glycolysis.

Mahajan et al. (1988) also investigated the influx of $^{22}Na^+$ into *A. laidlawii* cells which have depleted the glucose initially present in the isotonic assay buffer. At 37°C, Na^+ enters the cell down its concentration gradient at a much slower rate than it was pumped out, with about half the maximum intracellular Na^+ entering at 30–40 minutes. This rate of Na^+ entry is, nevertheless, much faster than that expected for the passive diffusion of Na^+ across the lipid bilayer. However, the rate of Na^+ entry

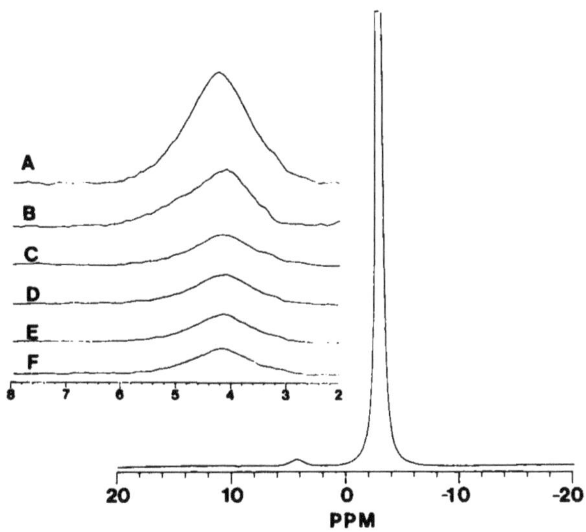

Figure 2. Effect of added glucose on the levels of sodium in *A. laidlawii* B cells as monitored by ^{23}Na-NMR spectroscopy. The changes in the intensity of the NMR signals of the intracellular sodium as a function of time after the addition of glucose are shown in the inset. Times were 0, 5, 10, 20, 30, and 40 min (**A** to **F**, respectively; from Mahajan et al., 1988, with permission).

is markedly temperature-dependent and, at temperatures below about 20°C, becomes too slow to measure. Using ^{31}P-NMR spectroscopy, these workers were able to show that in glucose-deprived cells maintained at 37°C, the entry of Na^+ ions is accompanied initially only by the loss of intracellular sugar phosphates and then by a loss of nucleoside triphosphates. If glucose is added before the nucleoside triphosphate level falls, Na^+ efflux is again observed. However, the addition of glucose to nucleoside triphosphate-depleted cells (1 hour more of glucose starvation) does not result in Na^+ efflux or an increase in intracellular sugar phosphate or nucleoside triphosphate levels.

Mahajan et al. (1988) also studied the temperature and lipid phase-state dependence of Na^+ extrusion. Cells enriched in oleic acid, whose membrane lipids are exclusively in the liquid–crystalline state throughout the physiological temperature, exhibit increasing rates of Na^+ extrusion with temperatures from 20–37°C. However, no decrease in intracellular Na^+ levels with time is detected at 15°C or below. In contrast, cells enriched in palmitic acid, which exhibit a gel to liquid–crystalline membrane lipid-phase transition over the temperature range 15–37°C, exhibit glucose-dependent Na^+ efflux at 37°C, but not at 20°C. The lack of Na^+ efflux in palmitate-enriched cells at 20°C is not due to a disruption in energy metabolism, since normal levels of sugar phosphates and nucleotide triphosphates are present. The absence of net Na^+ efflux at 20°C in palmitate-enriched cells could be due either to an inhibition of the function of the glucose transporter by gel–state lipid or to a leakage of Na^+ ions at the boundaries of gel and liquid–crystalline lipid domains, or both. However, the latter suggestion is unlikely, since palmitate-enriched, glucose-depleted cells maintained in Na^+-containing buffer at 20°C do not exhibit a detectable inward movement of Na^+. In fact, extensive evidence for inhibition of the function of the (Na^++Mg^{2+})-ATPase in this organism by gel-state lipid is available and will be discussed subsequently.

Lelong et al. (1989) have provided evidence for a monovalent cation/proton antiport activity in *A. laidlawii*. They showed that an artificially created ΔpH (inside acid) in intact cells can be dissipated more rapidly when Na^+ is added, and that the addition of K^+ after Na^+ further stimulates the collapse of the pH gradient, although the reverse is not true. However, the addition of K^+ is without effect in resealed membrane vesicles, as is the addition of HCl. The addition of nigericin, a K^+/H^+ antiporter, collapses the imposed ΔpH when added along with KCl, but the addition of high amounts of tetraphenyl phosphonium, which is known to collapse the transmembrane electrical potential, has no effect in intact cells. These workers suggest that a K^+-dependent Na^+/H^+ exchanger may exist in intact cells of *A. laidlawii* and this exchange may function both in pH homeostasis in the alkaline pH range as well as in the extrusion of excess intracellular Na^+. They further suggest that the osmotic balance may be maintained by the activity of a H^+-ATPase and a secondary Na^+/H^+ exchanger, rather than by the (Na^++Mg^{2+})-ATPase. However, as the existence and function of a H^+-ATPase has not been conclusively demonstrated in the organism while the existence of a Na^+-transporting ATPase has, it is likely that the (Na^++Mg^{2+})-ATPase is the primary system for osmo-regulation with the Na^+/H^+ exchanger playing a secondary role in maintaining cell volume and intracellular pH.

V. MODULATION OF THE ACTIVITY OF THE *A. LAIDLAWII* (Na^++Mg^{2+})-ATPASE BY THE HOST LIPID BILAYER

A. Lipid Fatty Acid, Polar Headgroup, and Cholesterol Compositional Variations in Native Membranes

The first investigation of the lipid and temperature dependence of the major ATPase of the *A. laidlawii* B membrane was that of de Kruijff et al. (1973), who studied the temperature-dependence of three membrane-bound enzymes in cells whose fatty acid and sterol compositions were widely varied. For NADH oxidase and p-nitrophenylphosphatase activities, Arrhenius plots are linear over the temperature range 5°C to 35°C and no breaks or slope changes are observed. Moreover, the absolute activity of these enzymes in isolated membranes does not vary with membrane lipid fatty acid or sterol composition. de Kruijff and co-workers thus concluded that NADH oxidase and p-nitrophylphosphatase, despite being integral membrane proteins, are unaffected by the fluidity and phase-state of the membrane lipids. In contrast, the Mg^{2+}-ATPase activity seems to exhibit biphasic linear Arrhenius plots in membranes enriched in relatively high-melting fatty acids (see Figure 3). The Arrhenius plot break temperatures always appeared to fall a few degrees above the lower boundary of the gel to liquid–crystalline lipid phase transition as measured by differential scanning calorimetry (DSC). When grown in the presence of cholesterol, the Arrhenius break temperatures are reduced by 6°C to 7°C, as is the lower boundary of the lipid-phase transition (see Figure 4). Treatment of cholesterol-enriched membranes with the polyene antibiotic, filipin, which specifically complexes cholesterol and withdraws it from interaction with the membrane glycerolipids, reverses the effect of cholesterol incorporation on both ATPase activity and the lipid-phase transition. Moreover, the incorporation of epicholesterol instead of cholesterol has no effect on either the ATPase activity-temperature profile or the membrane lipid-phase transition. Finally, the absolute activity of this enzyme does appear to vary significantly with alterations in membrane lipid fatty acid composition and cholesterol content. de Kruijff et al. (1973) concluded that the activity of the *A. laidlawii* B Mg^{2+}-ATPase is markedly influenced by the phase-state of the membrane lipids and that, within the phase-transition temperature range at least, this enzyme preferentially associates with the lipid molecular species having the lowest phase-transition temperatures. Qualitatively similar results were subsequently presented by Hsung et al. (1974), who used electron spin resonance (ESR) spectroscopy to monitor membrane lipid phase behavior.

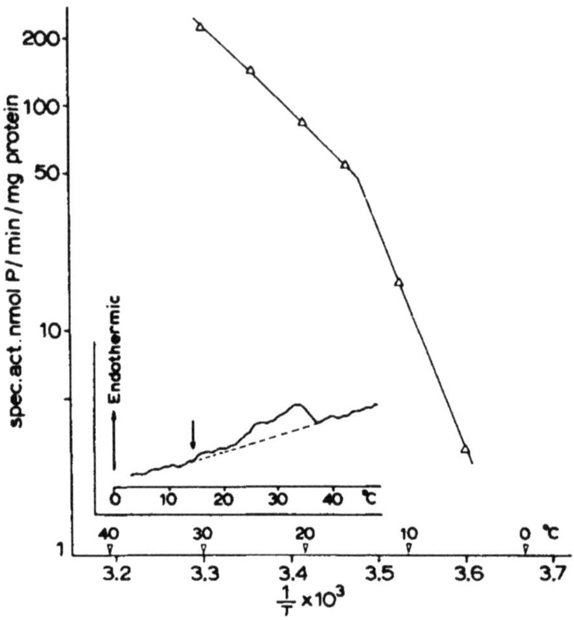

Figure 3. Arrhenius plots of the ATPase activity and lipid-phase transition determined by DSC in membranes from *A. laidlawii* cells grown on 18:1t determined by DSC. The arrow in the calorimetric scan indicates the temperature of the break in the Arrhenius plot of the ATPase activity (from de Kruijff et al., 1973, with permission).

Bevers et al. (1977) investigated the phospholipid requirement of the same three membrane-bound enzymes studied earlier by de Kruijff et al. (1973) using various phospholipases to selectively degrade the phosphatidylglycerol (PG) in *A. laidlawii* B membranes. These workers found that the complete hydrolysis of PG, the only phosphatide present in this *A. laidlawii* strain, by phospholipases A_2, C, or D has no effect on the activity of NADH oxidase or of *p*-nitrophenyl-phosphatase. Similarly, phospholipase-A_2 and -C hydrolysis of about 90% of the membrane PG could occur without effect on Mg^{2+}-ATPase activity, but hydrolysis of the final 10% of this phospholipid, which proceeds at a slower rate, results in a marked and progressive loss of activity. The complete conversion of PG to phosphatidic acid (PA) by phospholipase D treatment does not affect Mg^{2+}-ATPase activity, indicating that the glycerol portion of the PG headgroup is not required for activity. The inactivated Mg^{2+}-ATPase in PG-depleted membranes could be reactivated by adding PG, PA, or phosphatidylserine (PS), but not by phosphatidylcholine (PC), phosphatidylethanolamine (PE), nor any of the *A. laidlawii* glycolipids. These results indicate that this enzyme requires small amounts of a diacyl-phosphatide bearing a net negative charge for optimal activity. Phospholipid reconstitution experiments demonstrated that the fatty acid compositions of both the residual PG

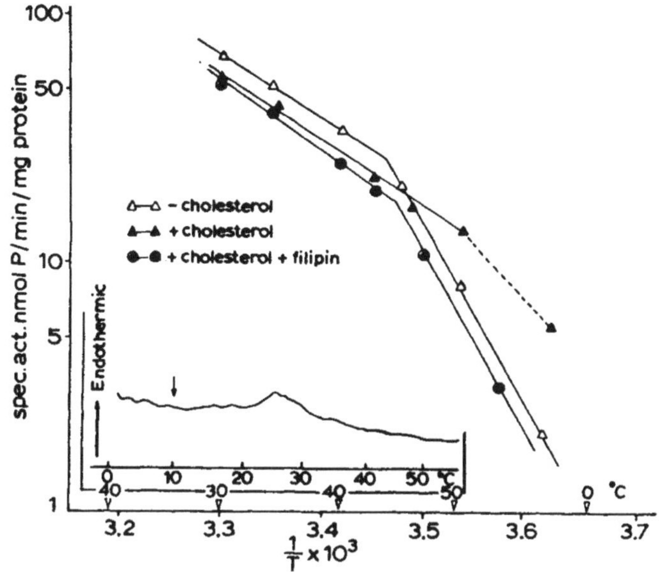

Figure 4. Arrhenius plot of the ATPase activity and lipid phase transition in membranes from *A. laidlawii* cells grown on 18:1t with or without cholesterol, or with cholesterol and filipin (from de Kruijff et al., 1973, with permission).

present in the membrane and of the added phospholipid determine the activation energy of the Mg^{2+}-ATPase and the Arrhenius plot break temperature. In these reconstitution experiments, a preferential association of this enzyme with the more fluid phospholipid species does not seem to occur, in contrast to the suggestion made earlier by de Kruijff et al. (1973). It is noteworthy that the ability to restore Mg^{2+}-ATPase by adding exogenous phospholipid to membranes containing less than 2% of their original PG is gradually lost with time. This result probably means that a certain minimal amount of phospholipid is required for stabilization of this enzyme in an active conformation in the native membrane, such as appears to be the case for the isolated enzyme (Lewis and McElhaney, 1983).

The dependence of the kinetic parameters of the *A. laidlawii* ATPase on the fluidity and phase-state of the membrane lipids have been investigated in more detail by McElhaney and coworkers. These investigators showed that the K_m for this enzyme for ATP is markedly temperature dependent, the K_m value increasing roughly 10-fold as the temperature is increased from 5°C to 40°C (Jinks et al., 1978). This temperature dependence of substrate-binding affinity appears to be an intrinsic property of the enzyme molecule, since it is not greatly affected by the fatty acid composition, cholesterol content, or phase-state of the membrane lipids. Moreover, Silvius et al. (1978) demonstrated that this and other enzymes with a temperature-dependent K_m can yield a variety of Arrhenius plot artifacts, most notably erroneous "breaks," if enzyme activity is assayed at a fixed substrate concentration; in fact, the positions of the Arrhenius plot "breaks" previously reported by de Kruijff et al. (1973) were actually significantly underestimated (by 7°C to 10°C) because ATPase activity was becoming substrate limited at the higher assay temperatures. These investigators also demonstrated that the Mg^{2+}-ATPase of *A. laidlawii* B is strongly and specifically stimulated by Na^+ and, therefore, that this enzyme is actually a (Na^++Mg^{2+})-ATPase.

The results of initial studies of the activity-temperature profile of the *A. laidlawii* (Na^++Mg^{2+})-ATPase, under conditions where the true maximum velocity of the enzyme was being measured, appeared to produce the classic biphasic linear Arrhenius plots for membranes enriched with fatty acid giving a lipid-phase transition in the physiological temperature range, except that the break temperatures now occur slightly below, but generally near, the phase-transition midpoint temperature instead of at the lower boundary of the transition (Jinks et al., 1978; Silvius et al., 1979). Interestingly, the Arrhenius plots of ATPase activity in membranes

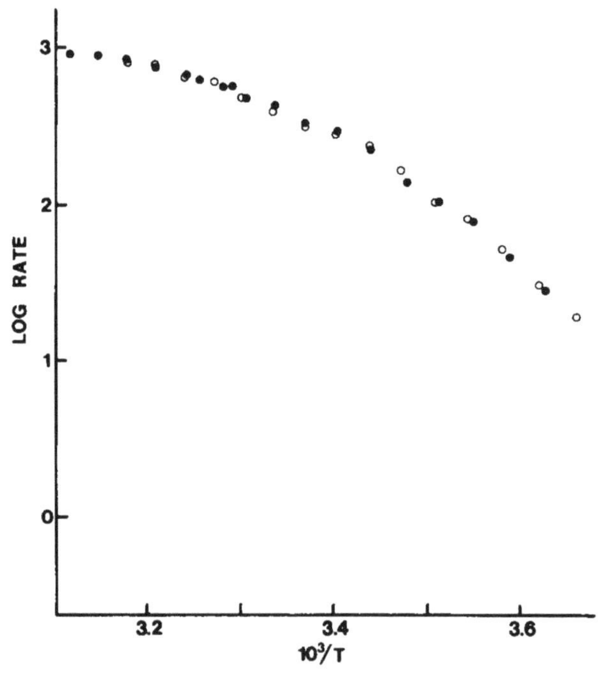

Figure 5. Arrhenius plots of the (Na^++Mg^{2+})-ATPase activity in membranes of *A. laidlawii* B grown with avidin plus either anteisopentadecanoic acid (•) or *cis*-vaccenic acid (o). The absolute ATPase activities of these two membranes are not equal, but have been normalized in order to facilitate comparison of their temperature (T) dependencies (from Silvius and McElhaney, 1980, with permission).

enriched in low-melting fatty acids appear to be curved downward instead of being linear as previously reported. Subsequent studies by Silvius and McElhaney (1980, 1982) and by Silvius et al. (1980), in which more accurate ATPase activity values at additional temperature points were determined, confirmed that in membranes containing exclusively liquid–crystalline lipid in the physiological temperature range, Arrhenius plots of (Na^++Mg^{2+})-ATPase activity are clearly nonlinear (slope gently downward; see Figure 5). The *temperature dependence* of the ATPase activity is not dependent on membrane lipid fatty acid composition as long as the lipids exist in the fluid state. The *absolute activity* of this enzyme, however, does vary significantly with fatty acid composition, but there is no discernible relation between enzyme activity and lipid fluidity *per se*. If a gel to liquid–crystalline phase transition occurs within the physiological range, however, a gently curving biphasic or triphasic Arrhenius plot is observed, in which the (Na^++Mg^{2+})-ATPase activity falls off more steeply with decreasing temperature than would otherwise be the case (see Figure 6). No effect of the lipid-phase transition on the ATPase activity is noted until about half of the membrane lipid is converted to the solid state, and some

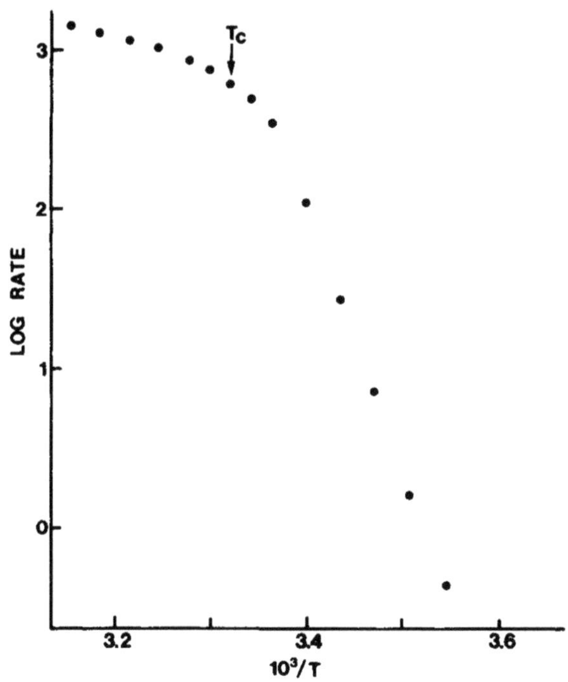

Figure 6. Arrhenius plot of the (Na^++Mg^{2+})-ATPase activity in membranes of *A. laidlawii* B grown with avidin plus isoheptadecanoic acid. The membrane lipid phase transition midpoint temperature (T_c) of 28.8°C is indicated (from Silvius and McElhaney, 1980, with permission).

ATPase activity remains at temperatures considerably *below* the *lower boundary* of the lipid-phase transition, although eventually all ATPase seems to be lost. These results suggest that the *A. laidlawii* (Na^{+}+Mg^{2+})-ATPase is active only in association with liquid-like boundary lipids, and that the ATPase hydrolytic reaction exhibits a significant heat capacity of activation in this case. This enzyme appears to become progressively inactivated when its boundary lipids undergo a liquid-like to solid-like "phase transition" that is driven by the liquid–crystalline to gel-phase transition of the bulk membrane lipid phase, but that is less cooperative and that takes place over a lower temperature range than does the bulk lipid transition. The lateral aggregation of intramembranous protein particles normally observed as the bulk lipid enters the gel-state is apparently not responsible for the loss of ATPase activity, since membranes enriched in methyl-isobranched fatty acids give Arrhenius plots indistinguishable from those described above, despite the fact that no significant clustering of intramembranous particles occurs in isobranched acid-enriched membranes. These results suggest that the familiar "biphasic linear" Arrhenius plots commonly reported for many membrane enzymes and transport systems may, in fact, have a more complex shape, analysis of which can furnish useful information regarding the behavior of the enzyme molecule in its membrane environment (Silvius and McElhaney, 1981).

The effect of pressure and of pentanol on the activity and temperature dependence of the ATPase of the *A. laidlawii* plasma membrane was studied by MacNaughton and Macdonald (1982). Increases in hydrostatic pressure increase the temperature at which the break in the Arrhenius plot of ATPase activity versus temperature is observed. Since this Arrhenius plot break is known to be induced by a gel to liquid–crystalline phase-transition of the membrane lipids, this implies that the effect of hydrostatic pressure is mediated predominantly through its action on the membrane lipid bilayer. Moreover, the increases in the Arrhenius plot slopes (apparent activation energies) observed at temperatures above the break temperature with increases in hydrostatic pressure can also be explained by changes in the physical properties of the lipid bilayer, since hydrostatic pressure increases the order of liquid–crystalline lipid bilayers (Macdonald and McLeod, 1980; Macdonald and Cossins, 1983). Pentanol, however, which is known to decrease the lipid-phase transition temperature and to increase the fluidity of liquid–crystalline bilayers of this organism (Macdonald and McLeod, 1980; Macdonald and Cossins, 1983), inhibits ATPase activity without affecting the Arrhenius break temperature at 37°C and atmospheric pressure, suggesting that it can act directly on the enzyme to inhibit its function. However, pentanol can partially offset the inhibitory effects of high hydrostatic pressure, which is consistent with a bilayer-fluidizing effect of this alcohol. It thus appears that pentanol can exert two opposite effects on the activity of this ATPase, with the net effect being determined by the temperature and pressure, and by the alcohol concentration employed.

B. Lipid Fatty Acid Compositional Variations in Reconstituted Membranes

With the development of a successful reconstitution procedure for the purified *A. laidlawii* B (Na^++Mg^{2+})-ATPase (Jinks and McElhaney, 1987; George et al., 1987a), McElhaney and co-workers have recently investigated, in more detail than is possible *in vivo*, the lipid- and temperature-dependence of the ATP hydrolytic activity of this enzyme (George et al., 1987b, 1989). In order to investigate the influence of lipid fluidity and phase-state on the temperature-dependence of the purified *A. laidlawii* (Na^++Mg^{2+})-ATPase, this enzyme was reconstituted in large, unilammelar vesicles of PCs containing saturated, branched-chain, or unsaturated fatty acids having a wide range of chain lengths. Arrhenius plots of ATPase activity vs. assay temperature for vesicles containing diisomyristoyl-, dioleoyl-, or diphytanoyl-PC were constructed. These three phospholipids have gel to liquid–crystalline phase transition temperatures below 0°C so that they exist entirely in the fluid state over the physiological temperature range. In all three vesicles, almost identical, smoothly curving Arrhenius plots are observed, indicating that the nature of the fatty acid has little or no effect on the absolute activity or temperature dependence of this enzyme (see Figure 7). In particular, ATPase activity is not very sensitive to the "fluidity" of the host lipid bilayer so long as it remains in the liquid–crystalline state. These Arrhenius plots are essentially identical to those observed in isolated *A. laidlawii* B membranes whose lipids have been highly enriched in various low-melting fatty acids (Silvius and McElhaney, 1980, 1982).

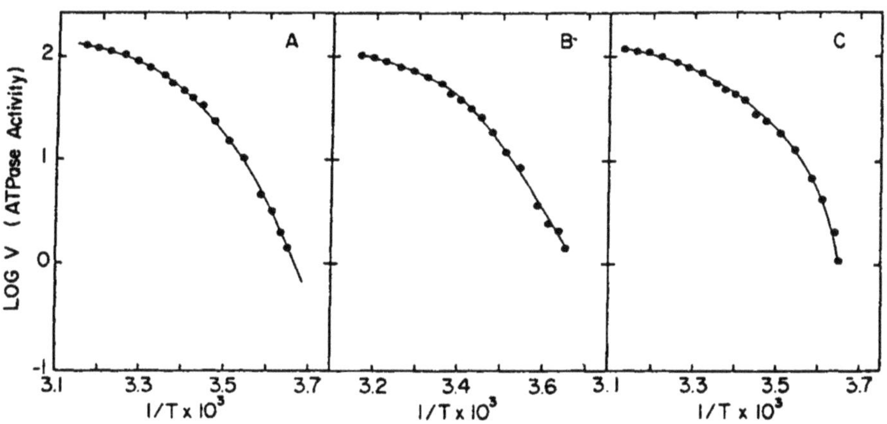

Figure 7. Arrhenius plots of the purified *A. laidlawii* B (Na^++Mg^{2+})-ATPase reconstituted into: (**A**) diisomyristoylphosphatidylcholine (PC), (**B**) dileoylPC, and (**C**) diphytanoylPC. Each of these ATPase-containing PC vesicles exhibits a gel to liquid–crystalline phase-transition temperature below 0°C (from George et al., 1987b, with permission).

Arrhenius plots of (Na^{+}+Mg^{2+})-ATPase activity in vesicles composed of PCs containing saturated or unsaturated fatty acids and that exhibit gel to liquid–crystalline phase transitions in the physiological temperature range were also constructed. In all cases, these Arrhenius plots are clearly biphasic (see Figure 8). The high-temperature portion of each plot exhibits a gentle downward slope, as observed for vesicles whose lipids exist entirely in the liquid–crystalline state above 0°C. The low-temperature portions of each plot are more nearly linear and exhibit much greater slopes. The temperature at which the change in slope occurs correlates well with the midpoint of the gel to liquid–crystalline lipid phase-transition temperature of the ATPase-containing vesicle as determined by DSC. These results indicate that this enzyme is fully active only when most of the phospholipid is in the liquid–crystalline state. However, at temperatures above the phospholipid phase-transition temperature, the absolute activities obtained by the ATPase are fairly similar. This indicates again that moderate alterations in phospholipid fatty acid structure, and thus membrane lipid fluidity, have only a small effect on enzyme activity. These Arrhenius plots are also essentially identical to those observed in isolated membranes whose lipids have been highly enriched in various higher-melt-

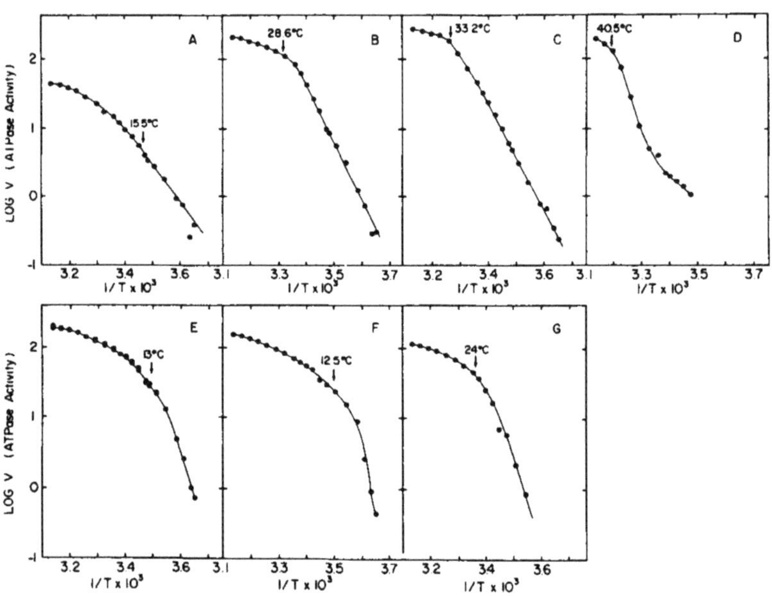

Figure 8. Arrhenius plots of the purified *A. laidlawii* B (Na^{+}+Mg^{2+})-ATPase reconstituted into: (**A**) ditridecanoylphosphatidylcholine (PC), (**B**) dimyristoylPC, (**C**) depentadecanoylPC, (**D**) dipalmitoylPC, (**E**) dielaidoylPC, (**F**) dierucoylPC, and (**G**) dinervonoylPC. The gel to liquid–crystalline phase-transition temperature of each of these ATPase-containing PC vesicles is indicated on the appropriate panel (from George et al., 1987b, with permission).

ing fatty acids (Silvius and McElhaney, 1980, 1982). These results also indicate that the wide variations in ATPase activity observed in *A. laidlawii* B membranes of differing fatty acid composition are due primarily to variations in the number, rather than to the specific activity, of the ATPase molecules present (Silvius and McElhaney, 1980, 1982; Silvius et al., 1980).

The effect of variations in the hydrocarbon chain length of a series of saturated, branched-chain and monounsaturated PCs on the activity of the purified (Na^++Mg^{2+})-ATPase was also studied (George et al., 1987b, 1989). The shorter-chain PCs (those containing fewer than 13 carbon atoms) either do not enhance the ATP hydrolytic activity of the lipid-depleted enzyme or actually inhibit it. This result was shown not to be due to a failure of this enzyme to associate with these short-chain PCs or to an inability of the enzyme–PC complexes to form bilayer assemblies. Interestingly, ATPase activity depends only weakly on acyl chain length for PCs containing 14–24 carbon atoms, and no clear optimal PC hydrocarbon chain length could be identified, in contrast to the results for several other ATPases, where enzyme activity is optimal for PC chain lengths of 16–18 or 18–20 carbons (Johannson et al., 1981a, b; Montecucco et al., 1982). Thus, once the minimal phospholipid hydrocarbon chain length requirement is met, the *A. laidlawii* (Na^++Mg^{2+})-ATPase can apparently function well in bilayers of considerably varying thickness, provided of course that an increase in acyl-chain length does not result in the formation of gel-state lipid at the assay temperature.

C. Cholesterol Structure and Content Variations in Reconstituted Membranes

In order to investigate the effect of cholesterol and epicholesterol (the α-epimer of cholesterol) on the activity and temperature-dependence of the purified (Na^++Mg^{2+})-ATPase, this enzyme was reconstituted into large, unilamellar vesicles formed from dimyristoylphosphatidylcholine (DMPC) and varying amounts of cholesterol or epicholesterol. The ATP hydrolytic activity of the reconstituted enzyme was then determined over a range of temperatures and the phase-state of the DMPC in the ATPase-containing vesicles was characterized by high-sensitivity DSC. In the vesicles containing only DMPC, the ATPase activity is higher in association with lipids in the liquid–crystalline state than with gel-state phospholipids, resulting in a curvilinear, biphasic Arrhenius plot with a pronounced change in slope at the elevated gel to liquid–crystalline phase-transition temperature of the DMPC (see Figure 9). The incorporation of increasing amounts of cholesterol into the DMPC vesicles results in a progressively greater degree of inhibition of ATPase activity at higher temperatures, but a stimulation of activity at lower temperatures, thus producing Arrhenius plots with progressively less curvature and without an abrupt change in slope at physiological temperatures. As cholesterol concentration in the ATPase–DMPC vesicles increases, the calorimetric phase-transition of the phospholipid is further broadened and eventually abolished.

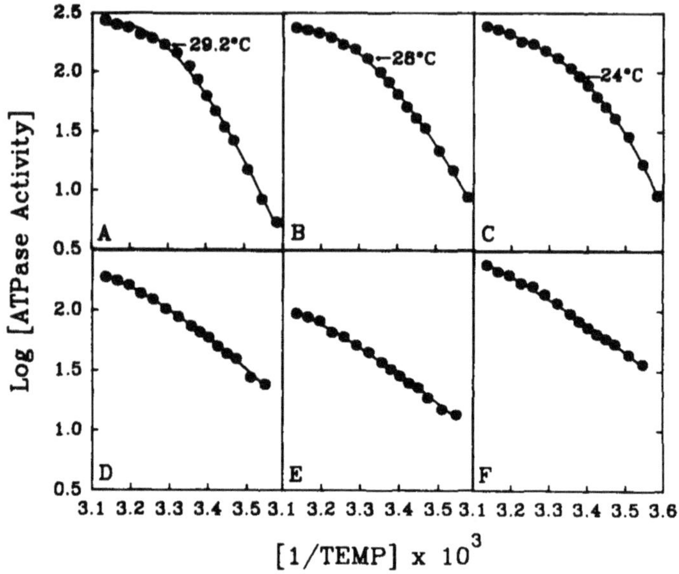

Figure 9. Temperature-dependence of the purified *A. laidlawii* B (Na^{+}+Mg^{2+})-ATPase reconstituted with DMPC and various concentrations of cholesterol or epicholesterol. Purified ATPase was reconstituted with either DMPC (2–3 mg/ml) alone or in combination with varying concentrations of cholesterol. ATPase activity was estimated at specified temperatures and the results represented as Arrhenius plots. Arrows indicate the DMPC phase-transition temperature determined by DSC. (**A**) DMPC alone, (**B**) DMPC + cholesterol (13 mol%), (**C**) DMPC + cholesterol (24 mol%), (**D**) DMPC + cholesterol (33 mol%), (**E**) DMPC + cholesterol (45 mol%), (**F**) DMPC + epicholesterol (50 mol%; from George and McElhaney, 1992, with permission).

The incorporation of epicholesterol into the DMPC proteoliposomes results in similar, but less pronounced effects on ATPase activity, and its effect on the phase behavior of the DMPC–ATPase vesicles is also similarly attenuated in comparison with cholesterol. Moreover, cholesterol added to the purified enzyme in the absence of phospholipid does not show any significant effect on either the activity or the temperature-dependence of the detergent-solubilized ATPase. These findings are consistent with the suggestion that cholesterol exerts its effect on the ATPase activity by altering the physical state of the phospholipid, since the ordering effect of cholesterol (or epicholesterol) on liquid–crystalline lipid results in a reduction of ATPase activity while the disordering of gel-state lipid results in an increase in activity (George et al., 1987b; George and McElhaney, 1992).

D. Lipid Polar Headgroup Variations in Reconstituted Membranes

In order to investigate the effect of the nature of the phospholipid polar headgroup on ATPase activity, this enzyme was reconstituted into vesicles composed of one of four different phospholipids and its activity assayed at 37°C, a temperature at which all these phospholipids exist in the liquid–crystalline state (George et al., 1987b, 1989). The two zwitterionic phospholipids tested, PC and PE, both support high levels of activity, while the anionic PS is much less effective; PG alone, an anionic lipid that is the major phosphatide of the *A. laidlawii* B membrane, does not support any ATPase activity. Moreover, when reconstituted with mixtures of PC plus PS or PG, enzyme activity is progressively diminished when the anionic phospholipid constitutes more than 20 mol% of the binary mixture. Interestingly, however, when increasing quantities of PG are combined with equimolar amounts of monoglucosyl-diacylglycerol (MGDG) and diglucosyl-diacylglycerol (DGDG), the major glycolipids of the *A. laidlawii* B membrane, the added PG actually *stimulates* ATPase activity up to 15 to 20 mol% and does not begin to inhibit ATP hydrolysis until levels of 35 to 40 mol% are reached. It is noteworthy, however, that even in the complete absence of PG or other anionic phospholipids, these neutral glycolipids support a relatively high enzyme activity. This latter finding is surprising in view of the study of Bevers et al. (1977) discussed earlier, which seemed to show an absolute requirement for small amounts (about 10 mol%) of PG for optimal Mg^{2+}-ATPase activity in *A. laidlawii* B membranes.

Hwang et al. (1992) have also studied the lipid requirements of the isolated and purified (Na^++Mg^{2+})-ATPase from the *A. laidlawii* membrane. These workers report a significant enhancement of enzyme activity upon reconstitution of the enzyme with liposomes composed of PC or PE, while PG, cardiolipin (diphosphatidylglycerol), and a mixture of soybean phospholipids were reported to actually inhibit basal enzyme activity. The incorporation of the (Na^++Mg^{2+})-ATPase into liposomes composed of mixtures of MGDG and DGDG also results in a substantial increase in enzyme activity. These results are very similar to those obtained previously by George et al. (1987b, 1989).

VI. CALORIMETRIC STUDIES OF THE INTERACTIONS OF THE PURIFIED (Na^++Mg^{2+})-ATPASE WITH LIPID BILAYERS

The nature of the interactions between cytoplasmic or various types of membrane-bound proteins and lipid bilayers can be investigated by monitoring the effect of these proteins on the thermotropic phase behavior of anionic or zwitterionic lipid vesicles by DSC (see Papahadjopoulos et al., 1975; McElhaney, 1986). Proteins can generally be put into one of three classes with respect to their effects on the temperature, enthalpy, and cooperativity of the lipid gel to liquid–crystalline phase transition and the change in these thermodynamic parameters in response to changes in the protein/lipid ratio. One class of proteins (designated Type I proteins), consisting of cytoplasmic or extrinsic membrane proteins, interacts only with anionic lipids exclusively by electrostatic forces. Interactions of this type usually result in little change or a small increase in the phase-transition temperature, a progressive increase in transition enthalpy, and a slight increase or little change in the cooperativity of the lipid chain-melting phase transition. For another class of water-soluble, but integral membrane proteins (designated Type II), interactions with lipid bilayers are thought to again involve an initial electrostatic interaction with the anionic lipid polar headgroups followed by the development of hydrophobic interactions with a portion of the nonpolar interior of the lipid bilayer. The effects of such proteins are to decrease the temperature slightly, decrease the enthalpy modestly, and decrease the cooperativity of the gel to liquid–crystalline phase transition of the host lipid bilayer. Finally, for the water-insoluble integral transmembrane proteins, which interact with lipid bilayers primarily by hydrophobic interactions with the nonpolar hydrocarbon core of the lipid bilayer (designated Type III), their interactions with either zwitterionic or anionic lipids result in no change in temperature, a marked and progressive linear decrease in enthalpy, and a decrease in the cooperativity of the lipid gel to liquid–crystalline phase transition with increasing protein incorporation. The progressive decrease in transition enthalpy observed with the incorporation of Type III proteins is thought to be the result of the removal of a defined number of protein-associated lipid molecules from the cooperative chain-melting process of the bulk lipid phase. Generally the number of lipid molecules perturbed by the protein (usually 20–100, depending on the size of the protein) correlates reasonably well with the number of boundary lipid molecules required to solvate the transbilayer region of the protein.

The purified (Na^{+}+Mg^{2+})-ATPase from *A. laidlawii* B plasma membranes was reconstituted with DMPC and the lipid thermotropic phase behavior of the proteoliposomes formed was investigated by DSC by George et al. (1990). The effect of this ATPase on the host lipid phase transition is markedly dependent on the amount of protein incorporated. At low protein/lipid ratios, the presence of increasing quantities of ATPase in the proteoliposomes increases the temperature and enthalpy while decreasing the cooperativity of the phospholipid gel to liquid–crystalline phase transition (see Figure 10). At higher protein/lipid ratios, the incorporation of increasing amounts of this enzyme does not further alter the temperature and cooperativity of the phospholipid chain-melting transition, but progressively and markedly decreases the transition enthalpy. Plots of lipid phase transition enthalpy versus protein concentration suggest that at the higher lipid/protein ratios, each ATPase molecule removes approximately 1,000 DMPC molecules from participation in the cooperative gel to liquid–crystalline phase transition of the bulk lipid phase. This number is much higher than would be required to form a single boundary lipid layer around the transmembrane region of even this very large enzyme complex. Despite being an integral transmembrane protein, the incorporation of the (Na^{+}+Mg^{2+})-ATPase into DMPC bilayers produces a complex mixture of Type I and Type III behavior and perturbs an unexpectedly large number of lipid molecules. These results indicate that this protein interacts in a complex, concentration-dependent manner with its host phospholipid and that such interactions probably involve both hydrophobic interactions with the lipid bilayer core and electrostatic interactions with the lipid polar headgroups at the bilayer surface.

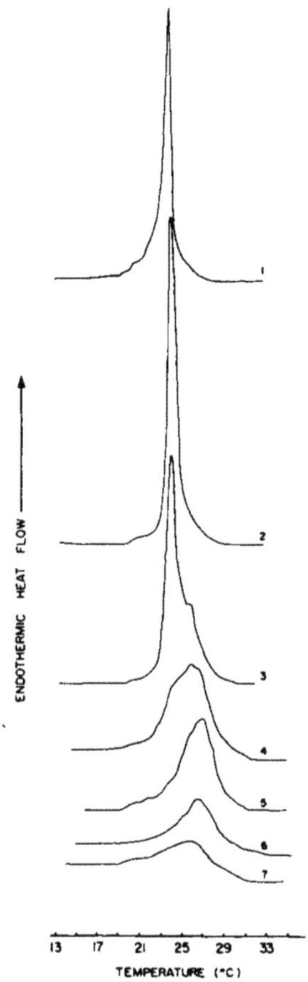

Figure 10. DSC thermograms of DMPC vesicles containing varying amounts of the *A. laidlawii* B (Na^++Mg^{2+})-ATPase. The thermograms shown have been normalized to a DMPC sample size of 2.5 mg/mL and a DSC heating rate of 35°C per hour. The protein concentrations (μg protein/mg DMPC) were as follows: curve 1, 0; curve 2, 19.8; curve 3, 49; curve 4, 79; curve 5, 148; curve 6, 403; curve 7, 608 (from George et al., 1989, with permission).

Hwang et al. (1992) have also studied the effect of the incorporation of the purified (Na^++Mg^{2+})-ATPase on the thermotropic phase behavior of phospholipid vesicles by DSC. These authors report that the incorporation of low concentrations of this enzyme into DMPC vesicles increases the enthalpy and decreases the cooperativity of the gel to liquid–crystalline phase transition while having little effect on the phase-transition temperature. These results are similar to those reported by George et al. (1990) for this same phospholipid, except that a small increase in the phase transition temperature was observed in the earlier study. In contrast, Hwang et al. (1992) report that the incorporation of low concentrations of $(Na^{+}+Mg^{2+})$-ATPase decreases the enthalpy as well as the cooperativity of the gel to liquid–crystalline phase transition of DMPG vesicles, again without changing the phase-transition temperature. The molecular basis for the opposite effects of the presence of this enzyme on the chain-melting phase transition enthalpy of DMPC and DMPG liposomes is unclear and its elucidation will require further study.

VII. CALORIMETRIC AND OTHER STUDIES OF THE THERMAL STABILITY OF THE $(Na^{+}+Mg^{2+})$-ATPASE IN NATIVE MEMBRANES

Chen et al. (1993) have recently employed high-sensitivity DSC to study the thermotropic properties of the lipids and proteins of the *A. laidlawii* membrane. These authors report the presence of five major endothermic transitions, whose midpoint temperatures range from about 37°C to 67°C in the initial heating scan. The lowest temperature endotherm (peak A) was shown to be fully reversible and to correspond to the gel to liquid–crystalline transition of the membrane lipids, while the four higher temperature endotherm (peaks B–E) were irreversible and resulted from the thermal denaturation of membrane proteins (see Figure 11). Further study of these membrane preparations by thermal gel analysis revealed that the electrophoretic band corresponding to the $(Na^{+}+Mg^{2+})$-ATPase is altered in position over the temperature range 48–65°C, presumably resulting from the change in conformation due to thermal denaturation, with about one-half of the enzymatic activity being lost at 53 ± 1°C. These authors thus concluded that the C

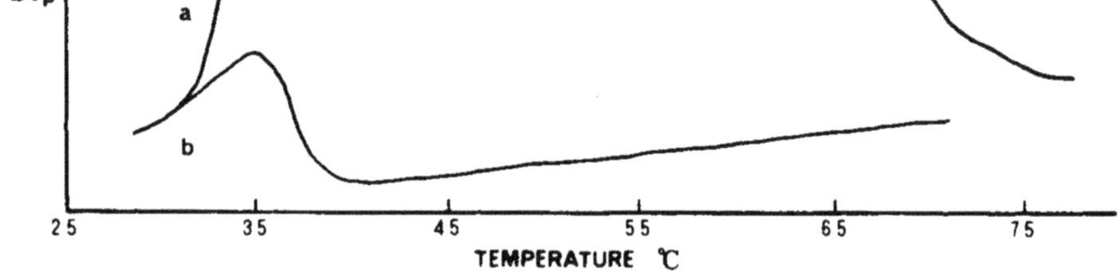

Figure 11. DSC traces of *A. laidlawii* membranes in 50 mM Tris-HCl/10 mM β-mercaptoethanol, pH 7.4 (**a**) and second scan of the same membrane preparation after denaturation of protein (**b**) (from Chen et al., 1993, with permission).

peak observed by DSC, which was centered at about 52°C, is due primarily to the thermal denaturation of the (Na^++Mg^{2+})-ATPase of the membrane of this organism.

Although the study of Chen et al. (1993) summarized above is a valuable one, I believe that it is unlikely that peak C observed in the initial DSC heating scan is due *primarily* to the thermal denaturation of the *A. laidlawii* (Na^++Mg^{2+})-ATPase, or indeed, to any single membrane protein, for several reasons. First, unlike the human erythrocyte membrane, which contains a relatively small number of proteins of which four or five are quantitatively predominant, the *A. laidlawii* membrane contains at least 200 polypeptides, none of which makes up more than a small fraction of the total (Nystrom et al., 1992). Thus, the description of the broad overlapping DSC protein endotherms stretching over the temperature range 40–80°C as consisting of four "distinct" peaks is not justified. In fact, when run under conditions of higher sensitivity, many more DSC peaks can be observed, as would be expected. Second, the asymmetric shape of peak C itself suggests that it consists of more than one component. Third, the (Na^++Mg^{2+})-ATPase accounts for no more than about 5 wt% of the total membrane protein (Lewis and McElhaney, 1983; Chen et al., 1984), yet "peak C" would appear to account for a substantial fraction of the total enthalpy of protein denaturation. Thus, although the thermal denaturation of the (Na^++Mg^{2+})-ATPase doubtlessly contributes to the overlapping endotherms making up peak C, it is unlikely to be the major component of this broad, multicomponent endothermy.

VIII. CONCLUSION

The general properties of the *A. laidlawii* (Na^++Mg^{2+})-ATPase in its native membrane environment, and when purified and reconstituted into lipid vesicles, are relatively well described. Similarly, the modulation of the activity of this enzyme by the phase-state, fluidity, surface charge density, and thickness of its host lipid bilayer are also relatively well characterized in both native and reconstituted membrane systems. More, however, needs to be learned about the structure and function of this (Na^++Mg^{2+})-ATPase and about the details of its interaction with its host lipid bilayer. Also, the detailed reaction mechanism of this enzyme is almost unknown.

ACKNOWLEDGMENTS

The work described in this chapter which emanated from my laboratory was supported by operating and major equipment grants from the Medical Research Council of Canada and by major equipment and personnel support grants from the Alberta Heritage Foundation for Medical Research.

REFERENCES

Bevers, E. M., Snoek, G. T., Op den Kamp, J. A. F., & van Deenen, L. L. M. (1977). Phospholipid requirement of the membrane-bound Mg^{2+}-dependent adenosine triphosphatase in *Acholeplasma laidlawii*. Biochim. Biophys. Acta 436, 346–356.

Chen, J-W., Sun, Q., & Hwang, F. (1984). Properties of the membrane-bound Mg^{2+}-ATPase isolated from *Acholeplasma laidlawii*. Biochim. Biophys. Acta 777, 151–154.

Chen, J-W., Hu, L-Y., & Hwang, F. (1993). Phase transitions of *Acholeplasma laidlawii* membranes. The involvement of Mg^{2+}-ATPase in the C transition. FEBS Lett. 322, 253–256.

de Kruijff, B., van Dijck, P. W. M., Goldback, R. W., Demel, R. A., & van Deenen, L. L. M. (1973). Influence of fatty acid and sterol composition on the lipid phase transition and activity of membrane-bound enzymes in *Acholeplasma laidlawii*. Biochim. Biophys. Acta 330, 269–282.

George, R., and McElhaney, R. N. (1985). Affinity labeling of the (Na^{+}+Mg^{2+})-affinity labeling of the (Na^{+}+Mg^{2+})-ATPase from *Acholeplasma laidlawii* B membranes by the 2′,3′-dialdehyde derivative of adenosine 5′-triphosphate. Biochim. Biophys. Acta 813, 161–166.

George, R., & McElhaney, R. N. (1992). The effect of cholesterol and epicholesterol on the activity and temperature dependence of the purified, phospholipid-reconstituted (Na^{+}+Mg^{2+})-ATPase from *Acholeplasma laidlawii* B membranes. Biochim. Biophys. Acta 1107, 111–118.

George, R., Lewis, R. N. A. H., Mahajan, S., & McElhaney, R. N. (1989). Studies on the purified, lipid-reconstituted (Na^{+}+Mg^{2+})-ATPase from *Acholeplasma laidlawii* B membranes: Dependence of enzyme activity on lipid headgroup and hydrocarbon chain structure. J. Biol. Chem. 264, 11598–11604.

George, R., Lewis, R. N. A. H., & McElhaney, R. N. (1985). Reconstitution and photolabeling of the purified (Na^{+}+Mg^{2+})-ATPase from the plasma membrane of *Acholeplasma laidlawii* B with phospholipids containing a photosensitive fatty acyl group. Biochim. Biophys. Acta 821, 253–258.

George, R., Lewis, R. N. A. H., & McElhaney, R. N. (1987a). Reconstitution of the purified (Na^{+}+Mg^{2+})-ATPase from *Acholeplasma laidlawii* B membranes into lipid vesicles and a characterization of the resulting proteoliposomes. Biochim. Biophys. Acta 903, 282–291.

George, R., Lewis, R. N. A. H., & McElhaney, R. N. (1987b). Lipid modulation of the activity and temperature dependence of the purified (Na^{+}+Mg^{2+})-ATPase from *Acholeplasma laidlawii* B membranes. Isr. J. Med. Sci. 23, 374–379.

George, R., Lewis, R. N. A. H., & McElhaney, R. N. (1990). Studies on the purified (Na^{+}+Mg^{2+})-ATPase from *Acholeplasma laidlawii* B membranes. A differential scanning calorimetric study of the ATPase-dimyristoylphosphatidylcholine interactions. Biochem. Cell Biol. 68, 161–168.

Hsung, J.-C., Huang, L., Hoy, D. J., & Haug, A. (1974). Lipid and temperature dependence of membrane-bound ATPase activity of *Acholeplasma laidlawii*. Can. J. Biochem. 52, 974–980.

Hwang, F., Chen, J. W., Hu, L.-Y., & Zhang, L. (1992). Molecular properties of the membrane-bound Mg^{2+}-ATPase of *Acholeplasma laidlawii*. IOM Lett. 1, 74.

Jinks, D. C., & McElhaney, R. N. (1987). Method for exchange of the lipid environment of the membrane-bound (Na^{+}+Mg^{2+})-ATPase of *Acholeplasma laidlawii* B. Anal. Biochem. 164, 331–335.

Jinks, D. C., Silvius, J. R., & McElhaney, R. N. (1978). Physiological role and membrane lipid modulation of the membrane-bound (Na^{+}+Mg^{2+})-adenosine triphosphatase activity in *Acholeplasma laidlawii*. J. Bacteriol. 136, 1027–1036.

Johannson, A., Keightley, C. A., Smith, G. A., Richards, C. D., Hesketh, T. R., & Metcalfe, J. C. (1981a). The effect of bilayer thickness and *n*-alkanes on the activity of the (Ca^{2+}+Mg^{2+})-dependent ATPase of sarcoplasmic reticulum. J. Biol. Chem. 256, 1643–1650.

Johannson, A., Smith, G. A., & Metcalfe, J. C. (1981b). The effect of bilayer thickness on the activity of (Na^{+}+K^{+})-ATPase. Biochim. Biophys. Acta 641, 416–421.

Lelong, I., Shirvan, M. H., & Rottem, S. (1989). A cation/proton antiport activity in *Acholeplasma laidlawii*. FEMS Microbiol. Lett. 59, 71–76.

Lewis, R. N. A. H., George, R., & McElhaney, R. N. (1986). Structure–function investigations of the membrane (Na^++Mg^{2+})-ATPase from *Acholeplasma laidlawii* B; studies of reactive amino acid residues using group-specific reagents. Arch. Biochem. Biophys. 247, 201–210.

Lewis, R. N. A. H., & McElhaney, R. N. (1983). Purification and characterization of the membrane (Na^++Mg^{2+})-ATPase from *Acholeplasma laidlawii*. Biochim. Biophys. Acta 735, 113–122.

Macdonald, A. G., & McLeod, K. T. (1980). The effect of pressure and of alcohols on the phase transition in the cell membrane of *Acholeplasma laidlawii* B. J. Physiol. Lond. 310, 19P.

Macdonald, A. G., & Cossins, A. R. (1983). Effects of pressure and pentanol on the phase transition in the membrane of *Acholeplasma laidlawii* B. Biochim. Biophys. Acta 730, 239–244.

MacNaughton, W., & Macdonald, A. G. (1982). Effects of pressure and pressure antagonists on the growth and membrane-bound ATPase of *Acholeplasma laidlawii* B. Comp. Biochem. Physiol. 72A, 405–414.

Mahajan, S., Lewis, R. N. A. H., George, R., Sykes, B. D., & McElhaney, R. N. (1988). Characterization of sodium transport in *Acholeplasma laidlawii* B cells and in lipid vesicles containing purified *A. laidlawii* (Na^++Mg^{2+})-ATPase using nuclear magnetic resonance spectroscopy and ^{22}Na tracer techniques. J. Bacteriol. 170, 5739–5746.

Mahajan, S., Lewis, R. N. A. H., & McElhaney, R. N. (1990). ATP-driven sodium transport in proteoliposomes reconstituted with the *Acholeplasma laidlawii* B (Na^++Mg^{2+})-ATPase. Zentralbl. Bakteriol. Suppl. 20, 670–673.

McElhaney, R. N. (1974a). The effect of alterations in the physical state of the membrane lipids on the ability of *Acholeplasma laidlawii* B to grow at various temperatures. J. Mol. Biol. 84, 145–157.

McElhaney, R. N. (1974b). The effect of membrane lipid phase transitions on membrane structure and on the growth of *Acholeplasma laidlawii* B. J. Supramol. Struct. 2, 617–628.

McElhaney, R. N. (1986). Differential scanning calorimetric studies of lipid-protein interactions in model membrane systems. Biochim. Biophys. Acta 864, 361–421.

McElhaney, R. N. (1992a). Membrane structure. In: Mycoplasmas: Molecular Biology and Pathogenesis (Maniloff, J., McElhaney, R. N., Finch, L. R., & Baseman, J. B., eds.), pp. 113–155. American Society for Microbiology, Washington, D.C.

McElhaney, R. N. (1992b). Membrane structure. In: Mycoplasmas: Molecular Biology and Pathogenesis (Maniloff, J., McElhaney, R. N., Finch, L. R., & Baseman, J. B., eds.), pp. 259–287. American Society for Microbiology, Washington, D.C.

Montecucco, C., Smith, G. A., Dabbeni-sala, F., Johannsson, A., Galante, Y. M., & Bisson, R. (1982). Bilayer thickness and enzymatic activity in the mitochondrial cytochrome c oxidase and ATPase complex. FEBS Lett. 144, 145–148.

Ne'eman, Z., Kahane, I., & Razin, S. (1971). Characterization of the *Mycoplasma* membrane proteins. II. Solubilization and enzyme activities of *Acholeplasma laidlawii* membrane proteins. Biochim. Biophys. Acta 249, 169–176.

Ne'eman, Z., Kahane, I., Kovartovsky, J., & Razin, S. (1972). Characterization of the *Mycoplasma* membrane proteins. III. Gel filtration and immunological characterization of *Acholeplasma laidlawii* membrane proteins. Biochim. Biophys. Acta 266, 255–268.

Nystrom, S., Wallbrandt, P., & Wieslander, A. (1992). Membrane protein acylation. Preference for exogenous myristic or endogenous saturated fatty acids in *Acholeplasma laidlawii*. Eur. J. Biochem. 204, 231–240.

Papahadjopoulos, D., Moscarello, M., Eylar, E. H., & Isac, T. (1975). Effects of proteins on thermotropic phase transitions of phospholipid membranes. Biochim. Biophys. Acta 401, 317–335.

Pollock, J. D., Razin, S., & Cleverdon, R. C. (1965). Location of enzymes in *Mycoplasma*. J. Bacteriol. 90, 617–622.

Razin, S. (1982). The mycoplasma membrane. In: Organization of Prokaryotic Cell Membranes (Ghosh, B. K., ed.), pp. 165–250. CRC Press, Boca Raton, FL.

Rottem, S., & Razin, S. (1966). Adenosine triphosphatase activity of mycoplasma membranes. J. Bacteriol. 105, 323–330.

Silvius, J. R., & McElhaney, R. N. (1978). Lipid compositional manipulation in *Acholeplasma laidlawii* B. Effect of exogenous fatty acids on fatty acid composition and cell growth when endogenous fatty acid production is inhibited. Can. J. Biochem. 56, 462–469.

Silvius, J. R., & McElhaney, R. N. (1980). Membrane lipid physical state and modulation of the $(Na^{+}+Mg^{2+})$-ATPase activity in *Acholeplasma laidlawii* B. Proc. Natl. Acad. Sci. USA 77, 1255–1259.

Silvius, J. R., & McElhaney, R. N. (1981). Non-linear Arrhenius plots and the analysis of reaction and motional rates in biological membranes. J. Theor. Biol. 88, 135–152.

Silvius, J. R., & McElhaney, R. N. (1982). Molecular properties of membrane lipids and activity of a membrane adenosine triphosphatase from *Acholeplasma laidlawii* B. Rev. Infect. Dis. 4, S50–S58.

Silvius, J. R., Read, B. D., & McElhaney, R. N. (1978). Membrane enzymes: Artifacts in Arrhenius plots due to temperature dependence of substrate-binding affinity. Science 199, 902–904.

Silvius, J. R., Jinks, D. C., & McElhaney, R. N. (1979). Effect of membrane lipid fluidity and phase state on the kinetic properties of the membrane-bound adenosine triphosphatase of *Acholeplasma laidlawii* B. In: Microbiology (Schlessinger, D., ed.), pp. 10–13. American Society for Microbiology, Washington, D. C.

Silvius, J. R., Mak, N., & McElhaney, R. N. (1980). Why do prokaryotes regulate membrane lipid fluidity? In: Control of Membrane Fluidity (Kates, M., and Kuksis, A., eds.), pp. 213–221. Humana Press, Clifton, NJ.

VACUOLAR H^+-ATPASE

Nathan Nelson

I. INTRODUCTION

The vacuolar system of eukaryotic cells consists of all of the internal membrane network of the cell excluding semiautonomous organelles such as mitochondria and chloroplasts (Mellman et al., 1986). The membranes of the vacuolar system are

Biomembranes
Volume 5, pages 317–341.

ISBN: 1-55938-662-2.

rich in transporters, carriers, exchangers, and so forth, most of which require energy for their function. Most of this energy is provided by an ATP-dependent proton pump called vacuolar H^+-ATPase or V-ATPase (Nelson, 1989). The enzyme pumps protons from the cytoplasm to the internal space of the organelles. Consequently, the pH inside the organelles is more acidic than that of the cytoplasm, a proton gradient across the membranes is established and a membrane potential (positive inside) is generated. Numerous organelles of the vacuolar system are energized by V-ATPases and each organelle has special requirements for its internal pH and membrane potential. Thus, endosomes operate with a smaller ΔpH than lysosomes and synaptic vesicles that accumulate glutamate maintain a larger membrane potential than chromaffin granules which take up catecholamines. These special requirements may be fulfilled by secondary transport systems such as chloride carriers and regulation of the V-ATPase activity. Figure 1 depicts some of the organelles and membranes in which V-ATPases are operating in a model eukaryotic cell. The puzzle is even more complicated because V-ATPase may function in dozens of other organelles. The specific assembly and differential activity in each

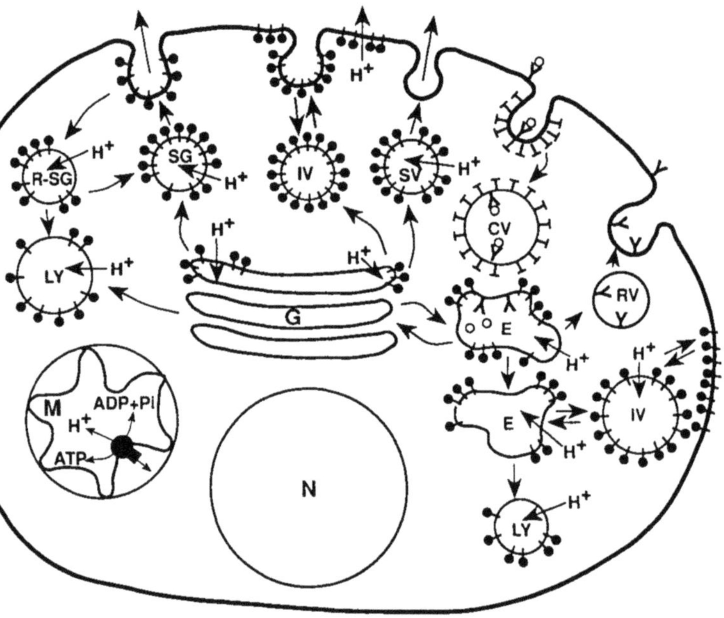

Figure 1. Schematic representation of the eukaryotic cell emphasizing organelles and membranes containing V-ATPases. •, F-ATPase; †, V-ATPase; T, clathrin; Y, receptor; o, ligand; N, nucleus; M, mitochondria; G, Golgi complex; LY, lysosome; E, endosome; IV, internal storage vesicle; RV, receptor recycling vesicle; CV, clathrin-coated vesicle; SV, synaptic vesicle; SG, secretory granule; R-SG, recycling secretory granule. Directions of proton pumping are shown by H^+ and arrow.

of these organelles is subject to extensive research in several laboratories. In this chapter, I shall review some of the recent advancements in understanding the structure, function, molecular biology, biogenesis, and evolution of V-ATPases.

II. BIOENERGETICS OF PROTON PUMPS

The electrochemical energy formed by proton pumps was termed proton-motive force (pmf), and according to the Mitchell hypothesis is the universal high energy intermediate across biological membranes (Mitchell, 1968). The pmf can be utilized for driving numerous secondary transport systems that maintain the solute composition inside the vacuolar system different from that of the cytoplasm. The following equations express the thermodynamics of pmf:

$$\text{ADP} + \text{P}_i + n\text{H}^+_{out} \leftrightarrow \text{ATP} + n\text{H}^+_{in} \quad (1)$$

where n stands for the number of protons involved in a single phosphorylation or hydrolysis step. The relationship between the electrochemical gradient of protons and the formation of ATP is described by the relation:

$$\Delta G = n\Delta\mu\text{H}^+ + \Delta G_p \quad (2)$$

where ΔG is the Gibbs free energy, ΔG_p the free energy of phosphorylation, and $\Delta\mu H^+$ the H^+ electrochemical proton gradient (pmf). When $\Delta G < 0$, ATP synthesis takes place; when $\Delta G > 0$, ATP hydrolysis takes place; and when $\Delta G = 0$, the reaction is at equilibrium. Proton-motive force is a combination of the proton gradient and the electric potential across the membrane. The relationship between the two is expressed as follows:

$$\Delta\mu\text{H}^+ = F\Delta\psi - 2.3\ RT\Delta\text{pH} \quad (3)$$

where $\Delta\psi$ represents the electrical potential across the membrane, ΔpH the difference in H^+ ion concentration on the two sides of the membrane, R the gas constant, T the absolute temperature, and F the Faraday constant. Dividing equation 3 by F expresses the pmf in millivolts (mV) and a simple equation of pmf = $\Delta\psi$ + 58.8 ΔpH is thereby obtained (2.3 RT/F is equal to 58.8 mV at room temperature).

Eubacteria, chloroplasts, and mitochondria contain an H^+-ATPase (F-ATPase) that utilizes the pmf generated by respiratory and photosynthetic electron transport chains for phosphorylating ADP into ATP. This enzyme operates at thermodynamic equilibrium (ΔG=0) most of the time, and depending on the levels of ATP and pmf, may work in a mode of ATP-synthesis (ΔG<0) or ATP-dependent proton pumping (ΔG>0). On the other hand, V-ATPase functions in eukaryotic cells exclusively as a proton pump presumably at ΔG>0 (see equation (2)). In archaebacteria, V-ATPase function in ATP-synthesis. The two families of proton pumps, F- and V-ATPases, have structural and mechanistic similarities and evolved from a common ancestral proton pump. Sequencing the genes encoding subunits of F- and V-ATPases from various sources sheds light on the structure, function, mechanism of action, and evolution of these two fundamental families of proton pumps.

III. STRUCTURE OF V-ATPASES

The molecular mass of V-ATPase is over 500 kD and is composed of two distinct sectors, each of which contains several subunits (Nelson, 1989). The catalytic sector, which protrudes from the membrane into the cytoplasm, is composed of five different subunits denoted as subunits A to E (Figure 2). They are present in an unequal stoichiometry of 3A, 3B, 1C, 1D, and 1E (Arai et al., 1988). The membrane sector contains a few subunits, some of which have not yet been identified (Nelson, 1992c). In mammalian V-ATPase, at least three subunits were isolated and studied by immunological methods and chemical modification. These subunits were denoted as Ac 115, Ac 39, and proteolipid, and the genes encoding these polypeptides were cloned and sequenced. More subunits are likely to comprise this sector. The structure of V-ATPases was studied by electron microscopy (Bowman et al., 1989; Taiz and Taiz, 1991). The general structure of V-ATPases resembles that of

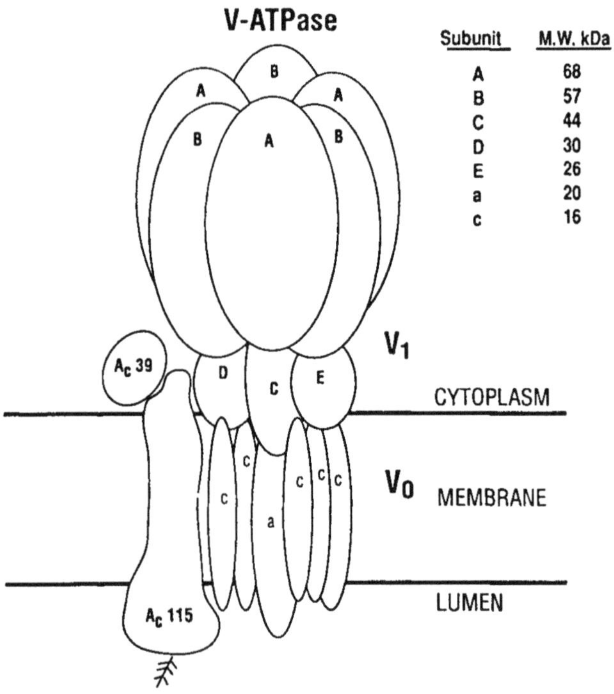

Figure 2. Schematic structure of V-ATPases. Subunits A to E are the catalytic sector. Subunits a and c are the functional membrane sector and Ac 115 and Ac 39 are accessory subunits associated with the membrane sector.

mitochondrial and chloroplast F-ATPases with a notable exception of appendices protruding from the periphery of the catalytic sector. The catalytic sector can be readily dissociated from the membranes by incubating the preparation on ice in the presence of MgATP and 0.2 M NaCl (Moriyama and Nelson, 1989a, b, c). This suggests that in one of the conformations of the catalytic cycle, the contact between the catalytic and the membrane sectors is hydrophobic in nature (hydrophobic interactions are weakened at low temperature). The exact organization of the catalytic sector is not known, but due to the homology and sequence conservation between F- and V-ATPases, it is logical to assume that their general structure is similar. Moreover, it was demonstrated that the A- and B-subunits of V-ATPase from the archaebacterium, *Sulfolobus acidocaldarius*, are alternating similarly to the α- and β-subunits of F-ATPase from *E. coli* (Schafer and Meyering-Vos, 1992). Since we are not expecting the three-dimensional structure of V-ATPases to be resolved in the near future, we have to resort to the general structure depicted in Figure 2. Ac 115 is the most peculiar subunit of V-ATPases as it has no homology nor analogy in F-ATPases. This subunit protrudes to the lumen of the vacuolar system and is the only glycosylated subunit of the enzyme (Perin et al., 1991). Its function is not known, but it is logical to assume that it may play a role in the recognition of V-ATPases at the cell exterior or in the lumen of the vacuolar system. This accessory subunit transverses the membrane several times and, therefore, is a *bona fide* membrane protein. The other accessory subunit, associated with the membrane sector, is the Ac 39. However, it does not contain an apparent transmembrane segment (Wang et al., 1989). Its function is not known and it is located on the cytoplasmic side of the membrane. Sequencing of the cDNA encoding this subunit revealed no sequence homology with any other protein and gave no clues to its function. All other subunits depicted in Figure 2 presumably are integral parts of the enzyme which could not function or assemble without each of them.

Subunit A (69 kD) contains the catalytic ATP-binding site of the enzyme. This was demonstrated by chemical modification with NEM, ATP binding by UV irradiation, and sequence homology with the β subunit of F-ATPases (Moriyama and Nelson, 1987b; Arai et al., 1987; Bowman et al., 1988a; Zimniak et al., 1988). The amino acid sequence of this subunit contains a "glycine-rich motif" common among ATP-binding proteins. This motif contains two cysteine residues, the modification of which causes inactivation of the enzyme (Feng and Forgae, 1992). Moreover, modification of a single cysteine on subunit A prevents the dissociation of the catalytic sector from the membrane by cold treatment (Moriyama and Nelson, 1989a). These and other observations leave little doubt that subunit A provides the catalytic ATP-binding site.

Subunit B (57 kD) contains an ATP-binding site that most probably is not the catalytic one. This subunit does not react with NEM and binds ATP analogs only under restricted conditions (Manolson et al., 1988). It does not bind ATP by UV irradiation (Moriyama and Nelson, 1987b). Sequence analysis revealed a potential ATP-binding domain and reasonable homology to the α-subunit of F-ATPases

(Bowman et al., 1988). These and other observations suggest that subunit B may function in regulating the activity of V-ATPases, but a direct role in the catalytic activity is not ruled out.

Subunit C (41 kD) is necessary for the assembly of V-ATPase in yeast cells (Beltrán et al., 1992). Its function in the catalytic activity of the enzyme is not known and it has no sequence homology with any subunit of F-ATPases. The amino acid sequence of this subunit is poorly conserved and exhibits only 37% identity between the subunits of yeast and bovine V-ATPases.

The function of subunit D (34 kD) is not known. It was identified on SDS gels in preparations from a variety of sources that ranged from yeast to mammals. Since the gene encoding this subunit has never been cloned, it is not certain that this polypeptide is an integral part of the enzyme. Immunological studies indicated that this polypeptide copurifies with the enzyme and, therefore, is likely to be a subunit of V-ATPases (Nelson, 1989).

Subunit E (29 kD) was one of the first to be identified in kidney V-ATPase by immunological studies, and the cDNA encoding this subunit was cloned and sequenced (Hirsch et al., 1988). The function of this subunit is not known but it was implicated in controlling the biogenesis of the kidney V-ATPase (Gluck et al., 1992). The gene encoding this subunit in yeast cells was cloned, sequenced, and interrupted (Foury, 1990). It was shown that subunit E is necessary for the functional assembly of the enzyme.

The proteolipid is a highly hydrophobic subunit of 16 kD that is the main functional subunit of the membrane sector of eukaryotic V-ATPases (Nelson, 1989). It binds DCCD which inactivates the proton pumping and ATPase activities of the enzyme (Arai et al., 1987a; Sze et al., 1992). The cDNA and gene encoding this subunit in mammals and yeast were cloned and sequenced (Mandel et al., 1988; Nelson and Nelson, 1989). The sequences revealed that the proteolipid evolved by gene duplication and fusion of an 8 kD ancestral gene homologous to that present in F-ATPases. Disruption of the gene encoding the proteolipid in yeast cells showed that it is necessary for the assembly of all other subunits of the enzyme (Nelson and Nelson, 1990). The proteolipid is likely to be involved in the process of proton translocation across the membrane.

IV. MOLECULAR BIOLOGY OF V-ATPASES

It was only since 1988 that cDNAs and genes encoding subunits of V-ATPases were cloned and sequenced (Bowman et al., 1988a, 1988b; Mandel et al., 1988; Manolson et al., 1988; Zimniak et al., 1988). The sequences revealed valuable information on the structure, function, and evolution of the various subunits as well as the evolution of F- and V-ATPases (Nelson, 1989; Gogarten et al., 1989; Nelson and Taiz, 1989). It became apparent that subunits A and B of V-ATPases, and subunits α and β of F-ATPases evolved from a common ancestral gene. The proteolipids of F- and V-ATPases also evolved from a common ancestral gene. All other subunits of F- and V-ATPases have no apparent homology and may have evolved independently of each other. In line with this evolutionary pattern, subunits A, B, and proteolipid are the most conserved subunits in V-ATPases and are likely to be the most fundamental subunits of the enzyme. In all of these subunits, there are sequences or amino acids indicative of their function. No such landmarks could be identified in all other subunits.

The cDNA and gene encoding subunit A were first cloned from plants (Zimniak et al., 1988), fungi (Bowman et al., 1988a), and the archaebacterium, *Sulfolobus acidocaldarius* (Denda et al., 1988b). It immediately became apparent that the enzyme that functions in ATP-synthesis in archaebacteria is V-ATPase and that subunit A is homologous to the β subunit of F-ATPases. It was also revealed that a yeast gene involved in trifluoperazine resistance and cloned the same year encodes subunit A of V-ATPase (Shih et al., 1988). This gene encodes a larger protein that undergoes protein splicing to produce the mature subunit A (Hirata et al., 1990; Kane et al., 1990). Aligning the amino acid sequences of A and β subunits from various sources produced a wealth of information. The glycine-rich loop in which the sequence GXXXXGKT/S is conserved in most nucleotide-binding proteins is situated in subunit A. This sequence of amino acids was shown to bind ATP analogs and NBD-Cl in F-ATPases (Penefsky and Cross, 1991), and NEM in V-ATPases (Feng and Forgac, 1992). The presence of ATP reduced the chemical modification of this region, and site-directed mutagenesis in F-ATPase from *E. coli* showed high sensitivity to amino acid changes. A single NEM bound to C^{254} in the bovine subunit A is sufficient for inhibiting the activity of the enzyme. This and other cysteine residues render V-ATPases from eukaryotic cells sensitive to oxygen. It is interesting that archaebacteria exchanged the corresponding cysteine residue to serine and their V-ATPase is not sensitive to NEM (Denda et al., 1988b). The conserved glycine-rich loop in nucleotide binding proteins was implicated as a primordial common structure for nucleotide binding. All of these properties indicate that the A subunit of V-ATPases and the β subunit of F-ATPase are the catalytic subunits of these proton pumps.

Subunit B of V-ATPases is most homologous to the α-subunit of F-ATPases. Significant homology in the amino acid sequences among the A and B subunits of V-ATPases and the β and α subunits of F-ATPases, left little doubt that all of these subunits evolved from a common ancestral gene (Gogarten et al., 1989). Subunit B is the most conserved subunit in V-ATPases, and it was argued that the current subunit is the closest to the common ancestor of the others (Nelson and Nelson, 1989). The function of subunit B is not clear, but is likely to have a similar function in regulating the enzyme as the α-subunit of F-ATPases. Unlike for the α-subunit, there is no wealth of experiments showing nucleotide binding to the B-subunit. There is only one report showing the photoaffinity binding of the ATP analog, 3-O-(4-benzoyl)benzoyladenosine 5′-triphosphate to the B-subunit of plant V-ATPase (Manolson et al., 1985). Moreover, in contrast to the α-subunit of F-ATPases that contains the consensus ATP-binding motif GXXXXGKT, in the B-subunit of V-ATPases, it is not present. Therefore, it would not be a surprise to find that V-ATPases contain only three nucleotide-binding sites. While yeast contains a single gene encoding the B-subunit, bovine contains two genes encoding very similar proteins (Puopolo et al., 1992). It is not known if each of these genes confers different properties to the enzyme.

Subunit C was cloned from the bovine adrenal library (Nelson et al., 1990). The deduced amino acid sequence showed no homology with any other proteins in the GenBank. It was concluded that this subunit evolved independently of any F-ATPase subunit. The gene encoding subunit C in yeast exhibits only 37% identity with the bovine protein, and except for a small stretch of highly conserved amino acids, the remainder of the conserved residues are scattered all over the protein. A mutant was generated in which the gene encoding the C subunit was interrupted (Beltran et al., 1992). This mutant fails to grow in a medium buffered at pH 7.5. A chimeric gene comprised of the yeast and bovine sequences successfully replaced the yeast gene and permitted growth at pH 7.5. The amino acid sequence of bovine or yeast subunits C shows no sequence homology with the known V-ATPase subunits from archaebacteria (Denda et al., 1990). The function of subunit C is not known, but its requirement for the functional assembly of the yeast enzyme is well established. It may be analogous to the γ-subunit of F-ATPases that is also poorly conserved and may function in coupling the ATPase activity on the catalytic sector with the proton translocation across the membrane sector.

The cDNA encoding subunit E of bovine kidney microsomes was cloned, sequenced and analyzed (Hirsch et al., 1988; Gluck et al., 1992). It exhibits no significant sequence homology with any of the subunits of F-ATPases. It also has no sequence homology with the predicted gene products of the archaebacterial operon encoding their V-ATPases. Studies with monoclonal antibodies, supported by partial DNA sequencing, revealed the existence of at least two isoforms of subunit E in the kidney. While V-ATPase isolated from kidney microsomes contains one form of subunit E, the enzyme from the kidney brush-border contains at least one additional form of subunit E. It was suggested that subunit E plays a role in the biogenesis of the enzyme and in regulating its polar assembly in types A and B of intercalated cells (Gluck et al., 1992). The gene encoding subunit E in yeast cells was cloned, sequenced and interrupted (Foury, 1990). The disruptant mutant exhibited similar phenotype to all the other V-ATPase disruptant mutants. While the proteolipid assembled into the membrane, all of the subunits of the catalytic sector did not assemble. Consequently, the mutant was not able to grow in a medium buffered at pH 7.5 (Ho et al., 1993).

A cDNA that encodes the proteolipid of V-ATPase was first cloned from the bovine adrenal library (Mandel et al., 1988). The predicted amino acid sequence revealed its relation to the proteolipids of F-ATPases and suggested that the proteolipid of eukaryotes evolved by gene duplication and fusion of an ancestral proteolipid gene. Subsequently, the gene encoding the yeast proteolipid was cloned and sequenced as well as genes and cDNAs encoding proteolipids from a variety of sources (Nelson, 1992b). Alignment of the predicted amino acid sequences of proteolipids of various eukaryotes demonstrated that this protein is one of the most conserved hydrophobic proteins in nature. The third and fourth transmembrane helices were almost totally conserved in proteolipids from yeast to mammals (Nelson, 1992b). So far, there is no other example in which the amino acid sequences of transmembrane helices are identical in yeast and mammalian proteins. Studies in yeast mutants showed that the proteolipid is necessary for the assembly of all other subunits (Nelson and Nelson, 1990). However, the proteolipid can assemble into the vacuolar membrane independently of the other subunits. The fourth transmembrane helix contains the glutamyl residue that presumably binds DCCD, resulting in inactivation of the ATPase and proton transport activities of the enzyme. Numerous mutations generated by site-directed mutagenesis of the yeast proteolipid were reported (Noumi et al., 1991). Displacement of E^{137} by any amino acid except D inactivate the enzyme. Several other displacements rendered the enzyme inactive, and it was concluded that water molecules coordinated to amino acids in the transmembrane helices of the proteolipid may play a role in proton translocation across the membrane. Analysis of the second-site suppressor mutant in the yeast proteolipid suggests a very tight organization of the proteolipids in the vacuolar membrane (Supek et al., 1992).

Except for archaebacteria and plants, a polypeptide of about 100 kD is present in every preparation of V-ATPase (Nelson, 1992c). The cDNA encoding this subunit in bovine brain was cloned and sequenced (Perin et al., 1991). It encodes a protein of 99 kD with potential glycosylation sites and transmembrane helices. A yeast gene encoding a protein with considerable similarity was cloned and sequenced (Manolson et al., 1992). Disruption of this gene did not abolish the activity of V-ATPase, supporting the notion that this is an accessory polypeptide that does not directly participate in the ATPase or proton translocation activities of the enzyme (Nelson, 1989). The function of this subunit is not known, but since it is the only glycosylated protein in the enzyme, it is likely to play a role in sensing the lumen or cell exterior.

A polypeptide (Ac 39) with an apparent molecular weight of 39 kD is present in every preparation of V-ATPase from a mammalian source (Moriyama and Nelson, 1989a, 1989b, 1989c). The cDNA encoding this polypeptide was cloned from the bovine adrenal library (Wang et al., 1989). It encodes a protein of 32 kD with no apparent transmembrane segments. Since it is fractionating with the membrane sector of V-ATPases, it is likely to be assembled as a complex with another membrane protein. The function of this polypeptide is not known.

A yeast gene (*VMA11*) encoding a hydrophobic protein with 57% identity at the amino acid level to the yeast proteolipid was cloned and studied by genetic analysis (Umemoto et al., 1991). Disruption of this gene prevented the assembly of other subunits of the V-ATPase including the proteolipid. Attempts to identify the gene product of *VMA11* in yeast vacuoles were not successful. The function of the

VMA11 gene product may be to promote the assembly of the membrane sector of the enzyme.

The complexity of the yeast V-ATPase is further demonstrated by the emerging gene encoding proteins participating in the assembly or function of the enzyme. In yeast, the genes denoted as *VMA 6*, 12, 13, 21, 22, and 23 were identified to encode potential subunits or organizers of the V-ATPase (Stevens, 1992). In mammals, potential specific inhibitors and enhancers were identified (Gluck, 1992). The involvement of more than 15 gene products in ATP-dependent pumping of the smallest ion in nature leaves one wondering about its marvels.

V. EVOLUTION OF V-ATPASES—FROM ARCHAEBACTERIA TO HUMAN BRAIN

Analysis of evolutionary trees demonstrated that subunits A and B of V-ATPases and subunits α and β of F-ATPases are paralogous polypeptides that evolved from a common ancestor and conserved a similar function through evolutionary time (Gogarten et al., 1992b). The fundamental function of F- and V-ATPases is a consequence of their early appearance in the most primitive life forms on earth. We assume that these primordial organisms evolved under the stress of high temperatures and low environmental pH (Nelson, 1992a). Two independent proton pumps took shape in the plasma membrane of these primitive cells; one is an ATP-dependent pump (the ancestor of F- and V-ATPases) and the second is a redox potential-driven quinone pump (photo- or chemo-synthetic electron transport). Both pumped protons from the cytoplasm to the acidic environment for maintaining the cytoplasmic pH neutral. While the reversibility of the electron transport-driven pump was not very useful, the reversibility of the ATP-driven pump was crucial to the evolution of more advanced life forms. When the environment became less acidic, the two proton pumps started working in concert in ATP formation, where the electron transport chain provided the pmf and the ATPase utilized it for the phosphorylation of ADP to ATP. It is assumed that the primordial atmosphere was anaerobic and that enzymes were not equipped with protective mechanisms against oxidative damage (Harold, 1986). Amino acids such as cysteine and tryptophan were more abundant in the primordial proteins. The amino acid sequences of subunits A and B of V-ATPases subscribe to such proteins, and therefore, it was proposed that the catalytic sector of the primordial proton pump was closer to the current V-ATPases (Nelson and Nelson, 1989). It was also proposed that F-ATPases evolved as a consequence of increasing oxygen tension concomitantly with the evolution of oxygen evolving photosystem II. As a result of this event, cysteine and tryptophan residues are no longer present in the catalytically crucial positions of F-ATPases. The archaebacterial V-ATPase still contains many more tryptophans than F-ATPases, but the cysteine residue at the ATP-binding site in the A subunit was substituted to a serine residue. Today, the V-ATPases of eukaryotic cells are protected against oxidation damage by their presence in the cytoplasm that main-

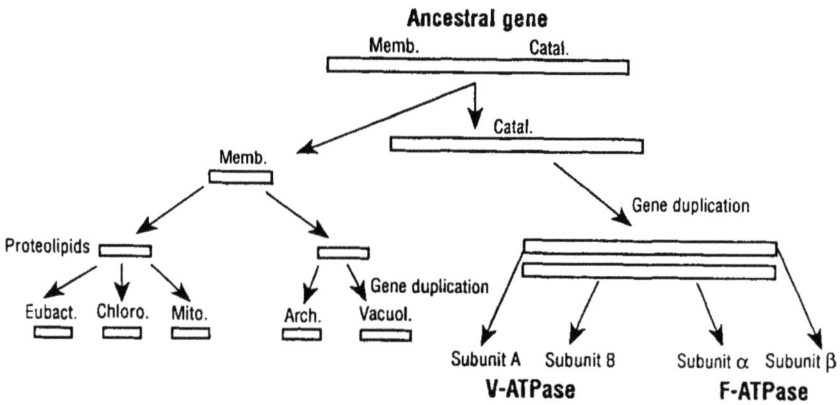

Figure 3. Schematic proposal for the early evolution of the catalytic and membrane sectors of F- and V-ATPases.

tains a relatively reduced environment. A schematic proposal for the main evolutionary events that progresses from the primordial enzyme to the current F- and V-ATPases is depicted in Figure 3. Apparently, most of the fundamental events in the evolution of F- and V-ATPases took place over three billion years ago. It was proposed that these enzymes evolved from a single ancestral subunit proton pump containing both catalytic and membrane sectors, acting as a homohexamer (Nelson, 1992a). Very early in the evolution of the system, the gene encoding this enzyme was separated into two genes encoding the catalytic and membrane sectors. Soon after, the gene encoding the catalytic sector was duplicated to give a heterohexamer composed of three A subunits and three B subunits. Concomitantly with the introduction of oxygen to the atmosphere by oxygenic photosynthesis, the β-subunit of F-ATPase evolved from the A-subunit and the α-subunit evolved from the B-subunit. V-ATPases of archaebacteria and eukaryota maintained the A- and B-subunits. In the remaining three billion years, the F- and V-ATPases were busy adding more and more subunits to the enzyme, increasing its complexity and obtaining a highly specialized enzyme to act efficiently in diverse organelles, membranes, and tissues.

The membrane sector underwent a separate evolution even more detrimental to the specific function of F- and V-ATPases in different living creatures and in various cells and organelles. It is assumed that the ancestral membrane sector had three transmembrane segments, two of which evolved into the ancestral proteolipid (Nelson, 1992a). The proteolipid maintained this structure in the two bacterial kingdoms as well as in chloroplasts and mitochondria. In all of these systems the main function of the F- or V-ATPase is to generate ATP at the expense of pmf. Very early on, a gene duplication and fusion of the ancestral proteolipid took place. This form of proteolipid containing four transmembrane helices is present in all V-ATPases of eukaryotic cells. Concomitantly, or even consequently, these enzymes function exclusively as proton pumps and can no longer generate ATP at the expense of pmf. Thermodynamically, this event may also be responsible for the difference in stoichiometry of H^+/ATP, which may be 2 in eukaryotic V-ATPases and 3 in F-ATPases (Nelson, 1989; Cross and Taiz, 1990). The size of the proteolipid is also correlated with the potential generation of proton leaks through the membrane sector in the absence of the catalytic sector. While most membrane sectors of F-ATPases and archaebacterial V-ATPases, when not gated by the catalytic sector, conduct protons down electrochemical gradients, the membrane sector of eukaryotic V-ATPases by itself is impermeable to protons (Beltrán and Nelson, 1992). All of these properties set up the proteolipids in the junction of energy transduction, and their evolution and conservation suggest a major role in the mechanism of ATP formation and proton pumping. The third transmembrane helix evolved into, or was displaced by, subunit a of F-ATPases and may be the 20 kD polypeptide of V-ATPases.

As with the evolution of the catalytic sector, these events took place at the dawn of life on earth. In the subsequent three billion years, more subunits were added to the various membrane sectors to optimize the function of these proton pumps in the numerous organelles and different membranes of the current eukaryotic cells.

VI. FUNCTION OF V-ATPASES IN DIFFERENT ORGANELLES AND MEMBRANES

A. Chromaffin Granules

Chromaffin granules are catecholamine storage vesicles of the adrenal medulla cells (Njus et al., 1981). They contain up to 0.5M catecholamines (adrenaline, noradrenaline, etc.), about 120 mM ATP, 20 mM calcium, 5 mM magnesium, 20 mM ascorbate, as well as several proteins and enzymes. Early studies of the energetization of catecholamine accumulation indicated that it is provided by an H^+-ATPase that generates pmf across the chromaffin granule membrane. This was the first demonstration of the presence of V-ATPase in the vacuolar system of eukaryotic cells. Subsequently, the V-ATPase of chromaffin granules was isolated and reconstituted into phospholipid vesicles. Upon the addition of MgATP, proton uptake into the reconstituted vesicles could be observed (Moriyama and Nelson, 1987a). The subunit structure of this enzyme was unraveled following the discovery of its cold lability (Moriyama and Nelson, 1989a). It was demonstrated that incubation of chromaffin granules on ice in the presence of MgATP resulted in the dissociation of the catalytic sector from the membrane. The released catalytic sector contained five polypeptides denoted as subunits A to E in order of decreasing molecular weights (Figure 2 and Moriyama and Nelson, 1989a). Similar cold inactivation was demonstrated with every membrane containing eukaryotic V-

ATPase. The ready dissociation of the catalytic sector presented a bioenergetic enigma. It is well known that upon removal of the catalytic sector of F-ATPases, their membrane sector conducts protons at high rates. Had similar phenomena occurred with the membrane sector of V-ATPase, chromaffin granules would loose all their catecholamines when even one of their enzymes lost its catalytic sector. Indeed it was not allowed by nature and it was demonstrated that the membrane sector of V-ATPases is impermeable to protons (Beltrán and Nelson, 1992). All of the above properties make the chromaffin granules a nearly perfect bioenergetic machine. Their V-ATPase provides pmf by hydrolyzing ATP. The internal pH is about 5.5, and together with the ascorbate provides an ideal environment to prevent oxidation of the catecholamines. The pH is maintained at 5.5 due to the proton slip

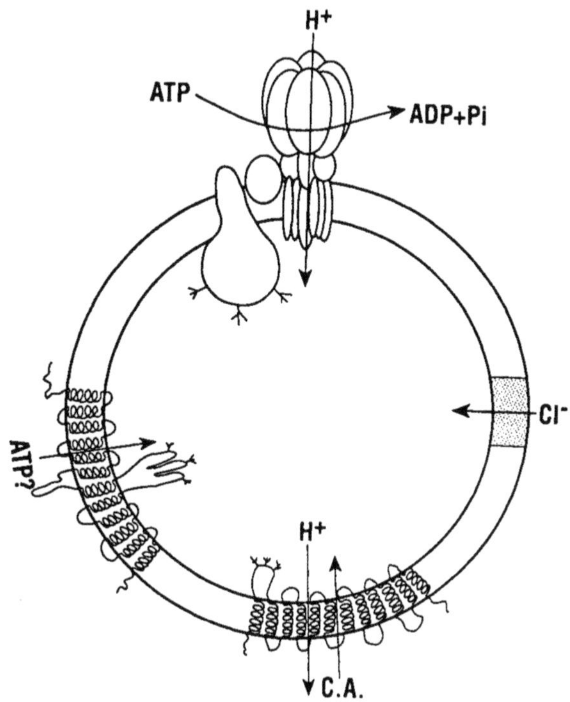

Figure 4. Four principal membrane proteins involved in ATP-dependent catecholamine and ATP uptake into chromaffin granules or synaptic vesicles. The V-ATPase provides the proton-motive force by ATP-dependent proton pumping. Cl^- is transported into the granule towards the positive potential by a specific channel. Catecholamines (C.A.) are taken up by exchanging with protons pumped inside by the V-ATPase. ATP is accumulated (driven by pmf) inside the granule via a specific carrier in the membrane.

that prevents overacidification of their interior (Nelson, 1991). The pH gradient generated by the V-ATPase is utilized for uptake and storage of the catecholamines. The catecholamines are taken up from the cytoplasm by a specific membrane protein that exchanges two protons for one catecholamine (Kanner, 1989). Figure 4 depicts the principal membrane proteins involved in this process. The cDNA encoding the catecholamine transporter was cloned and sequenced (Y. Liu et al., 1992). It is a hydrophobic protein containing 12 potential transmembrane helices. Similar vesicles storing catecholamines are present in various parts of the brain. While adrenaline secreted from the chromaffin granules acts on targets at long distance from the gland, in the brain the catecholamines are secreted into synaptic clefts and interact as neurotransmitters with specialized receptors on the post-synaptic cells. The structure and properties of the brain synaptic granule are similar to that of the chromaffin granules. This makes chromaffin granules an excellent model system for adrenergic granules in the brain.

B. Synaptic Vesicles

Synaptic vesicles accumulate and store neurotransmitters in presynaptic cells and release them following the arrival of an action potential. The driving force for the accumulation of neurotransmitters is the pmf generated by V-ATPase (Nelson, 1991). As shown in Figure 5, the neurotransmitter cycle is energized by two distinct ATPases. While V-ATPase functions on the synaptic vesicles membrane, a Na^+/K^+-ATPase functions in energizing the plasma membrane. Consequently the specific transporters for accumulation of neurotransmitters into the synaptic vesicles are driven by an electrochemical gradient of protons, and the specific transporters that function in the re-uptake of neurotransmitters from the synaptic cleft are driven by an electrochemical gradient of sodium (Kanner, 1989). The subunits of the catalytic sector of V-ATPases in synaptic vesicles are similar if not identical to those of chromaffin granules and other organelles of the vacuolar system (Cidon and Sihra, 1989). The subunit structure of the membrane sector was not studied, but is also likely to be similar to other membrane sectors of V-ATPases. The accumulation of glutamate into the synaptic vesicles was studied in more detail than the other neurotransmitters, and it appears to utilize primarily the membrane potential provided by the V-ATPase rather than the ΔpH driving the uptake of catecholamines (Moriyama et al., 1990; Tabb et al., 1992). The genes encoding glutamate, glycine, acetylcholine, and GABA transporters of the synaptic vesicles have not yet been cloned. The cDNAs encoding the sodium dependent neurotransmitter transporters of the plasma membrane were recently cloned and studied in detail (Liu et al., 1992, 1993). The neurotransmission cycle is one of the fine examples of Mitchelian coupling between ATPases, the electrochemical gradient, and a secondary process utilizing the electrochemical gradient.

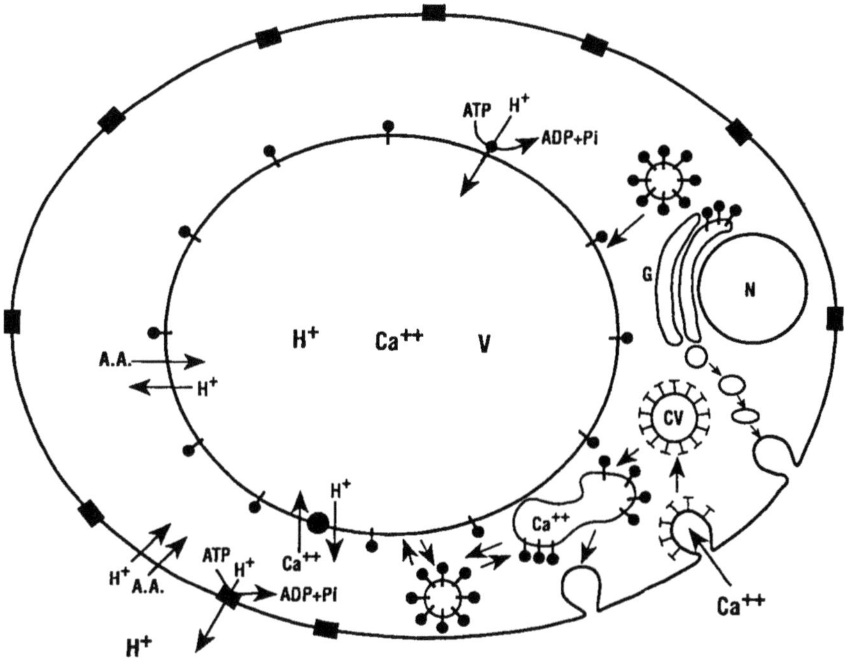

Figure 5. Schematic proposal of the principal driving forces of calcium homeostasis in plants and fungal cells. Symbols not common to Figure 1: V, vacuole; ■, P-type H^+-ATPase; ●, Ca^{2+}/H^+ exchanger; A.A., amino acid.

C. Lysosomes and Plant or Fungal Vacuoles

Most catalytic activities inside lysosomes require a low pH. Therefore, lysosomes have an internal pH of about 5 which is two pH units below that of the cytoplasm (Ohkuma et al., 1982). The acidification of lysosomes is catalyzed by an ATP-dependent proton pump that was shown to be V-ATPase (Schneider, 1981; Reeves and Reames, 1981; Moriyama and Nelson, 1989b). In addition, the V-ATPase provides most of the energy required for secondary transport systems in the lysosome membrane, some of which are vital for mammals and other higher eukaryotes. One of the possible functions of lysosomal and endosomal V-ATPases is in modulating drug influx into the vacuolar system by the multidrug resistance (MDR) P-glycoprotein transporter. Multi-drug resistance in several cell lines is related to increased levels of the MDR-transporter (Gottesman and Pastan, 1988). This transporter utilizes ATP for transporting the drugs, but the rates of drug transport and accumulation in reconstituted systems are not sufficient to explain the transport phenomenon *in vivo*. It was assumed that the P-glycoprotein functions at the plasma membrane and directly secrets the drugs to the extracellular space.

Recently it was suggested that the P-glycoprotein may function in the vacuolar system and secret a portion of the drugs into its organelles. Due to a much larger membrane area, the drug transport may be facilitated and eventually secreted by exocytosis. If correct, the low pH inside the vacuolar system may facilitate the release of the drug from the P-glycoprotein. Thus the V-ATPase that provides the low pH may indirectly be involved in drug efflux from mammalian cells. Moreover, it was demonstrated that the level of mRNA encoding subunit C of the V-ATPase increased several-fold in drug resistant cells in comparison with control cells (Ma and Center, 1992). This was not due to gene amplification but rather specific enhancement of subunit C mRNA transcription. This correlation between drug resistance and subunit C mRNA is only a first step in the study of gene expression of V-ATPases and malignancy.

The vacuoles of plants and fungi play a major role in their osmotic regulation, storage of metabolites, calcium homeostasis, and many other metabolic processes (Klionsky et al., 1990; Taiz, 1992). In most plant and fungal cells the vacuole occupies over 50% of the cell volume and the membrane contains numerous carriers, transporters, channels, and enzymes. Here, too, the V-ATPase is the major energy provider and most secondary transport processes are driven by the pmf generated by the enzyme. In plant vacuoles there exists a backup system of proton pyrophosphatase that also generates pmf by hydrolyzing pyrophosphate (Sarafian et al., 1992). A pivotal function of plant and fungal vacuoles is their function in calcium homeostasis. In contrast to mammalian cells that contain Ca^{2+}-ATPases and Ca^{2+}/Na^{+} exchangers in their plasma membrane, most calcium transport into and out of the cytoplasm of plant and fungal cells is mediated by the vacuolar membrane (Anraku et al., 1992). Therefore, calcium transport of plants and fungi is driven by pmf generated by V-ATPase or H^{+}-pyrophosphatase in plants. A Ca^{2+}/H^{+} exchanger functions in the vacuole membrane in Ca^{2+} transport from the cytoplasm to the vacuole. Most studies on vacuolar calcium transport were conducted in yeast. It was shown that calcium is taken into isolated yeast vacuoles utilizing the ΔpH generated by their V-ATPase. Consequently, disruptant mutants in which the yeast V-ATPase was inactivated are sensitive to high calcium concentrations in the medium (Noumi et al., 1991; Ohya et al., 1991). Though presumably the pmf in the vacuolar system is maintained in these mutants by fluid phase endocytosis (Nelson and Nelson, 1990), it is not sufficient to counteract calcium flow into the cytoplasm at high external calcium concentrations (see Figure 5). These yeast mutants are also sensitive to low calcium obtained by including 25 mM EGTA in the medium (Noumi et al., 1991). This may result from a higher than normal pH inside the vacuolar system that is not sufficient for dissociating the CaEGTA inside the vacuole (Figure 5). These studies demonstrated the crucial role of V-ATPase in calcium homeostasis in fungi and presumably in plants. It is tempting to suggest that a similar Ca^{2+}/H^{+} exchanger plays a role in calcium homeostasis in mammalian cells by controlling the calcium fluxes across membranes of organelles derived from the vacuolar system.

D. Endosomes and Receptor Recycling

The vacuolar system is very complicated, but highly organized machinery. Despite extensive flow of membranes among the intracellular organelles and membranes, the composition of each compartment is strictly preserved. Endocytosis and exocytosis are among the hallmarks of eukaryotic cells and are necessary processes for their life (Mellman et al., 1986). The exocytic pathway starts in the Golgi apparatus and is divided into two processes of constitutive and regulated secretions (Kelly, 1985). As shown in Figure 6, the regulated secretion contains V-ATPases that provide energy for their uptake systems and may play a role in the sorting of proteins into the system (Klionsky et al., 1990). The constitutive secretory pathway contains no V-ATPase and is used for secretion of several proteins and maintenance of the cell membrane. Synaptic vesicles and chromaffin granules are some of the organelles of the regulated pathway, and the function of the V-ATPase in these organelles has been discussed before. The existence of exocytosis, that adds large amounts membrane material to the cell membrane, required the development of endocytosis that is utilized for a variety of processes. Among the best known

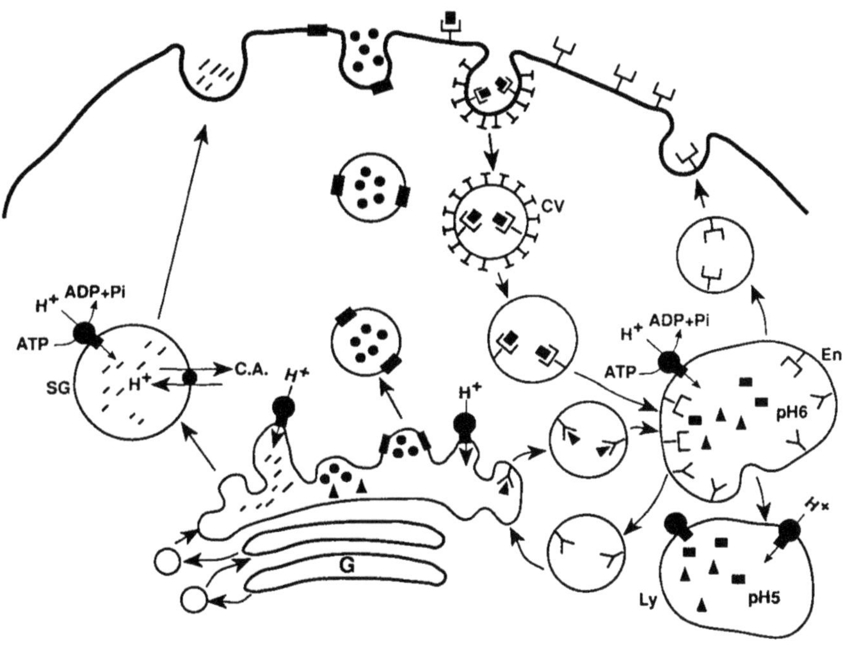

Figure 6. Involvement of V-ATPase in sorting and secretion. G, Golgi; Ly, lysosome; En, endosome; CV, clathrin-coated vesicle; ⧫, V-ATPase; ▬, Membrane protein; ●, secreted protein; ■, external ligand; ▲, ligand of the lumen of the vacuolar system; Y, receptor functions in the sorting of proteins into the vacuolar system.

processes of endocytosis is the receptor-mediated endocytosis (Brown and Goldstein, 1986). A receptor present at the cell surface binds its specific ligand. The binding of the ligand is pH-dependent. At neutral pH it exhibits high affinity to the ligand and at low pH its affinity is markedly lower. Following binding of its ligand, the receptor is internalized by clathrin-coated pits into clathrin-coated vesicles. In contrast to early observations, clathrin-coated vesicles are probably devoid of V-ATPase (Mellman, 1992). After the clathrin coat is removed, the uncoated vesicles fuse with endosomes containing V-ATPases. At the internal low pH of the endosome the ligand is dissociated from its receptor and the receptor can be either recycled to the plasma membrane or delivered into the lysosomes for disruption. This process of receptor-mediated endocytosis and pH-dependent release of the ligand is involved in numerous processes ranging from iron intake into cells through cholesterol uptake via low density lipoproteins to the action of several receptors on the surface of eukaryotes. Inhibition of V-ATPase results in interference of all those processes.

E. The Golgi Apparatus

The Golgi apparatus functions in sorting proteins of the vacuolar system and transports secreted proteins into specific delivery vesicles. Specific receptors function in the targeting of luminal and secreted proteins into their respective vesicles (see Figure 6). Some, but not all vesicles contain V-ATPases, and the trans-Golgi itself is furnished by the enzyme (Chanson and Taiz, 1985; Young et al., 1988; Moriyama and Nelson, 1989c). The possible involvement of V-ATPase in the sorting of some proteins was recognized by studying the effect of amines on the targeting of vacuolar proteins (Klionsky et al., 1990). It was shown that agents able to neutralize acidic pH in yeast vacuoles bring about the mistargetting of vacuolar enzymes. Mistargetting is also the result of treating yeast cells with specific V-ATPase inhibitors as well as mutant yeast cells in which genes encoding V-ATPase subunits were interrupted (Klionsky et al., 1992). In addition, the processing of the precursors of those vacuolar proteins is inhibited in the V-ATPase disruptant mutants. These mutants accumulate the intracellular precursors within the secretory pathway at some point before delivery to the vacuole and after transit to the Golgi complex (Yaver et al., 1993). These precursors are accumulated at the trans-Golgi complex or in post-Golgi vesicles. Thus, a picture emerges suggesting that V-ATPase plays a key role in several activities involving the Golgi apparatus including receptor-mediated targeting of luminal proteins as depicted in Figure 6. Studies involving antisense mRNA to subunit A of plant V-ATPases suggest the presence of two isozymes, one of which is localized to the Golgi apparatus (Gogarten et al., 1992a). This observation poses an important question about the sites and mechanism of biogenesis and assembly of V-ATPases in eukaryotic cells.

F. V-ATPases that Function on the Plasma Membrane of Eukaryotic Cells

There is only a single gene encoding each of the V-ATPase subunits in yeast cells. Therefore, it is likely that in yeast an identical enzyme functions in the Golgi apparatus, the vacuole, endosomes, and several other internal organelles of the vacuolar system. In mammalian cells, there are two or more genes encoding for some of the subunits of V-ATPase. Recently, it was shown that specialized subunit A may function in osteoclasts and two variants of subunit E are present in different kidney cells (Hemken et al., 1992; van Hille et al., 1993). In these two systems, V-ATPase functions not only in the vacuolar system, but also in the plasma membrane of the specialized cells. A third known cell in which the V-ATPase has an important role secreting protons outward is the sperm during fertilization. Acidification of the interior of the acrosome is vital for the function of the sperm (Wassarman, 1987). Upon binding of the sperm to the unfertilized egg, an acrosome reaction takes place. The acrosome reaction results in the secretion of the acrosome contents as well as exposure of the inner acrosomal membrane to the egg zona pellucida. Consequently, a part of the acrosomal membrane rich in V-ATPases is exposed to a cleft between the membrane and the zona pellucida to which low pH-dependent hydrolytic enzymes were secreted. Thus, the acrosomal V-ATPase acidifies the external cleft, providing a suitable environment for the activity of the hydrolytic enzymes. Neutralization of the cleft's low pH, generated by the V-ATPase, prevents fertilization—an action that is as essential for continuation of the human race as is fertilization.

The kidney plays a vital role in the acid–base balance of mammals. Hydrogen ion excretion involves several processes including bicarbonate reabsorption, carbonic anhydrase activity, and regulated pumping of protons across the plasma membrane by V-ATPase (Brown et al., 1992). In epithelial cells of the proximal urinary tubule, V-ATPase is present in the apical membrane and functions in proton secretion. In the collecting duct, V-ATPase may be found either in apical or basolateral membranes of specialized intercalated cells. These cells shuttle V-ATPase between intracellular vesicles and the plasma membrane in response to changes in the acid–base balance of the animal. It was shown that the distribution of V-ATPase in apical or basolateral membranes of intercalated cells changes during adaptation to acidosis or alkalosis (Bastani et al., 1991). The cells increase the number of V-ATPase enzymes in their apical membrane during acidosis and decrease their number during alkalosis. These observations suggest that V-ATPase plays a major role in maintaining pH homeostasis in mammals and other animals.

Bone resorption is necessary for bone growth, remodeling, and repair. Osteoclasts are multinucleated and highly motile cells that migrate between the bone and bone marrow and function in bone resorption. They attach to the mineralized bone matrix forming a close space in which hydrolytic enzymes are secreted (Chatterjee et al., 1992a). These enzymes require low pH for their optimal activity and the low pH is provided by V-ATPase located in the part of the plasma membrane in contact with the bone (Blair et al., 1989; Bekker and Gay, 1990; Vaananen et al., 1990). The principal bone mineral is hydroxylapatite, and protons are required for the release of each calcium ion from the mineral. The osteoclast V-ATPase provides all protons necessary for calcium resorption. As in kidney intercalated cells, osteoclasts shuttle V-ATPases between intracellular vesicles and the plasma membrane in response to attachment onto the bone matrix. This action renders an amoebae-like cell into a polar cell. The properties of the osteoclast V-ATPase are somewhat different from those of other V-ATPases in its greater sensitivity to vanadate, as well as the possible presence of specialized subunit A in the enzyme (Chatterjee et al., 1992b). The pharmacological value of studying the osteoclast V-ATPase is apparent because a specific slow down in its activity may prevent the onset of osteoporosis (Nelson, 1991). If, indeed, an enzyme with unique properties functions in osteoclasts, it would be more likely to discover a drug that will specifically inhibit the osteoclast V-ATPase.

VII. CONCLUDING REMARKS

Fundamental enzymes are highly conserved and maintain a high degree of amino acid sequence homology through evolution. Sequence conservation reveals not only the evolution of the given enzyme, but also indicates events that led to the evolution of eubacteria, archaebacteria, and eukaryota. Fundamental enzymes function in every known living cell and, therefore, nature did not permit eliminating them. Consequently a mutation leading to their inactivation is not permitted and if it occurs it causes lethality. This limits the diseases in which they may be involved and makes them less appealing for the less informed public. On the other hand, genuine scientists were always attracted to fundamental processes because they are the clue for understanding nature. The excitement in working with a fundamental enzyme such as V-ATPase is unsurpassed.

NOTE ADDED IN PROOFS

Since this chapter was written new yeast genes and mammalian cDNAs encoding V-ATPase subunits were cloned and sequenced. At the catalytic sector the yeast gene (*VMA8*) and bovine cDNA encoding subunits D were cloned and sequenced (Nelson et al., 1995; Graham et al., 1995). It was proposed that this subunit is analogous to the γ subunit of F-ATPases. A novel sixth subunit was identified and denoted as subunit F. The yeast gene (*VMA7*) and insect cDNA encoding this subunit were cloned and sequenced (Graf et al., 1994; Graham et al., 1994; Nelson et al., 1994). A yeast gene (*VMA13*) involved in the assembly of the catalytic sector was identified (Ho et al., 1993). For the membrane sector the yeast gene (*VMA6*) encoding Ac39 subunit was cloned and sequenced (Bauerle et al., 1993). A yeast gene (*VMA10*) and bovine cDNA encoding a V-ATPase subunit homologous to the b subunit of F-ATPases membrane sector were discovered and studied (Supekova et al., 1995, 1996). This is the fourth subunit that is homologous in F- and V-ATPases. A mammalian cDNA encoding organelle-specific subunit of V-ATPase was cloned from a bovine adrenal cDNA library (Supek et al., 1994). It was proposed that such accessory subunits will induce specific properties for V-ATPases located in specific organelles and membranes. Finally the hypothesis that yeast can survive, at low pH, the inactivation of V-ATPase by fluid-phase endocytosis was proven to be correct (Nelson and Nelson, 1990; Munn et al., 1994).

REFERENCES

Anraku, Y., Hirata, R., Wada, Y., & Ohya, Y. (1992). Molecular genetics of the yeast vacuolar H^+-ATPase. J. Exp. Biol. 172, 67–81.

Arai, H., Berne, M., & Forgac, M. (1987a). Inhibition of the coated vesicle proton pump and labeling of a 17,000-dalton polypeptide by *N,N'*-dicyclohexylcarbodiimide. J. Biol. Chem. 262, 11006–11011.

Arai, H., Berne, M., Terres, G., Terres, H., Puopolo, K., & Forgac, M. (1987b). Subunit composition and ATP site labeling of the coated vesicle proton-translocating adenosinetriphosphatase. Biochemistry 26, 6632–6638.

Arai, H., Terres, G., Pink, S., & Forgac, M. (1988). Topography and subunit stoichiometry of the coated vesicle proton pump. J. Biol. Chem. 263, 8796–8802.

Bastani, B., Purcell, H., Hemken, P., Trigg, D., & Gluck, S. (1991). Expression and distribution of renal vacuolar proton-translocating adenosine triphosphatase in response to chronic acid and alkali loads in the rat. J. Clin. Invest. 88, 126–136.

Bauerle, C., Ho, M. N., Lindorfer, M. A., & Stevens, T. H. (1993). The *Saccharomyces cerevisiae VMA6* gene encodes the 36-kDa subunit of the vacuolar H^+-ATPase membrane sector. J. Biol. Chem. 268, 12749–12757.

Bekker, P. J., & Gay, C. V. (1990). Biochemical characterization of an electrogenic vacuolar proton pump in purified chicken osteoclast plasma membrane vesicles. J. Bone Mineral Res. 5, 569–579.

Beltrán, C., & Nelson, N. (1992). The membrane sector of vacuolar H^+-ATPase by itself is impermeable to protons. Acta Physiol. Scand. 146, 41–47.

Beltrán, C., Kopecky, J., Pan, Y.-C. E., Nelson, H., & Nelson, N. (1992). Cloning and mutational analysis of the gene encoding subunit C of yeast V-ATPase. J. Biol. Chem. 267, 774–779..

Blair, H. C., Teitelbaum, S. L., Ghiselli, R., & Gluck, S. (1989). Osteoclastic bone resorption by a polarized vacuolar proton pump. Science 245, 855–857.

Bowman, E. J., Tenney, K., & Bowman, B. J. (1988a). Isolation of genes encoding the *Neurospora* vacuolar ATPase. J. Biol. Chem. 263, 13994–14001.

Bowman, B. J., Allen, R., Wechser, M. A., & Bowman, E. J. (1988b). Isolation of genes encoding the *Neurospora* vacuolar ATPase. J. Biol. Chem. 263, 14002–14007.

Bowman, B. J., Dschida, W. J., Harris, T., & Bowman, E. J. (1989). The vacuolar ATPase of *Neurospora crassa* contains an F_1-like structure. J. Biol. Chem. 264, 15606–15612.

Brown, D., Sabolic, I., & Gluck, S. (1992). Polarized targeting of V-ATPase in kidney epithelial cells. J. Exp. Biol. 172, 231–243.

Brown, M. S. & Goldstein, J. L. (1986). A receptor-mediated pathway for cholesterol homeostasis. Science 232, 34–47.

Chanson, A., & Taiz, L. (1985). Evidence for an ATP-dependent proton pump on the Golgi of corn coleoptiles. Plant Physiol. 78, 232–240.

Chatterjee, D., Chakraborty, M., Leit, M., Neff, L., Jamsakellokumpu, S., Fuchs, R., Bartkiewicz, M., Hernando, N., & Baron, R. (1992a). The osteoclast proton pump differs in its pharmacology and catalytic subunits from other vacuolar H^+-ATPases. J. Exp. Biol. 172, 193–204.

Chatterjee, D., Chakraborty, M., Leit, M., Neff, L., Samsa-Kellokumpu, S., Fuchs, R., & Baron, R. (1992b). Sensitivity to vanadate and isoforms of subunits A and B distinguish the osteoclast proton-pump from other vacuolar H^+-ATPases. Proc. Natl. Acad. Sci. USA 89, 6257–6291.

Cidon, S., & Sihra, T. (1989). Characterization of a H^+-ATPase in rat brain synaptic vesicles. Coupling to L-glutamate transport. J. Biol. Chem. 264, 8281–8288.

Cross, R. L., & Taiz, L. (1990). Gene duplication as a means for altering H^+/ATP ratios during the evolution of F_0F_1 ATPases and synthases. FEBS Lett. 259, 227–229.

Denda, K., Konishi, J., Oshima, T., Date, T., & Yoshida, M. (1988a). Molecular cloning of the β-subunit of a possible non-F_0F_1 type ATP synthase from the acidothermophilic archaebacterium, *Sulfolobus acidocaldarius*. J. Biol. Chem. 263, 17251–17254.

Denda, K., Konishi, J., Oshima, T., Date, T., & Yoshida, M. (1988b). The membrane-associated ATPase from *Sulfolobus acidocaldarius* is distantly related to F_1-ATPase as assessed from the primary structure of its α-subunit. J. Biol. Chem. 263, 6012–6015.

Denda, K., Konishi, J., Hajiro, K., Oshima, T., Date, T., & Yoshida, M. (1990). Structure of an ATPase operon of an acidothermophilic archaebacterium, *Sulfolobus acidocaldarius*. J. Biol. Chem. 265, 21509–21513.

Feng, Y., & Forgac, M. (1992). Cysteine 254 of the 73-kDa A subunit is responsible for inhibition of the coated vesicle (H^+)-ATPase upon modification by sulfhydryl reagents. J. Biol. Chem. 267, 5817–5822.

Foury, F. (1990). The 31-kDa polypeptide is an essential subunit of the vacuolar ATPase in *Saccharomyces cerevisiae*. J. Biol. Chem. 265, 18554–18560.

Gluck, S. (1992). V-ATPases of the plasma membrane. J. Exp. Biol. 172, 29–37.

Gluck, S. L., Nelson, R. D., Lee, B. S., Wang, Z.-Q., Guo, X.-L., Fu, J.-Y., & Zhang, K. (1992). Biochemistry of the renal V-ATPase. J. Exp. Biol. 172, 219–229.

Gogarten, J. P., Kibak, H., Dittrich, P., Taiz, L., Bowman, E. J., Bowman, B. J., Manolson, M. F., Poole, R. J., Date, T., Oshima, T., Konishi, J., Denda, K., & Yoshida, M. (1989). Evolution of the vacuolar H^+-ATPase: Implications for the origin of eukaryotes. Proc. Natl. Acad. Sci. USA 86, 6661–6665.

Gogarten, J. P., Fichmann, J., Braun, Y., Morgan, L., Styles, P., Taiz, S. L., DeLapp, K., & Taiz, L. (1992a). The use of antisense mRNA to inhibit the tonoplast H^+-ATPase in carrot. The Plant Cell 4, 851–864.

Gogarten, J. P., Starke, T., Kibak, H., Fishmann, J., & Taiz, L. (1992b). Evolution and isoforms of V-ATPase subunits. J. Exp. Biol. 172, 137–147.

Gottesman, M. M., & Pastan, I. (1988). The multidrug transporter, a double-edged sword. J. Biol. Chem. 263, 12163–12166.

Gräf, R., Lepier, A., Harvey, W. R., & Wieczorek, H. (1994). A novel 14-kDa V-ATPase subunit in the tobacco hornworm midgut. J. Biol. Chem. 269, 3767–3774.

Graham, L. A., Hill, K. J., & Stevens, T. H. (1994). VMA7 encodes a novel 14-kDa subunit of the *Saccharomyces cerevisiae* vacuolar H^+-ATPase complex. J. Biol. Chem. 269, 25974–25977.

Graham, L. A., Hill, K. J., & Stivens, T. H. (1995). *VMA8* encodes a 32-kDa V_1 subunit of the *Saccharomyces cerevisiae* vacuolar H^+-ATPase required for function and assembly of the enzyme complex. J. Biol. Chem. 270, 15037–15044.

Harold, F. M. (1986). The Vital Force: A Study of Bioenergetics. W.H. Freeman, New York.

Hemken, P., Guo, X.-L., Wang, Z.-Q., Zhang, K., & Gluck, S. (1992). Immunologic evidence that vacuolar H^+ ATPases with heterogeneous forms of $M_r = 31{,}000$ subunit have different membrane distributions in mammalian kidney. J. Biol. Chem. 267, 9948–9957.

Hirata, R., Ohsumi, Y., Nakano, A., Kawasaki, H., Suzuki, K., & Anraku, Y. (1990). Molecular structure of a gene, *VMA1*, encoding the catalytic subunit of H^+-translocating adenosine triphosphatase from vacuolar membranes of *Saccharomyces cerevisiae*. J. Biol. Chem. 265, 6726–6733.

Hirsch, S., Strauss, A., Masood, K., Lee, S., Sukhatme, V., & Gluck, S. (1988). Isolation and sequence of a cDNA clone encoding the 31-kDa subunit of bovine kidney vacuolar H^+-ATPase. Proc. Natl. Acad. Sci. USA 85, 3004–3008.

Ho, M. N., Hill, K. J., Lindorfer, M. A., & Stevens, T. H. (1993a). Isolation of vacuolar membrane H^+-ATPase-deficient yeast mutants; the *VMA5* and *VMA4* genes are essential for assembly and activity of the vacuolar H^+-ATPase. J. Biol. Chem. 268, 221–227.

Ho, M. N., Hirata, R., Umemoto, N., Ohya, Y., Takatsuki, A., Stevens, T. H., & Anraku, Y. (1993b). *VMA13* encodes a 54-kDa vacuolar H^+-ATPase subunit required for activity but not assembly of the enzyme complex in *Saccharomyces cerevisiae*. J. Biol. Chem. 268, 18286–18292.

Kane, P. M., Yamashiro, C. T., Wolczyk, D. F., Neff, N., Goebl, M., & Stevene, T. H. (1990). Protein splicing converts the yeast *TFP1* gene product to the 69-kD subunit of the vacuolar H^+-adenosine triphosphatase. Science 250, 651–657.

Kanner, B. I. (1989). Ion-coupled neurotransmitter transport. Curr. Opin. Cell Biol. 1, 735–738.

Kelly, R. B. (1985). Pathways of protein secretion in eukaryotes. Science 230, 25–32.

Klionsky, D. J., Herman, P. K., & Emr, S. D. (1990). The fungal vacuole: composition, function, and biogenesis. Microbiol. Rev. 54, 266–292.

Klionsky, D. J., Nelson, H., & Nelson, N. (1992). Compartment acidification is required for efficient sorting of proteins to the vacuole in *Saccharomyces cerevisiae*. J. Biol. Chem. 267, 3416–3422.

Liu, Q.-R., Mandiyan, S., Nelson, H., & Nelson, N. (1992). A family of genes encoding neurotransmitter transporters. Proc. Natl. Acad. Sci. USA 89, 6639–6643.

Liu, Q.-R., López-Corcuera, B., Mandiyan, S., Nelson, H., & Nelson, N. (1993). Molecular characterization of four pharmacologically distinct γ-aminobutyric acid transporters in mouse brain. J. Biol. Chem. 268, 2106–2112.

Liu, Y., Peter, D., Roghani, A., Schuldiner, S., Privé, G. G., Eisenberg, D., Brecha, N., & Edwards, R. H. (1992). A cDNA that suppresses MPP^+ toxicity encodes a vesicular amine transporter. Cell 70, 539–551.

Ma, L., & Center, M. S. (1992). The gene encoding vacuolar H^+-ATPase subunit C is overexpressed in multidrug-resistant HL60 cells. Biochem. Biophys. Res. Commun. 182, 675–681.

Mandel, M., Moriyama, Y., Hulmes, J. D., Pan, Y.-C. E., Nelson, H., & Nelson, N. (1988). Cloning of cDNA sequence encoding the 16-kDa proteolipid of chromaffin granules implies gene duplication in the evolution of H^+-ATPases. Proc. Natl. Acad. Sci. USA 85, 5521–5524.

Manolson, M. F., Rea, P. A., & Poole, R. J. (1985). Identification of 3-O-(4-benzoyl)benzoyladenosine 5′-triphosphate- and *N,N′*-dicyclohexylcarbodiimide-binding subunits of a higher plant proton-translocating tonoplast. J. Biol. Chem. 260, 12273–12279.

Manolson, M. F., Ouellette, B. F. F., Filion, M., & Poole, R. J. (1988). cDNA sequence and homologies of the "57-kDa" nucleotide-binding subunit of the vacuolar ATPase from *Arabidopsis*. J. Biol. Chem. 263, 17987–17994.

Manolson, M. F., Proteau, D., Preston, R. A., Stenbit, A., Roberts, T., Hoyt, M. Andrew, Preuss, D., Mulholland, J., Botstein, D., & Jones, E. W. (1992). The *VPH1* gene encodes a 95-kDa integral membrane polypeptide required for *in vivo* assembly and activity of the yeast vacuolar H^+-ATPase. J. Biol. Chem. 267, 14294–14303.

Manolson, M. F., Wu, B., Proteau, D., Taillon, B. E., Roberts, B. T., Hoyt, M. A., & Jones, E. W. (1994). *STV1* gene encodes functional homologue of 95-kDa yeast vacuolar H^+-ATPase subunit Vph1p. J. Biol. Chem. 269, 14064–14074.

Mellman, I., Fuchs, R., & Helenius, A. (1986). Acidification of the endocytic and exocytic pathways. Annu. Rev. Biochem. 55, 663–700.

Mellman, I. (1992). The importance of being acid: the role of acidification in intracellular membrane traffic. J. Exp. Biol. 172, 39–45.

Mitchell, P. (1968). Chemiosmotic Coupling and Energy Transduction, Bodmin.,

Moriyama, Y., & Nelson, N. (1987a). The purified ATPase from chromaffin granule membranes is an anion-dependent proton pump. J. Biol. Chem. 262, 9175–9180.

Moriyama, Y., & Nelson, N. (1987b). Nucleotide binding sites and chemical modification of the chromaffin granule proton ATPase. J. Biol. Chem. 262, 14723–14729.

Moriyama, Y., & Nelson, N. (1989a). Cold inactivation of vacuolar H^+-ATPases. J. Biol. Chem. 264, 3577–3582.

Moriyama, Y., & Nelson, N. (1989b). Lysosomal H^+-translocating ATPase has a similar subunit structure to chromaffin grauune H^+-ATPase complex. Biochim. Biophys. Acta 980, 241–247.

Moriyama, Y., & Nelson, N. (1989c). H^+-translocating ATPase in Golgi apparatus: Characterization as vacuolar H^+-ATPase and its subunit structures. J. Biol. Chem. 264, 18445–18450.

Moriyama, Y., Maeda, M., & Futai, M. (1990). Energy coupling of L-glutamate transport and vacuolar H^+-ATPase in brain synaptic vesicles. J. Biochem. 108, 689–693.

Munn, A. L., & Riezman, H. (1994). Endocytosis is required for the growth of vacuolar H^+-ATPase-defective yeast. Identification of six new *END* genes. J. Cell Biol. 127, 373–386.

Nelson, H., & Nelson, N. (1989). The progenitor of ATP synthases was closely related to the current vacuolar H^+-ATPase. FEBS Lett. 247, 147–153.

Nelson, H., & Nelson, N. (1990). Disruption of genes encoding subunits of yeast vacuolar H^+-ATPase causes conditional lethality. Proc. Natl. Acad. Sci. USA 87, 3503–3507.

Nelson, H., Mandiyan, S., Noumi, T., Moriyama, Y., Miedel, M. C., & Nelson, N. (1990). Molecular cloning of cDNA encoding the C subunit of H^+-ATPase from bovine chromaffin granules. J. Biol. Chem. 265, 20390–20393.

Nelson, H., Mandiyan, S., & Nelson, N. (1994). The *Saccharomyces cerevisiae VMA7* gene encodes a 14-kDa subunit of the vacuolar H^+-ATPase catalytic sector. J. Biol. Chem. 269, 24150–24155.

Nelson, H., Mandiyan, S., & Nelson, N. (1995). A bovine cDNA and a yeast gene-*VMA8* encoding subunit D of the vacuolar H^+-ATPase. Proc. Natl. Acad. Sci. USA 92, 497–501.

Nelson, N. (1989). Structure, molecular genetics and evolution of vacuolar H^+-ATPases. J. Bioenerg. Biomembr. 21, 553–571.

Nelson, N. (1991). Structure and pharmacology of the proton-ATPases. Trends Pharmac. Sci. 12, 71–75.

Nelson, N. (1992a). Evolution of organellar proton-ATPases. Biochim. Biophys. Acta 1100, 109–124.

Nelson, N. (1992b). Structural conservation and functional diversity of V-ATPases. Bioenerg. Biomembr. 24, 407–414.

Nelson, N. (1992c). Organellar proton-ATPases. Curr. Opin. Cell Biol. 4, 654–660.

Nelson, N., & Taiz, L. (1989). The evolution of H^+-ATPases. Trends Biochem. Sci. 14, 113–116.

Njus, D., Knoth, J., & Zallakian, M. (1981). Proton-linked transport in chromaffin granules. Curr. Topics Bioenergetics 11, 107–147.

Noumi, T., Beltrán, C., Nelson, H., & Nelson, N. (1991). Mutational analysis of yeast vacuolar H^+-ATPase. Proc. Natl. Acad. Sci. USA 88, 1938–1942.

Ohkuma, S., Moriyama, Y., & Takano, T. (1982). Identification and characterization of a proton pump on lysosomes by fluorescein isothiocyanate dextran fluorescence. Proc. Natl. Acad. Sci. USA 79, 2758–2762.

Ohya, Y., Umemoto, N., Tanida, I., Ohta, A., Iida, H., & Anraku, Y. (1991). Calcium-sensitive *cls* mutants of *Saccharomyces cerevisiae* showing a pet^- phenotype are ascribable to defects of vacuolar membrane H^+-ATPase activity. J. Biol. Chem. 266, 13971–13977.

Penefsky, H. S., & Cross, R. L. (1991). Structure and mechanism of F_0F_1-type ATP synthases and ATPases. In: Advances in Enzymology and Related Areas of Molecular Biology (Meister, A., ed.), Vol. 64, pp. 173–214. John Wiley & Sons, Inc., New York.

Perin, M. S., Fried, V. A., Stone, D. K., Xie, X.-S., & Südhof, T. C. (1991). Structure of the 116-kDa polypeptide of the clathrin-coated vesicle/synaptic vesicle proton pump. J. Biol. Chem. 266, 3877–3881.

Reeves, J. P., & Reames, T. (1981). ATP stimulates amino acid accumulation by lysosomes incubated with amino acid methyl esters. J. Biol. Chem. 256, 6047–6053.

Sarafian, V., Kim, Y., Poole, R. J., & Rea, P. A. (1992). Molecular cloning and sequence of cDNA encoding the pyrophosphate-energized vacuolar membrane proton pump of *Arabidopsis thaliana*. Proc. Natl. Acad. Sci. USA 89, 1775–1779.

Schneider, D. L. (1981). ATP-dependent acidification of intact and disrupted lysosomes. Evidence for an ATP-driven proton pump. J. Biol. Chem. 256, 3858–3864.

Schäfer, G., & Meyering-Vos, M. (1992). The plasma membrane ATPase of archaebacteria: a chimeric energy converter. In: Ion-Motive ATPases: Structure, Function, and Regulation (Scarpa, A., Carafoli, E., & Papa, S., eds.), Vol. 671, pp. 293–309. The New York Academy of Sciences, New York.

Shih, C.-K., Wagner, R., Feinstein, S., Kanik-Ennulat, C., & Neff, N. (1988). A dominant trifluoperazine resistance gene from *Saccharomyces cerevisiae* has homology with F_0F_1 ATP synthase and confers calcium-sensitive growth. Mol. Cell. Biol. 8, 3094–3103.

Stevens, T. H. (1992). The structure and function of the fungal V-ATPase. J. Exp. Biol. 172, 47–55.

Supek, F., Supekova, L., Beltrán, C., Nelson, H., & Nelson, N. (1992). Structure, function, and mutational analysis of V-ATPases. Ann. NY Acad. Sci. 671, 284–292.

Supek, F., Supekova, L., Mandiyan, S., Pan, Y.-C. E., Nelson, H., & Nelson, N. (1994). A novel accessory subunit for vacuolar H^+-ATPase from chromaffin granules. J. Biol. Chem. 269, 24102–24106.

Supekova, L., Supek, F., & Nelson, N. (1995). The *Saccharomyces cereviviae VMA10* is an intron-containing gene encoding a novel 18-kDa subunit of vacuolar H^+-ATPase. J. Biol. Chem. 270, 13726–13732.

Supekova, L., Sbia, M., Supek, F., Ma, Y-M., & Nelson, N. (1996). A novel subunit of vacuolar H^+-ATPase related to the *b* subunit of F-ATPases. J. Exp. Biol. 199, 1147–1156.

Sze, H., Ward, J. M., & Lai, S. (1992). Vacuolar H^+-translocating ATPases from plants: structure, function, and isoforms. J. Bioenerg. Biomembr. 24, 371–381.

Tabb, J. S., Kish, P. E., Van Dyke, R., & Ueda, T. (1992). Glutamate transport into synaptic vesicles. J. Biol. Chem. 267, 15412–15418.

Taiz, L. (1992). The plant vacuole. J. Exp. Biol. 172, 113–122.

Taiz, S. L., & Taiz, L. (1991). Ultrastructural comparison of the vacuolar and mitochondrial H^+-ATPases of *Daucus carota*. Bot. Acta 104, 85–168.

Umemoto, N., Ohya, Y., & Anraku, Y. (1991). MA*11*, a novel gene that encodes a putative proteolipid, is indispensable for expression of yeast vacuolar membrane H^+-ATPase activity. J. Biol. Chem. 266, 24526–24532.

van Hille, B., Richener, H., Evans, D. B., Green, J. R., & Bilbe, G. (1993). Identification of two subunit A isoforms of the vacuolar H^+-ATPase in human osteoclastoma. J. Biol. Chem. 268, 7075–7080.

Väänänen, H. K., Karhukorpi, E.-K., Sundquist, K., Wallmark, B., Roininen, I., Hentunen, T., Tuukkanen, J., & Lakkakorpi, P. (1990). Evidence for the presence of a proton pump of the vacuolar H^+-ATPase type in the ruffled borders of osteoclasts. J. Cell Biol. 111, 1305–1311.

Wang, S.-Y., Moriyama, Y., Mandel, M., Hulmes, J. D., Pan, Y.-C. E., Danho, W., Nelson, H., & Nelson, N. (1989). Cloning of cDNA encoding a 32-kDa protein: an accessory polypeptide of the H^+-ATPase from chromaffin granules. J. Biol. Chem. 263, 17638–17642.

Wassarman, P. M. (1987). The biology and chemistry of fertilization. Science 235, 553–560.

Yaver, D. S., Nelson, H., Nelson, N., & Klionsky, D. J. (1993). Vacuolar ATPase mutants accumulate precursor proteins in a pre-vacuolar compartment. J. Biol. Chem. 268, 10564–10572.

Young, G. P.-H., Qiao, J.-Z., & Al-Awqati, Q. (1988). Purification and reconstitution of the proton-translocating ATPase of Golgi-enriched membranes. Proc. Natl. Acad. Sci. USA 85, 9590–9594.

Zimniak, L., Dittrich, P., Gogarten, J. P., Kibak, H., & Taiz, L. (1988). The cDNA sequence of the 69 kDa subunit of the carrot vacuolar H^+-ATPase. J. Biol. Chem. 263, 9102–9112.

THE F_0F_1 ATP SYNTHASE: STRUCTURES INVOLVED IN CATALYSIS, TRANSPORT, AND COUPLING

Robert K. Nakamoto and Masamitsu Futai

Biomembranes
Volume 5, pages 343–367.

ISBN: 1-55938-662-2.

I. INTRODUCTION

The F_0F_1 ATP synthase has been the focus of extensive studies for the past thirty years. Such attention is not surprising considering its key role in cellular energetics and the complexity of its structure and mechanism. The multi-subunit complex, which has been largely conserved throughout evolution, is found in the inner membranes of mitochondria, chloroplast, and bacteria where it couples the downhill movement of protons (from the electrochemical gradient of protons, or proton motive force, generated by electron transport chains) to the synthesis of ATP from ADP and inorganic phosphate. In absence of oxidative conditions, many bacteria also use the F_0F_1 in the reverse reaction as a proton pump. In this case, hydrolysis of ATP is coupled to the transport of protons out of the cell producing an electrochemical gradient which provides energy for secondary systems such as proton-coupled transport and flagellar rotation. In this chapter, we will highlight molecular features of the ATP synthase, and discuss how its complex structure is integrated to link catalysis to transport.

A. General Features of the F_0F_1 Complex

The ATP synthase can be dissociated into two sectors: the water soluble F_1 sector which contains the catalytic sites, and the hydrophobic F_0 sector which conducts protons through the membrane. The F_1 sector consists of five types of subunits, α, β, γ, δ, and ε with a stoichiometry of 3:3:1:1:1, respectively (see Table 1; and for reviews Senior, 1988; Futai et al., 1989; Penefsky and Cross, 1991). Isolated F_1, which is released from the membrane in low ionic strength buffers in the absence of magnesium ion, catalyzes hydrolysis of ATP. A complex of α_3-β_3-γ (Futai et al., 1974) or in some cases, α_3–β_3 (Kagawa et al., 1989) is adequate for ATPase activity, however, binding of F_1 to F_0 and reconstitution of coupled transport requires all five subunits (Dunn and Futai, 1980). The F_0 sector makes up the integral membrane portion of the complex and consists of three or more types of hydrophobic subunits (see Fillingame, 1990 for a review). For example, the *Escherichia coli* F_0 has three subunits, *a*, *b*, and *c* in a stoichiometry of 1:2:10 ± 1. All three F_0 subunits are necessary for proton conductivity as well as binding of the F_1 sector (Schneider and Altendorf, 1987).

Since the complete sequences of the eight *E. coli* F_0F_1 subunits were reported (Futai and Kanazawa, 1983; Walker et al., 1984), primary structures for all subunits

Table 1. The *Escherichia coli* F_oF_1 Subunits Listed by Sector and Size

Subunit	*Amino Acid Residues*	*Mass*	*Stoichiometry[a] in Complex*
F_o *a*	271	30,285	1
F_o *b*	156	17,202	2
F_o *c*	79	8,264	10 ± 1
F_1 α	513	55,200	3
F_1 β	459	50,155	3
F_1 γ	286	31,428	1
F_1 δ	177	19,328	1
F_1 ε	132	14,920	1

Note: [a]Subunit stoichiometry was taken from Foster and Fillingame, 1982; Hermolin and Fillingame, 1989.

and from many sources have been derived from sequences of cDNAs and genes. This ever-growing data base has stimulated investigators to align sequences in order to identify conserved amino acids and structural motifs as an indicator of functionally important residues. As already expected from many physical and biochemical studies, the sequence analyses confirmed that ATP synthases from mitochondria, chloroplasts, and bacteria are closely related in structure as well as mechanism. It follows that, in most cases, basic findings from studies on the F_0F_1 from one species are relevant to others. In the following sections, we will see how many of these conserved residues have been found to play important roles in catalysis, transport, and the coupling between the two.

B. Organization of Subunits

The overall shape of F_0F_1 has been revealed by image analyses of electron micrographs of negatively stained or cryo-treated specimens (see Capaldi et al., 1992 for a review). The F_1 sector was seen as a globular sphere of approximately 100 Å diameter with a stem 45 Å long and 25–30 Å in diameter connecting it to the membranous portion (Gogol et al., 1987). The electron density of the hydrophobic F_0 is indistinct from the surrounding lipid, so little information about its shape has been obtained.

The subunits that make up the globular portion of F_1 were, in part, resolved by electron micrograph images of single complexes. The clarity of these images can be improved by classification according to orientation and averaging. Furthermore, in many methods for preparing the F_1 for imaging, the complex is found uniformly oriented on an axis normal to the bilayer (Akey et al., 1983; Tsuprun et al., 1984; Tiedge et al., 1985; Boekema et al., 1988; Gogol et al., 1989b). From this orientation, projection images are obtained which show six densities arranged in a hexameric ring around a 35 Å cavity. Monoclonal antibodies specific for α-subunits tag densities 120° apart, clearly showing that the three α- and three β-subunits alternate in the ring (Lündsdorf et al., 1984; Tiedge et al., 1985; Gogol et al., 1989a). Within the cavity, an asymmetrically placed seventh density is observed. The internal density is present even when δ- and ε-subunits are missing, indicating that the γ-subunit makes up most of this feature (Gogol et al., 1989a). As yet, the relationships between γ-, δ-, and ε-subunits and their interactions with α- and β-subunits are not known. Perhaps, the only reasonable method to reveal the relationships, as well as their fine structures, is by X-ray crystallography. Advances in this direction are described in the next section.

II. THE F_1 SECTOR AND CATALYSIS

A. The Structure of F_1 at High Resolution

The structure of F_1 has been obtained at much higher resolution by X-ray crystallography. Working with the soluble F_1 sector alone alleviates many of the problems encountered in attempts to crystallize membranous proteins. However, even though F_1 can be obtained highly purified and in adequate quantities, it presents a considerable crystallographic challenge owing to its very large size (~380 kD). Thus far, two groups have overcome some of the difficulties and have described crystals with different forms. First, the group of Amzel and Pedersen reported that their crystals of the rat liver F_1 were trigonal and belonged to space group *R*32 (Amzel and Pedersen, 1978). In contrast, the crystals of bovine heart F_1 made by Walker's group were orthorhombic of space group $P2_12_12_1$ (Lutter et al., 1993). These differences were manifest in the crystallographic results.

Bianchet et al. (1991) found that the rat liver F_1 sat on a crystallographic three-fold axis. Hence, each asymmetric unit contained only one-third of the F_1 complex and the resulting structure necessarily revealed a highly symmetric particle in which each α–β pair was identical to the other two. This crystal packing implies that the γ-, δ-, and ε-subunits, which appear only once in each F_1 complex and do not have internal three-fold symmetry, did not participate in packing contacts and were thus free to occupy random positions about the three-fold axis. As a result, their contributions to the structure were smeared around this three-fold axis. Nevertheless, the structure presented at 3.6 Å gave considerable detail of the α- and β-subunits. The β-subunits were found to be interdigitated between α-subunits and situated approximately 15 Å higher relative to the membrane surface. There is little β–β contact, but quite extensive α–α interactions at the bottom of the trimeric set.

In contrast, Lutter et al. (1993) found that their orthorhombic crystals of bovine heart F_1 diffracted with a unit cell that suggested an asymmetric unit of the size expected for the entire sector including γ, δ, and ε. Apparently a difference in the crystallization conditions fixed the small subunits in a single conformation. The structure presented at 6.5 Å resolution (Abrahams et al., 1993) was remarkable for the extent of asymmetry. Distinctly asymmetric features in the resulting structure were the 40 Å "stem" that extended from the bottom of the complex, a "pit" next to this stem, and a 15 Å depression with a trough at the top of the complex. In addition, a long, internal rod extended 90 Å along the pseudo-three-fold axis from the stem to the top of the complex. This rod, which is likely an α-helix, may be part of the δ- or γ-subunits, as both are predicted to have high α-helical contents. Other features appear to be related by three- or six-fold symmetry and thus are likely to belong to the α- and β-subunits. Identification of subunits responsible for the asymmetric distribution of mass will have important implications for the characteristics of the three potential active sites (see below), the role of the single-copy subunits, and the mechanism of coupling between transport and catalysis.

Because the bovine heart F_1 is essentially the same as the rat complex (α and β sequence identities are 98% and 95%, respectively), the discrepancy in the molecular symmetries from the two crystal forms must reflect the structural effects of the γ-, δ-, and ε-subunits on the crystal contacts, which in turn, must be due to differences in the crystallization conditions. Whereas the rat liver enzyme was crystallized from a mother liquor containing ATP and potassium phosphate, the bovine heart crystals were made in the presence of AMP–PNP, ADP, and Mg. A possible interpretation of the discrepancy is that ATP and phosphate seem to prevent the minor subunits from perturbing the three-fold symmetry of the α–β pairs, whereas AMP–PNP, ADP, and Mg force asymmetry which can only be accommodated if the asymmetric unit contains an entire F_1 molecule. This asymmetry could result either from an indirect effect on the hexameric ring of α- and β-subunits, or from direct participation of γ-, δ-, or ε-subunits in crystal contacts.

The influence of the nucleotides on the conformation of the complex suggests that the two crystal forms may have a functional significance. The differing patterns of crystallization are reminiscent of results from electron microscopy (summarized in Capaldi et al., 1992; see following), which showed that the minor subunits change conformation in response to the addition of nucleotides. These results are consistent with mechanistic studies, which established a kinetic asymmetry among the three potential catalytic sites on the β-subunits (reviewed in Boyer, 1993). Because the three α-subunits and the three β-subunits, respectively, have identical sequences, asymmetry must be imposed by the minor subunits. Furthermore, it is likely that the asymmetry changes during the catalytic cycle are an integral part of the catalytic and transport mechanism. Structures derived from the two different crystal forms may provide an opportunity to understand the nature of the conformational changes. As yet, the hydrolytic states of the nucleotides bound within the crystals are not known, so the published structures cannot be related to the catalytic cycle.

B. Subunits Involved in Catalysis

Catalytic turnover involves a highly cooperative mechanism which involves inter-subunit interactions between the three α-subunits and the three β-subunits. α and β are somewhat homologous to each other (~25% identical) and each contains a nucleotide-binding site which was recognized by consensus nucleotide-binding motifs (Walker et al., 1982). Furthermore, each site can be labeled by nucleotide analogs (see Cross, 1988; Senior, 1988; Penefsky and Cross, 1991; Boyer, 1993, for reviews). The three β-subunit sites readily exchange nucleotides and are potentially active. In contrast, the α-subunit sites, which appear to be at an α–β interface, are not catalytically active and bound nucleotides are not exchangeable. There is considerable debate about the location of the noncatalytic sites as well as their role, if any, in catalysis. Readers are referred to extensive discussions on this topic in the reviews mentioned previously.

Reconstitution of ATPase activity from subunits of the *E. coli* F_1 requires the minimum complex of α_3–β_3– (Futai, 1977; Dunn and Futai, 1980). However, the γ-subunit may not be strictly required because α_3–β_3 complexes reconstituted from subunits of the thermophilic bacterium PS3 F_1 (Kagawa et al., 1989; Miwa and Yoshida, 1989) or extracted from spinach chloroplast F_0F_1 (Avital and Gromet-Elhanan, 1991) had hydrolytic activity. These results suggest that α_3–β_3 contains the necessary structures and interactions for turnover; however, the addition of γ greatly stabilized the complexes and enhanced activity. Furthermore, several mutations in the γ-subunit cause greatly reduced catalytic activity (Iwamoto et al., 1990).

C. Catalytic Mechanism

Catalysis can be considered in two modes called uni-site and multi-site (see Penefsky and Cross, 1991 for a review). When measuring the uni-site ATP hydrolytic activity, an ATP:enzyme ratio well below 1:3 is used so that only one of the β-subunit catalytic sites is occupied (Cross et al., 1982). This first site binds ATP with very high affinity ($K_a \approx 10^9\ M^{-1}$) and ATP is hydrolyzed at about 0.1 s^{-1}. Significantly, the equilibrium between hydrolysis and synthesis ($ATP \leftrightarrow ADP + P_i$, K_2) is near unity. The products, ADP and P_i, are released slowly to give an overall $k_{cat} \approx 10^{-3}\ s^{-1}$. The K_d for P_i is greater than 2M so there is effectively no binding at physiological concentrations, and ADP binds with an affinity much lower than ATP ($K_d \approx 10^{-5}$ M).

Interactions between catalytic sites comes into play when the ATP concentration is raised in excess of enzyme, and one or two of the remaining sites are occupied. In a cooperative fashion, the release of products is accelerated to an overall rate $\geq 50\ s^{-1}$. K_2 remains unchanged indicating that the hydrolysis or synthesis rates are not affected by the cooperative mechanism and this step does not represent a significant free energy change. Instead, ATP is tightly bound in a manner which stabilizes the transition state and allows rapid catalytic rates (Senior et al., 1993). Overall turnover is increased by ATP binding at the second site which causes a change in the conformation of the first site, thereby increasing the release rate of products (Al-Shawi et al., 1990b). The cooperative release of hydrolysis products is known as the "binding change mechanism" (Boyer, 1989, 1993).

ATP synthesis is essentially the reverse of the hydrolysis reactions. Energization of the membrane (proton motive force or $\Delta\mu_{H^+}$) greatly increases the P_i association rate constant on the order of 10–100 s^{-1} (seven to eight orders of magnitude increase). In addition, the release rate of ATP is enhanced which also prevents product inhibition. Again, the equilibrium constant, K_2, has been found not to change significantly (Gräber and Labahn, 1992; O'Neal and Boyer, 1984), however, Zhou and Boyer demonstrated that energization of thylakoid membranes enhances ATP formation of the chloroplast enzyme and have argued that favoring formation of ATP is an essential part of $\Delta\mu_{H^+}$ driven synthesis (Zhou and Boyer, 1993).

Evidence for cooperative interactions between catalytic sites during ATP synthesis was obtained by Matsuno-Yagi and Hatefi (1990). They reported three apparent K_m values for ADP during oxidative phosphorylation. The first two sites had K_m values below micromolar while the third had a K_m well above micromolar; occupancy of the third site was required for rapid synthesis which demonstrated positive cooperativity among the catalytic sites. Upon filling of the third site, a 50-fold increase in the V_{max} was observed suggesting that all three catalytic sites participate during maximal activity.

D. Structure of the Catalytic Site

Sequence similarities between the β-subunit and other nucleotide-binding proteins have stimulated extensive modeling of the active site based on the crystal structures of adenylate kinase (Sachsenheimer and Schultz, 1977), elongation factor Ef-Tu (Clark et al., 1990), and the *ras* p21 protein (Pai et al., 1989; Milburn et al., 1990). These models incorporate residues β-137 to β-335 (*E. coli* numbering) into the active site and form a nucleotide-binding fold with six-parallel β-strands (see Duncan and Cross, 1992; Futai et al., 1992; and the next section). Considerable evidence suggests that residues β-137 to β-251 are near the β–γ phosphate bond cleavage site. Within this region is the glycine-rich sequence or P-loop (Saraste et al., 1990) between β-149 and β-156 whose structure, function, and interactions will be examined below.

E. Residues in the Catalytic Site

Adenosine triphosphopyridoxal (AP_3-PL) covalently modifies β-subunit residues βLys-155 and βLys-201 in an ATP-protectable manner (Noumi et al., 1987b; Ida et al., 1991), suggesting that the two lysine residues are near the γ-phosphate of ATP in the catalytic site. Interestingly, in the absence of Mg^{2+}, Lys-201 of the α-subunit is also labeled along with βLys-155 (Tagaya et al., 1988).

Mutagenesis was used to further study the role of βLys-155 and αLys-201. Mutants of αLys-201 had lower multi-site catalysis, but similar uni-site catalysis compared with wild-type. These data indicate the residue does not participate in catalysis, but may be important for catalytic cooperativity (Ida et al., 1991).

Supporting this notion, when the α-subunit is isolated, αLys-201 was labeled with adenosine diphosphopyridoxal (AP_2-PL), suggesting that α-subunits contain a complete nucleotide-binding site and the site is noncatalytic (Rao et al., 1988).

In contrast, substitution of βLys-155 caused drastic changes in catalytic properties. Lys-155 is a conserved residue in the glycine-rich sequence of the β-subunit (Gly-Gly-Ala-Gly-Val-Gly-**Lys**-Thr-Ala, *E. coli* positions 149–157, conserved residues underlined). The consensus sequence, Gly–X–X–X–X–Gly–Lys–Thr/Ser, is conserved in many nucleotide-binding proteins and the predicted secondary structure of the β-subunit glycine-rich sequence is similar to that of p21 *ras* protein and adenylate kinase (Figure 1). Consistent with the AP_3-PL labeling, the p21 *ras* structure indicates that the lysine residue interacts with the β- and γ-phosphate groups of ATP (Pai et al., 1989; Milburn et al., 1990). Not surprisingly, catalytic properties of βLys-155 to Ala, Ser, or Thr (Omote et al., 1992), or Glu or Gln (Senior et al., 1993) mutant enzymes were markedly changed. Mutant enzymes had greatly reduced or no multi-site activity. Uni-site measurements showed extremely slow hydrolysis rates and reduced affinity for ATP demonstrating that βLys-155 plays a critical role in catalysis.

The next residue in the glycine-rich sequence, βThr-156, corresponds to residue Ser-17 in the glycine-rich sequence of the p21 *ras* protein (Gly-Ala-Gly-Gly-Val-Gly-Lys-**Ser**-Ala, residues 10–18, see Reddy et al., 1982). Enzymes in which βThr-156 was replaced by serine or the entire glycine-rich sequence of the β-subunit was replaced by that of the *ras* protein, retained activity (Takeyama et al., 1990). In contrast, βThr-156→Cys, Ala or Asp, or βAla-156/βThr-157 mutants had no membrane ATPase activity (Omote et al., 1992). Consistent with these results, a glycine residue could not be inserted between βLys-155 and βThr-156 as in the case of adenylate kinase (Gly-Gly-Pro-Gly-Ser-Gly-Lys-**Gly**-Thr, residues 15–23), nor could the glycine-rich sequence of the β-subunit be replaced by that of adenylate kinase. These results indicate that residue βThr-156, although replaceable by a serine residue, is essential for catalysis, probably because of its hydroxyl moiety.

F. Residues Interacting with the Glycine-Rich Sequence

The glycine-rich sequence plays an important role in catalysis and does not function as an isolated domain. Functionally important interactions were recognized by isolation of pseudo-revertants. The negative phenotype of the defective βSer-174→Phe mutant (Noumi et al., 1984b; Parsonage et al., 1987) was suppressed by a second mutation in the glycine-rich sequence, βGly-149→Ser (Miki et al., 1990; Iwamoto et al., 1991). Changes of βGly-149 to Ala and Cys also suppressed the effect of the βPhe-174 mutation (Iwamoto et al., 1991). The ATPase activity of the βAla-149/βPhe-174 enzyme was similar to wild-type, and that of the βCys-149/βPhe-174 enzyme was about three-fold higher than the enzyme with βPhe-174 alone. The single mutation, βCys-149, resulted in a defective enzyme indicating that in the double mutant, βCys-149 and βPhe-174 mutually suppressed

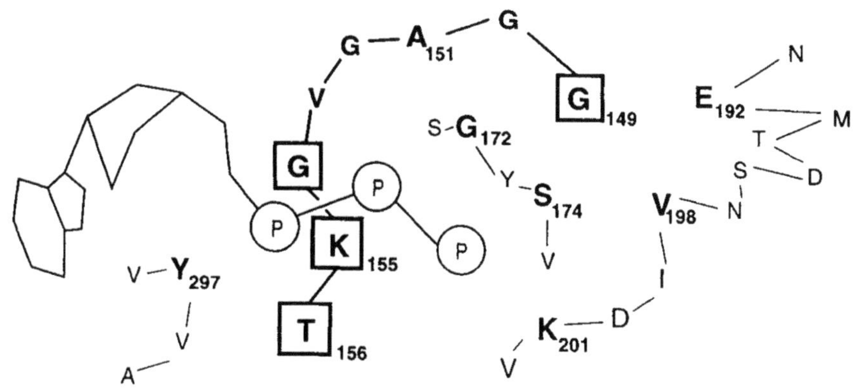

Figure 1. Model of the β-subunit in the region of the glycine-rich sequence (P-loop). The loop structure of the glycine-rich sequence of the β subunit is simulated from adenylate kinase (Sachsenheimer and Schultz, 1977) and p21 *ras* protein (Pai et al., 1989). Residues found to interact with the glycine-rich sequence by genetic studies are indicated (Iwamoto et al., 1993). βTyr-297 of mitochondrial F_1 (*E. coli* numbering) was suggested to be near βLys-155 by labeling studies (Andrews et al., 1984).

the effect of the other. The pseudo-revertant mutations at position β-149 appeared specific as shown by the inability of βGly-150→Ser or βGly-149→Thr changes to suppress the βPhe-174 mutation. Taken together, these results strongly suggest that residues βGly-149 and βSer-174 functionally interact with each other and both are located near the γ-phosphate moiety of ATP (see Figure 1).

In turn, pseudo-revertants of the deleterious βCys-149 mutation were found (Iwamoto et al., 1993). Four mutations conferred growth by oxidative phosphorylation: βGly-172→Glu, βSer-174→Phe, βGlu-192→Val, and βVal-198→Ala. These results suggested interactions between the glycine-rich sequence and βVal-198 which reinforced the finding that the AP_3-PL labeling site, βLys-201, is close to the γ-phosphate of ATP, and furthermore, brought βGlu-192 into the catalytic site which was already suggested by its reactivity with dicyclohexylcarbodiimide (DCCD, see Yoshida et al., 1982). The binding of DCCD to βGlu-192 in wild-type F_1 completely inhibited multi-site catalysis but only had a slight effect on uni-site catalysis (Tommasino and Capaldi, 1985), indicating that residue βGlu-192 is more involved in catalytic cooperativity.

G. Residues Involved in Cooperativity

We have now looked at residues which are located in the active sites. Inherent to the binding change mechanism model, these sites must communicate between each other to carry out the cooperative interactions required for multi-site catalysis. What parts of the complex are involved in the communication? Senior (1992) pointed out that most mutations in the β-subunit affect both multi- and uni-site catalysis, while mutations in the noncatalytic α-subunit affect only multi-site catalysis, suggesting that α-subunits are involved in the communication. For example, mutations between residues 347–375 of the α-subunit (*E. coli* numbering) caused extremely low multi-site activity, but retained normal uni-site hydrolysis (Wise et al., 1981, 1984; Kanazawa et al., 1984; Noumi et al., 1984a; Maggio et al., 1987; Soga et al., 1989). Furthermore, these mutant enzymes were no longer sensitive to sodium azide which inhibits by disrupting catalytic cooperativity (Noumi et al., 1987a). Kinetic and structural evidence also support the role of α-subunits in communication between the β-subunit sites. Catalytic activity of isolated β-subunits or α–β oligomers is quite different from uni-site activity of F_1 or α_3–β_3– indicating that characteristics of the mature active sites require interactions through α-subunits (Al-Shawi et al., 1990a). Furthermore, the crystallographic structure of Bianchet et al. (1991) indicates that the β-subunits interact intimately with α-subunits, but little or not at all with each other. Taken together, these results strongly suggest that the α-subunit is important for the cooperative interactions between active sites.

Not surprisingly, some residues of the β-subunit have been implicated in the cooperative mechanism as well. For example, we already mentioned βGlu-192 when modified by DCCD abolished multi-site activity while uni-site activity remained normal (Tommasino and Capaldi, 1985). Similarly, the substitution of βGlu-192 by Val had little effect on activity, but reduced sensitivity to sodium azide by a factor of 100 (Iwamoto et al., 1993). Furthermore, ATPase activity of the βGly-149→Ser mutant was insensitive to azide (Iwamoto et al., 1991). Taken together, the effects of many mutations near the catalytic site suggest that these residues are not only involved in catalysis, but also in the communication between sites.

III. THE F_0 SUBUNIT AND PROTON CONDUCTIVITY

A. F_0 Residues Involved in Proton Conductivity

Many experiments suggest that Asp-61 of subunit *c* (*E. coli* numbering), which is a completely conserved carboxylic acid residing in the middle of the membrane bilayer, participates in proton conduction. The requirement for Asp or Glu (most species have Glu at this position) seems to be very specific for the particular complex. For example, replacement of the *E. coli* *c*Asp-61 with Glu results in greatly reduced activity, which suggests strict chemical or structural requirements (Miller et al., 1990). In all wild-type F_0F_1, this residue is specifically modified by DCCD; the labeling results in blockage of proton conductivity, and inhibition of ATP-driven transport and ATP hydrolysis activity (reviewed in Sebald and Hoppe, 1981). Surprisingly, only one of the 10 ± 1 copies of subunit *c* need be modified to achieve maximal inhibition (Hermolin and Fillingame, 1989), which indicates a cooperative interaction between subunits (Hoppe and Sebald, 1984). Interestingly, when *c*Asp-61 is modified by mutagenesis, some amino acid replacements affect catalysis differently. Fillingame et al. (1984) reported that replacing the *E. coli* *c*Asp-61 with Asn not only blocked proton conduction, but also caused lower ATPase activity. The *c*Gly-61 mutant also blocked proton conductivity, but in contrast, had ATPase activity similar to wild-type.

In the hydrophilic loop of subunit *c*, the highly conserved residues *c*Arg-41, *c*Gln-42, and *c*Pro-43, which are known to be on the F_1 surface of the membrane (Girvin et al., 1989), are likely important for binding F_1 and coupling to catalysis. Substitutions in these three residues caused varying degrees of loss of ATP-driven proton pumping, loss of DCCD sensitivity of ATPase activity, and weaker binding of F_1 to F_0 (reviewed in Fillingame, 1992a).

In subunit *a*, identification of functionally important residues has relied completely upon mutagenesis studies. Strains with nonsense mutations causing truncated *a* subunits (*a*Trp-111→end, *a*Trp-231→end, or *a*Trp-252→end) had 50–70% of the normal membrane bound ATPase activity, but did not form active proton pathways (Eya et al., 1988). These results suggested that the carboxyl-terminus was important for proton conductivity. Indeed, characterizations of a large number of missense mutations bore out this conclusion. Three residues in the conserved carboxyl-terminal region were found to be particularly important. *a*Arg-210 appears to be the only essential residue of the subunit; proton conductivity is completely blocked by any substitution including the conservative replacement, Lys (Lightowelers et al., 1987; Cain and Simoni, 1989). Most replacements of *a*His-245 and *a*Glu-219 also block activity; however, some amino acid substitutions retained a trace of activity (Cain and Simoni, 1986, 1988; Lightowelers et al., 1988). Regardless of the degree of impaired proton conductivity, almost all subunit *a* mutations were capable of binding F_1 and retained ATPase activities sensitive to DCCD.

B. Structure of the F_0 Sector

As yet, little experimental evidence is available regarding the structure of the F_0 complex. The topography of subunits *b* and *c* appear to be relatively simple. *b* has a single hydrophobic segment near its amino-terminus that is long enough to span the membrane once. The hydrophilic carboxyl-terminus is oriented toward F_1 and is required for binding of F_1 to the membrane (Hoppe et al., 1983a, 1983b; Perlin et al., 1983; Schneider and Altendorf, 1985). As will be discussed in greater detail following, subunit *c* forms a hairpin loop with both termini on the surface of the membrane opposite F_1.

The topography of subunit *a* has been a matter of considerable debate. Various models based on hydropathy profiles (Senior, 1983; Walker et al., 1984; Cox et al., 1986; Hennig and Herrmann, 1986; Vik et al., 1988) and results of alkaline phosphatase gene fusion experiments (Bjørbæk et al., 1990; Lewis et al., 1990) propose anywhere from four to eight membrane spanning segments, and unfortunately, the predictions are most ambiguous towards the carboxyl-terminal end where the functionally important residues are located. Vik and Dao (1992) carried out an extensive analysis of possible topographies based on Fourier analysis of hydrophobicity, α-helical periodicity to search for possible amphipathic helices, and amino acid variation to identify helices that interact with membrane lipids on one face and with other helices on the opposite face. The authors settled upon a model with six-transmembrane crossings for subunit *a*. Two of the helices had variable faces which suggests contact between these two helices and the lipid, while the remaining four helices were predicted to entirely contact other helices. Fourier analysis of variation moments for subunits *b* and *c* found that each subunit had one variable face in contact with lipid. Based on these predictions and previous results from mutational studies, an arrangement of subunit *a* helices was proposed. Most of the conserved faces of subunit *a* helices were positioned to interact with each other. In addition, a single helix of subunit *a* contacts a cluster of 10 ± 1 *c* subunits and two other helices of *a* contact the two conserved helices from the *b* subunits.

Unlike the other F_0 subunits, there is direct structural information for the proteolipid subunit *c*. Girven and Fillingame (1993) examined the structure of the *E. coli* subunit *c* using multidimensional NMR. The purified subunit was labeled with a nitroxide derivative of DCCD (NCCD; N-[2,2,6,6-tetramethylpiperidyl-1-oxyl]-N′-[cyclohexyl]-carbodiimide) and studied in chloroform–methanol–water where it retained at least some characteristics of the native protein (i.e., DCCD reactivity and hairpin topology). A structural model was presented based on interactions between amino- and carboxyl-terminal residues and on resonance broadening of residues due to the nitroxide label within what would be the bilayer. A hairpin structure with two transmembrane helices was proposed with *c*Asp-61 on helix-2 placed within the bilayer between the side-chains of residues *c*Ala-24 and *c*Ile-28 on helix-1. The model was consistent with the effect of mutations at these positions, which greatly reduced sensitivity of Asp-61 to DCCD. This model also explained the functionality of a double mutant, *c*Asp-24-*c*Gly-61, which moves aspartic acid to helix-1. These results suggested that the two helices of subunit *c* interact to create an environment around *c*Asp-61 that makes the acidic group highly reactive to DCCD and most likely optimized to carry out the process of proton transport (Fillingame, 1990).

Genetic evidence suggests there are further helical–helical interactions between F_0 subunits that are important for proton conduction. Even though the enzyme with the double mutation, *c*Asp-24-*c*Gly-61, retains function, its activity is low enough to slow oxidative phosphorylation-dependent growth of the mutant strain. Fraga and Fillingame isolated several second-site mutations which suppressed the growth defect (described in Fillingame, 1992b). Thirteen of the mutations were found in subunit *a*, and 10 mapped to three positions, *a*Ala-217, *a*Ile-221, and *a*Leu-224. These residues are near the essential residue, *a*Arg-210, suggesting that *c*Asp-61 and *a*Arg-210, the only two residues clearly essential for proton conductivity, are located close to each other.

Similarly, the effects of a deleterious mutation in subunit *b*, Gly-9→Asp (Porter et al., 1985) in the middle of the single membrane spanning segment, were suppressed by mutations in subunit *a* and *c* (Kumamoto and Simoni, 1986, 1987). The subunit *a* mutations at position 240 (Pro→Ala or Leu) or subunit *c* mutation at position 62 (Ala→Ser) were again very close to the essential residues, *c*Asp-61 and *a*Arg-210. These results suggest that the original mutation, *b*Asp-9, may have affected the environment near *c*Asp-61 or *a*Arg-210, and implies that the hydrophobic segment of subunit *b* interacts with functionally critical helices of subunits *a* and *c*.

IV. ENERGY COUPLING—LINKING CATALYSIS TO TRANSPORT

A. Direct Versus Indirect Coupling

Two hypotheses for the mechanism of coupling the proton motive force to ATP synthesis have been discussed: the first, "direct coupling" as elaborated by Mitchell (Mitchell, 1974), utilizes protons energized by the $\Delta\mu_{H^+}$ directly in the formation of the phosphate-anhydride bond of ATP. The second, known as "indirect coupling" (Boyer, 1975), states that catalysis and transport take place in two separated domains of the enzyme complex and are linked via long-range conformational changes. Several lines of evidence argue against the former hypothesis. Direct coupling predicts that the equilibrium constant of the synthesis/hydrolysis reaction, $ADP + P_i + H^+ \leftrightarrow ATP + H_2O$ (K_2, see preceding) will be sensitive to $\Delta\mu_{H^+}$. Measurements done under conditions of energized membranes have shown that K_2 remains relatively constant regardless of the $\Delta\mu_{H^+}$ imposed across the membrane (O'Neal and Boyer, 1984; Gräber and Labahn, 1992). Furthermore, the F_oF_1 variant of *Propionigenium modestum* pumps Na^+ as well as H^+ (Laubinger and Dimroth, 1987, 1988; Laubinger et al., 1990), which eliminates the possibility that the transported or "vectorial" protons directly participate in the phosphate bond chemistry.

B. Transport Occurs in the F_o

Indirect coupling predicts that uphill transport can only be accomplished by carriers (transporters), not channels; therefore, the transport site must cycle through different states which switch accessibility from one side of the membrane to the other (Jencks, 1980; Tanford, 1983). An important question is where and how transport occurs. We have already discussed F_o residues which are involved in proton conductivity, but we do not know if these residues are involved in coupled transport, or if they passively guide protons to the transport machinery. Dimroth and co-workers took advantage of the sodium/proton transport properties of the *Propionigenium modestum* ATPase to demonstrate that the F_o sector alone has the properties of a carrier. Although the *P. modestum* enzyme has the characteristics of an F_0F_1 ATPase as denoted by biochemical and sequence similarities (Esser et al., 1990; Kaim et al., 1990; Ludwig et al., 1990; Krumholz et al., 1992), it normally transports sodium ions. However, in low sodium conditions, it will generate an ATP-dependent proton gradient (Laubinger and Dimroth, 1988, 1989). Using the *P. modestum* F_0 sector reconstituted into liposomes with the F_1 portion removed, Kluge and Dimroth showed that proton transport was inhibited by sodium ions from one side of the membrane only (Kluge and Dimroth, 1992). This behavior was indicative of a carrier and indirectly suggested that protons (or H_3O^+) and Na^+ are transported via the same pathway. Furthermore, sodium transport activity was reconstituted utilizing *E. coli* F_1 (Laubinger et al., 1990); therefore, ion translocation occurs within the F_0 sector, and F_1 is independent of the species of ion transported. The carrier behavior implies a conformational change occurs within the F_0 which must be coupled to the catalytic sites.

C. Transport is Linked to Catalysis Through Long-range Interactions

As we have seen, transport and catalytic functions are separated on two parts of the complex: transport in the F_0 sector and catalysis in the β-subunits of the F_1 sector. The indirect coupling hypothesis suggests that conformational changes, which occur as a part of the transport process of translocating protons from one side of the membrane to the other, cause corresponding conformational changes in the catalytic sites. The changes must be tightly linked to each other even though these sites are separated by 50–100 Å. Experimental evidence demonstrates that the conformation of F_0 subunits is linked to the catalytic sites. We already mentioned that mutations in subunit *c* disrupted not only passive proton conductivity, but in many cases lowered ATPase activity as well. Penefsky (1985) and Matsuno-Yagi et al. (1985) found that inhibitors bound specifically to the F_0 sector altered catalytic properties, and affected the fluorescence response of aurovertin which binds to β-subunits. In further investigations, Matsuno-Yagi and Hatefi characterized the effects of F_0-specific inhibitors on ATP hydrolysis and synthesis (Matsuno-Yagi and Hatefi, 1993a, 1993b). They found that oligomycin or DCCD completely blocked proton translocation and inhibited both multi- and uni-site ATP hydrolysis, whereas other compounds (i.e., venturicidin or tetracoordinate organotin compounds) attenuated rapid proton flux and inhibited only multi-site catalysis.

D. Subunits Involved in Coupling

Subunits directly involved in catalysis or transport must also be involved in coupling, and not surprisingly, a few mutations in subunits *c* and β affect coupling. We already discussed mutations in the polar loop of subunit *c* in this regard (see Section IIIA and Fillingame, 1992a). In the case of the β-subunit, Omote et al. (1994) recently showed that an interaction between the catalytic site and the α-subunit is important for coupling. They introduced a series of amino acid substitutions at position β-174 (βSer-174→Ala, Gly, Leu, Phe, and Thr) which is near the ATP γ-phosphate (see section on residues interacting with the glycine-rich sequence). Both βPhe-174 and βLeu-174 mutant enzymes were found defective in multi-site catalysis and had the same membrane ATP activities, but only the βPhe-174 enzyme was defective in coupling to transport. Significantly, the defective coupling of the βPhe-174 mutant enzyme was suppressed by a second-site mutation in the α-subunit, αArg-296→Cys. These results suggest the importance of interactions between a region near βSer-174 of the active site and αArg-296 in both catalytic cooperativity and coupling.

Consistent with, but not proving, their possible role in coupling, several observations suggest that the small F_1 subunits are conformationally sensitive to activity in both the catalytic and transport sites. The laboratories of Bragg and Capaldi have looked at conformational changes of γ and ε in response to ligand binding in the active sites (reviewed in Capaldi et al., 1992). Gogol et al. (1990) analyzed cryoelectron micrographs of immuno-decorated F_1 and observed γ- and ε-subunits in different positions when the enzyme was incubated with ATP (or ADP+P_i)+Mg^{2+} versus ATP + EDTA. Similarly, the same incubation conditions caused altered trypsin sensitivity of γ and ε, and altered cross-linking patterns between β and γ, α and ε, and γ and ε (Bragg and Hou, 1986, 1987; Gogol et al., 1990; Mendel-Hartvig and Capaldi, 1991a; Aggeler et al., 1992, 1993; Aggeler and Capaldi, 1993).

The ε-subunit has a strong effect on catalysis. The ε-subunit is a partial noncompetitive inhibitor of ATPase activity in isolated F_1 (Sternweis and Smith, 1980) and is believed to block product release (Dunn et al., 1987). Interestingly, inhibition is relieved when F_1 is bound to F_0 (Sternweis and Smith, 1980). The inhibition has been suggested to be a regulatory mechanism to prevent free F_1 from carrying out uncoupled ATP hydrolysis (Klionsky et al., 1984). Alternatively, the inhibition may indicate that ε plays a key role in coupling; the subunit may be involved in transmission of conformational information between the catalytic and transport sites. In the absence of F_0, the transmission is blocked which causes inhibition of the enzyme, while in the presence of F_0, the transmission is allowed to pass.

In fact, the conformations of ε as well as the γ-subunit are sensitive to the state of the F_0 sector. McCarty and co-workers demonstrated that protease sensitivity (Moroney and McCarty, 1982), reactivity with modifying reagents, and cross-linking within the chloroplast γ-subunit (Nalin and McCarty, 1984) were sensitive to energization of the thylakoid membrane by light. Likewise, ε became much more reactive to a polyclonal antibody upon exposure to light (Richter and McCarty, 1987). In the *E. coli* enzyme, Mendel-Hartvig and Capaldi (1991b) observed that the trypsin sensitivity of the *E. coli* ε-subunit was altered by DCCD labeling of subunit *c*.

In summary, there is abundant evidence that the catalytic and transport sites appear to be linked through conformational effects which are mediated through at least the γ- and ε-subunits. In the next section, we will analyze the involvement of specific residues of the γ-subunit in coupling.

E. Amino Acids of the γ-Subunit Involved in Coupling

The γ-subunit may be described as the core of the F_0F_1 ATPase. It has a role in catalysis because reconstitution of the minimum complex capable of ATPase activity (at least in *E. coli*) requires γ in addition to α and β (Futai, 1977; Dunn and Futai, 1980). Furthermore, mutations near the conserved carboxyl-terminus cause greatly reduced catalytic activity (Iwamoto et al., 1990). It also seems to have a role in transport because it has been reported to regulate proton flow through the chloroplast and mitochondrial F_0 (Schumann et al., 1985; Papa et al., 1990). These observations together with the demonstration of its conformational sensitivity to ligand binding or $\Delta\mu_{H^+}$ as described above, suggest that the γ-subunit plays an important role in linking catalysis to transport.

Experiments from the Capaldi laboratory demonstrated that the γ-subunit is linked to the catalytic sites. Aggeler and Capaldi (1992) introduced a series of Cys replacements throughout the *E. coli* γ-subunit. Using a novel bifunctional reagent, N-maleimido-N′-(4-azido-2,3,5,6-tetrafluorobenzamido)cystamine, they found that one of the Cys mutations, γCys-8, formed different cross-link products (Aggeler and Capaldi, 1993; Aggeler et al., 1993). If the γCys-8 mutant F_1 was reacted in the presence of ATP + Mg^{2+}, cross-linked products had an M_r of 102,000 and 84,000, whereas, if done in the presence of ADP + Mg^{2+}, an M_r = 108,000 product was observed. Both cross-linked enzymes had inhibited ATPase activity. When the cross-linker was reduced to break its disulfide bond, the 102,000 and 84,000 products regained activity, while the 108,000 product remained inhibited. Subsequently, the sites of cross-linking in the 108,000 product were identified as γCys-8 and a β-subunit sequence between βVal-145–βLys-155, which contains the glycine-rich sequence. This was compelling evidence that the γ-subunit does, in fact, have a role in the catalytic mechanism and undergoes a conformational change as a part of that mechanism.

Recent mutagenesis studies from the Futai laboratory linked residues of the γ-subunit to coupling. Shin et al. (1992) reported that replacement of the conserved γMet-23→Arg or Lys caused a slippage in the coupling between catalysis and transport. The mutant enzymes were extremely inefficient in both ATP-dependent proton transport as well as $\Delta\mu_{H^+}$-driven ATP synthesis. Interestingly, the effects of the γLys-23 mutation could be suppressed by second-site mutations in the carboxyl-terminal region of the γ-subunit. Nakamoto et al. (1993) identified seven mutations in the highly conserved region between γGln-269 and γVal-280; five mutations replaced residues identical in all the known γ-subunit sequences (Figure 2). The eighth, γArg-242→Cys, also changed a conserved residue. The effect of each second-site suppressor mutation was to restore efficient coupling to the γLys-23 mutant. These data strongly suggest that the two conserved portions of the γ-subunit, the carboxyl-terminal region, and the region near γMet-23, interact to mediate coupling between catalysis and transport.

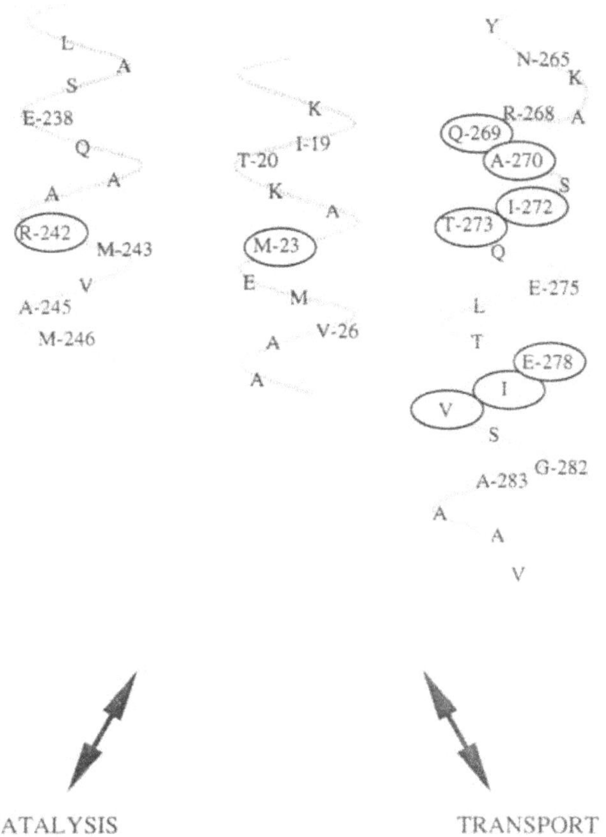

Figure 2. Interacting regions of the γ-subunit. Three sections of the *E. coli* γ-subunit are shown arranged in the predicted α-helices: residues Lys-18 to Ala-28 (center), Leu-265 to Met-246 (left), and Tyr-264 to Val-286 (right). The original deleterious mutation, γMet-23→Lys, was suppressed by each of the circled residues. Residues indicated by their position numbers are found conserved in all known γ-subunit sequences (see Nakamoto et al., 1993).

The γ-subunit interactions are reminiscent of subunit interactions at the hemoglobin α_1–β_2 interface which is critical for cooperativity between oxygen binding sites (Dickerson and Geis, 1983). This well-studied hemoglobin interface does not tolerate amino acid substitutions and even conservative replacements perturb the cooperative mechanism; it is not surprising that protein sequences in these regions are highly conserved. Similarly, the sequences in the termini of the F_1 γ-subunit are also highly conserved. Our next task is to identify coupling interactions of the γ-subunit with other F_oF_1 subunits that link catalysis to transport.

V. CONCLUSION

The experiments described above make clear that interactions between subunits are an essential part of catalytic cooperativity, ion translocation, and coupling between catalysis and transport. Future work will concentrate on revealing the structure of the complex at atomic resolution, and at the same time, mutagenesis and protein labeling experiments will identify functionally important amino acids. These two approaches will be used in concert to work towards an understanding of how the F_0F_1 carries out coupled transport.

ACKNOWLEDGMENTS

We would like to thank Dr. David Stokes of the University of Virginia for many useful discussions. This work was supported in part by research grants from the Ministry of Education, Science and Culture of Japan, and the Human Frontier Science Program. RKN was supported by a research fellowship from the Japan Society for the Promotion of Science.

REFERENCES

Abrahams, J. P., Lutter, R., Todd, R. J., van Raaij, M. J., Leslie, A. G. W., & Walker, J. E. (1993). Inherent asymmetry of the structure of F_1-ATPase from bovine heart mitochondria at 6.5 Å resolution. EMBO J. 12, 1775–1780.

Aggeler, R., & Capaldi, R. A. (1992). Cross-linking of the γ subunit of the *Escherichia coli* ATPase (ECF_1) via cysteines introduced by site-directed mutagenesis. J. Biol. Chem. 267, 21355–21359.

Aggeler, R., & Capaldi, R. A. (1993). ATP hydrolysis-linked structural changes in the N-terminal part of the γ subunit of *Escherichia coli* F_1-ATPase examined by cross-linking studies. J. Biol. Chem. 268, 14576–14579.

Aggeler, R., Chicas-Cruz, K., Cai, S.-X., Kenna, J. F. W., & Capaldi, R. A. (1992). Introduction of reactive cysteine residues in the ε subunit of *Escherichia coli* F_1 ATPase, modification of these sites with tetrafluorophenyl azido-maleimides, and examination of changes in the binding of the ε subunit when different nucleotides are in catalytic sites. Biochemistry 31, 2956–2961.

Aggeler, R., Cai, S. X., Keana, J. F. W., Koike, T., & Capaldi, R. A. (1993). The γ subunit of the *Escherichia coli* F_1-ATPase can be cross-linked near the glycine-rich loop region of a β subunit when ADP + Mg^{2+} occupies catalytic sites but not when ATP + Mg^{2+} is bound. J. Biol. Chem. 268, 20831–20837.

Akey, C. W., Crepeau, R. H., Dunn, S. D., McCarty, R. E., & Edelstein, S. J. (1983). Electron microscopy and single molecule averaging of subunit-deficient F_1-ATPases from *Escherichia coli* and spinach chloroplasts. EMBO J. 2, 1409–1415.

Al-Shawi, M. K., Parsonage, D., & Senior, A. E. (1990a). Adenosine triphosphatase and nucleotide binding activity of isolated β-subunit preparations from *Escherichia coli* F_1F_0-ATP synthase. J. Biol. Chem. 265, 5595–5601.

Al-Shawi, M. K., Parsonage, D., & Senior, A. E. (1990b). Thermodynamic analyses of the catalytic pathway of F_1-ATPase from *Escherichia coli*. J. Biol. Chem. 265, 4402–4410.

Amzel, L. M., & Pedersen, P. L. (1978). Adenosine triphosphatase from rat liver mitochondria: Crystallization and X-ray diffraction studies of the F_1-component of the enzyme. J. Biol. Chem. 253, 2067–2069.

Andrews, W. W., Hill, F. C., & Allison, W. S. (1984). Identification of the lysine residue to which the 4-nitrobenzofurazan group migrates after the bovine mitochondrial F_1-ATPase is inactivated with 7-chloro-4-nitro[^{14}C]benzofurazan. J. Biol. Chem. 259, 14378–14382.

Avital, S., & Gromet-Elhanan, Z. (1991). Extraction and purification of the β subunit and an active αβ-core complex from the spinach chloroplast CF_0F_1-ATPase synthase. J. Biol. Chem. 266, 7067–7072.

Bianchet, M., Ysern, X., Hullihen, J., Pedersen, P. L., & Amzel, L. M. (1991). Mitochondrial ATP synthase: quaternary structure of the F_1 moiety at 3.6 Å determined by X-ray diffraction analysis. J. Biol. Chem. 266, 21197–21201.

Bjørbæk, C., Försom, V., & Michelsen, O. (1990). The transmembrane topology of the *a* subunit from the ATPase in *Escherichia coli* analyzed by *PhoA* protein fusions. FEBS Lett. 260, 31–34.

Boekema, E. J., van Heel, M. G., & Gräber, P. (1988). Structure of the ATP synthase from chloroplasts studied by electron microscopy and image processing. Biochim. Biophys. Acta 933, 365–371.

Boyer, P. D. (1975). A model for conformational coupling of membrane potential and proton translocation to ATP synthesis and to active transport. FEBS Lett. 58, 1–6.

Boyer, P. D. (1989). A perspective of the binding change mechanism for ATP synthesis. FASEB J. 3, 2164–2178.

Boyer, P. D. (1993). The binding change mechanism for ATP synthase—Some probabilities and possibilities. Biochim. Biophys. Acta 1140, 215–250.

Bragg, P. D., & Hou, C. (1986). Effect of disulfide cross-linking between α and δ subunits on the properties of the F_1 adenosine triphosphatase of *Escherichia coli*. Biochim. Biophys. Acta 851, 385–394.

Bragg, P. D., & Hou, C. (1987). Ligand-induced conformational changes in the *Escherichia coli* F_1 adenosine triphosphatase probed by trypsin digestion. Biochim. Biophys. Acta 894, 127–137.

Cain, B. D., & Simoni, R. D. (1986). Impaired proton conductivity resulting from mutations in the *a* subunit of F_1F_0 ATPase in *Escherichia coli*. J. Biol. Chem. 261, 10043–10050.

Cain, B. D., & Simoni, R. D. (1988). Interaction between Glu-219 and His-245 within the *a* subunit of F_1F_0 ATPase in *Escherichia coli*. J. Biol. Chem. 263, 6606–6612.

Cain, B. D., & Simoni, R. D. (1989). Proton translocation by the F_1F_0 ATPase of *Escherichia coli*: Mutagenic analysis of the *a* subunit. J. Biol. Chem. 264, 3292–3300.

Capaldi, R. A., Aggeler, R., Gogol, E. P., & Wilkens, S. (1992). Structure of the *Escherichia coli* ATP synthase and role of the γ and ε subunits in coupling catalytic site and proton channeling functions. J. Bioenerg. Biomemb. 24, 435–439.

Clark, B. F. C., Kjeldgaard, M., la Cour, T. F. M., Thirup, S., & Nyborg, J. (1990). Structural determination of the functional sites of *E. coli* elongation factor Tu. Biochim. Biophys. Acta 1050, 203–208.

Cox, G. B., Fimmel, A. L., Gibson, F., & Hatch, L. (1986). The mechanism of ATP synthase: A reassessment of the functions of the *b* and *a* subunits. Biochim. Biophys. Acta 849, 62–69.

Cross, R. L. (1988). The number of functional catalytic sites on F_1-ATPase and the effects of quaternary structural asymmetry on their properties. J. Bioenerg. Biomemb. 20, 395–405.

Cross, R. L., Grubmeyer, C., & Penefsky, H. S. (1982). Mechanism of ATP hydrolysis by beef heart mitochondrial ATPase: Rate enhancements resulting from cooperative interactions between multiple catalytic sites. J. Biol. Chem. 257, 12101–12105.

Dickerson, R. E., & Geis, I. (1983). Hemoglobin: Structure, Function, Evolution and Pathology. Benjamin/Cummings, Menlo Park, NJ.

Duncan, T. M., & Cross, R. L. (1992). A model for the catalytic site of F_1-ATPase based on analogies to nucleotide binding domains of known structure. J. Bioenerg. Biomemb. 24, 453–461.

Dunn, S. D., & Futai, M. (1980). Reconstitution of a functional coupling factor from the isolated subunits of *Escherichia coli* F_1 ATPase. J. Biol. Chem. 255, 113–118.

Dunn, S. D., Zadorozny, V. D., Tozer, R. G., & Orr, L. E. (1987). ε subunit of *Escherichia coli* F_1-ATPase: Effects on affinity for aurovertin and inhibition of product release in unisite ATP hydrolysis. Biochemistry 26, 4488–4493.

Esser, W., Krumholz, L. R., & Simoni, R. D. (1990). Nucleotide sequence of the F_0 subunits of the sodium dependent F_1F_0 ATPase of *Propionigenium modestum*. Nucleic Acids Res. 18, 5887.

Eya, S., Noumi, T., Maeda, M., & Futai, M. (1988). Intrinsic membrane sector (F_0) H^+-ATPase (F_0F_1) from *Escherichia coli*: Mutations in the *a* subunit give F_0 with impaired proton translocation and F_1 binding. J. Biol. Chem. 263, 10056–10062.

Fillingame, R. H. (1990). Molecular mechanics of ATP synthesis by F_1F_0-type H^+-transporting ATP synthases. In: The Bacteria (Krulwich, T. A., ed.), pp. 345–391. Academic Press, New York.

Fillingame, R. H. (1992a). H^+ transport and coupling by the F_0 sector of the ATP synthase: Insights into the molecular mechanism of function. J. Bioenerg. Biomemb. 24, 485–491.

Fillingame, R. H. (1992b). Subunit *c* of F_1F_0 ATP synthase: Structure and role in transmembrane energy transduction. Biochim. Biophys. Acta 1101, 240–243.

Fillingame, R. H., Peters, L. K., White, L. K., Mosher, M. E., & Paule, C. R. (1984). Mutations altering aspartyl-61 of the omega subunit (*unc*E protein) of *Escherichia coli* H^+-ATPase differ in effect on coupled ATP hydrolysis. J. Bacteriol. 158, 1078–1083.

Foster, D. L., & Fillingame, R. H. (1982). Stoichiometry of subunits in the H^+-ATPase complex of *Escherichia coli*. J. Biol. Chem. 257, 2009–2015.

Futai, M. (1977). Reconstitution of the coupling factor, F_1, of *Escherichia coli*. Biochem. Biophys. Res. Commun. 79, 1231–1237.

Futai, M., & Kanazawa, H. (1983). Structure and function of proton-translocating adenosine triphosphatase (F_1F_0): Biochemical and molecular biological approaches. Microbiol. Rev. 47, 285–312.

Futai, M., Sternweis, P. C., & Heppel, L. A. (1974). Purification and properties of reconstitutively active and inactive adenosinetriphosphatase from *Escherichia coli*. Proc. Natl. Acad. Sci. USA 71, 2725–2729.

Futai, M., Noumi, T., & Maeda, M. (1989). ATP synthase (H^+-ATPase): Results by combined biochemical and molecular biological approaches. Annu. Rev. Biochem. 58, 111–136.

Futai, M., Iwamoto, A., Omote, H., & Maeda, M. (1992). A glycine-rich sequence in the catalytic site of F-type ATPase. J. Bioenerg. Biomemb. 24, 463–467.

Girvin, M. E., & Fillingame, R. H. (1993). Helical structure and folding of subunit *c* of F_1F_0 ATP synthase: ^1H NMR resonance assignments and NOE analysis. Biochemistry 32, 12167–12177.

Girvin, M. E., Hermolin, J., Pottorf, R., & Fillingame, R. H. (1989). Organization of the F_0 sector of *Escherichia coli* H^+-ATPase: The polar loop region of subunit *c* extends from the cytoplasmic face of the membrane. Biochemistry 28, 4340–4343.

Gogol, E. P., Lücken, U., & Capaldi, R. A. (1987). The stalk connecting the F_1 and F_0 domains of ATP synthase visualized by electron microscopy of unstained specimens. FEBS Lett. 219, 274–278.

Gogol, E. P., Aggeler, R., Sagermann, M., & Capaldi, R. A. (1989a). Cryoelectron microscopy of *Escherichia coli* F_1 adenosinetriphosphatase decorated with monoclonal antibodies to individual subunits of the complex. Biochemistry 28, 4717–4724.

Gogol, E. P., Lücken, U., Bork, T., & Capaldi, R. A. (1989b). Molecular architecture of *Escherichia coli* F_1 adenosinetriphosphatase. Biochemistry 28, 4709–4716.

Gogol, E. P., Johnston, E., Aggeler, R., & Capaldi, R. A. (1990). Ligand-dependent structural variations in *Escherichia coli* F_1 ATPase revealed by cryoelectron microscopy. Proc. Natl. Acad. Sci. USA 87, 9585–9589.

Gräber, P., & Labahn, A. (1992). Proton transport-coupled unisite catalysis by the H^+-ATPase from chloroplasts. J. Bioenerg. Biomemb. 24, 493–497.

Hennig, J., & Herrmann, R. G. (1986). Chloroplast ATP synthase of spinach contains nine nonidentical subunit species, six of which are encoded by plastid chromosomes in two operons in a phylogenetically conserved arrangement. Mol. Gen. Genet. 203, 117–128.

Hermolin, J., & Fillingame, R. H. (1989). H^+-ATPase activity of *Escherichia coli* F_1F_0 is blocked after reaction of dicyclohexylcarbodiimide with a single proteolipid (subunit *c*) of the F_0 complex. J. Biol. Chem. 264, 3896–3903.

Hoppe, J., & Sebald, W. (1984). The proton conducting F_0-part of bacterial ATP synthases. Biochim. Biophys. Acta 768, 1–27.

Hoppe, J., Friedl, P., Schairer, H. U., Sebald, W., von Meyenburg, K., & Jørgensen, B. B. (1983a). The topology of the proton translocating F_0 component of the ATP synthase from *E. coli* K12; studies with proteases. EMBO J. 2, 105–110.

Hoppe, J., Montecucco, C., & Friedl, P. (1983b). Labeling of subunit *b* of the ATP synthase from *Escherichia coli* with a photoreactive phospholipid analogue. J. Biol. Chem. 258, 2882–2885.

Ida, K., Noumi, T., Maeda, M., Fukui, T., & Futai, M. (1991). Catalytic site of F_1-ATPase of *Escherichia coli*: Lys-155 and Lys-201 of the β subunit are located near the γ-phosphate group of ATP in the presence of Mg^{2+}. J. Biol. Chem. 266, 5424–5429.

Iwamoto, A., Miki, J., Maeda, M., & Futai, M. (1990). H^+-ATPase γ subunit from *Escherichia coli*: Role of the conserved carboxylic terminal region. J. Biol. Chem. 265, 5043–5048.

Iwamoto, A., Omote, H., Hanada, H., Tomioka, N., Itai, A., Maeda, M., & Futai, M. (1991). Mutations in Ser^{174} and the glycine-rich sequence (Gly^{149}, Gly^{150}, and Thr^{156}) in the β subunit of *Escherichia coli* H^+-ATPase. J. Biol. Chem. 266, 16350–16355.

Iwamoto, A., Park, M.-Y., Maeda, M., & Futai, M. (1993). Domains near ATP γ-phosphate in the catalytic site of H^+-ATPase: Model proposed from mutagenesis and inhibitor studies. J. Biol. Chem. 268, 3156–3160.

Jencks, W. P. (1980). The utilization of binding energy in coupled vectorial processes. Adv. Enzymol. 51, 75–106.

Kagawa, Y., Ohta, S., & Otawara-Hamamoto, Y. (1989). $\alpha_3\beta_3$ complex of thermophilic ATP synthase: Catalysis without the γ-subunit. FEBS Lett. 249, 67–69.

Kaim, G., Ludwig, W., Dimroth, P., & Schleifer, K. H. (1990). Sequence of subunits *a* and *b* of the sodium ion translocating adenosine triphosphate synthase of *Propionigenium modestum*. Nucleic Acids Res. 18, 6697.

Kanazawa, H., Noumi, T., Matsuoka, I., Hirata, T., & Futai, M. (1984). F_1-ATPase of *Escherichia coli*: A mutation (*uncA*401) located in the middle of the α subunit affects the conformation essential for F_1 activity. Arch. Biochem. Biophys. 288, 258–269.

Klionsky, D. J., Brusilow, W. S. A., & Simoni, R. D. (1984). *In vivo* evidence for the role of the ε subunit as an inhibitor of the proton-translocating ATPase of *Escherichia coli*. J. Bacteriol. 160, 1055–1060.

Kluge, C., & Dimroth, P. (1992). Studies on Na^+ and H^+ translocation through the F_0 part of the Na^+-translocating F_1F_0 ATPase from *Propionigenium modestum*: Discovery of a membrane potential dependent step. Biochemistry 31, 12665–12672.

Krumholz, L. R., Esser, U., & Simoni, R. D. (1992). Characterization of the genes coding for the F_1F_0 subunits of the sodium dependent ATPase of *Propionigenium modestum*. FEMS Microbiol. Lett. 70, 37–41.

Kumamoto, C. A., & Simoni, R. D. (1986). Genetic evidence for interaction between the *a* and *b* subunits of the F_0 portion of the *Escherichia coli* proton translocating ATPase. J. Biol. Chem. 261, 10037–10042.

Kumamoto, C. A., & Simoni, R. D. (1987). A mutation of the *c* subunit of the *Escherichia coli* proton-translocating ATPase that suppresses the effects of a mutant *b* subunit. J. Biol. Chem. 262, 3060–3064.

Laubinger, W., & Dimroth, P. (1987). Characterization of the Na^+-stimulated ATPase of *Propionigenium modestum* as an enzyme of the F_1F_0 type. Eur. J. Biochem. 168, 475–480.

Laubinger, W., & Dimroth, P. (1988). Characterization of the ATP synthase *Propionigenium modestum* as a primary sodium pump. Biochemistry 27, 7531–7537.

Laubinger, W., & Dimroth, P. (1989). The sodium ion translocating adenosine triphosphatase of *Propionigenium modestum* pumps protons at low sodium ion concentrations. Biochemistry 28, 7194–7198.

Laubinger, W., Deckers-Hebestreit, G., Altendorf, K., & Dimroth, P. (1990). A hybrid adenosinetriphosphatase composed of F_1 of *Escherichia coli* and F_0 of *Propionigenium modestum* is a functional sodium ion pump. Biochemistry 29, 5458–5463.

Lewis, M. L., Chang, J. A., & Simoni, R. D. (1990). A topological analysis of subunit *a* from *Escherichia coli* F_1F_0-ATP synthase predicts eight transmembrane segments. J. Biol. Chem. 265, 10541–10550.

Lightowelers, R. N., Howitt, S. M., Hatch, L., Gibson, F., & Cox, G. B. (1987). The proton pore in the *Escherichia coli* F_0F_1-ATPase: A requirement for arginine at position 210 of the *a* subunit. Biochim. Biophys. Acta 894, 399–406.

Lightowelers, R. N., Howitt, S. M., Hatch, L., Gibson, F., & Cox, G. B. (1988). The proton pore in the *Escherichia coli* F_0F_1-ATPase: Substitution of glutamate by glutamine at position 219 of the *a*-subunit prevents F_0-mediated proton permeability. Biochim. Biophys. Acta 933, 241–248.

Ludwig, W., Kaim, G., Laubinger, W., Dimroth, P., Hoppe, J., & Schleifer, K. H. (1990). Sequence of subunit *c* of the sodium ion translocating adenosine triphosphate synthase of *Propionigenium modestum*. Eur. J. Biochem. 193, 395–399.

Lündsdorf, H., Ehrig, K., Friedl, P., & Schairer, H. U. (1984). Use of monoclonal antibodies in immuno-electron microscopy for the determination of subunit stoichiometry in oligomeric enzymes: there are three α-subunits in the F_1-ATPase of *Escherichia coli*. J. Mol. Biol. 173, 131–136.

Lutter, R., Abrahams, J. P., van Raaij, M. J., Todd, R. J., Lundqvist, T., Buchanan, S. K., Leslie, A., & Walker, J. E. (1993). Crystallization of F_1-ATPase from bovine heart mitochondria. J. Mol. Biol. 229, 787–790.

Maggio, M. B., Pagan, J., Parsonage, D., Hatch, L., & Senior, A. E. (1987). The defective proton-ATPase of *uncA* mutants of *Escherichia coli*: Identification by DNA sequencing of residues in the α-subunit which are essential for catalysis or normal assembly. J. Biol. Chem. 262, 8981–8984.

Matsuno-Yagi, A., & Hatefi, Y. (1990). Studies on the mechanism of oxidative phosphorylation: Positive cooperativity in ATP synthesis. J. Biol. Chem. 265, 82–88.

Matsuno-Yagi, A., & Hatefi, Y. (1993a). Studies on the mechanism of oxidative phosphorylation: ATP synthesis by submitochondrial particles inhibited at F_0 by venturicidin and organotin compounds. J. Biol. Chem. 268, 6168–6173.

Matsuno-Yagi, A., & Hatefi, Y. (1993b). Studies on the mechanism of oxidative phosphorylation: Different effects of F_0 inhibitors on unisite and multisite ATP hydrolysis by bovine submitochondrial particles. J. Biol. Chem. 268, 1539–1545.

Matsuno-Yagi, A., Yagi, T., & Hatefi, Y. (1985). Studies on the mechanism of oxidative phosphorylation: Effects of specific F_0 modifiers on ligand-induced conformation changes of F_1. Proc. Natl. Acad. Sci. USA 82, 7550–7554.

Mendel-Hartvig, J., & Capaldi, R. A. (1991a). Catalytic site nucleotide and inorganic phosphate dependence of the conformation of the ε subunit in *Escherichia coli* adenosinetriphosphatase. Biochemistry 30, 1278–1284.

Mendel-Hartvig, J., & Capaldi, R. A. (1991b). Nucleotide-dependent and dicyclohexylcarbodiimide-sensitive conformational changes in the ε subunit of *Escherichia coli* ATP synthase. Biochemistry 30, 10987–10991.

Miki, J., Fujiwara, K., Tsuda, M., Tsuchiya, T., & Kanazawa, H. (1990). Suppression mutations in the defective β subunit of F_1-ATPase from *Escherichia coli*. J. Biol. Chem. 265, 21567–21572.

Milburn, M. V., Tong, L., de Vos, A. M., Brunger, A., Yamaizumi, Z., Nishimura, S., & Kim, S-H. (1990). Molecular switch for signal transduction: Structural differences between active and inactive forms of protoncogenic *ras* proteins. Science 247, 939–945.

Miller, M. J., Oldenburg, M., & Fillingame, R. H. (1990). The essential carboxyl group in subunit *c* of the F_1F_0 ATP synthase can be moved and H^+-translocating function retained. Proc. Natl. Acad. Sci. USA 87, 4900–4904.

Mitchell, P. (1974). A chemiosmotic molecular mechanism for proton-translocating adenosine triphosphatases. FEBS Lett. 43, 189–194.

Miwa, K., & Yoshida, M. (1989). The $\alpha_3\beta_3$ complex, the catalytic core of F_1-ATPase. Proc. Natl. Acad. Sci. USA 86, 6484–6487.

Moroney, J. V., & McCarty, R. E. (1982). Light-dependent cleavage of the γ subunit of coupling factor 1 by trypsin causes activation of Mg^{2+}-ATPase activity and uncoupling of photophosphorylation in spinach chloroplasts. J. Biol. Chem. 257, 5915–5920.

Nakamoto, R. K., Maeda, M., & Futai, M. (1993). The γ subunit of the *Escherichia coli* ATP synthase: mutations in the carboxyl-terminal region restore energy coupling to the amino-terminal mutant, γMet-23→Lys. J. Biol. Chem. 268, 867–872.

Nalin, C. M., & McCarty, R. E. (1984). Role of a disulfide bond in the γ subunit in activation of the ATPase of chloroplast coupling factor 1. J. Biol. Chem. 259, 7275–7280.

Noumi, T., Futai, M., & Kanazawa, H. (1984a). Replacement of serine-373 by phenylalanine in the α subunit of *Escherichia coli* F_1-ATPase results in loss of steady state catalysis by the enzyme. J. Biol. Chem. 259, 10076–10079.

Noumi, T., Mosher, M. E., Natori, S., Futai, M., & Kanazawa, H. (1984b). A phenylalanine for serine substitution in the β subunit of *Escherichia coli* F_1-ATPase affects dependency of its activity on divalent cations. J. Biol. Chem. 259, 10071–10075.

Noumi, T., Maeda, M., & Futai, M. (1987a). Mode of inhibition of sodium azide on H^+-ATPase of *Escherichia coli*. FEBS Lett. 213, 381–384.

Noumi, T., Tagaya, M., Miki-Takeda, K., Maeda, M., Fukui, T., & Futai, M. (1987b). Loss of unisite and multisite catalysis by *Escherichia coli* F_1 through modification with adenosine tri- or tetraphosphopyridoxal. J. Biol. Chem. 262, 7686–7692.

O'Neal, C. C., & Boyer, P. D. (1984). Assessment of the rate of bound substrate interconversion and of ATP acceleration of product release during catalysis by mitochondrial adenosine triphosphatase. J. Biol. Chem. 259, 5761–5767.

Omote, H., Maeda, M., & Futai, M. (1992). Effects of mutations of conserved Lys-155 and Thr-156 residues in the phosphate-binding glycine-rich sequence of the F_1-ATPase β subunit of *Escherichia coli*. J. Biol. Chem. 267, 20571–20576.

Omote, H., Park. M.-Y., Maeda, M., & Futai, M. (1994). The α/β subunit interaction in H^+-ATPase (ATP synthase): An *Escherichia coli* α subunit mutation (αArg-296→Cys) restored coupling efficiency to the deleterious β subunit mutant (βSer-174→Phe). J. Biol. Chem. 269, 10265–10269.

Pai, E. F., Kabsch, W., Krengel, U., Holmes, K. C., John, J., & Wittinghofer, A. (1989). Structure of the guanine-nucleotide-binding domain of the Ha-*ras* oncogene product p21 in the triphosphate conformation. Nature (London) 341, 209–214.

Papa, S., Guerrieri, F., Zanotti, F., Fiermonte, M., Capozza, G., & Jirillo, E. (1990). The γ subunit of F_1 and the PVP protein of F_0 (F_0I) are components of the gate of the mitochondrial F_0F_1 H^+-ATP synthase. FEBS Lett. 272, 117–120.

Parsonage, D., Duncan, T. M., Wilke-Mounts, S., Kironde, F. A. S., Hatch, L., & Senior, A. E. (1987). The defective proton-ATPase of *uncD* mutants of *Escherichia coli*: Identification by DNA sequencing of residues in the β-subunit which are essential for catalysis or normal assembly. J. Biol. Chem. 262, 6301–6307.

Penefsky, H. S. (1985). Mechanism of inhibition of mitochondrial adenosine triphosphatase by dicyclohexylcarbodiimide and oligomycin: Relationship to ATP synthesis. Proc. Natl. Acad. Sci. USA 82, 1589–1593.

Penefsky, H. S., & Cross, R. L. (1991). Structure and mechanism of F_0F_1-type ATP synthases and ATPases. Adv. Enzymol. 64, 173–213.

Perlin, D. S., Cox, D. N., & Senior, A. E. (1983). Integration of F_1 and the membrane sector of the proton-ATPase of *Escherichia coli*: Role of subunit "*b*" (*uncF* protein). J. Biol. Chem. 258, 9793–9800.

Porter, A. C. G., Kumamoto, C. A., Aldape, K., & Simoni, R. D. (1985). Role of the *b* subunit of the *Escherichia coli* proton-translocating ATPase: A mutagenic analysis. J. Biol. Chem. 260, 8182–8187.

Rao, R., Cunningham, D., Cross, R. L., & Senior, A. E. (1988). Pyridoxal 5′-diphospho-5′-adenosine binds at a single site on isolated α-subunit from *Escherichia coli* F_1-ATPase and specifically reacts with lysine 201. J. Biol. Chem. 263, 5640–5645.

Reddy, E. P., Reynolds, R. K., Santos, E., & Barbacid, M. (1982). A point mutation is responsible for the acquisition of transforming properties by the T24 human bladder carcinoma oncogene. Nature 300, 149–152.

Richter, M. L., & McCarty, R. E. (1987). Energy-dependent changes in the conformation of the ε subunit of the chloroplast ATP synthase. J. Biol. Chem. 262, 15037–15040.

Sachsenheimer, W., & Schultz, G. E. (1977). Two conformations of crystalline adenylate kinase. J. Mol. Biol. 114, 23–36.

Saraste, M., Sibbald, P. R., & Wittinghofer, A. (1990). The P-loop—a common motif in ATP- and GTP-binding proteins. Trends Biochem. Sci. 15, 430–434.

Schneider, E., & Altendorf, K. (1985). Modification of subunit *b* of the F_0 complex from *Escherichia coli* ATP synthase by a hydrophobic maleimide and its effects on F_0 functions. Eur. J. Biochem. 153, 105–109.

Schneider, E., & Altendorf, K. (1987). Bacterial adenosine 5′-triphosphate synthase (F_1F_0): Purification and reconstitution of F_0 complexes and biochemical and functional characterization of their subunits. Microbiol. Rev. 51, 477–497.

Schumann, J., Richter, M. L., & McCarty, R. E. (1985). Partial proteolysis as a probe of the conformation of the γ subunit in activated soluble and membrane-bound chloroplast coupling. J. Biol. Chem. 260, 11817–11823.

Sebald, W., & Hoppe, J. (1981). On the structure and genetics of the proteolipid subunit of the ATP synthase complex. Curr. Top. Bioenerg. 12, 1–64.

Senior, A. E. (1983). Secondary and tertiary structure of membrane proteins involved in proton translocation. Biochim. Biophys. Acta 726, 81–95.

Senior, A. E. (1988). ATP synthesis by oxidative phosphorylation. Physiol. Rev. 68, 177–231.

Senior, A. E. (1992). Catalytic sites of *Escherichia coli* F_1-ATPase. J. Bioenerg. Biomemb. 24, 479–484.

Senior, A. E., Wilke-Mounts, S., & Al-Shawi, M. K. (1993). Lysine 155 in β-subunit is a catalytic residue of *Escherichia coli* F_1 ATPase. J. Biol. Chem. 268, 6989–6994.

Shin, K., Nakamoto, R. K., Maeda, M., & Futai, M. (1992). F_0F_1-ATPase γ subunit mutations perturb the coupling between catalysis and transport. J. Biol. Chem. 267, 20835–20839.

Soga, S., Noumi, T., Takeyama, M., Maeda, M., & Futai, M. (1989). Mutational replacements of conserved amino acid residues in the α subunit change the catalytic properties of *Escherichia coli* F_1-ATPase. Arch. Biochem. Biophys. 268, 643–648.

Sternweis, P. C., & Smith, J. B. (1980). Characterization of the inhibitory (ε) subunit of the proton-translocating adenosine triphosphatase from *Escherichia coli*. Biochemistry 19, 526–531.

Tagaya, M., Noumi, T., Nakano, K., Futai, M., & Fukui, T. (1988). Identification of α-subunit Lys-201 and β-subunit Lys-155 at the ATP-binding sites in *Escherichia coli* F_1-ATPase. FEBS Lett. 233, 347–351.

Takeyama, M., Ihara, K., Moriyama, Y., Noumi, T., Ida, K., Tomiyoka, N., Itai, A., Maeda, M., & Futai, M. (1990). The glycine-rich sequence of the β subunit of *Escherichia coli* H^+-ATPase is important for activity. J. Biol. Chem. 265, 21279–21284.

Tanford, C. (1983). Mechanism of free energy coupling in active transport. Annu. Rev. Biochem. 52, 379–409.

Tiedge, H., Lunsdorf, H., Schafer, G., & Schairer, H. U. (1985). Subunit stoichiometry and juxtaposition of the photosynthetic coupling factor 1: Immunoelectron microscopy using monoclonal antibodies. Proc. Natl. Acad. Sci. USA 82, 7874–7878.

Tommasino, M., & Capaldi, R. A. (1985). Effect of dicyclohexyl-carbodiimide on unisite and multisite catalytic activities of the adenosinetriphosphatase of *Escherichia coli*. Biochemistry 24, 3972–3976.

Tsuprun, V. L., Mesyanzhinova, I. V., Kozlov, I. A., & Orlova, E. V. (1984). Electron microscopy of beef heart mitochondrial F_1-ATPase. FEBS Lett. 167, 285–290.

Vik, S. B., & Dao, N. N. (1992). Prediction of transmembrane topology of F_0 proteins from *Escherichia coli* F_1F_0 ATP synthase using variational and hydrophobic moment analyses. Biochim. Biophys. Acta 1140, 199–207.

Vik, S. B., Cain, B. D., Chun, K. T., & Simoni, R. D. (1988). Mutagenesis of the *a* subunit of the F_1F_0-ATPase from *Escherichia coli*: Mutations at Glu-196, Pro-190, and Ser-199. J. Biol. Chem. 263, 6599–6605.

Walker, J. E., Saraste, M., Runswick, M. J., & Gay, N. J. (1982). Distantly-related sequences in the alpha and beta subunits of ATP synthase, myosin, kinases and other ATP-requiring enzymes and a common nucleotide binding fold. EMBO J. 1, 945–951.

Walker, J. E., Saraste, M., & Gay, N. J. (1984). The *unc* operon: Nucleotide sequence, regulation and structure of ATP-synthase. Biochim. Biophys. Acta 768, 164–200.

Wise, J. G., Latchney, L. R., & Senior, A. E. (1981). The defective proton-ATPase of *uncA* mutants of *Escherichia coli*: Studies of nucleotide binding sites, bound aurovertin fluorescence, and labeling of essential residues of the purified F_1-ATPase. J. Biol. Chem. 256, 10383–10389.

Wise, J. G., Latchney, L. R., Ferguson, A. M., & Senior, A. E. (1984). Defective proton ATPase of *uncA* mutants of *Escherichia coli*. 5′-adenylyl imidodiphosphate binding and ATP hydrolysis. Biochemistry 23, 1426–1432.

Yoshida, M., Allison, W. S., Esch, F. S., & Futai, M. (1982). The specificity of carboxyl group modification during the inactivation of the *Escherichia coli* F_1-ATPase with dicyclohexyl $[C^{14}]$carbodiimide. J. Biol. Chem. 257, 10033–10037.

Zhou, J.-M., & Boyer, P. D. (1993). Evidence that energization of the chloroplast ATP synthase favors ATP formation at the tight binding catalytic site and increases the affinity for ADP at another catalytic site. J. Biol. Chem. 268, 1531–1538.

ATP-DIPHOSPHOHYDROLASES, APYRASES, AND NUCLEOTIDE PHOSPHOHYDROLASES:
BIOCHEMICAL PROPERTIES AND FUNCTIONS

Adrien R. Beaudoin, Jean Sévigny, and Maryse Picher

Biomembranes
Volume 5, pages 369–401.

ISBN: 1-55938-662-2.

I. INTRODUCTION

The terms ATP-diphosphohydrolase and apyrase refer to a family of enzymes which split the γ- and β-phosphate residues of triphospho- and diphosphonucleosides (Meyerhoff, 1945). Kalckar was the first to describe the properties of apyrases in potato tubers. Indeed, in a paper submitted in January, 1944 to the *Journal of Biological Chemistry*, he demonstrated that a single enzyme was involved in the hydrolysis of both diphospho- and triphosphonucleosides. In the years that followed, the potato apyrase was extensively studied, but its function was never clearly demonstrated. It is noteworthy that in Kalckar's studies on potato apyrase, adenylate kinase was excluded to explain the production of phosphate when ADP was used as the substrate. The latter enzyme can convert two molecules of ADP to ATP and AMP, which could then be degraded by an ATPase and a 5′-nucleotidase, respectively. Research work of Lee and Eiler (1951) and the confrontation of their results with those of Kalckar (1944) and Krishnan (1949a, 1949b) suggested that the potato apyrase activity was attributable to a mix of enzymes. Partial purification led the group of Liébecq et al. (1962, 1963) to propose that their fraction contains an ATPase and an apyrase. However, in the same period, Molnar and Lorand (1961) confirmed that potato extract contains two distinct apyrases, one of them having much higher affinity for ATP than ADP. These two enzymes were later purified and characterized by Traverso-Cori and others (Kalckar, 1944; Krishnan, 1949a, 1949b; Lee and Eiler, 1951; Szekely, 1951; Kiessling, 1956; Cohn and Meek, 1957; Molnar and Lorand, 1961; Liébecq et al., 1962, 1963; Traverso-Cori and Cori, 1962; Cori et al., 1965; Traverso-Cori et al., 1965, 1970; Miller and Westheimer, 1966; Valenzuela et al., 1973, 1988, 1989; Del Campo et al., 1977; Kettlun et al., 1981, 1982; Mancilla et al., 1984, 1987; Anich et al., 1990). Following the pioneer work on potato tubers, presence of the enzyme was demonstrated in several other plant species, namely the "chick-pea" *Cicer arietinum* (Vara and Serrano, 1981), the "Alaska" pea (Tognoli and Marrè, 1981; Tognoli, 1984), the cabbage (Mazelis, 1959), the motile plant, *Mimosa pudica* pulvinius (Ishikawa et al., 1984) and the yellow lupin, *Lupinus luteus* (Guranovski et al., 1991).

Even though animal apyrases have been occasionally mentioned very early in the literature, no definite proof for their existence was obtained until Ribeiro and

Garcia (1979, 1980, 1981) found the enzyme in the salivary gland and saliva of *Rhodnius prolixus*. Presence of apyrase activity was confirmed in the same species by Smith et al. (1980). The activity was later reported in the salivary gland and saliva of a variety of blood-unrelated arthropods such as the tsetse fly, *Glossina morsitans* (Mant and Parker, 1981), the mosquitoes, *Aedes aegypti* (Ribeiro et al., 1984a) and three species of *Anopheles* (Ribeiro et al., 1985a), the ticks, *Ixodes dammini* (Ribeiro et al., 1985b) and *Ornithodoros moubata* (Ribeiro et al., 1991), the sandfly, *Lutzomyia longipalpis* (Ribeiro et al., 1986), three species of *Phlebotomus* (Ribeiro et al., 1989), and three rodent flea species (Ribeiro et al., 1990). The detection of the enzyme in the medicinal leech (Rigbi et al., 1987) leads one to believe that apyrase is a common feature of the saliva in these hematophagous animals. An ATPDase has recently been found on the external surface of the tegument of *Schistosoma mansoni*, an intestinal worm (Vasconcelos et al., 1993). Intringuingly, this enzyme activity was also related to blood coagulation.

An ATPDase was identified in vertebrate tissues in 1980 by LeBel et al. and its kinetic properties were described in detail two years later by Laliberté et al. (1982) and Laliberté and Beaudoin (1983). Following these reports, the ATPDase was found in several tumor cells (Knowles et al., 1983) and in a variety of cells and tissues (Knowles et al., 1983; Miura et al., 1987; Schadeck et al., 1989; Valenzuela et al., 1989, 1992; Yagi et al., 1989, 1991, 1992; Papamarcaki and Tsolas, 1990; Battastini et al., 1991; Côté et al., 1991, 1992a,b; Moodie et al., 1991; Pieber et al., 1991; Sarkis and Salto, 1991; Strobel and Rosenberg, 1992; Picher et al., 1993, 1994). In addition to these studies where the enzyme has been clearly identified, a great number of reports deal with various types of activities, namely Ca^{2+}, Mg^{2+}-ATPases, ecto-ATPases, ecto-ADPases, ectonucleotidases, and nucleotide triphosphatases that exhibit some characteristics of the ATPDases. In this review, the general properties of ATPDases from plants, invertebrates, and vertebrates have been analyzed. Problems associated with identification and characterization of ATPDases are discussed. A section discussing some vertebrate nucleotide phosphohydrolase activities, which could potentially be attributable to ATPDases, is also included. There are a few reports on ATPDases in prokaryotic cells that are not discussed in this review (Curdova et al., 1982; Jechova et al., 1982). An overview of the possible roles played by this enzyme in the various types of organisms is given.

II. PROPERTIES OF ATP-DIPHOSPHOHYDROLASES (APYRASES)

A. Plants

Plant apyrases are readily soluble either in dilute or concentrated salt solutions, indicating the electrostatic nature of the bond with the cell membranes. As shown in Table 1, these enzymes are generally more active at acid pH, the optimum pH

Table 1. Kinetic Properties of Plant ATP-diphosphohydrolases (Apyrases)

Source of ATPDase (apyrase)	Optimum pH (substrate)	$K_{m,app.}$ (μM) ATP	$K_{m,app.}$ (μM) ADP	MW (kD)	Ratio ATP/ADP
Potato tubers (*Solanum tuberosum*)					
variety Pimpernel (soluble) (1,2)[a]	6.0 (ATP) 8.0 (ADP)	60	250	49[b]	10–12[d]
variety Desirée (soluble) (1)	6.0 (ATP & ADP)	25	100	48–49[b]	1.0
variety Desirée (microsomal) (3)	6.0 (ATP & ADP)	28	26	52[b]	0.9
Cabbage leaves					
soluble apyrase (4)	5.0				~1.4
particulate apyrase (4)	8.7				~2.5
"Chick-pea" roots					
Cicer arietinum (5)	6.5 (ATP & ADP)	200	500	50[c] 55[b]	0.5
"Alaska" pea stems (6)	6.0 (ATP) 7.0 (ADP)	70–80	21–35	30 & 48[c] 70–80[b]	0.7
Pulvinius					
Mimosa pudica (7)					0.1–0.2
Yellow Lupin cotyledons					
Lupinus luteus (8)	6–8	~20	~20	51[b,c]	~1.3

Notes: [a](1) Kettlun et al., 1982; (2) Del Campo et al., 1977; (3) Valenzuela et al., 1989; (4) Mazelis, 1959; (5) Vara and Serrano, 1981; (6) Tognoli and Marrè, 1981; (7) Ishikawa et al., 1984; (8) Guranovski et al., 1991.

[b]molecular weight estimated by gel filtration.

[c]molecular weight estimated by SDS-PAGE (sodium dodecyl sulfate polyacrylamide gel electrophoresis.

[d]The ATP/ADP ratio can be influenced by the incubation buffer and pH.

varying from 5.0 to 7.0. With two exceptions, diphosphonucleosides are better substrates than triphosphonucleosides (ATP/ADP < 1.0). By definition, these enzymes do not hydrolyze monophosphoric esters or pyrophosphates. The fact that, in most instances, these ATPDases have not been purified, somewhat complicates the analysis of their kinetic parameters, namely K_m and V_{max}. However, extensive studies have provided a more defined picture in the case of potato apyrases. There are two homogeneous isozymes in potato tubers which show marked differences in substrate specificity (ATP vs ADP). Both isoapyrases have the same molecular weight ~50 kD, but differ in their isoelectric point (pI). The K_m for ATP and ADP are about the same for the membrane-bound enzyme, on the order of 25 μM, whereas it is slightly different for the soluble enzyme (25 μM for ATP and 100 μM for ADP). Presence of a heat-stable activator of the potato apyrase is an interesting feature of plant apyrases (Mancilla et al., 1987). This activator seems to increase animal apyrase activity, although its effect remains relatively modest (Valenzuela et al., 1989). The identity of this activator and the modalities of its interaction have not yet been described.

B. Invertebrates

As summarized in Table 2, apyrase activities were measured in saliva and salivary glands of a variety of hematophagous invertebrates including insects (Ribeiro et al., 1989, 1990 and reviewed by Ribeiro, 1987, 1989), ticks (Ribeiro et al., 1985b, 1991), and leeches (Rigbi et al., 1987). In one species of parasitic worm (Vascon-

Table 2. Kinetic Properties of Invertebrate ATP-diphosphohydrolases (Apyrases)

Source of ATPDase (apyrase)	*Optimum pH (substrate)*	*$K_{m,app.}$ (μM)*		*MW (kD)*	*Ratio ATP/ADP*
		ATP	*ADP*		
Bug					
Rhodnius prolixus (1,2,3)[a]	7.5–8.5	—	500	83[b]	1.7–1.8[d]
(saliva & salivary gland)	(ATP & ADP)	229	291		
Mosquitoes					
Aedes aegypti (4)	9.0			65[b]	1.4
(salivary gland)	(ATP & ADP)				
Anophele freeborni (5)	9.0				1.3
(salivary gland)					
Anophele stephensi (5)	9.0				1.8
(salivary gland)					
A. sp. nr. salbaii (5)	9.0				1.4
(salivary gland)					
Sand flies					
Lutzomyia longipalpis (6)	8.0				0.6
(salivary gland)	(ATP & ADP)				
Phlebotomus papatasi (7)	7.0–8.5				0.8
(salivary gland)					
Phlebotomus argentipes (7)	8.0–9.0				1.0
(salivary gland)					
Phlebotomus perniciosus (7)	7.0–8.5				0.8–1.3
(salivary gland)					

(continued)

Table 2. (Continued)

Source of ATPDase (apyrase)	Optimum pH (substrate)	$K_{m,app.}$ (μM) ATP	$K_{m,app.}$ (μM) ADP	MW (kD)	Ratio ATP/ADP
Fleas					
Orchopea howardii (8)	7.0–8.5				1.1
(salivary gland)	(ATP & ADP)				
Oropsylla bacchi (8)	7.0–8.5				1.2
(salivary gland)	(ATP & ADP)				
Xenopsylla cheopis (8)	7.0–8.0				1.8
(salivary gland)	(ATP & ADP)				1.1
Ticks					
Ixodes dammini (9)	8.5–9.0				1.1
(saliva)	(ATP & ADP)				
Ornithodoros Moubata (10)	7.0 (ADP)				1.3
(saliva)	8.0 (ATP)				
Leech					
Hirudo medicinalis (11) (saliva)	7.5 (ATP & ADP)			form I: >400 form II: 45[b]	~1.3
Worm					
Schistosoma mansoni (12) (tegument)		~450	~250	49–58[c]	1.1

Notes: [a](1) Ribeiro and Garcia, 1980; (2) Smith et al., 1980; (3) Sarkis et al., 1986; (4) Ribeiro et al., 1984a; (5) Ribeiro et al., 1985a; (6) Ribeiro et al., 1986; (7) Ribeiro et al., 1989a; (8) Ribeiro et al., 1990; (9) Ribeiro et al., 1985b; (10) Ribeiro et al., 1991; (11) Rigbi et al., 1987; (12) Vasconcelos et al., 1993.

[b]Molecular weight estimated by gel filtration.

[c]Molecular weight estimated by SDS-PAGE (sodium dodecyl sulfate polyacrylamide gel electrophoresis.

[d]The ATP/ADP ratio can be influenced by the incubation buffer and pH.

celos et al., 1993), the enzyme is found in the tegument, its catalytic site being exposed to the external *milieu*. In these invertebrates, it is not always clear whether the enzyme is soluble or not. Its location in saliva does not necessarily mean that it is in a soluble form. In this respect, ATPDases has been reported in gland secretions where they were found to be associated with microvesicles (Beaudoin et al., 1986; Beaudoin and Grondin, 1991). In contrast to plants, the enzymes from invertebrates exhibit an optimum pH for catalysis in the neutral or alkaline ranges, and most of them can function or are activated by divalent cations. Monovalent cations do not appear to have any significant effect. The K_m values for ADP and

ATP are relatively high (200 μM) and comparable for the two substrates, whereas similar rates of hydrolysis of these two substrates were obtained. Sarkis et al. (1986) purified the *Rhodnius prolixus* salivary apyrase and described its kinetic properties in detail. The K_m for ATP and ADP were 229 μM and 291 μM, respectively, and the K_i values for ATP and ADP inhibition were of the same order, confirming competition for the same catalytic site. Specific activity of the purified fraction was very high, with 580- and 335-μmoles phosphate/min/mg of protein (units/mg of protein) for ATP and ADP, respectively. The kinetics of ATP hydrolysis are similar to the pig pancreas ATPDase (Laliberté and Beaudoin, 1983). The enzyme appears to remove both phosphate groups with a single substrate contact, with no accumulation of ADP in the reaction medium. Although there is still a limited number of studies, the general properties of invertebrate ATPDases reveal some important differences to the plant ATPDases, from the point of view of solubility and kinetic properties.

C. Vertebrates

Ectonucleotidases have been known for many years in vertebrate tissues, but identification of ATPDases is relatively recent. In 1980, LeBel et al. demonstrated this enzyme in the pig pancreas. Laliberté et al. (1982) and Laliberté and Beaudoin (1983) later purified the enzyme to a high degree (68 units/mg of protein) and studied its kinetic properties in detail. Cytochemical and biochemical observations have shown that the enzyme is associated with the zymogen granule and plasma membrane (Beaudoin et al., 1980). In certain conditions, one can also find the enzyme associated with microvesicles in secretions of the gland (Beaudoin and Grondin, 1991). The optimum pH of catalysis is in the alkaline range (pH of 8.0 for ATP and 9.0 for ADP) (Table 3). It is activated by Ca^{2+} or Mg^{2+}, but can use

Table 3. Kinetic Properties of Vertebrate ATP-diphosphohydrolases

Source of ATPDase (apyrase)	Optimum pH (substrate)	$K_{m,app.}$ (μM) ATP	$K_{m,app.}$ (μM) ADP	MW (kD)	Ratio ATP/ADP
Pig pancreas (1,2,3,4)[a]	8.0 (ATP)	7.3	7.4	58 and/or 28[b]	1.1[e]
	9.0 (ADP)	3.1	5.6	132 ± 19[c]	
Bovine aorta (4,5,6,7,8)	7.5 (ATP & ADP)	23 ± 3		110[b] 189 ± 30[c]	1.1
Bovine spleen (9)	7.5–8.0 (ATP) 8.0–8.5 (ADP)		9	100[b]	1.0

(continued)

Table 3. (Continued)

Source of ATPDase (apyrase)	*Optimum pH (substrate)*	*$K_{m,app.}$ (μM)* ATP	ADP	*MW (kD)*	*Ratio ATP/ADP*
Bovine lung (10)	7.5 (ATP & ADP)	7 ± 2		70 ± 3[c]	1.4
Bovine trachea (11) smooth muscle	7.3 (ATP & ADP)	15 ± 2		71 ± 5[c]	1.5
Human umbilical vessels (12)				75[b]	1.1
Human term placenta (13)	8.0 (ADP) 8.5 (ATP)				0.8
Rat placental tissue (14)	8.0 (ATP & ADP)	50	33	67[d]	1.1
Rat salivary glands (15)	8.0 (ATP) 9.0 (ADP)	13	15		1.0
Rat mammary glands (15,16)	8.0 (ATP) 9.0 (ADP)	12	16		0.9 1.3
Rat uterus (15,16)	8.0 (ATP) 9.0 (ADP)	12	27	67[d]	1.4 1.5
Rat synaptosome from hypothalamus (17)	8.0 (ATP & ADP)				3.7–5.6
Rat synaptosome from adult cerebral cortex (18)	8.0–8.5 (ATP & ADP)	80	45		2.3
Ray *Torpedo marmorata* (19) synap. from the electric organ	7.5–8.0 (ATP & ADP)	117 ± 11	123 ± 14		1.4
Sarcoma Li-7(m) (20)	8.0 (ADP) ≥8.0 (ATP)	k_{m1}: 87 k_{m2}: 1610	k_{m1}: 44 k_{m2}: 1100		1.7
Chicken liver (21,22)				85[b]	~4
Chicken oviduct (21,22)				80[b]	~4

Notes: [a](1) Lebel et al., 1980; (2) Laliberté et al., 1982; (3) Laliberté and Beaudoin, 1983; (4) Côté et al., 1991; (5,6) Côté et al., 1992a, b; (7) Miura et al., 1987; (8) Yagi et al., 1989; (9) Moodie et al., 1991; (10) Picher et al., 1993; (11) Picher et al., 1994; (12) Yagi et al., 1992; (13) Papamarcaki and Tsolas, 1990; (14) Pieber et al., 1991; (15,16) Valenzuela et al., 1989, 1992; (17) Schadeck et al., 1989; (18) Battastini et al., 1991; (19) Sarkis and Salto, 1991; (20) Knowles et al., 1983; (21) Strobel and Rosenberg, 1992; (22) Strobel, 1992.

[b]molecular weight estimated by SDS-PAGE (sodium dodecyl sulfate polyacrylamide gel electrophoresis).

[c]molecular weight estimated by irradiation-inactivation technique.

[d]molecular weight estimated by gel filtration.

[e]The ATP/ADP ratio can be influenced by the incubation buffer and pH.

ATP^{4-} as a substrate as well. A slight preference for triphosphonucleosides was noted. The $K_{m,app.}$ for ATP and ADP are in the low micromolar range. Competition experiments have shown a simultaneous breakdown of ATP and ADP with initial relative rates of 2.1- and 3.8-units/mg of protein, respectively. The V_{max} for ADP hydrolysis is twice that of ATP. The two are competitive inhibitors of each other, and their K_i values are comparable to their $K_{m,app.}$ (10 μM). The following sequential scheme was proposed: ATP binds to the enzyme, its γ-phosphate group is hydrolyzed, resulting in an enzyme–ADP complex which either breaks down to free enzyme and ADP, or is further processed via hydrolysis of the β-phosphate group, releasing free enzyme, AMP and P_i. Experimental data showed that the processing step is favored.

The common occurrence of ATPDase in mammalian plasma membrane was demonstrated by Knowles et al. (1983). Plasma membranes from the following tissues and cells were examined: mouse liver, mouse brain, dog kidney, mouse sarcomas, human astrocytoma, oat cell carcinoma, and melanoma. ATPDase activities were particularly high in mouse sarcoma plasma membrane. As for the pancreas enzyme, the enzyme from these different sources were all inhibited by sodium azide. There was also a preference for triphospho- as compared to diphosphonucleosides. However, the $K_{m,app.}$ for ATP was a little higher than for ADP. With the cell line, Li-7(m), they noticed a drop of ATPDase activity in cell culture. Interestingly, the activity reappeared in the tumors grown from these same cells after they were injected in nude mice, thereby implying some kind of regulation in the *in vivo* system.

The presence of both ectoATPases and ectoADPases in the vascular system has been known for many years, and up until the work of Yagi et al. (1989, 1991), they were attributed to two distinct enzymes. They purified these activities and showed that in bovine aorta, a single enzyme was responsible for the sequential hydrolysis of ATP and ADP. Like the pancreas and other mammalian ATPDases, the enzyme shows a slight preference for triphospho—as compared to diphosphonucleosides, and is stimulated by Ca^{2+} or Mg^{2+}. Purification to homogeneity was demonstrated by SDS-polyacrylamide gel electrophoresis and silver staining. The apparent molecular mass of the pure enzyme was estimated at 110 kD. The existence of the ATPDase in bovine aorta was corroborated by Côté et al. (1991) who showed identical heat and irradiation inactivation curves, with ATP and ADP as substrates. A comparison of the biochemical properties led them to propose that the bovine aorta enzyme was different from the pancreas ATPDase. Indeed, the enzymes have different native molecular weights, optimum pH and sensitivities to inhibitors. They proposed to identify pancreas enzyme as type I, and the bovine aorta enzyme as type II. In the bovine aorta, the enzyme was found to be associated with smooth muscle cells and endothelial cells and could inhibit platelet aggregation induced by ADP (Miura et al., 1987; Yagi et al., 1991; Côté et al., 1992a).

More recently, Yagi et al. (1992) reported an ATPDase in human umbilical vessels. The human enzyme showed broad substrate specificity and sensitivity to various inhibitors and calcium ions. The purified umbilical enzyme was inhibited by only 20% with an antiserum which inhibited bovine aorta enzyme by about 80% at a dilution of 1/250. This suggests some important structural differences between bovine aorta and human umbilical ATPDases. The molecular weight of the purified protein was estimated at 75 kD as compared to 110 kD for the bovine enzyme. However, the specific activities were comparable (37- and 58-units/mg of protein, respectively).

A bovine spleen ADPase was purified by Moodie et al. (1991) and was shown to be also an ATPDase. Like the pancreas, the optimum pH of catalysis is slightly different for ATP and ADP. The K_m values for ATP and ADP are in the low micromolar range and the enzyme requires Ca^{2+}. SDS electrophoretic gels showed the presence of a major polypeptide with an estimated molecular mass of 100 kD, in close agreement with the ATPDase from the bovine aorta (Yagi et al., 1989). The specific activity of the purified enzyme from the bovine spleen was significantly higher (115- vs. 58-units/mg of protein) than the bovine aorta, with ADP as substrate. Papamarcaki and Tsolas (1990) found an ATPDase in human placenta. The properties of the enzyme were essentially similar to the previously reported mammalian ATPDase. Intriguingly, CTP seemed to be a poor substrate. In another study, Pieber et al. (1991) described ATPase–ADPase activities associated with a microsomal fraction of rat placental tissues. They mentioned that the enzyme shares many characteristics with ATPDases.

The detection of ATPDase in blood vessels led Picher et al. (1993) to look for ATPDase activity in the lung, a highly vascularized organ. In bovine lung, they found a distinct type of ATPDase. By irradiation inactivation curves, the native molecular mass of the lung enzyme was estimated at 70 kD as compared to 132 kD for the pancreas and 189 kD for the aorta enzymes (Côté et al., 1991). Migration patterns after polyacrylamide gel electrophoresis under nondenaturing conditions were slightly different. These differences in molecular mass estimation and migration pattern on acrylamide gels could be explained by the association of the aorta ATPDase with another protein. Picher et al. (1994) also described an ATPDase in bovine trachea smooth muscles that shows the same properties as the lung enzyme.

The presence of the ATPDase in the nervous system, more specifically at nerve endings, may have significant physiological meaning. As for the vascular system, ATPDase activities have been described in various brain samples (Knowles et al., 1983; Schadeck et al., 1989; Battastini et al., 1991), and in synaptosomes of the electric organ of the ray, *Torpedo marmorata* (Sarkis and Salto, 1991). In 1989, Schadeck et al. demonstrated the presence of an ATPDase on the outer surface of the synaptosomal membrane isolated from the hypothalamus of adult rats. A unique feature of this enzyme is a clear preference for ATP over ADP (4- to 6-fold). Two years later, Sarkis and Salto (1991) characterized an ATPDase associated with the synaptosomes isolated from the electric organ of *Torpedo marmorata*. In contrast to the enzyme isolated from rat synaptosomes, the enzyme hydrolyses almost equally well different nucleoside di- and triphosphates. The enzyme was insensitive to many inhibitors of other ATPDases including sodium azide (5.0 mM), which appears to be an exceptional case. Battastini et al. (1991) further characterized the ATPDase from synaptosomes of the rat cerebral cortex. The enzyme was only slightly inhibited by sodium azide (5 mM), but chlorpromazine and fluoride were powerful inhibitors. In addition, the enzyme exhibited a clear preference for triphospho- versus diphosphonucleosides.

Only one study has reported ATPDases in birds. Indeed, Strobel and Rosenberg (1992) have purified two different ATPDases from chicken liver and chicken oviductal secretions. The specific activities of their preparation far exceeded any purified ATPDases previously described. This is attributable to a very efficient immunoaffinity chromatography technique using specific monoclonal antibodies. Both enzymes are very active with ATP and ADP. Surprisingly, there is a small but significant hydrolysis of AMP (Strobel, 1992). Molecular weights of the purified enzymes were 80- and 85-kD for the oviduct and the liver, respectively. The authors did not examine the kinetic properties of their enzyme, which would have been very informative. The monoclonal antibodies do not seem to react with bovine aorta enzyme (unpublished results). A common feature to all these vertebrate ATPDases is their firm association with membranes. Moreover, in the cases tested, the enzyme is bound by concanavalin A columns, thereby indicating the presence of glycosyl residues in the molecule (LeBel et al., 1980; Moodie et al., 1991; Yagi et al., 1992; Strobel and Rosenberg, 1992).

IV. PHYSIOLOGICAL ROLES OF ATP-DIPHOSPHOHYDROLASES

A. Plants

Although there is a considerable amount of work devoted to the characterization of plant ATPDases, especially potato apyrases, there is still much to be learned about the functions of the enzyme. The approach taken by Anich et al. (1990) to resolve this question was to follow the changes in metabolites associated with the enzyme and its activity in relation to tuber and shoot development in potatoes. For this purpose, levels of adenine nucleotides, inorganic phosphate, starch, and apyrase activity were measured. Although their results did not clarify the role of the apyrase, they could propose a function related to starch synthesis. Indeed, several glycosyltransferases of sucrose and starch synthesis involve reversible reactions producing UDP and ADP, inhibitory products of this enzyme (Sadler et al., 1982) which can be removed by ATPDase. Such a role has been suggested for the nucleoside diphosphatase associated with gluconeogenesis, glycogen synthesis, glycolysis (Plaut, 1955; Ernster and Jones, 1962; Yamazaki and Hayaishi, 1965, 1968; Novikoff et al., 1971), and lactose synthesis (Kuhn and White, 1977). As mentioned by Sarkis et al. (1986), this effect would be similar to that of inorganic pyrophosphatase which drives the formation of activated amino acids in protein synthesis and of carbamoyl phosphate in the urea cycle. In addition, the enzyme might have a similar role in the biosynthesis of many other sugar nucleotides, including those that are involved in the synthesis of the cell wall. In the latter case, transglycosylases also generate diphosphonucleosides as products (Bonner and Varner, 1976). The detection of apyrase in the "chick-pea" (Vara and Serano, 1981), and in the "Alaska" pea (Tognoli and Marrè, 1981), two plants that accumulate starch, gives some support to these proposed functions for the enzyme.

B. Invertebrates

The role of ATPDase in the saliva of blood-feeding arthropods has been reviewed by Ribeiro (1987, 1989) in the wider context of the role of saliva in hematophagy. As these hematophagous animals probe and salivate into their host tissues, blood vessels and other tissues are lacerated and a massive concentration of cytoplasmic nucleotides is released by the damaged cells. This, joined to the exocytic release of ATP and ADP from activated platelets, favor a rapid platelet aggregation. Apyrase, converting extracellular pro-aggregatory ADP into the anti-aggregatory AMP, joined to other factors found in saliva, keep the blood in a liquid state, favoring its accumulation into hemorrhagic pools and thereby facilitating the feeding process. In addition, the ATP, released by platelets and damaged cells, enhances local inflammation by inducing mast cell degranulation (Coutts et al., 1981) which leads to release of thromboxane A2 platelet activating factor, PAF, and other arachidonic derivatives having vasoactive properties, in addition to serotonin and histamine (Bach, 1982). ATP also contributes to inflammation by other mechanisms which involve neutrophils, causing their aggregation (Ford-Hutchinson, 1982). By converting ATP into AMP, the ATPDase counteracts these protective mechanisms of the host. Taking these functions of the enzyme into account, one is led to conclude that apyrase is a major player in the feeding of these animals.

Because these invertebrates evolved independently of hemophagy, and ATPDase is not found at such high activity in nonhematophagous animals, including the male partners of mosquitoes (Rossignol et al., 1984), it was proposed that such activities were a case of convergent evolution. This enzyme probably had a "domestic" metabolic role in salivary glands of nonhematophagous species and evolved to the status of a secretory product upon development of the blood feeding habit (Ribeiro et al., 1984b).

In the case of *Schistosoma mansoni*, the enzyme is found at the external surface of the tegument (Vasconcelos, 1993). It was shown that the enzyme could inhibit ADP induced platelet aggregation. They proposed that the enzyme degrades nucleotides in the bloodstream of the host in the surroundings of the parasite. It would be one of the mechanisms developed by the parasite to ensure its survival in the circulation.

C. Vertebrates

The definition of the physiological role played by the ATPDase is both simple and complex. It is simple in the sense that the enzyme, localized either on the outer cell surface or in extracytoplasmic spaces, converts triphospho- and diphosphonucleosides into their monophosphate derivatives. However, it is more complex as one refers to the *in vivo* conditions. One must then take into consideration tissue localization, availability of substrates, presence of enzymes which compete for the same substrate (protein kinases for ATP), other enzymes that catalyse the hydrolysis of the reaction products (5′-nucleotidase and adenosine deaminase for AMP), and finally the receptors, which are the ultimate targets of nucleotides and their dephosphorylated derivatives.

The role of ATPDase is perhaps best illustrated by the control of platelet reactivity in hemostasis. As described by Marcus and Safier (1993), there are at least three thrombo-regulatory mechanisms associated with endothelial cells: eicosanoids, endothelium-dependent relaxing factor (EDRF/NO), and the ecto-ATPDase. They demonstrated the role of vessel ATPDase by blocking the effects of EDRF with hemoglobin, and the effects of PGI_2 with aspirin, simultaneously (Marcus et al., 1991; Marcus and Safier, 1993).

The nervous system is another example which illustrates the potential role of ATPDase. It is well recognized that ATP is released by the peripheral nervous system at both pre-ganglionic and post-ganglionic levels. More specifically, it is released as a cotransmitter from cholinergic and adrenergic nerve terminals and as a neurotransmitter from non-adrenergic and non-cholinergic nerve terminals. ATP can also act as a fast excitatory transmitter at synapses between neurons (Edwards et al., 1992; Evans et al., 1992). The presence of an ATPDase within or in the vicinity of these nerve terminals may modulate the action of the neurotransmitter. The released ATP would have a short period of time to interact with P2-purinoceptors on the target cells, before being converted to adenosine by the combined actions of the enzyme and 5′-nucleotidase. The latter reaction product, adenosine, can then interact with its P1-purinoceptors either on the target cell or on the neuron itself, to inhibit further release of the neurotransmitter. Such a type of interaction occurs at the level of smooth muscle cells in the arteries, where ATPDase has been localized. It leads us to believe that ATPDase may represent an underestimated partner in the control of the arterial pressure. Since the enzyme is present in nonvascular smooth muscles of the trachea, and probably of many other organs, it is difficult to realize all the physiological roles of this enzyme. Apart from the smooth muscles of the cardiovascular and respiratory systems, the enzyme may intervene with many other cellular targets where extracellular ATP can strike. This would include secretory glands and different parts of the gastrointestinal system.

Perhaps the role of the enzyme can be better visualized if one looks at the distribution of purine receptors throughout the body and the specific effect associated with their activation by extracellular nucleotides. For an exhaustive review see the recent papers of Olsson and Pearson (1990), El-Moatassim et al. (1992), and Burnstock (1993). To summarize, let us state that ATPDase, by converting ATP and ADP into AMP, could convert a P2 effect (mediated by ATP and ADP) into a P1 effect (mediated by AMP and adenosine), effects that are often opposite. Therefore, when ATP is released in the external *milieu*, a P2 effect preceedes a P1 effect. This P1 effect could be further delayed if the P1-purinoceptor is only activated by adenosine. Indeed, it is becoming recognized that AMP is inactive on adenosine receptors (Ragazzi et al., 1991). Since 5′-nucleotidase is inhibited by high concentrations of ATP and ADP, this P1 effect would only emerge after concentrations of ATP and ADP have been sufficiently reduced by the action of ATPDase. Finally, an additional function of the enzyme may be to degrade extracellular nucleotides, allowing the recovery of the base by a nucleotide salvage pathway.

V. PROBLEMS ASSOCIATED WITH THE IDENTIFICATION AND CHARACTERIZATION OF ATP-DIPHOSPHOHYDROLASES

The presence of other enzymes somewhat complicates the tasks of identifying and characterizing ATPDases. In intact cells, one may find alkaline phosphatases and other nonspecific phosphohydrolases on the cell surface. The former can be efficiently inhibited by tetramisole. Any residual activity can be measured with or eliminated by p-nitrophenylphosphate in the case of radioactive substrates. At this stage, it would be advisable to carry out a comparative study of nucleotide specificity and if possible, competition studies, although the latter type of study may be problematic, as we will see below.

If one starts with a homogenate, enzymes that are inaccessible to the substrate when intact cells are used become accessible, namely adenylate kinase, nucleotide hydrolyzing enzymes of the mitochondria (ATPases, ADPases), Golgi apparatus (nucleotide diphosphatase), endosomes (vacuolar ATPase), cytoplasm (protein kinases), as well as plasma membrane Na^+, K^+-ATPase and Ca^{2+}-stimulated Mg^{2+}-dependent ATPase (Ca^{2+}-pumps), which considerably complicate the task. Partial purification therefore becomes a prerequisite for enzyme characterization. Since vertebrate ATPDases are particulate enzymes, the most simple way to start purification is to isolate the most active fractions by differential and density gradient ultracentrifugations. This step offers the advantage of eliminating any degradation by soluble proteases. Addition of protease inhibitors can also considerably reduce any residual protease associated with membranes. Further purification requires the solubilization of the particulate enzyme. For this purpose, some non-ionic detergents: Triton X-100, NP-40, Tween 20, and CHAPSO have been used with variable success. The solubilized enzyme appears to be much more unstable, as occasionally mentioned in the literature. At this stage, one may find it useful to look for other nucleotide phosphohydrolases by polyacrylamide gel electrophoresis under non-denaturing conditions. For this purpose, the procedure of Côté et al. (1991) ensures a good separation of the proteins while preserving the activity of the enzyme. Different substrates can be tested like ATP, ADP, and AMP, thereby evaluating the contribution of other nucleotide phosphohydrolases. If there is a single band that hydolyzes both ATP and ADP, there is a good chance that apyrase is the enzyme responsible for their hydrolysis. This may be confirmed by comparing heat and/or γ-irradiation inactivation curves with diphospho- and triphosphonucleosides (Côté et al., 1991; Picher et al., 1993). Ideally, one should carry out an analysis of the kinetic properties of an enzyme on the purest preparation possible. As mentioned, the use of intact cells or semi-purified fraction may lead to some erroneous results. In these preparations, one finds nucleotide receptors, nucleotide transporters, protein kinases, and other enzymes that compete for the substrate. In this respect, analysis of the binding of 5′-p-fluorosulfonylbenzoyl adenosine (FSBA), an analog of ATP, to proteins of a crude fraction (30-fold purification) of vascular smooth muscle cells have revealed more than 30 bands on gel electrophoresis. This labeling could be prevented by adding ATP or ADP to the incubation medium (unpublished results).

The role of metal ions is another parameter which has not always been taken into account when the kinetic parameters of ATPDases were defined. Analysis of the literature indicates that Ca^{2+} and Mg^{2+} are generally required for the activity of ATPDase. In fact, Ca^{2+} and Mg^{2+} form complexes with the nucleotide and these complexes are generally the true substrates of the enzyme. In the determination of K_m, one should take into account the concentration of these complexes, instead of the free nucleotide. Otherwise, there may be an excess of free nucleotides, or its Na^+ or K^+ salts, which are not hydrolyzed by the enzyme, and may behave like competitive inhibitors.

There are other parameters that one must pay attention to, namely the chemical stability of the reactants (substrates and inhibitors), and the method of phosphorus determination. The classical method of Fiske and Subbarow (1925) can not be recommended because of its low sensitivity, the instability of the reagents, and the interference by ATP and other nucleotides.

Perhaps the most difficult step of the purification process is to identify the catalytic subunit of the enzyme. The finding of a unique band after polyacrylamide gel electrophoresis is very strong and convincing evidence. There is still, however, a certain level of subjectivity in this type of demonstration. The time allowed for the reaction of detection may totally change the apparent homogeneity of the preparation. However, if monospecific antibodies are raised against this purified band and these antibodies can immunoprecipitate the ATPDase, one may be confident that the right protein has been identified. Additional evidence for the identity of the protein could come from labeling experiments with substrate analogs such as azido- or etheno-derivatives of nucleotides, or with other analogs such as FSBA. Obviously, the final proof for the identity of the enzyme will be its expression by recombinant DNA techniques in cells which are devoid of ATPDase activity.

VI. POTENTIAL ATP-DIPHOSPHOHYDROLASES IN MAMMALIAN TISSUES

Analysis of the literature reveals a surprising number of enzymes which exhibit some of the properties of ATPDases, namely specificity towards triphospho- and diphosphonucleosides, sensitivity to Ca^{2+} or Mg^{2+}, inhibition of ATP hydrolysis by ADP and *vice versa*, and inhibition by sodium azide. In this section, we describe some of those plasma membrane ATPases and/or ADPases that could potentially be ATPDases, although substrate specificity was not always fully addressed. One finds them on cells lining blood vessels or in the circulation, in cardiac tissues, in skeletal muscles, and in many organs and glands.

A. Blood Vessels

A number of biochemical and cytochemical studies have shown that nucleoside phosphohydrolase activities are present as ectoenzymes on intact endothelial and vascular smooth muscle cells of large vessels (Heyns et al., 1974; Dieterle et al., 1978; Dosne et al., 1978; Habliston et al., 1978; Cooper et al., 1979; Pearson et al., 1980; Wilson et al., 1982; Ogawa et al., 1986), as well as on cultured fibroblasts (Dosne et al., 1978; Glasgow et al., 1978). Pearson and coworkers have conducted a number of experiments to identify the enzymes responsible for hydrolysis of extracellular nucleotides in blood vessels (for a review see Pearson, 1986). By studying patterns of degradation of radioactive nucleotides perfused in piglet lung, they have demonstrated that ATP hydrolysis involves the sequential loss of two terminal phosphate groups, so that ATP is converted to ADP, and then to AMP (Pearson and Gordon, 1979; Carleton et al., 1979; Hellewell and Pearson, 1987). They also studied nucleotide hydrolysis at the surface of vascular endothelial and smooth muscle cells in culture (Pearson et al., 1980, 1985; Gordon et al., 1986, 1989). Based on inhibition by analogs and substrate specificity, they suggested that hydrolysis of ATP to adenosine is attributable to three Mg^{2+}-stimulated enzymes: nucleoside triphosphatase (E. C. 3.6.1.15), nucleoside diphosphatase (E. C. 3.6.1.6), and 5′-nucleotidase (E. C. 3.1.3.5).

Lieberman and coworkers studied the properties and subcellular localization of an ADPase activity in pig aorta smooth muscle cells. Subcellular distribution of this enzyme follows that of 5′-nucleotidase, a plasma membrane marker, but not of other organelles (Lieberman and Lewis, 1980; Lieberman et al., 1982). Simultaneous addition of ATP and ADP did not result in an increase in the rate of hydrolysis, suggesting that they both compete for the same catalytic site. Optimum pH was estimated at 7.3 and K_m for ATP at 10.3 μM. Sun et al. (1990) presented a more exhaustive characterization of a pig plasma membrane Ca^{2+}, Mg^{2+}-ATPase having affinities for both Ca^{2+} and Mg^{2+}. They successfully eliminated a Ca^{2+}-pump by calmodulin-affinity chromatography. This Ca^{2+}-pump was ATP-specific, whereas the purified enzyme could use all triphospho- and diphosphonucleosides as substrates. It was activated by cations, and insensitive to vanadate. Contribution of adenylate kinase was eliminated by using Mg^{2+} concentrations below the millimolar range (Noda, 1973). In light of the recent work of Miura et al. (1987) and Côté et al. (1991), it is highly probable that the enzyme activities described above are attributable, at least in great part, to an ATPDase.

ATPase and ADPase activities have also been localized in plasma membranes of small blood vessels in lung, kidney, small intestine, and heart tissues (Marchesi and Barnett, 1963, 1964; Crutchley et al., 1978; Cooper et al., 1979; Chelliah and Bakhle, 1983; Ryan, 1986; Grantham and Bakhle, 1988; Hulstaert et al., 1991) by cytochemistry and perfusion studies. These activities were characterized in mesenteric arteries by Plesner and coworkers (Juul et al., 1991; Plesner et al., 1991). They showed that sections of small mesenteric arteries incubated in bicarbonate buffer hydrolyze all triphospho- and diphosphonucleosides at similar rates, with K_m values in the micromolar range (2.5 μM for Ca^{2+}-ATP and 10 μM for Mg^{2+}-ATP). These activities are insensitive to P-type, F-type, and V-type ATPase inhibitors. Functional removal of the endothelium did not reduce these activities, thereby indicating that ectonucleotidases are also present on smooth muscle cells of small vessels, as described earlier for large vessels.

B. Heart

Many studies have shown that adenine nucleotides are rapidly dephosphorylated on a single passage through the coronary bed (Baer and Drummond, 1968; Paddle and Burnstock, 1974, 1987; Ronca-Testoni and Borghini, 1982; Belardinelli et al., 1984; Fleetwood et al., 1989). As reported earlier for large vessels, patterns of catabolites are consistent with the sequential dephosphorylation of ATP to ADP, ADP to AMP, and then AMP to adenosine. K_m values for ATP, ADP, and AMP were estimated at 450-, 300-, and 93-μM, respectively.

Ectonucleotidases have been reported at the surface of intact cardiomyocytes (Bowditch et al., 1985), but neither the nature of the enzymes nor their kinetic properties have been studied in detail. Meghji et al. (1992) recently demonstrated the sequential hydrolysis of ATP to adenosine at the surface of rat ventricular myocytes in suspension. K_m values for ATP and ADP are in the micromolar range. With ATPase-resistant analogs, AMP–PCP and AMP–PNP, ATP-pyrophosphohydrolase activity was evaluated to be less than 5% of total ATPase activity. β-glycerophosphate and p-nitrophenylphosphate did not inhibit nucleotide hydrolysis, suggesting that the contribution of nonspecific phosphatases was negligible.

An ADPase was purified from rat ventricle homogenate (De Vente et al., 1984). Similar subcellular distributions were obtained for the enzyme and a plasma membrane marker, 5′-nucleotidase. This enzyme activity was inhibited when another nucleoside diphosphate was added along with ADP. However, this inhibition was overcome by raising ADP concentration, thereby suggesting that they might all be competing for the same catalytic site. $K_{m,app.}$ for ADP is about 20 μM. As for cardiomyocyte ectoenzymes, non-specific nucleotidase activities were ruled out using a series of inhibitors.

C. Organs Containing Nonvascular Smooth Muscles

Kwan et al. (1984) published a review in which they compared ATPase and ADPase properties of vascular and nonvascular smooth muscle plasma membrane and microsomal fractions. Ratios of ATP/ADP hydrolysis ranged from 1.5 to 1.9 in rat vas deferens, gastric fundus, and mesenteric arteries, and in dog aorta and mesenteric arteries (Kwan et al., 1981, 1982a, 1982b, 1983a, 1983b; Kwan and Ramlal, 1982). They demonstrated that sodium azide, a known inhibitor of mitochondrial ATPase, also inhibits ATPase and ADPase activities located in the plasma membrane. These activities were inhibited by at least 30% with ATP and 50% with ADP. They were all activated by cations, preferentially Ca^{2+} or Mg^{2+}.

Ectonucleotidases have been found in nonvascular smooth muscles of guinea pig taenia coli (Cusack and Hourani, 1984; Welford et al., 1986) and urinary bladder (Cusack and Hourani, 1984). Welford et al. (1987) showed that sections of guinea pig urinary bladder in suspension completely dephosphorylated ATP, GTP, CTP, ADP, and AMP, and at similar rates. K_m values for ATP and ADP are 532 μM and 864 μM, respectively. Rat intestinal basal membranes also show some Ca^{2+}-ATPase activity having properties distinct from those of a Ca^{2+}-pump (Moy et al., 1986). The enzyme has higher activity and is insensitive to vanadate. It hydrolyzes all triphospho- and diphosphonucleosides, but not AMP or p-nitrophenylphosphate, whereas the pump is specific for ATP. The pump is activated by Mg^{2+}, and the Ca^{2+}-ATPase is activated by either Mg^{2+} or Ca^{2+}. Increasing concentrations of Mg^{2+} inhibits the Ca^{2+}-ATPase activity, suggesting competion for the same binding site.

Rat myometrium plasma membranes contain two different ATPases (Enyedi et al., 1988). They were later identified as two Ca^{2+} or Mg^{2+} nucleotide phosphohydrolase activities localized at the surface of intact cells (Magocsi and Penniston, 1991). One of them is labile, disappearing rapidly during enzyme assays (half-life of about 2 min), while the other is stable. The labile component has a K_m for Ca^{2+} of nearly 1 mM, cleaves all triphosphates but not diphosphates, is inhibited by p-chloromercuriphenylsulfonate and inorganic phosphate, but not by sodium azide. The stable component is more sensitive to Ca^{2+} (K_m of about 0.1 mM). It accepts all tri- and diphosphates as substrates, is not inhibited by p-chloromercuriphenylsulfonate or inorganic phosphate, but is inhibited by 20 mM sodium azide. A similar biphasic character was observed by Missiaen et al. (1988a,b) in rat myometrium microsomal fraction.

D. Muscles

Skeletal muscles also bear ectonucleotidases at their surface. Dunkley et al. (1966) incubated frog leg muscles with ATP and noticed the presence of ADP, AMP, and IMP in the incubation medium. This enzyme was later localized in the chicken transverse tubule membrane, and was shown to exhibit a number of properties distinct from those of transport ATPases present in the sarcolemma and in the sarcoplasmic reticulum (Moulton et al., 1986). The K_m for ATP was estimated at 14 μM. The enzyme requires either Mg^{2+} or Ca^{2+} in millimolar concentrations as a co-substrate, displays a broad pH optimum and nucleotide specificity, and is insensitive to a variety of inhibitors known to affect transport ATPases, such as oligomycin, ouabain, and vanadate. Molecular composition seems to differ among species, but in general, a glycoprotein of 102–105 kD is reported (Hidalgo et al., 1983; Okamoto et al., 1985; Moulton et al., 1986; Damiani et al., 1987; Kirley, 1988; Horgan and Kuypers, 1988). Saborido et al. (1991) recently provided evidence for the extracellular orientation of this enzyme. They purified the enzyme from the free portion of the transverse tubule. In the microvesicle fraction, its activity was oriented in the same direction as other ectoenzymes such as acetylcholinesterase. Since the transverse tubules constitute the major fraction of the total surface membranes in the muscle fiber, this Mg^{2+}-ATPase may represent an effective protection mechanism against cellular effects of ATP and other extracellular nucleotides. Finally, Alertsen et al. (1958) have shown that in the medium surrounding rat diaphragm, breakdown of ATP proceeded via ADP, AMP, IMP, and inosine.

E. Other Organs and Glands

Liver

Nucleoside diphosphatases have been localized in rat liver plasma membranes (Emmelot et al., 1964, 1968). This enzyme was later shown to follow the same subcellular distribution as 5′-nucleotidase, a plasma membrane marker (Wattiaux-de-Coninck and Wattiaux, 1969). It hydrolyzes all diphosphonucleosides, and a 5- to 10-fold increase in activity was obtained with Mg^{2+} and Ca^{2+}, respectively. The same optimum pH was obtained with both cations. The possibility that this activity was attributable to adenylate kinase is weak since Novikoff and Hues (1963) have demonstrated that in liver, this enzyme is mainly found in mitochondrial and soluble fractions of the homogenate. The authors suggested that this enzyme was an ATPDase. Loterztajn et al. (1981, 1984) later purified a plasma membrane Ca^{2+}, Mg^{2+}-ATPase. This enzyme is capable of hydrolyzing ADP as well as all triphosphonucleosides (ATP, GTP, CTP, UTP, and ITP). It is insensitive to calmodulin, but requires either Ca^{2+} or Mg^{2+} to function at maximal rate. Alkaline phosphatase represents less than 5% of this hydrolytic activity. Aware of the presence of a Ca^{2+}-pump in this fraction, they eliminated its possible contribution by using low concentrations of nucleotides (0.25 mM) and by omitting Mg^{2+} in their assays. $K_{0.5}$ for ATP was estimated at 25 μM, and V_{max} at 2 units/mg of protein. An activator protein sensitive to trypsin was eliminated by chromatography on a DE-52 column.

These results, along with kinetic studies, suggest the presence of two sites for the substrates, one for hydrolysis and another controlling the enzyme activity.

Lin and coworkers were later able to eliminate a Ca^{2+}-pump with a concanavalin A-sepharose 4B chromatography column (Lin and Fairn, 1984). The Ca^{2+}-pump did not bind to the matrix, while a Ca^{2+}, Mg^{2+}-ATPase was retained. Both enzymes are insensitive to calmodulin, activated by cations, and have $K_{0.5}$ in the micromolar range (Lin, 1985). However, they differ in molecular weight evaluated from SDS-PAGE, with 118 kD for the Ca^{2+}-pump and 70 kD for the other nucleotidase. Moreover, the Ca^{2+}-pump is ATP-specific, whereas the nucleotidase hydrolyzes all triphospho- and diphosphonucleosides. They later demonstrated that this enzyme is located at the surface of hepatocytes in primary cultures (Lin and Russell, 1988), and they cloned a gene encoding for a 100 kD ecto-ATPase (Lin and Guidotti, 1989).

Kidney

Renal plasma membranes have been shown to possess Ca^{2+}-stimulated ATPase activities measurable in the absence of added Mg^{2+} (Berger and Sacktor, 1970; Rorive and Kleinzeller, 1972; Kinne-Saffran and Kinne, 1974). In the tubular basolateral membrane, this enzyme hydrolyzes all triphospho- and diphosphonucleosides (Kinne-Saffran and Kinne, 1974), and the K_m for ATP is around 200 μM. Culic et al. (1990) demonstrated that purified rat renal brush-border membranes hydrolyze all triphospho- and diphosphonucleosides at similar rates. Simultaneous additions of ATP and ADP produce similar rates of hydrolysis than with ATP alone. These activities are stimulated by a variety of metal ions, and K_m is estimated at 380 μM for ADP. Again, inhibitors were used to make sure no other enzymes were significantly involved in the nucleotides' hydrolysis. The authors consider that these enzyme activities are due to an ATPDase (Culic, personal communication).

Secretory Glands and Other Cells

A number of glands and other cells have been identified as ectonucleotidase-bearing tissues. Pituitary plasma membranes contain a calcium-sensitive ATPase activity that hydrolyzes ATP and other triphosphonucleosides at slightly lower rates (Lorenson et al., 1981). Parotid plasma membranes have been extensively studied for their ATPase activities other than the Ca^{2+}-pumps (Gutman and Glushevitzky-Strachman, 1973; Gantzer and Grisham, 1979; Knowles and Leng, 1984; Teo et al., 1988; Matsukawa, 1990). Based on their requirements for Ca^{2+} and/or Mg^{2+}, in many cases, more than one ATPase has been demonstrated. As summarized by Teo et al. (1988), there are two Ca^{2+}-stimulated ATPase activities in the plasma membrane of rat parotid: (1) an ATPase with high affinity for free Ca^{2+} ($K_{m,app.} = 0.21$ μM) that requires micromolar concentrations of Mg^{2+}, and (2) an ATPase with relatively low affinity for Ca^{2+} ($K_{0.5} = 23$ μM) or Mg^{2+} ($K_{0.5} = 26$ μM). Addition of both ions did not produce superior rates of hydrolysis. Oligomycin and Ruthenium red, inhibitors of mitochondrial ATPase, as well as calmodulin, ouabain, and vanadate, inhibitors of ion-transport ATPases, had no effect on these enzyme activities. The high-affinity enzyme dephosphorylated all triphosphonucleosides at similar rates and ADP at a lower rate. The fact that trifluoperazine inhibits the low-affinity Ca^{2+}-ATPase much more effectively (50%) than the high-affinity Ca^{2+}-ATPase further supports the concept of two different enzymes.

More recently, a Ca^{2+}-ATPase, not stimulated by Mg^{2+}, was separated from a Mg^{2+}-ATPase by papain treatment of a plasma membrane-rich fraction of bovine parotid gland (Matsukawa, 1990). Following purification on DEAE-cellulose, gel filtration on HPLC, and ion-exchange on HPLC, a protein of 100 kD was recovered. The enzyme does not hydrolyze p-nitrophenylphosphate and is not inhibited by transport or subcellular ATPase inhibitors. All triphospho- and diphosphonucleosides are hydrolyzed at different rates. An ATPase with a high affinity for Ca^{2+} (K_m = 0.23 μM, molecular weight = 100 kD) was isolated from rat parotid plasma membranes and immunoprecipitated with an antibody which was raised against a rat liver plasma membrane ecto-ATPase (Cheung et al., 1992). These results are strong indications that parotid plasma membranes possess an ATPDase.

ATP released from chromaffin cells during their secretory response can be hydrolyzed by ectonucleotidases found on these cells (Torres et al., 1990). K_m values for ecto-ATPase, ecto-ADPase, and ecto-AMPase activities were estimated at 250 μM, 365 μM, and 55 μM, respectively, in the presence of 1 mM Mg^{2+} and 2.5 mM Ca^{2+}.

A number of blood cells are known to hydrolyze extracellular nucleotides. Engelhardt (1957) reported that most of the ATPase activity of nucleated erythrocytes seemed to be associated with the cell surface. Clearance of ADP in an erythrocyte suspension has been attributable to two different enzymes, an adenylate kinase that leaked from the cells and an ecto-ADPase (Lüthje et al., 1988). Presence of adenylate kinase has been ascertained with a specific inhibitor, adenosine(5′)pentaphospho(5′) adenosine (Ap_5A). This ADPase activity was membrane-bound and the main reaction product was AMP. Studies with various inhibitors revealed that degradation of ADP is not due to a nonspecific phosphatase. K_m for ADP was estimated at 28 μM with 5×10^{12} cells, in the presence of Ca^{2+} and Mg^{2+}. This enzyme was able to inhibit ADP-induced platelet aggregation.

In the immune system, granulocytes and leukocytes were also shown to hydrolyze extracellular nucleotides (DePierre and Karnovsky, 1974a, b; Weiss and Sachs, 1977; Smolen and Weissman, 1978; Medzihradsky et al., 1980; Smith and Peters, 1981; Smith et al., 1981; Wilson et al., 1981; Ochs and Reed, 1984). An ATPase activated by Ca^{2+} or Mg^{2+} was identified on mast cells, macrophages, mononuclear cells, and lymphocytes (Dornand et al., 1974, 1986; Cooper and Stanworth, 1977; Chakravarty and Echetebu, 1978; Kragballe and Ellegaard, 1978; Chakravarty and Nielsen, 1980). Such activities were recently characterized on cytolytic T-lymphocytes (Filippini et al., 1990). This enzyme hydrolyzes all triphosphonucleosides and ADP, but not AMP. According to these authors, the fact that all triphosphonucleosides are hydrolyzed at higher rates than ATP indicates the presence of other nucleotide phosphatases on these cells: a Ca^{2+}/Mg^{2+}-ectoATPase with higher selectivity toward ATP and ADP, and another ectonucleotidase having a broad substrate specificity. No increase in total activity was observed when both Ca^{2+} and Mg^{2+} were present in the assay medium. This ectonucleotidase was shown to protect other cells against the lytic effects of extracellular ATP. On mast cells, ADP is a competitive inhibitor for ATP dephosphorylation. It is insensitive to Na^+ or K^+, and to ouabain (Pepys and Edwards, 1979). An ectoATPase has also been characterized on human natural killer cells (Dombrowski et al., 1993). The enzyme requires both Ca^{2+} and Mg^{2+}, and purine and pyrimidine nucleotides are competitive inhibitors of ATP hydrolysis. The K_m for ATP was estimated at 41 μM. An ATP-binding protein of 68–80 kD was obtained by photoaffinity labeling of intact NK3.3 cells with [α-^{32}P]-8-azidoATP.

F. Nervous System

In Section IIC we mentioned the detection of ATPDases in the brain and nervous system. There is an extensive literature reporting some ectoATPase and ectoADPase activities in these tissues (Agren et al., 1971; Stefanovic et al., 1974, 1976; Trams and Lauter, 1974, 1978; Rosenblatt et al., 1976). An ectoenzyme chain consisting of ATPase, ADPase, and 5′-nucleotidase activities has been suggested (Rosenblatt et al., 1976; Zimmermann et al., 1979; Shubert et al., 1981). White (1978) was the first to observe the very rapid hydrolysis of ATP by hypothalamic synaptosomes. Enzyme localization on the presynaptic plasma membrane was later confirmed by Sorensen and Mahler (1982). Nagy and collaborators characterized these ectonucleotidase activities in chicken brain synaptosomes (Nagy and Rosenberg, 1981; Nagy et al., 1983), results they later presented in an extensive review on synaptic ectonucleotidases and nucleotide recycling (Nagy, 1986). These tissues possess an ectoATPase activity that requires either Ca^{2+} or Mg^{2+}, with an optimum pH of 7.4–7.8. The K_m for ATP was estimated at 25 μM, either with Ca^{2+} or Mg^{2+}. Nucleotide triphosphates (GTP, UTP, ITP) were hydrolyzed to a similar extent as ATP, and ADP was hydrolyzed at 30% of their rates. Similar substrate specificity was obtained with neuroblastoma cells (Stefanovic et al., 1974). Inhibitors of ion-transporting ATPases, nonspecific alkaline phosphatases and kinases had no appreciable effect on this ATPase activity.

Zimmermann and coworkers, reported the hydrolysis of ATP to adenosine at the surface of intact synaptosomes of the electric organ of the electric ray, *Torpedo marmorata* (Keller and Zimmermann, 1983; Zimmermann et al., 1986). Kinetic and biochemical properties of these ectoenzymes corresponded to those of the chicken brain nucleotidases described above (Nagy, 1986), in terms of cation requirements and K_m for ATP (Ca^{2+}-ATP = 73 μM, Mg^{2+}-ATP = 53 μM). However, substrate specificity was not fully addressed. Battastini et al. (1991) later demonstrated that an ATPDase was responsible for the conversion of ATP to AMP, at least in brain tissues.

VII. CONCLUSION

From our analysis, it appears that ATPDases from plants are quite different from those of invertebrates and vertebrates. In invertebrates, the enzyme is associated with parasites and would be part of their strategy to counteract their host defense mechanisms. The role of the ATPDase in vertebrates is far more complex. It is very closely associated with the main control systems of the organism. There is still much confusion about the identification of this enzyme. Cloning the encoding genes will probably be the most appropriate strategy to define its nature and its expression in normal and pathological situations.

DEDICATION

We dedicate this review to a great and generous scientist, Dr. Murry Rosenberg from the University of Minnesota, with whom we had the privilege of sharing our ideas and views, and most importantly, our friendship. He died in 1993 of a malign form of parotid cancer.

ACKNOWLEDGMENTS

This work has been supported by grants from Heart and Stroke Foundation of Quebec, F.C.A.R. (Fonds pour la Formation des Chercheurs et l'Aide à la Recherche), Québec. Jean Sévigny is a recipient of a Research Traineeship from F.R.S.Q. (Fonds de la Recherche en Santé du Québec).

ENDNOTE

1. The terms apyrase and ATP-diphosphohydolase have been used indiscriminately in this text.

REFERENCES

Agren, G., Ponten, J., Ronquist, G., & Westmark, B. (1971). Demonstration of an ATPase at the cell surface of intact normal and neoplastic tumor cells in culture. J. Cell. Physiol. 78, 171–176.

Alertsen, A. R., Walaas, O., & Walass, E. (1958). Enzymic conversion of mononucleotides by rat diaphragm. Acta Physiol. Scand. 43, 122–134.

Anich, M., Fanta, N., Mancilla, M., Kettlun, A. M., Valenzuela, M. A., & Traverso-Cori, A. (1990). Apyrase activity and changes in metabolites during germination and tuberization of *Solanum tuberosum*. Phytochemistry 29, 1411–1415.

Bach, M. K. (1982). Mediators of anaphylaxis and inflammation. Ann. Rev. Microbiol. 36, 371–413.

Baer, H. P., & Drummond, G. I. (1968). Catabolism of adenine nucleotides by the isolated perfused rat heart. Proc. Soc. Exp. Biol. Med. 127, 33–36.

Battastini, A. J., Da Rocha, J. B., Barcellos, C. K., Dias, R. D., & Sarkis, J. J. (1991). Characterization of an ATP diphosphohydrolase (EC 3.6.1.5) in synaptosomes from cerebral cortex of adult rats. Neurochem. Res. 16, 1303–1310.

Beaudoin, A. R., & Grondin, G. (1991). Shedding of vesicular material from the cell surface of eukaryotic cells: Different cellular phenomena. Biochim. Biophys. Acta 1071, 203–219.

Beaudoin, A. R., Laliberté, J.-F., LeBel, D., Lord, A., Grondin, G., Phaneuf, S., & St-Jean, P. (1980). Localization and physiological role of the ATP-diphosphohydrolase in the pancreatic acinal cell. In: Biology of Normal and Cancerous Exocrine Pancreatic Cells (Ribet, A., Pradayrol, L., & Susini, C., eds.) INSERM Symposium No. 15, pp. 273–280. Elsevier, North Holland Press.

Beaudoin, A. R., Vachereau, A., Grondin, G., Rosenberg, M. D., & Strobel, R. (1986). Microvesicular secretion, a mode of cell secretion associated with the presence of an ATP-diphosphohydrolase. FEBS Lett. 203, 1–2.

Belardinelli, L., Shryock, J., West, G. A., Clemo, H. F., Dimarco, J. P., & Berne, R. M. (1984). Effects of adenosine and adenine nucleotides on the atrioventricular node of isolated guinea pig hearts. Circulation 70, 1083–1091.

Berger, S. J., & Sacktor, B. (1970). Isolation and biochemical characterization of brush borders from rabbit kidney. J. Cell Biol. 47, 637–645.

Bonner, J., & Vanner, J. E. (1976). Plant Biochemistry, 3rd Ed. Academic Press, New York.

Bowditch, J., Nigdikar, S., Brown, A. K., & Dow, J. W. (1985). 5'-nucleotidase activity of isolated mature rat cardiac myocytes. Biochim. Biophys. Acta 845, 21–26.

Burnstock, G. (1993). Physiological and pathological roles of purines: An update. Drug. Dev. Res. 28, 195–206.

Carleton, J. S., Gordon, J. L., Hutchings, A., & Pearson, J. D. (1979). Secretion and extracellular metabolism of adenine nucleotides by endothelial cells in culture. J. Physiol. (London), 291–40P.

Chakravarty, N., & Echetebu, Z. (1978). Plasma membrane adenosine triphosphatases in rat mast cells and macrophages: the relation of the mast cell enzyme to histamine release. Biochem. Pharmacol. 27, 1561.

Chakravarty, N., & Nielsen, H. (1980). Ca^{2+}-Mg^{2+}-activated adenosine triphosphatase in plasma and granule membranes in non-secreting and secreting mast cells. An electron microscopic histochemical study. Exper. Cell Res. 130, 175–184.

Chelliah, R., & Bakhle, Y. S. (1983). The fate of adenine nucleotides in the pulmonary circulation of isolated lung. Q. J. Exp. Physiol. 68, 289–300.

Cheung, P. H., Dowd, F. J., Porter, J. E., & Li, L.-S. (1992). A Ca^{2+}-ATPase from rat parotid gland plasma membranes has the characteristics of an ecto-ATPase. Cell. Sign. 4, 25–35.

Cohn, M., & Meek, G. A. (1957). The mechanism of hydrolysis of adenosine di- and triphosphate catalysed by potato apyrase. Biochem. J. 66, 128–132.

Cooper, D. R., Lewis, G. P., Lieberman, G. E., Webb, H., & Westwick, J. (1979). ADP metabolism in vascular tissue, a possible thrombo-regulatory mechanism. Thromb. Res. 14, 901–914.

Cooper, P. H., & Stanworth, D. R. (1977). Effect of removal of calcium-activated adenosine triphosphatase from rat mast cells by treatment with sodium glycocholate. Biochem. J. 168, 583–585.

Cori, O., Traverso-Cori, A., Tetas, M., & Chaimovich, H. (1965). Substrate specificity and inhibition studies on potato apyrase. Biochem. Z. 342, 345–358.

Côté, Y. P., Picher, M., St-Jean, P., Béliveau, R., Potier, M., & Beaudoin, A. R. (1991). Identification and localization of ATP-diphosphohydrolase (apyrase) in bovine aorta: Relevance to vascular tone and platelet aggregation. Biochim. Biophys. Acta 1978, 187–191.

Côté, Y. P., Filep, J. G., Battistini, B., Gauvreau, J., Sirois, P., & Beaudoin, A. R. (1992a). Characterization of ATP-diphosphohydrolase activities in the intima and media of the bovine aorta: evidence for a regulatory role in platelet activation in vitro. Biochim. Biophys. Acta 1139, 133–142.

Côté, Y. P., Ouellet, S., & Beaudoin, A. R. (1992b). Kinetic properties of type-II ATP diphosphohydrolase from the tunica media of the bovine aorta. Biochim. Biophys. Acta 1160, 246–250.

Coutts, A. A., Jorrizo, J. L., Eady, R. A. J., Greaves, M. W., & Burnstock, G. (1981). Adenosine triphosphate-evoked vascular changes in human skin: Mechanism of action. Eur. J. Pharmacol. 76, 391–401.

Crutchley, D. J., Eling, E. E., & Anderson, M. W. (1978). ADPase activity of isolated perfused rat lung. Life Sci. 22, 1413–1420.

Culic, O., Sabolic, I., & Zanic-Grubisic, T. (1990). The stepwise hydrolysis of adenine nucleotides by ectoenzymes of rat renal brush-border membranes. Biochim. Biophys. Acta 1030, 143–151.

Curdova, E., Jechova, V., & Hostalek, Z. (1982). Properties of apyrase and inorganic pyrophosphatase in *Streptomyces aurofaciens*. Folia Microbiol. 27, 159–166.

Cusack, N. J., & Hourani, S. M. O. (1984). Some pharmacological and biochemical interactions of the enantiomers of adenylyl 5′-(β,γ-methylene)-diphosphonate with the guinea-pig urinary bladder. Br. J. Pharmacol. 82, 155–160.

Damiani, E., Margreth, A., Furlan, A., Dahms, A. S., Arnn, J., & Sabbadini, R. A. (1987). Common structural domains in the sarcoplasmic reticulum calcium ATPase and the transverse tubule magnesium-ATPase. J. Cell. Biol. 104, 461–472.

De Vente, J., Velema, J., & Zaagsma, J. (1984). Properties and subcellular localization of adenosine diphosphatase in rat heart. Arch. Biochem. Biophys. 233, 180–187.

Del Campo, G., Puente, J., Valenzuela, M. A., Traverso-Cori, A., & Cori, O. (1977). Hydrolysis of synthetic pyrophosphoric esters by an isoenzyme of apyrase from *Solanum tuberosum*. Biochem. J. 167, 525–529.

DePierre, J. W., & Karnowsky, M. I. (1974a). Ectoenzymes of the guinea-pig polymorphonuclear leukocyte. I. Evidence for an ecto-adenosine monophosphatase, -adenosine triphosphatase, and -p-glycerophenylphosphatase. J. Biol. Chem. 249, 7111–7120.

DePierre, J. W., & Karnowsky, M. L. (1974b). Ecto-enzymes of the guinea-pig polymorphonuclear leukocyte. II. Properties and suitability as marker for the plasma membrane. J. Biol. Chem. 249, 7121–7129.

Dieterle, Y., Ody, C., Ehrensberger, A., Stalder, H., & Junod, A. F. (1978). Metabolism and uptake of adenosine triphosphate and adenosine by porcine aortic and pulmonary endothelial cells and fibroblasts in culture. Circ. Res. 42, 869–876.

Dombrowski, K. E., Trevillyan, J. M., Cone, J. C., Lu, Y., & Phillips, C. A. (1993). Identification and partial characterization of an ectoATPase expressed by human natural killer cells. Biochemistry 32, 6515–6522.

Dornand, J., Mani, J. C., Mousseron-Canet, M., & Pau, B. (1974). Propriétés d'une ATPase Ca^{2+}- ou Mg^{2+}-dépendante des membranes plasmiques de lymphocytes. Effet de la concanavaline A sur les ATPases membranaires. Biochimie (Paris) 56, 851–857.

Dornand, J., Bonnafous, J. C., Gartner, A., Favero, J., & Mani, J. C. (1986). Sudy of ecto-5′-nucleotidase, ecto-ATPase, and adenosine deaminase activities in relation to lymphoid cell differentiation. In: Cellular Biology of Ectoenzymes (Kreutzberg, et al., eds.) pp. 72–85. Springer-Verlag, Berlin, Heidelberg.

Dosne, A. M., Legrand, C., Bauvois, B., Bodevin, E., & Caen, J. P. (1978). Comparative degradation of adenylnucleotides by cultured endothelial cells and fibroblasts. Biochem. Biophys. Res. Commun. 85, 183–189.

Dunkley, C. R., Manery, J. F., & Dryden, E. E. (1966). The conversion of ATP to IMP by muscle surface enzymes. J. Cell Physiol. 68, 241–248.

Edwards, F. A., Gibb, A. J., & Colquhoun, D. (1992). ATP receptor-mediated synaptic currents in the central nervous system. Nature 359, 144–147.

El-Moatassim, C., Dornand, J., & Mani, J-C. (1992). Extracellular ATP and cell signalling. Biochim. Biophys. Acta 1134, 31–45.

Emmelot, P., Bos, E. L., Benedetti, E. L., & Rumke, P. (1964). Studies on plasma membranes. I. Chemical composition and enzyme content of plasma membranes isolated from rat liver. Biochim. Biophys. Acta 90, 126–145.

Emmelot, P., Bos, E. L., Benedetti, E. L., & Peeters, H. (1968). Protides of the Biological Fluids, 15th Colloquium Bruges, Vol. 15. Elsevier, Amsterdam.

Engelhardt, W. A. (1957). Enzymes as structural elements of physiological mechanisms. Proc. Int. Symp. Enzyme Chem. (Tokyo and Kyoto) 2, 163–166.

Enyedi, A., Minami, J., Caride, A. J., & Penniston, J. T. (1988). Characteristics of the Ca^{2+} pump and Ca^{2+}-ATPase in the plasma membrane of rat myometrium. Biochem. J. 252, 215–220.

Ernster, L., & Jones, L. C. (1962). A study of nucleoside tri- and diphosphate activities of rat liver microsomes. J. Cell Biol. 15, 563–578.

Evans, R. J., Derkach, V., & Surprenant, A. (1992). ATP-mediated fast synaptic transmission in mammalian neurons. Nature 357, 503–505.

Filippini, A., Taffs, R. E., Agui, T., & Sitkovsky, M. V. (1990). Ecto-ATPase activity in cytolytic T-lymphocytes. Protection from the cytolytic effects of extracellular ATP. J. Biol. Chem. 265, 334–340.

Fiske, C. H., & Subbarow, Y. (1925). The colorimetric determination of phosphorus. J. Biol. Chem. 66, 375–400.

Fleetwood, G., Coade, S. B., Gordon, J. L., & Pearson, J. D. (1989). Kinetics of adenine nucleotide catabolism in the coronary circulation of rats. Am. J. Physiol. 256, H1565–H1572.

Ford-Hutchinson, A. W. (1982). Aggregation of rat neutrophils by nucleotide triphosphates. Br. J. Pharmacol. 76, 467–471.

Gantzer, M. L., & Grisham, C. M. (1979). Characterization of magnesium-ATPase from sheep kidney medulla: Purification. Arch. Biochem. Biophys. 198, 263–267.

Glasgow, J. G., Schade, R., & Pitlick, F. A. (1978). Evidence that ADP hydrolysis by human cells is related to thrombogenic potential. Thromb. Res. 13, 255–266.

Gordon, E. L., Pearson, J. D., & Slakey, L. L. (1986). The hydrolysis of extracellular adenine nucleotides by cultured endothelial cells from pig aorta. Free-foward inhibition of adenosine production at the cell surface. J. Biol. Chem. 261, 15496–15504.

Gordon, E. L., Pearson, J. D., Dickinson, E. S., Moreau, D., & Slakey, L. L. (1989). The hydrolysis of extracellular adenine nucleotides by arterial smooth muscle cells. Regulation of adenosine production at the cell surface. J. Biol. Chem. 264, 18986–18992.

Grantham, C. J., & Bakhle, Y. S. (1988). Effect of acute lung injury on metabolism of adenine nucleotides in rat perfused lung. Br. J. Pharmacol. 94, 1029–1036.

Guranowski, A., Starzynska, E., Rataj-Guranowska, M., & Gunther Sillero, M. A. (1991). Purification of apyrase from yellow Lupin cotyledons after extraction with perchloric acid. Prot. Exp. Purif. 2, 235–239.

Gutman, Y., & Glushevitzky-Strachman, D. (1973). Effect of dehydration, food deprivation, saline and adrenalectomy on microsomal (Na^{+}+K^{+})-dependent ATPase in the salivary glands and intestinal mucosa. Biochim. Biophys. Acta 304, 533–540.

Habliston, D. L., Ryan, U. S., & Ryan, J. W. (1978). Endothelial cells degrade adenosine-5′-diphosphate. J. Cell. Biol. 79, 206A.

Hellewell, P. G., & Pearson, J. D. (1987). Nucleotides and pulmonary endothelium. In: Pulmonary Endothelium in Health and Disease (Ryan, U.S., ed.), pp. 327–348.

Heyns, A. du P., Van den Berg, D. J., Potgieter, G. M., & Retief, F. P. (1974). The inhibition of platelet aggregation by an aorta intima extract. Thrombos. Diathes. Haemorrh. 32, 417–431.

Hidalgo, C., Gonzalez, M. E., & Lagos, R. (1983). Characterization of the Ca^{2+}- or Mg^{2+}-ATPase of transverse tubule membranes isolated from rabbit skeletal muscles. J. Biol. Chem. 258, 13937–13945.

Horgan, D. J., & Kuypers, R. (1988). Biochemical properties of purified transverse tubules isolated from skeletal muscle triads. Arch. Biochem. Biophys. 260, 1–9.

Hulstaert, C. E., Kalicharan, D., Poelstra, K., Bakker, W. W., & Hardonk, M. J. (1991). Survey of the occurrence of adenosine polyphosphatase in extracellular matrix of rat tissues. A cytochemical investigation. Histochemistry 96, 441–447.

Ishikawa, H., Tamiya, T., Tsuchiya, T., & Matsumoto, J. J. (1984). A novel ATP–ADPase from *Mimosa* pulvinus. Comp. Biochem. Physiol. 78B, 59–61.

Jechova, V., Curdova, E., & Hostalek, Z. (1982). Role of phosphatase in the biosynthesis of chlortetracycline in *Streptomyces aureofaciens*. Folia Microbiol. 27, 153–158.

Juul, B., Lüscher, M. E., Aalkjaer, C., & Plesner, L. (1991). Nucleotide hydrolytic activity of isolated intact rat mesenteric small arteries. Biochim. Biophys. Acta 1067, 201–207.

Kalckar, H. M. (1944). Adenylpyrophosphatase and myokinase. J. Biol. Chem. 153, 355–367.

Keller, F., & Zimmermann, H. (1983). Ecto-adenosine triphosphatase activity at the cholinergic nerve endings of the *Torpedo* electric organ. Life Sci. 33, 2635–2641.

Kettlun, A. M., Uribe, L., Silva, S., Rivera, J., Valenzuela, M. A., & Traverso-Cori, A. (1981). Chemical modification of aromatic acid and basic amino-acids of 2 isoenzymes of apyrase EC-3.6.1.5. from *Solanum tuberosum*. Arch. Biol. Med. Exp. 14, 171–176.

Kettlun, A. M., Uribe, L., Calvo, V., Silva, S., Rivera, J., Mancilla, M., Valenzuela, M. A., & Traverso-Cori, A. (1982). Properties of two apyrases from *Solanum tuberosum*. Phytochemistry 21, 551–558.

Kiessling, K. H. (1956). Studies on thiamine triphosphate. Biochim. Biophys. Acta 20, 293–298.

Kinne-Saffran, E., & Kinne, R. (1974). Localization of a calcium-stimulated ATPase in the basal-lateral plasma membranes of the proximal tubule of rat kidney cortex. J. Membr. Biol. 17, 263–274.

Kirley, T. L. (1988). Purification and characterization of the Mg^{2+}-ATPase from rabbit skeletal muscle transverse tubule. J. Biol. Chem. 263, 12682–12689.

Knowles, A. F., & Leng, L. (1984). Purification of a low affinity $Mg^{2+}(Ca^{2+})$-ATPase from the plasma membranes of a human oat cell carcinoma. J. Biol. Chem. 259, 10919–10924.

Knowles, A. F., Isler, R. E., & Reece, J. F. (1983). The common occurrence of ATP diphosphohydrolase in mammalian plasma membranes. Biochim. Biophys. Acta 731, 88–96.

Kragballe, K., & Ellegaard, J. (1978). ATPase activity of purified human normal T- and B-lymphocytes. Scand. J. Haematol. 20, 271–279.

Krishnan, P. S. (1949a). Studies on apyrases. I. Purification of potato apyrase by fractional precipitation with ammonium sulfate. Arch. Biochem. 20, 261–271.

Krishnan, P. S. (1949b). Studies on apyrase. II. Some properties of potato apyrase. Arch. Biochem. 20, 272–283.

Kuhn, N. J., & White, A. (1977). The role of nucleoside diphosphatase in a uridine nucleotide cycle associated with lactose synthesis in rat mammary-gland Golgi apparatus. Biochem. J. 168, 423–433.

Kwan, C. Y., & Ramlal, T. (1982). On the inhibition of smooth muscle membrane ATPase by sodium azide. Biochem. Intern. 4, 439–449.

Kwan, C.-Y., Lee, R. M. K. W., & Daniel, E. E. (1981). Isolation of plasma membranes from mesenteric veins: A comparison of their physical and biochemical properties with arterial membranes. Blood Vessels 18, 171–186.

Kwan, C.-Y., Sakai, Y., Grover, A. K., & Lee, R. M. L. W. (1982a). Isolation and characterization of plasma membrane fraction from gastric fundus smooth muscle of the rat. Mol. Physiol. 2, 107–120.

Kwan, C.-Y., Triggle, C. R., Grover, A. K., Lee, R. M. K. W., Ramlal, T., & Daniel, E. E. (1982b). Canine aortic microsomes: Isolation and characterization. Can. Fed. Proc. 25, 98.

Kwan, C.-Y., Lee, R. M. K. W., & Grover, A. K. (1983a). Isolation and characterization of plasma membrane fraction from gastric fundus smooth muscle of the rat. Mol. Physiol. 2, 107–120.

Kwan, C.-Y., Triggle, C. R., Grover, A. K., Lee, R. M. K. W., & Daniel, E. E. (1983b). An analytical approach to the preparation and characterization of subcellular membranes from canine mesenteric arteries. Prep. Biochem. 13, 275–314.

Kwan, C.-Y., Kostka, P., & Ramlal, T. (1984). Properties of a widespread azide-sensitive adenosine diphosphohydrolase in smooth muscle. Molec. Physiol. 6, 99–114.

Laliberté, J.-F., & Beaudoin, A. R. (1983). Sequential hydrolysis of the γ and β-phosphate groups of ATP by the ATP diphosphohydrolase from pig pancreas. Biochim. Biophys. Acta 742, 9–15.

Laliberté, J.-F., St-Jean, P., & Beaudoin, A. R. (1982). Kinetic effects of Ca^{2+} and Mg^{2+} on ATP hydrolysis by the purified ATP-diphosphohydrolase. J. Biol. Chem. 257, 3869–3874.

LeBel, D., Poirier, G. G., Phaneuf, S., St-Jean, P., Laliberté, J.-F., & Beaudoin, A. R. (1980). Characterization and purification of a calcium-sensitive ATP diphosphohydrolase from pig pancreas. J. Biol. Chem. 255, 1227–1233.

Lee, K. H., & Eiler, J. J. (1951). Temperature-dependent characteristics of an adenylpyrophosphatase preparation from potatoes. Science 114, 393–395.

Liébecq, C., Lallemand, A., & Degueldre-Guillaume, M. J. (1962). The apyrase activity of potato extract. Arch. Biochem. Biophys. 97, 609–610.

Liébecq, C., Lallemand, A., & Degueldre-Guillaume, M. J. (1963). Purification partielle et propriétés de l'apyrase de la pomme de terre. Bull. Soc. Chim. Biol. 45, 573–594.

Lieberman, G. E., & Lewis, G. P. (1980). Properties of vascular ADPase. Artery 8, 374–380.

Lieberman, G. E., Leake, D. S., & Peters, T. J. (1982). Subcellular localization of adenosine diphosphatase in cultured pig arterial endothelial cells. Thromb. Haemostas. (Stittgart) 47, 249–253.

Lin, S.-H. (1985). The rat liver plasma membrane high affinity (Ca^{2+}-Mg^{2+})-ATPase is not a calcium pump. Comparison with ATP-dependent calcium transporter. J. Biol. Chem. 260, 10976–10980.

Lin, S.-H., & Fairn, J. N. (1984). Purification of (Ca^{2+}-Mg^{2+})-ATPase from rat liver plasma membranes. J. Biol. Chem. 259, 3016–3020.

Lin, S.-H., & Guidotti, G. (1989). Cloning and expression of a cDNA coding for a rat liver plasma membrane ecto-ATPase. The primary structure of the ecto-ATPase is similar to that of the human biliary glycoprotein I. J. Biol. Chem. 264, 14408–14414.

Lin, S.-H., & Russell, W. E. (1988). Two Ca^{2+}-dependent ATPases in rat liver plasma membrane. The previously purified (Ca^{2+}-Mg^{2+})-ATPase is not a Ca^{2+}-pump but an ecto-ATPase. J. Biol. Chem. 263, 12253–12258.

Lorenson, M. Y., Lee, Y.-C., & Jacobs, L. S. (1981). Identification and characterization of an anion-sensitive Mg^{2+}-ATPase in pituitary secretory granule membranes. J. Biol. Chem. 256, 12802–12810.

Lotersztajn, S., Hanoune, J., & Pecker, F. (1981). A high affinity calcium-stimulated magnesium-dependent ATPase in rat liver plasma membranes. Dependence on an endogenous protein activator distinct from calmodulin. J. Biol. Chem. 256, 11209–11215.

Lotersztajn, S., Epand, R. M., Mallat, A., & Pecker, F. (1984). Inhibition by glucagon of the calcium pump in liver plasma membranes. J. Biol. Chem. 259, 8195–8201.

Lüthje, J., Schomburg, A., & Ogilvie, A. (1988). Demonstration of a novel ecto-enzyme on human erythrocytes, capable of degrading ADP and of inhibiting ADP-induced platelet aggregation. Eur. J. Biochem. 175, 285–289.

Magocsi, M., & Penniston, J. T. (1991). Ca^{2+} or Mg^{2+} nucleotide phosphohydrolases in myometrium: two ecto-enzymes. Biochim. Biophys. Acta 1070, 163–172.

Mancilla, M., Kettlun, A. M., Valenzuela, M. A., & Traverso-Cori, A. (1984). Structural studies of two apyrases from *Solanum tuberosum*. Phytochemistry 23, 1397–1400.

Mancilla, M., Valenzuella, M. A., Anich, M., Kettlun, A. M., Jara, O., & Traverso-Cori, A. (1987). Identification of an apyrase activating protein and of calmodulin in *Solanum tuberosum*. Phytochemistry 26, 24721–2474.

Mant, M. J., & Parker, K. R. (1981). Two platelet aggregation inhibitors in tsetse (*Glossina*) saliva with studies of roles of thrombin and citrate in *in vitro* platelet aggregation. Br. J. Haematol. 48, 601–608.

Marchesi, V. T., & Barnett, R. J. (1963). The localization of nucleoside-phosphatase activity in different types of small blood vessels. J. Ultrastruct. Res. 10, 103–115.

Marchesi, V. T., & Barrnett, R. J. (1964). The localization of nucleosidephosphatase activity in different types of small blood vessels. J. Ultrastruct. Res. 10, 103–115.

Marcus, A. J., & Safier, L. B. (1993). Thromboregulation: Multicellular modulation of platelet reactivity in haemostasis and thrombosis. FASEB J. 7, 516–522.

Marcus, A. J., Safier, L. B., Hajjar, K. A., Ullman, H. L., Islam, N., Broekman, M. J., & Eiroa, A. M. (1991). Inhibition of platelet function by an aspirin-insensitive endothelial cell ADPase. Thromboregulation by endothelial cells. J. Clin. Invest. 88, 1690–1696.

Matsukawa, R. (1990). Separation of Ca^{2+}-ATPase from Mg^{2+}-ATPase in plasma membrane-rich fraction of bovine parotid gland. Arch. Biochem. Biophys. 280, 362–368.

Mazelis, M. (1959). Enzymatic degradation of adenosine triphosphate to adenine by cabbage leaf preparations. Plant Physiol. 34, 153–158.

Medzihradsky, F., Cullen, E. I., Lin, H. L., & Bole, G. G. (1980). Drug-sensitive ecto-ATPase in human leucocytes. Biochem. Pharmacol. 29, 2285–2290.

Meghji, P., Pearson, J. D., & Slakey, L. L. (1992). Regulation of extracellular adenosine production by ectonucleotidases of adult rat ventricular myocytes. Am. J. Physiol. 263, H40–H47.

Meyerhof, O. (1945). The origin of the reaction of harden and young in cell-free alcoholic fermentation. J. Biol. Chem. 157, 105–119.

Miller, D. L., & Weistheimer, F. H. (1966). The enzymic hydrolysis of γ-phenylpropyl di- and triphosphate. J. Am. Chem. Soc. 88, 1511–1513.

Missiaen, L., Wuytack, F., & Casteels, R. (1988a). Effect of ovarian steroids on membrane ATPase activities in microsomes (microsomal fractions) from rat myometrium. Inhibition of a component of the Mg^{2+}-activated ATPase by Ca^{2+}-calmodulin and by oxytocin. Biochem. J. 250, 571–577.

Missiaen, L., Wuytack, F., & Casteels, R. (1988b). Characterization of the Mg^{2+}-activated ATPase activity in smooth-muscle membranes. NADH oxidase and adenylate kinase interfere with NADH-coupled enzyme assay. Biochem. J. 250, 579–588.

Miura, Y., Hirota, K., Arai, Y., & Yagi, K. (1987). Purification and partial characterization of adenosine diphosphatase activity in bovine aorta microsomes. Thromb. Res. 46, 685–695.

Molnar, J., & Lorand, L. (1961). Studies on apyrases. Arch. Biochem. Biophys. 93, 353–363.

Moodie, F. D. L., Baum, H., Butterworth, P. J., & Peters, T. J. (1991). Purification and characterization of bovine spleen ADPase. Eur. J. Biochem. 202, 1209–1215.

Moulton, M. P., Sabbadini, R. A., Norton, K. C., & Dahms, A. S. (1986). Studies on the transverse tubule membrane Mg-ATPase. Lectin-induced alterations of kinetic behavior. J. Biol. Chem. 261, 12244–12251.

Moy, T. C., Walters, R. F., & Weiser, M. M. (1986). Intestinal basolateral membrane Ca^{2+}-ATPase activity with properties distinct from those of the Ca^{2+}-pump. Biochem. Biophys. Res. Commun. 141, 979–985.

Nagy, A. (1986). Enzymatic characteristics and possible role of synaptosomal ecto-adenosine triphosphatase from mammalian brain. In: Cellular Biology of Ectoenzymes (Kreutzberg, et al., eds.), pp. 49–59. Springer-Verlag, Berlin, Heidelberg.

Nagy, A., & Rosenberg, M. D. (1981). Adenosine triphosphate activity at the external surface of synaptosomes. Eighth Meet. Int. Soc. Neurochem., Nottingham.

Nagy, A., Shuster, T. A., & Rosenberg, M. D. (1983). Adenosine triphosphatase activity at the external surface of chicken brain synaptosomes. J. Neurochem. 40, 226–234.

Noda, L. (1973). Adenylate kinase. In: The Enzymes, Vol. 8, (Boyer, P. D., ed.), pp. 379–405. Academic Press, New York.

Novikoff, A. B., & Heus, M. (1963). A microsomal nucleoside diphosphatase. J. Biol. Chem. 238, 710–716.

Novikoff, P. M., Novikoff, A. B., Quintana, N., & Hauw, J. J. (1971). Golgi-apparatus, GERL, and lysosomes of neurons in rat dorsal root ganglia studied by thick section and thin section cytochemistry. J. Cell Biol. 50, 859–886.

Ochs, D. L., & Reed, P. W. (1984). Ca^{2+}-stimulated, Mg^{2+}-dependent ATPase activity in neutrophil plasma membrane vesicles. Coupling to Ca^{2+} transport. J. Biol. Chem. 259, 102–106.

Ogawa, K. S., Fujimoto, K., & Ogawa, K. (1986). Ultracytochemical studies of adenosine nucleotidases in aortic endothelial and smooth muscle cells Ca^{2+}-ATPase and Na^{+}, K^{+}-ATPase. Acta Histochem. Cytochem. 19, 601–620.

Okamoto, V. R., Moulton, M. P., Runte, E. M., Kent, C. D., Lebherz, H. G., Dahms, A. S., & Sabbadini, R. A. (1985). Characterization of transverse tubule membrane proteins: Tentative identification of the magnesium-ATPase. Arch. Biochem. Biophys. 237, 43–54.

Olsson, R. A., & Pearson, J. D. (1990). Cardiovascular purinoceptors. Physiol. Rev. 70, 761–845.

Paddle, B. M., & Burnstock, G. (1974). Release of ATP from perfused heart during coronay vasodilatation. Blood Vessels 11, 110–119.

Paddle, B. M., & Burnstock, G. (1987). Release of ATP from perfused heart during coronary vasodilatation. Blood Vessels 11, 110–119.

Papamarcaki, T., & Tsolas, O. (1990). Identification of ATP diphosphohydrolase activity in human term placenta using a novel assay for AMP. Molec. Cell Biochem. 97, 1–8.

Pearson, J. D. (1986). Ectonucleotidases of vascular endothelial cells: Characterization and possible physiological roles. In: Cellular Biology of Ectoenzymes (Kreutzberg, et al., eds.), pp. 16–26. Springer-Verlag Berlin, Heidelberg.

Pearson, J. D., & Gordon, J. L. (1979). Vascular endothelial and smooth muscle cells in culture selectively release adenosine nucleotides. Nature (London) 281, 384–386.

Pearson, J. D., Carleton, J. C., & Gordon, J. L. (1980). Metabolism of adenine nucleotides by ectoenzymes of vascular endothelial and smooth-muscle cells in culture. Biochem. J. 190, 421–429.

Pearson, J. D., Coade, S. B., & Cusack, N. J. (1985). Characterization of ectonucleotidases on vascular smooth-muscle cells. Biochem. J. 230, 503–507.

Pepys, J., & Edwards, A. M. (1979). The mast cells. Its role in health and disease. Proceedings of an International Symposium, Davos, Switzerland.

Picher, M., Côté, Y. P., Béliveau, R., Potier, M., & Beaudoin, A. R. (1993). Demonstration of a novel type of ATP-diphosphohydrolase (E.C.3.6.1.5.) in the bovine lung. J. Biol. Chem. 268, 4699–4703.

Picher, M., Béliveau, R., Potier, M., Savaria, D., Rousseau, É., & Beaudoin, A. R. (1994). Demonstration of an ectoATP-diphosphohydrolase (E.C.3.6.1.5.) in non-vascular smooth muscles of the bovine trachea. Biochim. Biophys. Acta 1200, 167–174.

Pieber, M., Valenzuela, M. A., Kettlun, A. M., Mancilla, M., Aranda, E., Collados, L., & Traverso-Cori, A. (1991). ATPase–ADPase activities of rat placental tissue. Comp. Biochem. Biophys. 100B, 281–285.

Plaut, G. W. E. (1955). An inosine diphosphatase from mammalian liver. J. Biol. Chem. 217, 235–245.

Plesner, L., Juul, B., Skriver, A., & Aalkjaer, C. (1991). Characterization of Ca^{2+}- or Mg^{2+}-dependent nucleoside triphosphatase from rat mesenteric small arteries. Biochim. Biophys. Acta 1067, 191–200.

Ragazzi, E., Wu, S. N., Shryock, J., & Belardinelli, L. (1991). Electrophysiological and receptor binding studies to assess activation of the cardiac adenosine receptor by adenine nucleotides. Circ. Res. 68, 1035–1044.

Ribeiro, J. M. C. (1987). Role of saliva in blood-feeding by arthropods. Ann. Rev. Entomol. 32, 463–478.

Ribeiro, J. M. C. (1989). Vector saliva and its role in parasite transmission. Exp. Parasitol. 69, 104–106.

Ribeiro, J. M. C., & Garcia, E. S. (1979). Apyrase activity in saliva and crop of *Rhodnius prolixus*. An. Acad. Bras. Ciênc. 51, 182.

Ribeiro, J. M. C., & Garcia, E. S. (1980). The salivary and crop apyrase activity of *Rhodnius prolixus*. J. Insect Physiol. 26, 303–307.

Ribeiro, J. M. C., & Garcia, E. S. (1981). Platelet antiaggregating activity in the salivary secretion of the blood sucking bug *Rhodnius prolixus*. Experimentia 37, 384–386.

Ribeiro, J. M. C., Sarkis, J. J. F., Rossignol, P. A., & Spielman, A. (1984a). Salivary apyrase of *Aedes aegypti*: Characterization and secretory fate. Comp. Biochem. Physiol. 79B, 81–86.

Ribeiro, J. M. C., Rossignol, P. A., & Spielman, A. (1984b). Role of mosquito saliva in blood vessel location. J. Exp. Biol. 108, 1–7.

Ribeiro, J. M. C., Sarkis, J. J. F., Rossignol, P. A., & Spielman, A. (1985a). Salivary gland apyrase determines probing time in anopheline mosquitoes. J. Insect Physiol. 31, 689–692.

Ribeiro, J. M. C., Makoul, G. T., Levine, J., Robinson, D. R., & Spielman, A. (1985b). Antihemostatic, antiinflammatory and immunosuppressive properties of the salivary of a tick, *Ixodes dammini*. J. Exp. Med. 161, 332–344.

Ribeiro, J. M. C., Rossignol, P. A., & Spielman, A. (1986). Blood-finding strategy of a capillary feeding sandfly *Lutzomyia longipalpis*. Comp. Biochem. Physiol. 83A, 683–686.

Ribeiro, J. M. C., Modi, G. B., & Tesh, R. B. (1989). Salivary apyrase activity of some Old World phlebotomine sandflies. Insect Biochem. 19, 409–412.

Ribeiro, J. M. C., Vaughan, J. A., & Azad, A. F. (1990). Characterization of the salivary apyrase of three rodent flea species. Comp. Biochem. Physiol. 95B, 215–219.

Ribeiro, J. M. C., Endris, T. M., & Endris, R. (1991). Saliva of the soft tick, *Ornithodoros moubata*, contains anti-platelet and apyrase activities. Comp. Biochem. Physiol. 100A, 109–112.

Rigbi, M., Levy, H., Iraqui, F., Teitelbaum, M., Orevi, M., Alajoutsijärvi, A., Horowitz, A., & Galun, R. (1987). The saliva of the medicinal leech *Hirudo medicinalis*. I. Biochemical characterization of the high molecular weight fraction. Comp. Biochem. Physiol. 87B, 567–573.

Ronca-Testoni, S., & Borghini, F. (1982). Degradation of perfused adenine compounds up to uric acid in isolated rat heart. J. Mol. Cell. Cardiol. 14, 177–180.

Rorive, K. G., & Kleinzeller, A. (1972). The effect of ATP and Ca^{2+} on the cell volume in isolated kidney tubules. Biochim. Biophys. Acta 274, 226–239.

Rosenblatt, D. E., Lauter, C. J., & Trams, E. C. (1976). Deficiency of a Ca^{2+}-ATPase in brains of seizure prone mice. J. Neurochem. 27, 1299–1304.

Rossignol, P. A., Ribeiro, J. M. C., & Spielman, A. (1984). Increased intradermal probing time in sporozoite-infected mosquitoes. Am. J. Trop. Med. Hyg. 33, 17–20.

Ryan, U. U. (1986). Metabolic activity of pulmonary endothelium: Modulations of structure and function. Ann. Rev. Physiol. 48, 263–277.

Saborido, A., Moro, G., & Megias, A. (1991). Transverse tubule Mg^{2+}-ATPase of skeletal muscle. Evidence for extracellular orientation of the chicken and rabbit enzymes. J. Biol. Chem. 266, 23490–23498.

Sadler, J. E., Beyer, T. A., Oppehheimer, C. L., Paulson, J. C., Prieels, J. I., Rearick, J. I., & Hill, R. L. (1982). Purification of mammalian glycosyltransferases. Meth. Enzymol. 83, 458–514.

Sarkis, J. J. F., & Salto, C. (1991). Characterization of synaptosomal ATP diphosphohydrolase from the electric organ *Torpedo marmorata*. Brain Res. Bull 26, 871–876.

Sarkis, J. J. F., Guimaraes, J. A., & Ribeiro, J. M. C. (1986). Salivary apyrase of *Rhodnius prolixus*: Kinetics and purification. Biochem. J. 233, 885–891.

Schadeck, R. J. G., Sarkis, J. J. F., Dias, R. D., Araujo, H. M. M., & Souza, D. O. G. (1989). Synaptosomal apyrase in the hypothalamus of adult rats. Braz. J. Med. Biol. Res. 22, 303–314.

Shubert, P., Lee, K., & Kreutzberg, G. W. (1981). Formation and function of adenosine in the CNS. I. Release and modulatory action. Eighth Meet. Soc. Neurochem., Nottingham.

Smith, G. P., & Peters, T. J. (1981). Subcellular localization and properties of adenosine diphosphatase activity in human polymorphonuclear leukocytes. Biochim. Biophys. Acta 673, 234–242.

Smith, G. P., Shah, T., Webster, A. D. B., & Peters, T. J. (1981). Studies on the kinetic properties and subcellular localization of adenosine diphosphatase in human peripheral blood lymphocytes. Clin. Exp. Immunol. 46, 321–326.

Smith, J. J. B., Cornish, R. A., & Wilkes, J. (1980). Properties of a calcium-dependent apyrase in the saliva of the blood-feeding bug, *Rhodnius prolixus*. Experimentia 36, 898–899.

Smolen, J. E., & Weissmann, G. (1978). Mg^{2+}-ATPase as a membrane ecto-enzyme of human granulocytes. Inhibitors, activators and response to phagocytosis. Biochim. Biophys. Acta 512, 525–538.

Sorensen, R. G., & Mahler, H. R. (1982). Localization of endogenous ATPase at the nerve terminal. J. Bioenerg. Biomembr. 14, 527–547.

Stefanovic, V., Ciesielski-Treska, J., Ebel, A., & Mandel, P. (1974). Nucleoside triphosphate activity at the external surface of neuroblastoma cells. Brain Res. 81, 427–441.

Stefanovic, V., Ledig, M., & Mandel, P. (1976). Divalent cation activated ecto-nucleoside triphosphatase activity of nervous cells in tissue culture. J. Neurochem. 27, 799–805.

Strobel, R. (1992). Purification and immunochemical characterization of ectoATP-diphosphohydrolase from chicken oviduct and liver. Thesis. University of Minnesota.

Strobel, R., & Rosenberg, M. (1992). Immunoaffinity chromatographic purification of chicken ectoATP diphosphohydrolase. Ann. N.Y. Acad. Sci. 671, 487–489.

Sun, H.-T., Yoshida, Y., & Imai, S. (1990). A Ca^{2+}-activated, Mg^{2+}-dependent ATPase with high affinity for both Ca^{2+} and Mg^{2+} in vascular smooth muscle microsomes: Comparison with plasma membrane Ca^{2+}-pump ATPase. J. Biochem. 108, 730–736.

Szekely, M. (1951). Purification of potato apyrase and investigation into its reaction mechanisms. Acta Chim. Acad. Sci. Hung. 1, 325–333.

Teo, T. S., Thiyagarajah, P., & Lee, M. K. (1988). Characterization of a high affinity Ca^{2+}-stimulated, Mg^{2+}-dependent ATPase in the rat parotid plasma membrane. Biochim. Biophys. Acta 945, 202–210.

Tognoli, L. (1984). Interaction of nucleoside triphosphate diphosphohydrolase from pea with a membrane protein factor. Phytochemistry 23, 515–518.

Tognoli, L., & Marrè, E. (1981). Purification and characterization of a divalent cation-activated ATPase-ADPase from pea stem microsomes. Biochim. Biophys. Acta 642, 1–14.

Torres, M., Pintor, J., & Miras-Portugal, T. (1990). Presence of ectonucleotidases in cultured chromaffin cells: Hydrolysis of extracellular adenine nucleotides. Arch. Biochem. Biophys. 279, 37–44.

Trams, E. G., & Lauter, C. J. (1974). On the sideness of plasma membrane enzymes. Biochim. Biophys. Acta 345, 180–197.

Trams, E. G., & Lauter, C. J. (1978). Ecto-ATPase deficiency in glia of seizure prone mice. Nature (London) 271, 270–271.

Traverso-Cori, A., & Cori, O. (1962). Splitting of the terminal phosphate group of adenosine triphosphate by potato apyrase. Biochim. Biophys. Acta 57, 158–160.

Traverso-Cori, A., Chaimovich, H., & Cori, O. (1965). Kinetic studies and properties of potato apyrase. Arch. Biochem. Biophys. 109, 173–184.

Traverso-Cori, A., Traverso, S., & Reyes, H. (1970). Different molecular forms of potato apyrase. Arch. Biochem. Biophys. 137, 133–142.

Valenzuela, M. A., Del Campo, G., Marin, E., & Traverso-Cori, A. (1973). Effects of protein-modifying reagents on an isoenzyme of potato apyrase. Biochem. J. 133, 755–763.

Valenzuela, M. A., Kettlun, A. M., Mancilla, M., Calvo, V., Fanta, N., & Traverso-Cori, A. (1988). The effect of bivalent metal ions on ATPase-ADPase activities of apyrase from *Solanum tuberosum*. Phytochemistry 27, 1981–1985.

Valenzuela, M. A., Lopez, J., Depix, M., Mancilla, M., Kettlun, A. M., Catalan, L., Chiong, M., Garrido, J., & Traverso-Cori, A. (1989). Comparative subcellular distribution of apyrase from animal and plant sources. Characterization and microsomal apyrase. Comp. Biochem. Physiol. 93B, 509–513.

Valenzuela, M. A., Collados, L., Kettlun, A. M., Mancila, M., Lara, H., Puente, J., Aranda, E., Chayet, L., Alvarez, A., & Traverso-Cori, A. (1992). Changes in apyrase activity in uterus and mammary gland during the lactogenic cycle. Comp. Biochem. Physiol. 103B, 113–118.

Vara, F., & Serrano, R. (1981). Purification and characterization of a membrane-bound ATP diphosphohydrolase from *Cicer arietinum* (chick-pea) roots. Biochem. J. 197, 637–643.

Vasconcelos, E. G., Nascimento, P. S., Mierelles, M. N. L., Verjovski-Almeida, S., & Ferreira, S. T. (1993). Characterization and localization of an ATP-diphosphohydrolase on the external surface of the tegument of *Schistosoma mansoni*. Mol. Biochem. Parasitol. 58, 205–214.

Wattiaux-de-Coninck, S., & Wattiaux, R. (1969). Nucleoside diphosphatase activity in plasma membrane of rat liver. Biochim. Biophys. Acta 183, 118–128.

Weiss, B., & Sachs, L. (1977). Differences in surface membrane ecto-ATPase and ecto-AMPase in normal and malignant cells. I. Decrease in ecto-ATPase in myeloid leukemic cells and the independent regulation of ecto-ATPase and ecto-AMPase. J. Cell Physiol. 93, 183–188.

Welford, L. A., Cusack, N. J., & Hourani, S. M. O. (1986). ATP analogues and the guinea-pig *taenia coli*: A comparison of the structure–activity relationships of ectonucleotidases with those of the P2-purinoceptor. Eur. J. Pharmacol. 129, 217–224.

Welford, L. A., Cusack, N. J., & Hourani, S. M. O. (1987). The structure-activity relationships of ectonucleotidases and of excitatory P2-purinoceptors: evidence that dephosphorylation of ATP analogues reduces pharmacological potency. Eur. J. Pharmacol. 141, 123–130.

White, T. D. (1978). Release of ATP from a synaptosomal preparation by elevated extracellular potassium and by veratridine. J. Neurochem. 30, 329–336.

Wilson, P. D., Rustin, G. J., Smith, G. P., & Peters, T. J. (1981). Electron microscopic cytochemical localization of nucleoside phosphatases in normal and chronic granulocytic human neutrophils. Histochem. 13, 73–84.

Wilson, P. D., Lieberman, G. E., & Peters, T. J. (1982). Ultrastructural localization of adenosine diphosphatase activity in cultured aortic endothelial cells. Histochem. J. 14, 215–219.

Yagi, K., Arai, Y., Kato, N., Hirota, K., & Miura, Y. (1989). Purification of ATP diphosphohydrolase from bovine aorta microsomes. Eur. J. Biochem. 180, 509–513.

Yagi, K., Shinbo, M., Hashizume, M., Shimba, L. S., Kurimura, S., & Miura, Y. (1991). ATP diphosphohydrolase is responsible for ecto-ATPase and ecto-ADPase activities in bovine aorta endothelial and smooth muscles cells. Biochem. Biophys. Res. Commun. 180, 1200–1206.

Yagi, K., Kato, N., Shinbo, M., Shimba, L. S., & Miura, Y. (1992). Purification and characterization of adenosine diphosphatase from human umbilical vessels. Chem. Pharm. Bull. 40, 2143–2146.

Yamazaki, M., & Hayaishi, O. (1965). The activation of nucleoside diphosphatase by nucleoside triphosphates and its possible metabolic significance. J. Biol. Chem. 240, 2761–2762.

Yamasaki, M., & Hayaishi, O. (1969). Allosteric properties of nucleoside diphosphatase and its identity with thiamine pyrophosphatase. J. Biol. Chem. 243, 2934–2942.

Zimmermann, H., Dowdall, M. J., & Lane, D. A. (1979). Purine salvage at the cholinergic nerve endings of the *Torpedo* electric organ: The central role of adenosine. Neuroscience 4, 979–993.

Zimmermann, H., Grondal, E. J. M., & Keller, F. (1986). Hydrolysis of ATP and formation of adenosine at the surface of cholinergic nerve endings. In: Cellular Biology of Ectoenzymes (Kreutzberg, et al., eds.), pp. 35–48. Springer-Verlag, Berlin, Heidelberg.

THE KDP-ATPASE OF *ESCHERICHIA COLI*

Karlheinz Altendorf and Wolfgang Epstein

Principles of
Volume 5, pages 403–420.

ISBN: 1-55938-662-2.

I. INTRODUCTION

Kdp is an active transport system that accumulates K^+ in *Escherichia coli*, where it was first identified, as well as in many other bacteria. Kdp is one of several systems that accumulate K^+ in bacteria. Kdp is the system with the highest affinity for K^+ and is necessary only when the external concentration of K^+ is low. The role of Kdp as a reserve or stand-by system is supported by its regulation. Kdp is expressed only when other systems are unable to meet the cells' needs for K^+. It is somewhat surprising that this regulation does not appear to be determined by K^+ per se, but rather by the effect of cytoplasmic K^+ in maintaining the level of turgor pressure needed for growth of bacteria.

Kdp is a P-type transport ATPase of unique structure, a complex of three membrane proteins ranging in size from 20 to 72 kD. A 3 kD peptide may also be part of the Kdp complex. The fact that Kdp is not necessary under most conditions allows it to be manipulated genetically at will. A variety of mutations, including deletions and point mutations that abolish or alter Kdp activity or regulation, are readily isolated. The ability to alter Kdp at will makes it considerably easier to study than most eukaryotic P-type ATPases that have an essential function in the cell.

The Kdp system has been the subject of several recent reviews (Epstein, 1990; Siebers and Altendorf, 1993; Altendorf and Epstein, 1993). Here we will stress the latest developments and the major questions that studies of this system are addressing.

II. GENETIC STRUCTURE OF THE *kdp* GENES

All of the genes for the components of Kdp as well as for their specific regulation are clustered (Polarek et al., 1992) in a single, continuous region (Figure 1) of almost 8 kb near min 16 of the map of *E. coli* (Bachmann, 1990). These genes constitute two transcription units or operons, both of which are read in the clockwise direction of the map. The *kdpFABC* operon begins at the upstream extremity and encodes the three large protein subunits of Kdp as well as the small KdpF peptide whose function is not known. The promoter region extends about 80 base pairs upstream of the transcription start site, and is the region where the KdpE regulatory protein binds (Sugiura et al., 1992). The promoter region is characterized by runs of T residues with a period of about 10 base pairs, a pattern that leads to bending of DNA (Tanaka et al., 1991). That the Kdp promoter region is a region of bending is also suggested by the fact that it was isolated in a search for sequences that resulted in bending of the DNA (Tanaka et al., 1991). The -10 and -35 regions of the promoter resemble the consensus for such sequences fairly well, but the promoter does not appear to be functional by itself, since absence of the regulatory proteins results in very low expression of Kdp (Polarek et al., 1992).

The first open reading frame of the *kdpFABC* transcript is that for the 29-residue KdpF peptide whose GUG start codon begins 28 bases from the start of the

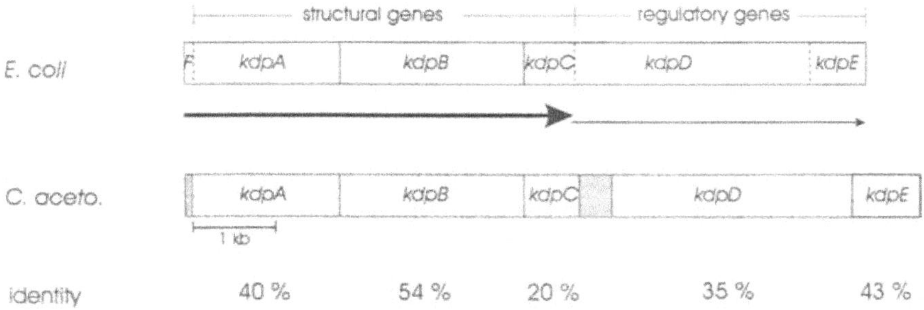

Figure 1. A diagram to scale of the *kdp* genes of *E. coli* (top) and of *C. acetobutylicum* (bottom). Dashed lines indicate gene junctions where stop and start codons overlap, or where a somewhat larger overlap occurs; solid lines and the shaded box show junctions at which genes are separated by a number of untranslated bases. The large arrow represents the major, controlled transcript that encodes the Kdp structural proteins; the thinner arrow shows the weaker constitutive transcript that encodes the two regulatory proteins.

transcript. Its UGA stop codon overlaps the AUG start codon of the 557 residue KdpA protein. The coding sequences for the 682 residue KdpB and the 190 residue KdpC proteins follow, separated from the preceding coding region by 22 and eight untranslated bases, respectively. The start codons of each of the four genes encoded by the *kdpFABC* transcript are preceded by Shine-Dalgarno sequences similar to the consensus for sequence and spacing. Most *kdpFABC* transcripts end early in the *kdpD* gene, but a fraction of them continue to transcribe both genes of the distal *kdpDE* operon (Polarek et al., 1992).

The *kdpDE* operon encodes regulatory proteins, the membrane-bound 98 kD KdpD sensor kinase and the soluble 26 kD KdpE response regulator (Walderhaug et al., 1992). The promoter of this operon is within the *kdpC* gene, the transcript beginning some 90 base pairs before the end of the gene. There are only fair matches of the -10 and -35 regions with the consensus for sigma-70 promoters of *E. coli*. Expression from this promoter is at a relatively low level and does not appear to be subject to any regulation. The start codon of the large and membrane-bound KdpD protein is five base pairs upstream of the stop codon of the KdpC protein. The UGA stop codon of KdpD overlaps the ATG start codon of KdpE. Distal to *kdpE* is a region of several hundred base pairs after which there are genes for ornithine decarboxylase and a putrescine transport protein (Kashiwagi et al., 1991).

Each of the three large subunits of the Kdp complex ought to be made in equimolar amounts, since they are so present in the Kdp complex. This is achieved by the small, untranslated regions between *kdpA* and *kdpB*, and between *kdpB* and *kdpC*. Curiously, overlap ('translational coupling'; Normark et al., 1983) occurs between the *kdpD* and *kdpE* genes where there is no need for stoichiometry production, and between *kdpF* and *kdpA* where this need might exist. The function that overlap of genes serves in the *kdp* cluster is unclear, but it seems that producing equimolar amounts of gene products is not involved.

The only other *kdp* genes sequenced to date are from the Gram-positive anaerobe, *Clostridium acetobutylicum* (Treuner and Dürre, personal communication). The clustering and order of genes are identical to that in *E. coli*, and in spite of the large evolutionary distance, the genes are over 40% identical in amino acid sequence (Figure 1).

III. STRUCTURE OF THE KDP COMPLEX

Kdp is a complex containing equimolar amounts of each of its three large subunits: KdpA, KdpB, and KdpC. Intracistronic complementation between some mutations affecting KdpA (Epstein and Davies, 1970) indicates that the complex is oligomeric

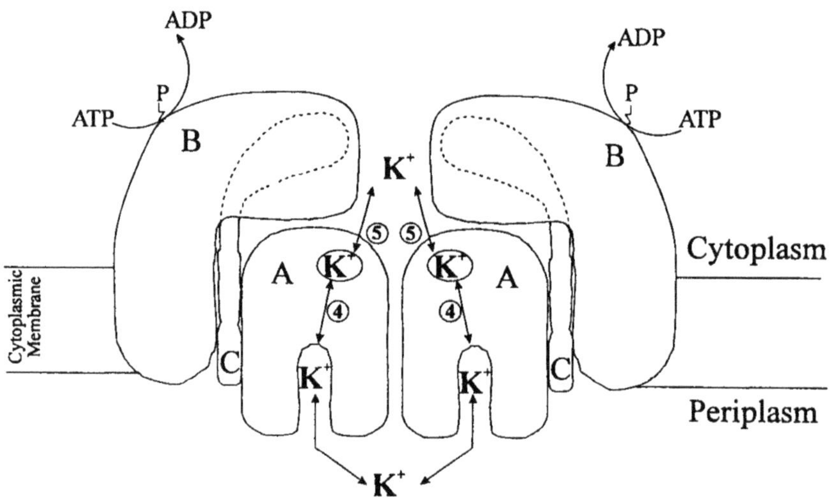

Figure 2. A schematic diagram indicating how Kdp may act to transport K^+. Kdp is shown as a dimer in which contact between the two halves is mediated by the KdpA subunit. KdpC is shown between the other two subunits because it appears to be necessary for assembly of KdpA and KdpB. Steps in the transport of K^+ are numbered to correspond with those of the kinetic scheme of Figure 4. Phosphorylation of KdpB on Asp 307 is an early step. Not shown are changes corresponding to steps 2 and 3 of Figure 4, in which a major conformational change and hydrolysis of the aspartyl-phosphate intermediate occur. Then, in step 4, binding of K^+ at the periplasmic binding site results in release of phosphate from the enzyme and movement of K^+ to a site within the cytoplasmic domain of KdpA. Finally, in step 5, binding of ATP results in release of K^+ to the cytoplasm.

and must have at least two copies of each type of subunit. Thus the minimal size of Kdp would be as an $A_2B_2C_2$ complex. The Kdp complex is stable to solubilization in a number of nonionic detergents, and has been purified to near homogeneity (Siebers et al., 1992). The largest subunit, the 72 kD KdpB protein, has extensive regions of homology with the large subunit of other P-type ATPases. The 59 kD KdpA subunit is very hydrophobic and has a large number of segments that are predicted to span the membrane. The 20.5 kD KdpC subunit is predicted to span the membrane only once. Neither KdpA nor KdpC has extensive regions of homology to other proteins in the Genebank data base. A suggested structure of Kdp is shown in Figure 2.

A. The Topology of KdpB

The topology of KdpB appears to conform to that of most other P-type ATPases, characterized by two sets of closely-spaced membrane-spanning regions in the N-terminal half of the protein and two to four membrane-spanning regions near the C-terminus. The rest of the protein is cytoplasmic, where the highly conserved DKTGT sequence that contains the phosphorylated Asp residue and other conserved sequences are found. Extensive cytoplasmic exposure of KdpB is consistent with its sensitivity to protease digestion in inside-out vesicle preparations (Altendorf et al., 1992). This structure matches the key features of all P-type ATPases (Epstein et al., 1990) and is compatible with the proposed head–piece–stalk model for this class of enzymes (Serrano, 1988; Taylor et al., 1986). The latter model has recently been supported by three-dimensional cryo-electron microscopy of the Ca^{2+}-ATPase of sarcoplasmic reticulum at 14 Å resolution (Toyoshima et al., 1993).

B. The Topology of KdpA

The topology of the KdpA subunit (Figure 3) has recently been studied by the use of protein fusions to alkaline phosphatase and to β-galactosidase and shown to have 10 membrane-spanning segments (Buurman et al., 1995). The region between residues 356 and 399 is quite hydrophobic and therefore predicted to form two membrane spans in most programs that predict membrane spans on the basis of hydrophobicity (Eisenberg et al., 1984; Rao and Argos, 1986). The fusions show that this region is cytoplasmic and thus analogous to the hydrophobic internal parts of soluble proteins. The C-terminus is periplasmic, while the N-terminus is predicted to be in the membrane, perhaps close to the periplasmic surface. All of the extramembranous loops, with the exception of the region from 356 to 399, are hydrophilic. The distribution of charged residues is typical for membrane proteins in bacteria where basic residues are found predominantly in cytoplasmic loops (von Heijne, 1986).

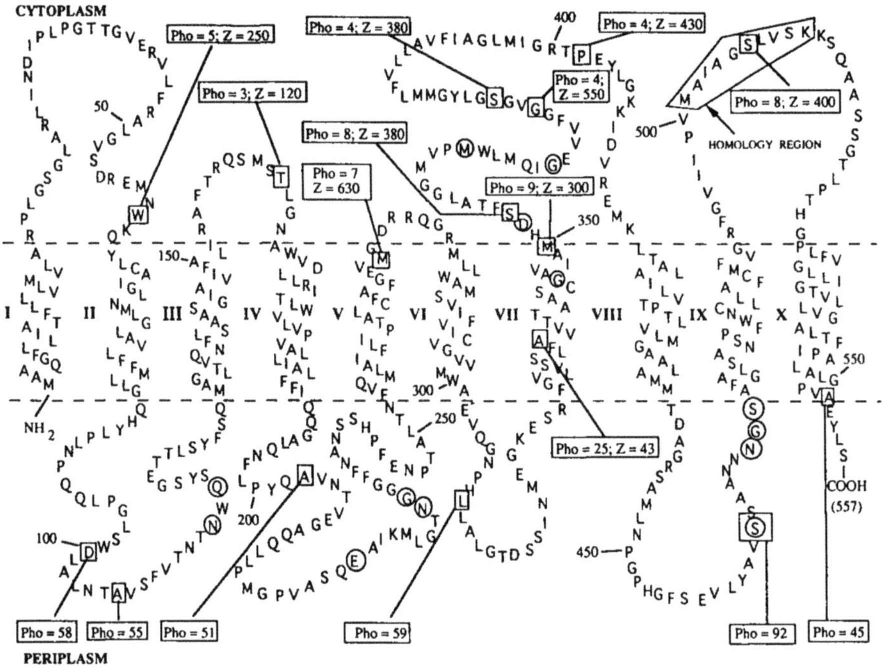

Figure 3. Topology of the KdpA protein as deduced from analysis of protein fusions and locations of residues that reduce affinity for K^+. Dashed lines separate the membrane from the cytoplasm at the top and the periplasm at the bottom. The amino- and carboxy-termini, and every 50th residue are labeled. Membrane spans, shown as helical barrels with roman numerals, are at the center. The pentagon at the top shows a small region of homology with the large subunit of other P-type ATPases. Small boxes are sites of fusions whose activities are in the larger, connected boxes. PhoA is alkaline phosphatase activity, Z is β-galactosidase activity. Circles are residues where alterations reduce affinity for K^+ (reproduced with permission from Buurman et al., 1995).

C. The Topology of KdpC

KdpC appears to have only a single membrane-spanning α-helix close to its N-terminus. The rest of the protein presumably is in the cell interior, but as indicated below is not extensively exposed to the cytoplasm in the Kdp complex. Predicted characteristics of its secondary structure are two antiparallel β-sheets and an amphipathic α-helix at the carboxy-terminus.

The predicted structure of KdpC is similar to that of the β1 and β2 isoforms of the small subunit of the Na^+, K^+-ATPase (Shull et al., 1986) and the β-subunit of the H^+, K^+-ATPase (Shull, 1990). All of these proteins have a single transmembrane helix near their N-termini followed by a large extramembraneous domain. However, this resemblance is probably fortuitous. There is very little similarity of their amino acid sequences, and the extramembraneous parts of the β-subunits of the eukaryotic enzymes are outside the cell and are heavily glycosylated. There may be a similarity in function, since the β-subunits (McDonough et al., 1990) and KdpC (see following) seem to be important in the assembly of a functional ATPase complex.

There is evidence for an ATP-binding site on KdpC that may have a regulatory function. Recent photoaffinity labeling of the Kdp complex with [^{32}P]-8-azido-ATP resulted in labeling of both the KdpB and KdpC, with greater incorporation of radioactivity into KdpC. Labeling in KdpC was in the N-terminal cyanogen bromide fragment, from Met1 to Met75 (Dröse and Altendorf, unpublished observations). The notion that ATP may play a regulatory role in addition to that of substrate was suggested earlier in enzymatic studies of Kdp (Siebers and Altendorf, 1989).

D. The KdpF Peptide

This 29-residue peptide was first identified as an open reading frame. Subsequently, it has been shown that the protein is synthesized when *kdpF* is selectively expressed in a minicell system (Möllenkamp and Altendorf, unpublished observations). A protein fusion joining the first 23 residues of KdpF to β-galactosidase has β-galactosidase activity of which over 80% is membrane bound (Covello and Epstein, unpublished observations). Since β-galactosidase is not readily exported (Silhavy and Beckwith, 1985), this result suggests that the N-terminus of this very hydrophobic peptide is near the outside of the membrane while the C-terminus is in or near the cytoplasm.

E. Assembly of the Kdp Complex

Information about how the three subunits of Kdp interact in the complex has been obtained by solubilizing membrane proteins with a nonionic detergent and examining the structure formed when only two of the three subunits are made (Siebers, Epstein, and Altendorf, unpublished observations). Strains with an *amber* mutation in one of the three Kdp subunits were found to produce only the other two subunits; not even an *amber* fragment was seen, suggesting that the fragments were rapidly degraded. When the Kdp proteins were solubilized in the strain lacking KdpC, the KdpA and KdpB subunits no longer comigrated during ion exchange or dye–ligand chromatography, suggesting they were not associated. Conversely, when only KdpA and KdpC were made they comigrated during separation. The same behavior was seen when only KdpB and KdpC were made although these results were complicated by considerable proteolytic degradation of KdpB. These results imply that only KdpC makes contact with each of the other two subunits. Thus KdpC probably is important either in forming a stable complex and/or in maintaining this structure. Accordingly, we have drawn the Kdp complex so that KdpC lies between the KdpA and KdpB subunits (Figure 2).

Mutants in which the KdpF peptide is made in reduced amounts, by changing the GUG start codon to CUG, or in which it is deleted appear to have reduced Kdp activity. It is clear that KdpF is not essential for Kdp function, since its deletion does not abolish the ability of constructs to complement *kdp* deletion mutants (Möllenkamp and Altendorf, unpublished observations). However, the peptide may be important in facilitating assembly and/or in altering the kinetic properties of Kdp.

IV. ENZYMOLOGY OF KDP

Kdp mediates a cycle of phosphorylation that resembles those of the other P-type ATPases. Formation of the phosphorylated intermediate of KdpB from ATP is extremely rapid and requires only μmolar concentrations of ATP (Siebers and Altendorf, 1989; Naprstek et al., 1992). The intermediate is relatively stable at acid pH and very labile at alkaline pH, as is typical of acyl phosphates (Post and Kume, 1973). Mutation of the Asp 307 in the conserved DKTGT sequence to glutamate or several other amino acids abolished transport activity *in vivo* and ATPase activity *in vitro* (Puppe et al., 1992). These results indicate that Asp 307 is probably the site of phosphorylation in Kdp.

Kinetic analysis of the formation and discharge of the phosphorylated intermediate is consistent with a reaction cycle similar to, but somewhat different from that of other P-type ATPases (Siebers and Altendorf, 1989; Naprstek et al., 1992). The kinetic model of Kdp shown in Figure 4 is based in large measure on data for other P-type enzymes, particularly on data for the Na^+, K^+-ATPase (Glynn and Karlish, 1975). Such models include two conformational forms of the enzyme, E_1 and E_2. Studies of Kdp have shown that no cation appears to be necessary to form E_1–P in step 1, and that E_1–P is of high energy since step 1 is readily reversible. The unusual finding is that there must be very little of the E_2–P form since it could not be detected as a kinetic intermediate.

After formation of the first intermediate, E_1–P, there is a conformational change to yield a low energy form, E_2–P. In Kdp, the latter must be very rapidly hydrolyzed to yield a form of E_2 which may retain phosphate but as a noncovalent ligand. Such an intermediate is known for bacterial alkaline phosphatase (Coleman and Gettins, 1983). By analogy with other ATPases, E_2 (or E_2–P) accepts K^+ for transport to form a conformation in which phosphate is released, but K^+ is bound (occluded). In step 5, K^+ is liberated in the cytoplasm in a reaction greatly stimulated by the binding of ATP. The binding of ATP in step 5 explains the paradox that the apparent affinity (K_m) of the free E_1 form of the enzyme for ATP is 0.001 mM or lower, while the affinity for the ATPase reaction is about 0.1 mM. Thus, there must be a step involving ATP in which the affinity is only modest. Low affinity binding of ATP in step 5 has been demonstrated in other ATPases (Post et al., 1972).

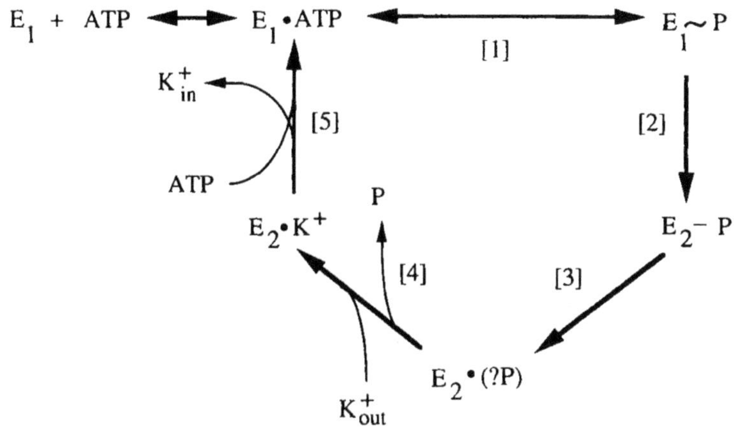

Figure 4. A proposed kinetic scheme of the steps in transport and ATP hydrolysis by Kdp. E_1 and E_2 represent two conformational states that have been well documented for several other P-type ATPases. All steps are presumed to be reversible; those rapidly and readily reversible have double ended arrows. Steps 1, 2, and 5 are very similar to those of other P-type ATPases. Since Kdp is not believed to export any ion, we show no ion dependence of step 1 (in P-type ATPases that export ions, this is the step dependent on and associated with occlusion of an exported ion). The conformational change from $E_1 \sim P$ to E_2 in step 2 is followed very rapidly by step 3 in which it is assumed that E_2 retains phosphate, but no longer in a covalently bound state. In the subsequent step 4, phosphate is released and K^+ is bound (occluded) in the enzyme. The final step is release of K^+ to the cytoplasm, a step that is postulated to be stimulated by ATP binding and is the step responsible for the low affinity of the enzyme for ATP.

V. MECHANISM OF TRANSPORT BY KDP

A. Stoichiometry of Transport

Major questions about Kdp that have yet to be answered are the nature and numbers of ions transported. *In vivo*, the system accumulates K^+ to achieve very high gradients which indicate that, at most, two K^+ ions can be transported per ATP hydrolyzed (Epstein et al., 1978). There is neither a Na^+ requirement for transport *in vivo* nor for ATPase activity *in vitro*, suggesting that Na^+ is not a substrate. There remains the possibility that Kdp exports protons coupled to uptake of K^+. Vesicle systems that could test this remain to be perfected. Kdp is active in native right-side-out vesicles when provided with an internal ATP generating system (Kollmann and Altendorf, 1993), but proton movement due to Kdp is difficult to measure in such vesicles. Reconstitution of the Kdp-ATPase in liposomes has been achieved, but the rate of K^+ transport is too low and permeability to protons is too high to allow a test of proton movement by Kdp (unpublished observations). In our diagram illustrating transport by Kdp (Figure 4), we assume that only K^+ is transported.

B. Initial Binding of K^+

The regions of Kdp that determine binding of K^+ were identified by the isolation and characterization of mutants in which the affinity of Kdp for K^+ was reduced by a factor ranging from 100-fold to more than 10,000-fold (Buurman et al., 1995). Thirty seven independent mutants of this type were examined; 33 of them alter the KdpA protein. These 33 represent only 16 different mutations, many having been isolated independently several times. The residues altered in the affinity mutants are circled in the diagram of Figure 3. Transport rates in all of the mutants were at least 30% of the wild-type, indicating that the change in affinity did not greatly affect rate-limiting steps in transport. The mutants also alter affinity for Rb^+, which is a poor substrate for wild-type Kdp, but is a fairly good substrate of many of the mutants.

Three of the mutants altered the KdpB subunit and one altered KdpC. The mutant altering KdpC and two of those altering KdpB resulted in a drastic reduction in the rate of transport. We believe that these mutations may alter affinity indirectly, by an allosteric effect in which a conformational change is transmitted to regions involved in binding.

The residues of KdpA altered by the affinity mutants are clustered in four well separated parts of the *kdpA* gene. Three of the clusters identified by mutations are in the first, second, and fourth periplasmic loops of KdpA, and the other cluster is a large cytoplasmic loop. The locations of the residues affected by these mutations suggest that K^+ first interacts with Kdp at a binding site formed by three of the periplasmic loops of KdpA. K^+ in the external medium would have free access to bind at this site.

C. Transmembrane Movement of K^+

The identification of two K^+-binding sites in KdpA, one in the periplasm and one in the cytoplasmic compartment, leads to the suggestion that this subunit forms the path for transmembrane movement of K^+ as well as the sites for K^+ binding. The cluster of residues that alter affinity in the fourth periplasmic loop ends at the beginning of membrane span IX. One of the residues implicated in the cytoplasmic binding site is in membrane span VII. Therefore, both membrane spans VII and IX are candidates to participate in the structure through which K^+ moves. Placing these features in the context of transport, K^+ binding in the periplasmic site leads to a conformational change which allows K^+ to move through the membrane.

After crossing the membrane, K^+ is bound at the other binding site in the large fourth cytoplasmic loop of KdpA. Since one of the residues identified is in membrane span VII, part of this site may be within the membrane. The movement of K^+ from the outside to the inside binding site must be reversible, since mutations at the inside site (as well as the outside site) affect affinity for K^+ as well as for one of its cation analogs, Rb^+. If this step were not reversible, alterations of the internal site would not alter apparent affinity.

Transmembrane movement of K^+ corresponds to step 4 in the kinetic scheme of Figure 4. K^+ in the cytoplasmic site is presumably occluded. K^+ occlusion by Kdp has not been demonstrated, but is inferred by analogy with other ATPases (Glynn and Karlish, 1990).

D. Release of K^+ to the Cytoplasm

In the final step of transport, corresponding to step 5 in Figure 4, K^+ is released into the cytoplasm. As indicated in the kinetic scheme, release is stimulated by binding of ATP and is associated with a change in conformation from the E_2 to the E_1 form of the enzyme which is then immediately phosphorylated to begin the cycle again. The cytoplasmic binding site is probably very close to parts of KdpB where ATP binds, since it is believed that most of the cytoplasmic domain of KdpA is covered by the large cytoplasmic domains of KdpB which binds ATP and is the site of phosphorylation. The suggested proximity of KdpB to the cytoplasmic binding site could explain the finding that some mutations altering KdpB reduce affinity without much effect on rate. This is true of one of the mutations isolated in the selection for reduced affinity (Buurman et al., 1995) and of several site-directed alterations of Asp 300 of KdpB (Puppe et al., 1992). It is possible that the cytoplasmic binding site is made up in part by residues of KdpB.

A comparison of amino acid sequences of K^+ channels of prokaryotes and eukaryotes revealed a motif H5, flanked by two hydrophobic segments M1 and M2, of low, but significant similarity (Jan and Jan, 1994). Based on mutational analyses, this motif has been suggested to confer ion specificity to those channels. It is interesting to note that KdpA may contain two sets of M1, H5, and M2 segments. Furthermore, defects leading to K_m mutants are found within or close to the H5 motifs, which are part of two periplasmic loops of KdpA (Buurman et al., 1995). The picture now emerging for KdpA is that of a hydrophobic core made up by membrane-spanning segments, in which hydrophilic loops, including those containing the H5 motifs, fold into, thereby determining the ion specificity of the transport system.

VI. REGULATION OF KDP EXPRESSION

Expression of the Kdp system is mediated by two regulatory proteins, KdpD and KdpE, members of the large family of sensor kinase/response regulator proteins that control a wide variety of genetic and physiological responses in bacteria and some eukaryotes (Stock et al., 1989; Parkinson and Kofoid, 1992; Maeda et al., 1994). In all cases subjected to biochemical analysis, the larger sensor kinase becomes phosphorylated when activated and transfers phosphate to the smaller response regulator. The phosphorylated form of the response regulator produces the ultimate effect, be it to turn on expression of some genes or alter a biochemical response such as rotation of the flagellar motor.

The sensor kinase KdpD has 894 amino acid residues and a calculated molecular weight of 98,700. The hydrophilic C- and N-terminal domains were shown to be cytoplasmic, the protein being anchored in the cytoplasmic membrane by four closely-spaced membrane spans near its middle (Zimmann et al., 1995). There is moderate homology of parts of the C-terminal domain with other sensor kinases; the homologous regions include the His residue which is the site of phosphorylation in other sensor kinases.

The KdpE protein is a soluble protein with 225 amino acid residues and a predicted molecular weight of 25,200. It is homologous to other response regulator proteins, with similarity greatest to the OmpR, CreB, and PhoB regulators (Walderhaug et al., 1992). The protein is stable in cytoplasmic extracts and readily purified.

The signal to turn on Kdp is postulated to produce a conformational change in KdpD that results in its autophosphorylation. Phosphorylation of KdpD is readily demonstrated *in vitro* by crude cell extracts (Nakashima et al., 1992; Voelkner et al., 1993), by a soluble form of KdpD, and by the intact partially purified protein after reconstitution into liposomes (Nakashima et al., 1993b). Phosphorylation proceeds most rapidly in the presence of Mg^{2+}, and at a lower rate in the presence of Ca^{2+} (Nakashima et al., 1992). Increasing salt concentration greatly stimulates phosphorylation of KdpD, an effect not specific to the ion used since NaCl and KCl have similar effects (Voelkner et al., 1993). Phospho-KdpD has stability properties consistent with phosphohistidine (Nakashima et al., 1993a). When His673 was replaced by glutamine, no phosphorylation was observed, supporting the evidence from homology that the site of phosphorylation is probably His673 (Voelkner et al., 1993).

The transfer of phosphate from phospho-KdpD to KdpE has been demonstrated with each of the preparations listed above in which formation of phospho-KdpD occurs. In a mixture containing γ-^{32}P-labeled ATP and KdpD, addition of KdpE results in rapid transfer of label to KdpE and such transfer is stimulated by high salt (Nakashima et al., 1992; Voelkner et al., 1993). Stimulation may reflect an increase in the kinase and/or phosphotransferase activity of KdpD. In the absence of KdpD, no phosphorylation of KdpE occurred. The total amount of phospho-protein when both KdpD and KdpE were present fell rather rapidly, in contrast to the situation where only KdpD was present. The instability of phospho-KdpD in the presence of KdpE is explained by the rapid transfer of phosphate to KdpE and the relative instability of phospho-KdpE. It was noted that replacing Mg^{2+} with Ca^{2+} increased the stability of phospho-KdpE *in vitro*.

KdpE resembles the OmpR protein sufficiently that it can be phosphorylated by EnvZ, the sensor kinase of OmpR (Nakashima et al., 1992). Such heterologous activity, known as 'cross-talk', is a common occurrence due to the similarity of many members of this family of regulators. KdpE is probably phosphorylated on Asp52, the homolog of the site of phosphorylation of homologous proteins (Stock et al., 1989). The stability properties of phospho-KdpE are consistent with those of an acylphosphate (Nakashima et al., 1993a).

The ultimate effect of regulation, increasing expression of the Kdp system, is mediated by the interaction of KdpE upstream of the promoter of the *kdpFABC* operon. Unphosphorylated KdpE binds to an activator site that spans the region from 50 to 72 base pairs upstream of the transcription start site (Sugiura et al., 1992). The DNA sequence of this region is AT rich and has a number of runs of A residues (on the template strand) separated by about 10 base pairs. This region exhibits inherent DNA bending that persists when it binds KdpE (Tanaka et al., 1991; Sugiura et al., 1993). Phospho-KdpE has increased affinity for binding to the activator site and stimulates transcription from the *kdpFABC* promoter *in vivo* (Nakashima et al., 1993a).

To terminate expression of Kdp, phospho-KdpE must be dephosphorylated. In a number of two-component systems, such as the EnvZ–OmpR pair, the sensor kinase also has phosphatase activity that acts on the phosphorylated form of its cognate response regulator and that is stimulated by ATP. We suggest that KdpD has phosphatase activity to dephosphorylate phospho-KdpE when the signal to turn on expression ceases.

VII. THE SIGNAL FOR KDP EXPRESSION

The signal that KdpD senses to phosphorylate itself is believed to be turgor pressure or some effect thereof, reflecting the role of K^+ as a cytoplasmic osmotic solute. A current model for regulation of Kdp expression is shown schematically in Figure 5. The key feature of this model is the dependence of the conformation and hence of the kinase activity of KdpD on membrane stretch. When turgor, and hence stretch, is high, KdpD kinase is inactive. When stretch falls, KdpD alters its conformation and the kinase becomes active, resulting in formation of phospho-KdpE and, thus, expression of Kdp.

The turgor control model was based on the pattern of expression of Kdp in response to changes in medium K^+ concentration and turgor (Laimins et al., 1981). The Kdp system can be turned on by reducing the K^+ concentration of the medium. However, it is neither the external or internal concentration of K^+ that is sensed, but what might be called the 'need' for K^+ to maintain turgor. In wild-type strains, Kdp is not expressed when medium K^+ is above 2 mM. In mutants lacking all saturable transport systems, Kdp is expressed in media containing 50 mM K^+ or less. There was no correlation of Kdp expression with internal K^+ concentration when this parameter was altered by changing medium osmolality. Control by turgor was supported by finding that a sudden increase in medium osmolality, a maneuver that reduces turgor, was able to turn on Kdp expression transiently without any

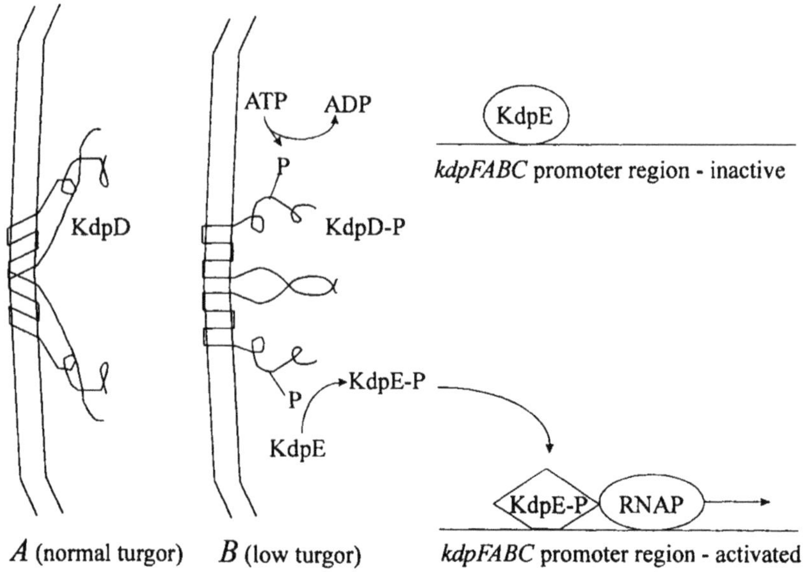

Figure 5. A schematic diagram of control of Kdp expression by turgor. Membrane-bound KdpD is shown as a dimer, extending into the cytoplasm from its four central membrane spans that anchor it in the membrane. In **A**, the KdpD protein is shown under conditions of high turgor, leading to a conformation without kinase activity. In **B**, turgor reduction alters the conformation of KdpD which catalyzes its own phosphorylation from ATP and phosphotransfer to KdpE. Phospho-KdpE in turn binds in the promoter region of the *kdpFABC* operon to stimulate transcription, here shown as promoting the binding of RNA polymerase (RNAP).

reduction of external or internal K^+ concentration. Expression was transient because K^+ uptake ultimately restored turgor.

This model has been challenged by more recent findings that do not, at first glance, appear to be consistent with control by turgor pressure, or by turgor alone. In steady-state growth at elevated osmolality in medium of intermediate K^+ concentration, Kdp is expressed if the osmotic solute is a salt, but not if it is a sugar present at the same osmotic concentration (Gowrishankar, 1985; Sutherland et al., 1986; Asha and Gowrishankar, 1993). Kdp can be turned on in medium with sugar if K^+ concentration in the medium is reduced, and conversely Kdp can be turned off in medium with salt by an increase in K^+ concentration. These results indicate that the critical medium K^+ concentration below which expression of Kdp begins is higher when osmolality is created by a salt than by a sugar. In addition, expression of Kdp in a number of situations (Asha and Gowrishankar, 1993) and in *kdpD* mutants that make expression partially constitutive don't fit expectations of the turgor model. To explain these findings, it has been suggested that expression of

Kdp is determined by a special pool of cytoplasmic K^+ or the rate of K^+ transport (Gowrishankar, 1987; Asha and Gowrishankar, 1993).

Analysis of mutant forms of KdpD that result in constitutive expression of *kdp* led Sugiura et al. (1994) to suggest that KdpD senses two signals, turgor and K^+. These mutants had altered amino acid residues in the distal two membrane spans or neighboring C-terminal part of KdpD. Expression of *kdp* in the mutants was stimulated from 6- to 25-fold when medium osmolality was increased by either salts or sugars. Increasing medium K^+ concentration had only a small effect, reducing *kdp* expression up to 3-fold in medium of low osmolality and having variable, small effects in medium of high osmolality.

The ability of a marked reduction in turgor to turn on Kdp indicates that turgor or its effects are probably sensed by KdpD. However, salts must also be sensed in some way to account for the quantitatively different effects of salts and sugars in stimulating expression of Kdp. The stimulation of KdpD kinase by salts *in vitro* may reflect sensing of salts.

VIII. CONCLUSIONS

The high ion selectivity and high rate of Kdp makes this complex especially useful for the analysis of the structure and regulation of transport. The high affinity and specificity of Kdp for K^+ has been exploited in generating mutants with a wide range of affinities for K^+. Further analyses of this type of mutation should illuminate mechanisms for discriminating between very similar ions. In concert with the topology of KdpA in the membrane these studies should also lead to an understanding of how K^+ ions move across the membrane. The role of K^+ as a preferred and essential osmoregulatory ion in bacteria explains the widespread distribution of the Kdp-type ATPase among distantly related eubacterial groups. However, a Kdp-like system has so far not been detected in archaebacteria.

K^+ translocation by the Kdp complex is the coordinated work of three (or even four) subunits. This makes the enzyme an excellent model system for the study of subunit assembly and of communication between components of a hetero-oligomeric enzyme complex. Another challenging feature of the Kdp system is its regulation by turgor pressure. An environmental stimulus or stimuli evoking mechanical forces in the membrane is transmitted into a signal acting at the level of gene expression and, at the same time, exerts control on the activity of already synthesized and membrane-integrated Kdp transport complexes. The analysis of this kind of control is facilitated by the ease of genetic manipulations and the over-expression of the proteins involved.

ACKNOWLEDGMENTS

This work was supported by grants from the Deutsche Forschungsgemeinschaft (SFB 171) and the Fonds der Chemischen Industrie to K.A., and grant GM22323 from the National

Institutes of Health to W.E.. We thank Dr. Kirsten Jung and Thomas Möllenkamp for drawing figures 1, 2, 3, and 5; Dr. Kirsten Jung for her comments; and Johanna Petzold for typing the manuscript.

REFERENCES

Altendorf, K., & Epstein, W. (1993). Kdp-ATPase of *Escherichia coli*. Cell. Physiol. Biochem. 4, 160–168.

Altendorf, K., Siebers, A., & Epstein, W. (1992). The Kdp ATPase of *Escherichia coli*. Ann. N. Y. Acad. Sci. 671, 228–243.

Asha, H., & Gowrishankar, J. (1993). Regulation of *kdp* operon expression in *Escherichia coli*: Evidence against turgor as signal for transcriptional control. J. Bacteriol. 175, 4528–4537.

Bachmann, B. J. (1990). Linkage map of *Escherichia coli* K-12, edition 8. Microbiol. Rev. 54, 130–197.

Buurman, E. T., Kim, K-T., & Epstein, W. (1995). Genetic evidence for two sequentially occupied K^+ binding sites in the Kdp transport ATPase. J. Biol. Chem. 270, 6678–6685.

Coleman, J. E., & Gettins, P. (1983). Alkaline phosphatase, solution, structure and mechanism. Adv. Enzymol. Relat. Areas Mol. Biol. 55, 381–452.

Eisenberg, D., Schwarz, E., Komaromy, M., & Wall, R. (1984). Analysis of membrane and surface protein sequences with the hydrophobic moment plot. J. Mol. Biol. 179, 125–142.

Epstein, W. (1990). Bacterial transport ATPases. In: The Bacteria: A Treatise on Structure and Function, Vol. 12, Bacterial Energetics (Krulwich, T. A., ed.), pp. 87–110. Academic Press, Orlando, FL.

Epstein, W., & Davies, M. (1970). Potassium-dependent mutants of *Escherichia coli* K-12. J. Bacteriol. 101, 836–843.

Epstein, W., Whitelaw, V., & Hesse, J. (1978). A K^+-transport ATPase in *Escherichia coli*. J. Biol. Chem. 253, 6666–6668.

Epstein, W., Walderhaug, M. O., Polarek, J. W., Hesse, J. E., Dorus, E., & Daniel, J. M. (1990). The bacterial Kdp K^+-ATPase and its relation to other transport ATPases, such as the Na^+/K^+- and the Ca^{2+}-ATPases in higher organisms. Phil. Trans. R. Soc. Lond. B. 326, 479–487.

Glynn, I. M., & Karlish, S. J. D. (1975). The sodium pump. Annu. Rev. Physiol. 37, 13–55.

Glynn, I. M., & Karlish, S. J. D. (1990). Occluded cations in active transport. Annu. Rev. Biochem. 59, 171–205.

Gowrishankar, J. (1985). Identification of osmoresponsive genes in *Escherichia coli*: Evidence for participation of potassium and proline transport systems in osmoregulation. J. Bacteriol. 164, 434–445.

Gowrishankar, J. (1987). A model for the regulation of expression of the potassium-transport operon, *kdp*, in *Escherichia coli*. J. Genet. 66, 87–92.

Jan, L. Y., & Jan, Y. N. (1994). Potassium channels and their evolving gates. Nature 371, 119–122.

Kashiwagi, K., Suzuki, T., Suzuki, F., Furuchi, T., Kobayashi, H., & Igarashi, K. (1991). Coexistence of the genes for putrescine transport protein and ornithine decarboxylase at 16 min on the *Escherichia coli* chromosome. J. Biol. Chem. 266, 20922–20927.

Kollmann, R., & Altendorf, K. (1993). ATP-driven potassium transport in right-side-out membrane vesicles via the Kdp system of *Escherichia coli*. Biochim. Biophys. Acta 1143, 62–66.

Laimins, L. A., Rhoads, D. B., & Epstein, W. (1981). Osmotic control of *kdp* operon expression in *Escherichia coli*. Proc. Natl. Acad. Sci. USA 78, 464–468.

Maeda, T., Wurgler-Murphy, S. M., & Saito, H. (1994). A two-component system that regulates an osmosensing MAP kinase cascade in yeast. Nature 369, 242–245.

McDonough, A. A., Geering, K., & Farley, R. A. (1990). The sodium pump needs its β subunit. FASEB J. 4, 1598–1605.

Nakashima, K., Sugiura, A., Momoi, H., & Mizuno, T. (1992). Phosphotransfer signal transduction between two regulatory factors involved in the osmoregulated *kdp* operon in *Escherichia coli*. Mol. Microbiol. 6, 1777–1784.

Nakashima, K., Sugiura, A., Kanamaru, K., & Mizuno, T. (1993a). Signal transduction between the two regulatory components involved in the regulation of the *kdpABC* operon in *Escherichia coli*: Phosphorylation-dependent functioning of the positive regulator, KdpE. Mol. Microbiol. 7, 109–116.

Nakashima, K., Sugiura, A., & Mizuno, T. (1993b). Functional reconstitution of the putative *Escherichia coli* osmosensor, KdpD, into liposomes. J. Biochem. (Tokyo) 114, 615–621.

Naprstek, J., Walderhaug, M. O., & Epstein, W. (1992). Purification and kinetic characterization of the Kdp-ATPase. Ann. N.Y. Acad. Sci. 671, 481–483.

Normak, S., Berström, S., Edlund, T., Grundström, T., Jaurin, B., Linberg, F. P., & Olsson, O. (1983). Overlapping Genes. Annu. Rev. Genet. 17, 499–525.

Parkinson, J. S., & Kofoid, E. C. (1992). Communication modules in bacterial signaling proteins. Annu. Rev. Genet. 26, 71–112.

Polarek, J. W., Williams, G., & Epstein, W. (1992). The products of the *kdpDE* operon are required for expression of the Kdp ATPase of *Escherichia coli*. J. Bacteriol. 174, 2145–2151.

Post, R. L., & Kume, S. (1973). Evidence for an aspartyl phosphate residue at the active site of sodium and potassium ion transport adenosine triphosphatase. J. Biol. Chem. 248, 6993–7000.

Post, R. L., Hegyvary, C., & Kume, S. (1972). Activation by adenosine triphosphate in the phosphorylation kinetics of sodium and potassium ion transport adenosine triphosphatase. J. Biol. Chem. 247, 6530–6540.

Puppe, W., Siebers, A., & Altendorf, K. (1992). The phosphorylation site of the Kdp-ATPase of *Escherichia coli*. Site-directed mutagenesis of aspartic acid residues 300 and 307 of the KdpB subunit. Mol. Microbiol. 6, 3511–3520.

Rao, J. K. M., & Argos, P. (1986). A conformational preference parameter to predict helices in integral membrane proteins. Biochim. Biophys. Acta 869, 197–214.

Serrano, R. (1988). Structure and function of proton translocating ATPase in plasma membranes of plants and fungi. Biochim. Biophys. Acta 947, 1–28.

Shull, G. E. (1990). cDNA cloning of the β-subunit of the rat gastric H,K-ATPase. J. Biol. Chem. 265, 12123–12126.

Shull, G. E., Lane, L. K., & Lingrel, J. B. (1986). Amino acid sequence of the β-subunit of the (Na^+ + K^+) ATPase deduced from a cDNA. Nature 312, 429–431.

Siebers, A., & Altendorf, K. (1989). Characterization of the phosphorylated intermediate of the K^+-translocating Kdp ATPase from *Escherichia coli*. J. Biol. Chem. 264, 5831–5838.

Siebers, A., & Altendorf, K. (1993). K^+-translocating Kdp-ATPases and other bacterial P-type ATPases. In: Alkali Cation Transport Systems in Prokaryotes (Bakker, E. P., ed.), pp. 225–252. CRC Press, Boca Raton, FL.

Siebers, A., Kollmann, R., Dirkes, G., & Altendorf, K. (1992). Rapid, high yield purification and characterization of the K^+-translocating Kdp-ATPase from *Escherichia coli*. J. Biol. Chem. 267, 12717–12721.

Silhavy, T. J., & Beckwith, J. R. (1985). Uses of *lac* fusions for the study of biological problems. Microbiol. Rev. 49, 398–418.

Stock, J. B., Ninfa, A. J., & Stock, A. M. (1989). Protein phosphorylation and regulation of adaptive responses in bacteria. Microbiol. Rev. 53, 450–490.

Sugiura, A., Nakashima, K., Tanaka, K., & Mizuno, T. (1992). Clarification of the structural and functional features of the osmoregulated *kdp* operon of *Escherichia coli*. Mol. Microbiol. 6, 1769–1776.

Sugiura, A., Nakashima, K., & Mizuno, T. (1993). Sequence-directed DNA curvature in activator-binding sequence in the *Escherichia coli kdpABC* promoter. Biosci. Biotech. Biochem. 57, 356–357.

Sugiura, A., Hirokawa, K., Nakashima, K., & Mizuno, T. (1994). Signal-sensing mechanisms of the putative osmosensor KdpD in *Escherichia coli*. Mol. Microbiol. 14, 929–938.

Sutherland, L., Cairney, J., Elmore, M. J., Booth, I. R., & Higgins, L. F. (1986). Osmotic regulation of transcription: Induction of the *proU* betaine transport gene is dependent on accumulation of intracellular potassium. J. Bacteriol. 168, 805–814.

Tanaka, K., Muramatsu, S., Yamada, H., & Mizuno, T. (1991). Systematic characterization of curved DNA segments randomly cloned from *Escherichia coli* and their functional significance. Mol. Gen. Genet. 226, 367–376.

Taylor, K. A., Dux, L., & Martonosi, A. (1986). Three-dimensional reconstruction of negatively-stained crystals of the Ca^{2+}-ATPase from muscle sarcoplasmic reticulum. J. Mol. Biol. 187, 417–427.

Toyoshima, C., Sasabe, H., & Stokes, D. L. (1993). Three-dimensional cryo-electron microscopy of the calcium ion pump in the sarcoplasmic reticulum membrane. Nature 362, 469–471.

Voelkner, P., Puppe, W., & Altendorf, K. (1993). Characterization of the KdpD protein, the sensor kinase of the K^+-translocating Kdp system of *Escherichia coli*. Eur. J. Biochem. 217, 1019–1026.

von Heijne, G. (1986). The distribution of positively charged residues in bacterial inner membrane proteins correlates with the trans-membrane topology. EMBO J. 5, 3021–3027.

Walderhaug, M. O., Polarek, J. W., Voelkner, P., Daniel, J. M., Hesse, J. E., Altendorf, K., & Epstein, W. (1992). KdpD and KdpE, proteins that control expression of the *kdpABC* operon, are members of the two-component sensor-effector class of regulators. J. Bacteriol. 174, 2152–2159.

Zimmann, P., Puppe, W., & Altendorf, K. (1995). Membrane topology analysis of the sensor kinase KdpD of *Escherichia coli*. J. Biol. Chem. 270, 28282–28288.

INDEX

www.ingramcontent.com/pod-product-compliance
Ingram Content Group UK Ltd.
Pitfield, Milton Keynes, MK11 3LW, UK
UKHW020229290926
15470UKWH00001B/425